# THE CARBON STAR PHENOMENON

INTERNATIONAL ASTRONOMICAL UNION

UNION ASTRONOMIQUE INTERNATIONALE

# THE CARBON STAR PHENOMENON

PROCEEDINGS OF THE 177TH SYMPOSIUM OF THE
INTERNATIONAL ASTRONOMICAL UNION,
HELD IN ANTALYA, TURKEY, MAY 27–31, 1996

EDITED BY

ROBERT F. WING

*Astronomy Department,
The Ohio State University,
Columbus OH, U.S.A.*

KLUWER ACADEMIC PUBLISHERS

DORDRECHT / BOSTON / LONDON

A C.I.P. Catalogue record for this book is available from the Library of Congress.

ISBN 0-7923-6346-9

*Published on behalf of
the International Astronomical Union
by
Kluwer Academic Publishers, P.O. Box 17, 3300 AA Dordrecht, The Netherlands.*

*Sold and distributed in North, Central and South America
by Kluwer Academic Publishers,
101 Philip Drive, Norwell, MA 02061, U.S.A.*

*In all other countries, sold and distributed
by Kluwer Academic Publishers,
P.O. Box 322, 3300 AH Dordrecht, The Netherlands.*

*Printed on acid-free paper*

*All Rights Reserved*
© 2000 International Astronomical Union

*No part of the material protected by this copyright notice may be reproduced or utilized in
any form or by any means, electronic or mechanical, including photocopying, recording
or by any information storage and retrieval system, without written permission from the
publisher.*

*Printed in the Netherlands.*

# CONTENTS

Preface .................................................. xvii
Tribute to Nüzhet Gökdoğan .............................. xxvi
Conference photo ....................................... xxviii
List of participants ..................................... xxxi

## Introduction

Introductory Remarks: Watch the Details ..................... 3
*How to Deal with Unexpected Scientific Results*
   William P. Bidelman (read *in absentia*)

What Theorists Think They Know about AGB Stars .............. 7
   John C. Lattanzio

## Session I — Surveys

Deep IV–N and Yellow-Red Spectral Surveys for Carbon Stars ....... 13
   O. M. Kurtanidze and M. G. Nikolashvili

Space Distribution of Carbon Stars in Our Galaxy ................. 21
   K. Noguchi, Z. Qian, J. Sun, G. Wang, J. Wang and Y. Rao

Understanding the Elusive Dwarf Carbon Star ..................... 27
   Paul J. Green

A Proper-Motion Search for Dwarf Carbon and S Stars .............. 37
   D. J. MacConnell, R. L. Williamson II and W. J. Roberts

Large Magellanic Cloud Carbon Stars: Reading the Rosetta Stone
of Stellar Evolution ............................................ 41
   Jay A. Frogel and Edgardo Costa

Surveys for Carbon Stars in External Galaxies ..................... 51
   Marc Azzopardi

A Photometric and Spectroscopic Survey of AGB Stars in M31 ....... 59
   James P. Brewer, Harvey B. Richer and Dennis R. Crabtree

## Session II — Models and Abundance Determinations

Model Atmospheres for Normal and Peculiar Red Giant Stars ........ 71
   Bertrand Plez

Quantitative Analysis of Carbon Isotope Ratios in N–, SC–, and
J–type Carbon Stars .................................................. 81
    K. Ohnaka and T. Tsuji

$^{12}C/^{13}C$ Ratios and Li Abundances in Low-Mass C Stars .......... 89
    Carlos Abia and Jordi Isern

Carbon and Hydrocarbon Molecules in Very Cool White Dwarfs ..... 97
    Turgut Aslan and Irmela Bues

The s-Process in the Yellow Symbiotic AG Draconis ................ 103
    Katia Cunha, Verne V. Smith and Alain Jorissen

The Chromospheres of Carbon Stars ................................ 105
    Donald G. Luttermoser

## Session III — Observed Spectra and Energy Distributions

Spectral Characteristics of RV Tauri Stars ....................... 117
    Sunetra Giridhar

Applications of Narrow-Band Photometry in the Study of Peculiar
Red Giants ........................................................ 127
    Robert F. Wing

Spectral Classification of Carbon-Peculiar G Stars ............... 141
    Tuba Koktay and R. F. Garrison

Mass-losing AGB Stars in the LMC ................................. 145
    Jacco Th. van Loon, Albert A. Zijlstra, Patricia A. Whitelock,
    Cecile Loup and L. B. F. M. Waters

## Session IV — Variability

Trend Analysis of 51 Carbon Long-Period Variables ................ 155
    Janet Akyüz Mattei and Grant Foster

Comparison of Mean Light Curve Parameters of M, S and C Mira
and Semi-Regular Variable Stars Using 75 Years of AAVSO Data .. 165
    Marie-Odile Mennessier, Hichame Boughaleb and
    Janet A. Mattei

The Variability of R, N, and C Stars from HIPPARCOS and AAVSO
Data .............................................................. 171
    Michel Grenon, Janet A. Mattei, Laurent Eyer and Grant Foster

Infrared Light Curves of Carbon-Rich Variables ................... 179
    Patricia Whitelock

R CrB: Photometric Evidence Regarding the Nature of its Pulsation, and a Putative Connection between Pulsation and Deep Declines .. 191
  *J. D. Fernie*

The Pulsations and Evolution of AGB Stars in the Large Magellanic Cloud ............................................................. 199
  *Z. G. Gong and Y. Li*

## Session V — Post–AGB Stars

Post–AGB Variables and Stellar Mass-Loss ........................ 207
  *M. W. Feast*

Lines of Circumstellar $C_2$, CN, and $CH^+$ in the Optical Spectra of Post–AGB Stars ................................................. 217
  *Eric J. Bakker, David L. Lambert and Ewine F. Van Dishoeck*

Chemical Composition and Evolution of Post–AGB Stars ........... 225
  *M. Parthasarathy*

Optical Spectroscopic Monitoring of the Carbon-Rich Post–AGB Star HD 56126: *Pulsation and Shock Waves* ......................... 237
  *Agnès Lèbre, Nicolas Mauron, Denis Gillet and Dominique Barthès*

Collimated Outflow from Stars: The Planetary Nebula Abell 78 ..... 245
  *Paris Pişmiş* (read *in absentia*)

## Session VI — Binarity

The Role of Binaries in the Carbon Star Phenomenon ............... 249
  *Robert D. McClure*

Barium Stars and Tc-Poor S Stars: Binary Masqueraders within the Carbon-Star Family .............................................. 259
  *A. Jorissen and S. Van Eck*

Binarity among Barium Dwarfs and CH Subgiants: Will They Become Barium Giants? ......................................... 269
  *Pierre North, Alain Jorissen and Michel Mayor*

The Chemical Composition and Orbital Parameters of Barium Stars.. 277
  *Laimons Začs*

Binary "Post–AGB" Stars ......................................... 285
  *Hans Van Winckel, Christoffel Waelkens and Laurens B. F. M. Waters*

Light and Velocity Variability of Post–AGB Stars .................... 293
    *Bruce J. Hrivnak and Wenxian Lu*

The Case for S Star Binaries ........................................ 299
    *Thomas B. Ake*

## Session VII — Mass–Loss, Winds, and Formation of Dust Grains

Inferring Mass Loss Rates for Cool Luminous Stars from
  High-Resolution GHRS Spectra .................................... 303
    *J. L. Linsky, G. M. Harper, J. Valenti, P. D. Bennett and
    A. Brown*

Dust Formation in Stellar Photospheres ............................ 313
*The Case of Carbon Stars from Dwarfs to AGB Stars*
    *Takashi Tsuji*

Dust-Driven Winds of Rotating Carbon Stars ....................... 325
    *Ernst A. Dorfi and Susanne Höfner*

Nucleating Dust in Carbon-Rich AGB Stars ........................ 331
    *I. Cherchneff*

Grain Formation in the Winds of Cool Red Giant Stars ............. 337
    *E. Sedlmayr and J. M. Winters*

Carbon Star Dust from Meteorites .................................. 349
    *Uffe Gråe Jørgensen and Anja C. Andersen*

## Session VIII — Circumstellar Shells

Circumstellar Dust around M, S and C Stars ....................... 361
    *Irene R. Little-Marenin*

Observations of Mass Loss and Circumstellar Matter around Cool
  Carbon Stars .................................................... 367
    *T. Lloyd Evans*

Dynamical Models of Circumstellar Dust Shells around Long-Period
  Variables ....................................................... 377
    *A. J. Fleischer, J. M. Winters and E. Sedlmayr*

Modelling the Spectral Energy Distributions of AGB Stars in
  the LMC ........................................................ 385
    *M. A. T. Groenewegen, J. Th. van Loon, P. A. Whitelock,
    P. R. Wood and A. A. Zijlstra*

IR Emission from Dusty Winds — Scaling and Self-Similarity
  Properties .................................................. 391
    Moshe Elitzur and Željko Ivezić

Dust Emission from IRC +10216 ................................. 399
    Željko Ivezić and Moshe Elitzur

## Session IX — Circumstellar Emission and Environment

Molecular Optical and Infrared Emission from the Red Rectangle ... 407
*Carriers of Diffuse Circumstellar, Nebular and Interstellar Bands*
    T. H. Kerr, J. R. Miles, M. E. Hurst, R. E. Hibbins and
    P. J. Sarre

Scattered Light from Envelopes around N–type Stars ............... 409
    Bengt Gustafsson, Kjell Eriksson, Dan Kiselman, Nils Olander,
    Hans Olofsson and Hugo E. Schwarz

The Neutral Envelopes around AGB and Post–AGB Objects ........ 413
*Their Structure and Kinematics*
    H. Olofsson

Extended Dust Shells around Carbon Stars in the Infrared and in
  Optical Light .................................................. 425
    Hideyuki Izumiura, L. B. F. M. Waters, T. de Jong, C. Loup
    and O. Hashimoto

Asymmetries around Luminous Red Variables ...................... 433
    Antonio Mário Magalhães and Kenneth H. Nordsieck

## Session X — Nucleosythesis and Evolution

Heavy-Element Nucleosynthesis in AGB Stars ...................... 443
    Verne V. Smith

Nucleosynthesis in Intermediate-Mass Stars ........................ 449
    John C. Lattanzio, Cheryl A. Frost, Robert C. Cannon and
    Peter R. Wood

Fluorine Production in Asymptotic Giant Branch Stars ............. 459
    N. Mowlavi, A. Jorissen and M. Arnould

Carbon Stars in the Early–AGB Stage of Evolution ................. 463
    Ju. L. Frantsman

From the Tip of the AGB Towards a Planetary: A Hydrodynamical
  Simulation ..................................................... 469
    D. Schönberner, M. Steffen, J. Stahlberg, K. Kifonidis and
    T. Blöcker

## Session XI — Galactic Chemical Evolution

Carbon Stars and Nucleosynthesis in Galaxies .................... 481
  Bengt Gustafsson and Nils Ryde

## Session XII — Observing Facilities

Variability Studies with International Networks ................. 499
  François R. Querci and Monique Querci

A New Optical Observatory in Turkey ............................ 507
  Zeki Aslan, Selim O. Selam and Akif Esendemir

**Abstracts of Posters**
  *(ordered alphabetically by first author)*

$V$ and $R$ Observations of Two Carbon Stars: UX Dra and RY Dra .. 515
  Hasan Ak, Berahitdin Albayrak, Zeki Aslan, Osman Demircan,
  Zekeriya Müyesseroğlu, Sacit Özdemir and Kutluay Yüce

Unusual Light Curves of Some Carbon Stars ...................... 516
  Andrejs Alksnis

Modelling the M-S-C Giants Spectral Sequence ................... 517
  France Allard, Peter H. Hauschildt, David R. Alexander,
  Martin Cohen and Gordon C. Augason

Carbon Isotope Ratios in Carbon Stars of the Galactic Halo ....... 518
  Wako Aoki and Takashi Tsuji

Carbon Stars in Open Clusters .................................. 519
  Bernhard Aringer

The SiO Molecule in the Atmospheres of Cool AGB Stars ........... 520
  Bernhard Aringer, Uffe G. Jørgensen, Stephen R. Langhoff
  and Josef Hron

The Eddington Limit, Radiative Instabilities and the Declines of
  R Coronae Borealis Stars ..................................... 521
  Martin Asplund

Modelling of Carbon-Rich Stars with Far Infrared Flux Excess ...... 522
  Stefano Bagnulo, Gerry Doyle, Chris Skinner and
  Vincenzo Andretta

Selective Depletion of Elements in Stellar Atmospheres: A Unified
  Picture? ..................................................... 523
  Eric J. Bakker, Guillermo Gonzalez and David L. Lambert

How to Make Carbon Stars: A New Approach to Model Boundaries
of Convective Regions .............................................. 524
    T. Blöcker, F. Herwig, D. Schönberner and M. El Eid

Opacities for Carbon Dwarfs and M Dwarfs ......................... 525
    Aleksandra Borysow and Uffe Gråe Jørgensen

A Kinematic Survey of Carbon Stars in the Small Magellanic Cloud.. 526
    Russell Cannon, Barry Croke, Despina Hatzidimitriou and
    David Morgan

Mid-Infrared Silicate Variation in Long-Period, Oxygen-Rich
Variable Stars ..................................................... 527
    M. J. Creech-Eakman and R. E. Stencel

LMC/SMC Outer Halo Carbon Star Survey: Radial Velocities of
500 Newly Identified Stars ......................................... 528
    Serge Demers, W. E. Kunkel and M. J. Irwin

The Backwarming Effect and Carbon Stars .......................... 529
    Dimitri N. Doikov and Ekaterina M. Doikova

Proposal for a Photometric System for the Classification of
Carbon Stars ...................................................... 530
    Uldis Dzervitis

Carbon to Oxygen Abundance Ratios in the Atmospheres of Carbon
Stars of the Orion and Perseus Galactic Arms ..................... 531
    I. Eglitis and M. Eglite

Condensation of SiC in Circumstellar Dust Shells of C-Rich
Red Giants ........................................................ 532
    Andreas Gauger, John J. Keady and Erwin Sedlmayr

New Results from the Modeling of the Shell around IRC +10216 .... 533
    M. A. T. Groenewegen

JHK Photometry of AGB Stars in the SMC ......................... 534
    M. A. T. Groenewegen and J. A. D. L. Blommaert

Frequency Sampling for Radiative Transfer Calculations in
Cool Stars ........................................................ 535
    Christiane Helling and Uffe Gråe Jørgensen

Macro-Molecules in Model Atmospheres ............................. 536
    Christiane Helling, Uffe Gråe Jørgensen, Bertrand Plez and
    Hollis R. Johnson

Molecular Hydrogen in the Circumstellar Shells of Carbon Stars .... 537
    Kenneth Hinkle and John Keady

Atmospheric Dynamics and Dust-Driven Winds of Carbon Stars .... 538
  *Susanne Höfner and Ernst A. Dorfi*

The Shape of Silicate Features in Semiregular and Mira Variables ... 539
  *Josef Hron, Bernhard Aringer and Franz Kerschbaum*

FG Sge as a New-Born Carbon Star .................................. 540
  *Takashi Iijima*

Experimental Gas-Phase Spectroscopy of Carbon Molecular
  Structures and Diffuse Interstellar Bands ........................ 541
  *J. Janča, M. Plonka, M. Šolc and M. Vetešník*

The Population of Red Giant Stars in Globular Clusters of the
  Fornax Dwarf Galaxy .............................................. 542
  *Uffe Gråe Jørgensen and Raul Jimenez*

Synthetic JHK Colors for M Dwarfs, M Giants, and Carbon Stars ... 543
  *Uffe Gråe Jørgensen and Robert F. Wing*

Near-Infrared Spectroscopy of High Galactic Latitude Carbon Stars –
  They Might Be Giants? ............................................ 544
  *Richard R. Joyce*

A Simulation for a Carbon Star .................................... 545
  *Gülçin Kandemir and Cem Güçlü*

On the Nature of Irregular Variables of Type Lb ................... 546
  *Franz Kerschbaum, Peter Habison and Josef Hron*

Comparison of C-Rich Mira, Semiregular, and Irregular Variables ... 547
  *Franz Kerschbaum, Peter Habison, Rita Loidl, Hans Olofsson
  and Josef Hron*

The Chemical Composition of the Halo Mira V CrB ................. 548
  *Tônu Kipper and Uffe Gråe Jørgensen*

A 200 km/sec Molecular Wind in the Carbon Star V Hya ........... 549
  *Gillian Knapp, Alain Jorissen and Ken Young*

Formation of Core–Mantle Type Grains Consisting of a SiC Core and
  a Carbon Mantle in Circumstellar Envelopes of Carbon Stars ..... 550
  *Takashi Kozasa and Hisato Sogawa*

Open and Globular Cluster Ages Using Theoretical Isochrones ...... 551
  *İbrahim Küçük*

Tien Shan Astronomical Observatory ................................ 552
  *Kenesken Kuratov*

Statistics of Carbon Stars in the Galaxy .......................... 553
  *Omar M. Kurtanidze*

A Close Association of Three Carbon Stars in the Direction of M92 .. 554
    O. M. Kurtanidze and M. G. Nikolashvili

Near IR (*JHK*) Observations of Selected Carbon Stars .............. 555
    Vladimir P. Kuz'kov

A Comparison between the Mira–OH/IR and the Carbon-Rich AGB
    Sequences ...................................................... 556
    Jacques R. D. Lépine

Interferometric Molecular Line Observations of RW LMi ............ 557
    Michael Lindqvist, Robert Lucas, Hans Olofsson, Fredrick
    Larsen, Alain Omont, Kjell Eriksson and Bengt Gustafsson

Are There Silicate – S Stars? ...................................... 558
    Irene R. Little-Marenin

Classification of Dust Emission Features in Carbon Stars ............ 559
    Irene R. Little-Marenin, Gregory C. Sloan and Stephan D. Price

Synthetic Spectra for Carbon-Rich Long-Period Variables ........... 560
    Rita Loidl, Bernhard Aringer, Uffe Jørgensen, Susanne Höfner
    and Josef Hron

Circumstellar Envelopes of Peculiar and Normal J–Type Stars ...... 561
    S. Lorenz Martins

Modelling of OH/IR Dust Envelopes ............................... 562
    S. Lorenz Martins and F. X. de Araújo

Southern Carbon Stars Found on Near–IR Objective-Prism Plates .. 563
    D. Jack MacConnell

Absolute Magnitude and Kinematics of Barium Stars ............... 564
    Marie-Odile Mennessier, Ana Gómez, Xavier Luri, Suzanne
    Grenier, Louis Prévot, Jordi Torra and Francesca Figueras

The Overtone Spectrum of Molecular Hydrogen and Methane in the
    Visible: Recent Measurements .................................. 565
    Michael Mickelson, Lee Larson, Lars English and David Ferguson

Multiwavelength Photometric Observations of Northern Carbon Stars 566
    Anatoly S. Miroshnichenko, Kenesken S. Kuratov, Željko Ivezić
    and Moshe Elitzur

AFGL 4106: Proto–Planetary Nebula or Post–Red Supergiant? ..... 567
    Frank Molster, Jacco van Loon, Rens Waters and
    Hans Van Winckel

The Detection of Low-Luminosity Carbon Stars in the Leo II and

Fornax Dwarf Galaxies .................................... 568
  Gérard Muratorio and Marc Azzopardi

High-Resolution Coudé–Echelle Spectrometer for the 1.5-m Kazan
  University Telescope at the Turkish National Observatory ......... 569
  Faig Musaev and Ilfan Bikmaev

Spectroscopic Analysis of Single-Lined Spectroscopic Binaries with
  Unseen Companions ................................................ 570
  Faig Musaev, Ilfan Bikmaev and Laimons Začs

Radiative Transfer and Dynamics of Stellar Outflows ................ 571
  Nathan Netzer

A Determination of the C/M5+ Ratio in the Galactic Plane ......... 572
  Maria G. Nikolashvili

On the Molecular Structure of Circumstellar Envelopes Surrounding
  C Stars .......................................................... 573
  A. Beate C. Patzer, Jan Martin Winters and Erwin Sedlmayr

Maser Mapping of Red Supergiants and the Onset of Bipolar Outflow . 574
  Anita M. S. Richards, Jeremy A. Yates and R. James Cohen

The Peculiar Object IRAS 06088+1909 ................................ 575
  A. Richichi, G. Calamai, F. Lisi, B. Stecklum, T. Herbst and
  E. Thamm

Observations, Assignments and Profiles of $SiC_2$ Absorption and
  Emission Bands in Carbon Stars ................................... 576
  Peter J. Sarre, Mark E. Hurst and Tom Lloyd Evans

The Dust around Cool Stars ........................................ 577
  Irakli Simonia and Tsitsino Simonia

Observations of the 11 $\mu$m Silicon Carbide Feature in Carbon Star
  Shells ........................................................... 578
  Angela K. Speck, M. J. Barlow and C. J. Skinner

Carbon– and Oxygen–Rich Stars in the IRAS Two-Color Diagram:
  Results from Hydrodynamical Models of AGB Winds ............. 579
  M. Steffen, R. Szczerba, A. Men'shchikov and D. Schönberner

Energy Distribution in the Spectra of Carbon Stars ................. 580
  Janis-Imants Straume

On Carbon Star Evolution in the IRAS Two-Color Diagram ........ 581
  Ryszard Szczerba, Matthias Steffen and Kevin Volk

Carbon Stars in the Galactic Halo ................................. 582
  Ed J. Totten and Michael J. Irwin

Infrared Observations of Peculiar Carbon Stars .................... 583
   *Ana Ulla, Peter Thejll, Tônu Kipper and Uffe Gråe Jørgensen*

The Frequency of Extrinsic and Intrinsic S Stars in the Henize
   Sample .................................................... 584
   *Sophie Van Eck, Alain Jorissen, Michel Mayor, Stephane Udry
   and Michel Burnet*

New Input Data for Synthetic AGB Evolution ..................... 585
   *J. Wagenhuber*

Silicon and Sulphur Chemistry in the Inner Envelopes of Carbon
   Stars ..................................................... 586
   *Karen Willacy and Isabelle Cherchneff*

Synthetic Colors of Carbon Stars ................................ 587
   *Walter Windsteig, Ernst A. Dorfi, Susanne Höfner, Josef Hron
   and Franz Kerschbaum*

Temperatures of Peculiar G-Type Stars from Narrow-Band
   Near-Infrared Photometry ..................................... 588
   *Robert F. Wing, Robert F. Garrison and Tuba Koktay*

A Photometric Search for Dwarf Carbon Stars .................... 589
   *Robert F. Wing and D. Jack MacConnell*

Optical Appearance of Dynamical Models for Circumstellar Dust
   Shells around Long-Period Variables: AFGL 3068 ................ 590
   *Jan Martin Winters, Axel J. Fleischer, Thibaut Le Bertre
   and Erwin Sedlmayr*

Envelope Pulsations of a $1\,M_\odot$ AGB Star During Thermal Pulses ... 591
   *A. Ya'ari, Y. Tuchman and J. Wagenhuber*

Motions of Carbon Stars ......................................... 592
   *Cahit Yeşilyaprak, Zeki Aslan, Orhan Gölbaşı and
   Tuncay Özdemir*

VRI Observations of S Stars ..................................... 593
   *Sandra B. Yorka and Tracy L. Huard*

Liquid and Solid Carbon Particles in Cool White Dwarf Atmospheres 594
   *Victor Zubko*

## Indexes

**Author Index** .................................................. 597
**Object Index** .................................................. 601
**Subject Index** ................................................. 607

View of the Antalya harbor. Visible in the rear center is the 13th century fluted brick Seljuk minaret which has become a landmark of the city.

# PREFACE

Symposium 177 of the International Astronomical Union was held in late May of 1996 in the coastal city of Antalya, Turkey. It was attended by 142 scientists from 32 countries. The purpose of the symposium was to discuss the causes and effects of the composition changes that often occur in the atmospheres of cool, evolved stars such as the carbon stars in the course of their evolution. This volume includes the full texts of papers presented orally and one-page abstracts of the poster contributions.

The chemical composition of a star's observable surface layers depends not only upon the composition of the interstellar medium from which it formed, but in many cases also upon the star's own history. Consequently, spectroscopic studies of starlight can tell us much about a star's origin, the path it followed as it evolved, and the physical processes of the interior which brought about the composition changes and made them visible on the surface. Furthermore, evolved stars are often surrounded by detectable shells of their own making, and the compositions of these shells provide additional clues concerning the star's evolutionary history.

It was Henry Norris Russell who showed in 1934 that the gross spectroscopic differences between the molecular spectra of carbon stars and M stars could be explained as due to a simple reversal of the abundances of oxygen and carbon. It was also realized early on that the interstellar medium is nowhere carbon-rich, and that changes in chemical composition must be the result of nuclear reactions that take place in the hot interiors of stars, not in their atmospheres. The very existence of carbon stars thus showed that at least a few stars have managed to change their atmospheric compositions by mixing processed material to the surface. Until recently, however, it has been widely supposed that the atmospheric compositions of most evolved stars are unchanged since their days on the main sequence.

Recent work has shown that composition changes are far more widespread than previously thought. Stars on the asymptotic giant branch of the HR diagram (AGB stars) often show changes in a whole gamut of elements including the heavy $s$-process nuclei, and even stars on the first ascent of the red giant branch may show evidence of changes in C, N, O, and their isotopes, i.e. the nuclei affected by hydrogen-burning reactions. Some stars have lost so much of their outer layers that all the material remaining has gone through hydrogen burning, and we see them as hydrogen-deficient. Still other stars show evidence of the composition changes associated with the AGB although they are not luminous enough to have reached that stage themselves; the mysteries of such stars are now being solved in terms of mass exchange in binaries systems, as more and more of them are found to have white-dwarf companions which must already have been through the AGB stage.

Work on these problems has seen enormous advances in recent years. Improved survey techniques are helping us discover and sort the many kinds of chemically-peculiar late-type stars, even revealing their presence in external galaxies. Advances in the modeling of stellar atmospheres, including the effects of opacity by millions of molecular lines, have made it possible to determine accurate abundances and isotope ratios for cool stars, while work on interior models has led to better understanding of energy production, mixing mechanisms, and composition changes. Infrared observations have provided data on circumstellar dust shells and have stimulated theoretical work on mass-loss mechanisms and grain formation, while radio observations reveal the composition of the molecular envelopes surrounding many cool stars. Imaging techniques are now being used to study the circumstellar environments of cool red giants and even to show the presence of spots on their surfaces. And because red giant stars return a good deal of processed material to the interstellar medium, it has become important to address the question of the effects of red giant stars on galactic evolution.

To bring together the researchers interested in these diverse problems relating to composition changes in cool, evolved stars, the IAU Working Group on Peculiar Red Giants proposed in 1994 to hold a meeting on "The Carbon Star Phenomenon." It was emphasized that the carbon stars themselves are only the most prominent manifestation of a widespread phenomenon affecting to some degree the evolution of essentially all stars of intermediate mass. The proposal was sponsored by IAU Commission 45 on Stellar Classification and co-sponsored by Commissions 27 (Variable Stars), 29 (Stellar Spectra), and 36 (Theory of Stellar Atmospheres), and we thank these Commissions and their Presidents (respectively Jack MacConnell, John Percy, David Lambert, and Wolfgang Kalkofen) for their support.

This Symposium may be considered a sequel to two others involving the same Working Group. A colloquium on "Cool Stars with Excesses of Heavy Elements" was organized by Carlos Jaschek and held in 1984 in Strasbourg, France; it was here that Jaschek proposed the formation of a working group to maintain the momentum generated by the conference and founded its "Newsletter on Chemically–Peculiar Red Giant Stars." This working group became the IAU–sponsored WG on Peculiar Red Giants at the General Assembly in New Delhi the following year, with Hollis Johnson as its Chairman. Johnson then organized the very successful IAU Colloquium 106 on "Evolution of Peculiar Red Giant Stars," which was held in 1988 in Bloomington, Indiana, U.S.A. When I took over as Chair of the WG in 1991, we were already starting to think about a third meeting along similar lines. I did, however, consider that it would be advantageous to hold our next meeting somewhere other than in Western Europe or North America, which have had more than their share of astronomy conferences.

Symposium 177 was held on May 27–31, 1996, in the city of Antalya, Turkey. It was the first IAU–sponsored conference to be held in Turkey, a country with more than 50 IAU members, several major departments and institutes of Astronomy, and a new national observatory. The choice of Turkey as the meeting site was also strongly influenced by my family's interest in that country: my daughter Sylvia had studied the Turkic languages, married a Turk, and had given me a guided tour of Istanbul during which I made the acquaintance of several Turkish astronomers. Organizing a conference far from one's home institution has obvious difficulties and could be carried out successfully only because of the enthusiastic support of many Turkish astronomers and graduate students. The key to this success was the hard work of Zeki Aslan, Head of the Physics Department at Akdeniz University in Antalya, who served as Chair of the Local Organizing Committee and took care of all local arrangements.

The Scientific Organizing Committee for Symposium 177 consisted of Zeki Aslan (Turkey), Hollis Johnson (U.S.A.), Uffe Jørgensen (Denmark), Tom Lloyd Evans (South Africa), Mário Magalhães (Brazil), Janet Mattei (U.S.A.), Monique Querci (France), Verne Smith (U.S.A.), Takashi Tsuji (Japan), and Robert Wing (Chair, U.S.A.). With two exceptions (Aslan and Mattei), these committee members constituted the Organizing Committee of the WG on Peculiar Red Giants. Aslan was added because of his role as LOC Chair and his interest in the photometric and kinematic properties of red giants. Janet Mattei was invited to join because of her Turkish background and the committee's desire to include a variable-star expert. Hollis Johnson resigned from the Committee shortly before the symposium when he realized he would not be able to attend, and has place was taken by John Lattanzio (Australia). All other SOC members did attend the symposium, except that Monique Querci was replaced by her husband François.

Antalya is a substantial, but relatively quiet, city on the southern coast of Turkey. It has a long and rich history, a fine archaeological museum, and a young university (Akdeniz Universitesi, named for the Mediterranean Sea, or "White Sea" in Turkish). It is the nearest city to Bakırlıtepe, a mountain in the Toros range that is the site of the new Turkish National Observatory, headed by Zeki Aslan. Antalya also boasts several first-rate hotels, some with enormous conference facilities, strung out along the Mediterranean coast. As our conference site we chose the more modest Talya Hotel because of its pleasant atmosphere and central location, within easy walking distance of shops, restaurants, and the lovely ancient harbor. The hotel sits on a cliff overlooking the sea (to get to the pier from the hotel lobby, one takes an elevator down 7 flights). Most participants stayed at the Talya and will remember its excellent buffet-style breakfasts and dinners, which gave us many good opportunities to interact socially.

The Local Organizing Committee included astronomers from each of Turkey's principal astronomy centers, as well as several of their graduate students. The original membership consisted of Zeki Aslan (Chair) and Orhan Gölbaşı from Antalya; Çetin Bolcal, Hülya Çalışkan, Levent Denizman, Dursun Koçer, Tuba Koktay, and Talat Saygaç from Istanbul; Cafer Ibanoğlu and Varol Keskin from İzmir; and Osman Demircan from Ankara. Most of the above attended the symposium and provided the essential support needed for the smooth running of the meeting. In addition, several other Turkish astronomers and students offered help at the symposium, and in particular we thank Cahit Yeşilyaprak and Şerafettin Yaltkaya, both of the Physics Department of Akdeniz University, for their assistance.

The final count of participants was 142, including 113 from outside Turkey. They came from 32 countries, representing all of the world's continents except Antarctica. In addition there were 20–25 accompanying persons from abroad, and an unknown number from the host country. A List of Participants follows this Preface. There I have given affiliations as of the time of the conference, since that is part of the symposium record, but I have given current (1999) email addresses as far as possible in order to make them more useful for communication. Many participants have changed affiliations since the symposium, as is often apparent from discrepancies between the listed affiliations and email addresses. If the reader needs the current postal address of a participant, I would suggest sending email and requesting it directly from the participant. In compiling the List of Participants, I discovered that alphabetizing the list of names was not a trivial matter. Accented letters in Turkish are considered separate letters, usually directly following their unaccented counterparts in the alphabet;

thus Çalışkan and Çay come after Cunha, Göğüş follows Gong, and so on. To be consistent, Jørgensen follows Jorissen and Joyce, since ø comes at the end of the Danish alphabet. But the French letter è is *not* a separate letter, but merely an accented *e*, so Lèbre come before LeSqueren. Is everyone with me?

The conference photo on p. xxviii was taken on an interior stairway of the Talya Hotel leading to the lower level where the conference room was located.

The SOC was particularly pleased by the attendance of a substantial number of young astronomers (students and post-docs), who for the most part were relatively unknown to the organizers before the symposium but who made great contributions to its success. On the other hand, we were unable to secure the attendance of the most senior members of our field, perhaps because of the distance most of them would have had to travel.

Here I would like to mention several of the real pioneers of our discipline, whose work has greatly influenced many of the topics we discussed. Much of our knowledge of the occurrence and statistics of peculiar red giants has come from surveys of our Galaxy and the Magellanic Clouds carried out by Bengt Westerlund, Victor Blanco, Martin McCarthy, and Bruce Stephenson. The recognition and classification of the various sub-groups of peculiar stars owes much to Philip Keenan, William P. Bidelman, Y. Yamashita, and Carlos Jaschek. Abundance determinations for cool stars got their start with the work of Yoshio Fujita, and our modern understanding of the interiors and evolution of red giants is largely based on the work of Icko Iben, Jr. Philip Keenan was almost persuaded to give a talk on the classification of carbon stars, 55 years after his classic paper with Morgan on the subject, but in the end he decided against making the trip. Billy Bidelman was invited to give an historical talk to put the symposium in perspective, and although he declined on his doctor's advice, he did send a short paper which I read to open the conference. These leaders of our field were all alive at the time of the symposium (Carlos Jaschek has since died) and sufficiently active to be likely to read these proceedings. Other names could of course be added, and I certainly don't wish to offend anyone by omission, but I want at least these gentlemen to know that although they did not attend Symposium 177, they were most certainly in the thoughts of its participants.

Two Turkish pioneers of astronomy played significant roles at Symposium 177 and should be mentioned here. One of them, Nüzhet Gökdoğan, received her Ph.D. in Astronomy at Istanbul University in 1937, at a time when the only astronomers there were foreign-born; the other, Paris Pişmiş, received her Ph.D. from the same university in the same year, but in Mathematics. Both were ground-breakers, being among the first women to study the sciences in the new Republic of Turkey at the time of Atatürk. After receiving their degrees, however, their careers diverged, at least geographically. Nüzhet Gökdoğan, an authority in solar physics, stayed in Turkey and was a leading force in the development of modern astronomy in that country; over the years she supervised the dissertations of many of today's leading Turkish astronomers, including members of our LOC. Paris Pişmiş, meanwhile, accepted an appointment as assistant at Harvard College Observatory shortly before the outbreak of World War II; at Harvard she met her husband, a Mexican graduate student, and moved with him to Mexico, where she had great impact on the development of that country's fledgling astronomy programs. We were delighted that Drs. Gökdoğan and Pişmiş both expressed great interest in Symposium 177. Dr. Gökdoğan had guided Turkey's admission to the IAU in 1961 and so was especially pleased to take part in Turkey's first IAU symposium. One of her former students, LOC member Dursun Koçer, made sure that she attended the welcoming reception in the gardens of the Antalya Archaeological Museum, giving us an opportunity to present her with a certificate of appreciation (see p. xxvi). Dr. Pişmiş had accepted our invitation to give a paper on planetary nebulae and their progenitors, but unfortunately she was prevented from traveling by a broken hip. When we reached her spot on the program, several participants spoke movingly about her influence and contributions. These Proceedings include the abstract of her intended talk, submitted just before the symposium. It is with deep sorrow that I must now add that Paris Pişmiş died in August 1999.

Several social activities were arranged for participants and their guests, and photos from these occasions will be found scattered throughout the volume. The first of these was the opening reception and cocktail, held in the gardens of the Antalya Archaeological Museum amid Roman statuary and (live) peacocks, at the end of the first day. It was attended by officials from Akdeniz University and the City of Antalya, and we were treated to a guided tour of the museum collection, an illustrated lecture by a local archaeologist, and snacks accompanied by the ever-present traditional Turkish *rakı*.

On the third day of the symposium, papers were scheduled for only half a day to allow time for a refreshing bus tour to the archaeological sites of Perge and Aspendos. Both ancient cities had been visited by Alexander the Great, although most of what one sees today is from a later (Roman) period. The large amphitheater at Aspendos is said to be the best-preserved theater of the classical world.

The conference banquet was held at a marina west of the city, outdoors but under cover. We were treated to drinks, a sumptuous spread of food, and belly-dancing lessons.

On the Saturday following the conclusion of the Symposium, participants were invited to a full-day excursion to Bakırlıtepe, the site of the new Turkish National Observatory. This peak, located about 50 km west of Antalya, had been selected by a site-survey team led by our LOC Chair Zeki Aslan (see p. 507), who continues as Head of the Observatory Project. Several photos from this excursion, including one of the mountain itself (p. 12), are included in this volume. At the time of our visit, construction was well along on the dormitory and the building for the smaller of two telescopes, and was about to begin on the building for the larger telescope. The smaller telescope, a 0.4–m photometric telescope installed in collaboration with the University of Utrecht, has now been in operation for some time. The larger facility, the 1.5–m Kazan University telescope employing a Russian-made mirror, is now in place and has recently seen first light at the Cassegrain focus. A description of the high-resolution spectrometer designed for the coudé focus is given on p. 569. The Observatory was officially opened on September 5, 1997, with Turkey's President in attendance.

It is my sad duty to report that two contributors to these Proceedings, in addition to Paris Pişmiş, have died since the symposium was held. Chris Skinner, an observer of circumstellar shells who did not attend the symposium but co-authored two of its posters, passed away tragically in October 1997 just as his career was starting to blossom. Then in July 1998 the Latvian theoretician Jurij Frantsman died suddenly while returning from a conference in Canada. The full text of his paper is included here.

Death has also struck the editor's family, not once but twice while work on this volume was in progress. My mother, Charlotte Wing, died in March 1997 at the age of 88. On behalf of the SOC, I would like to thank her for her support of the Symposium and for her financial assistance, which allowed us to provide small but important subsistence grants to 30 of the participants, many of whom would not otherwise have been able to attend. As my family was getting over that loss, my wife Ingrid — whom many participants met in Antalya — learned that her cancer had returned, this time in a more aggressive form. A two-year barrage of radiation and chemotherapy, including several experimental varieties, may have slowed the progress of the disease but did not change the inevitable outcome, which occurred in October 1999.

To say that these circumstances have interfered with the editorial work on this volume would be an understatement. But it would also be misleading to imply that these were the only factors that slowed the work. Of the many other contributors, I should mention my own serious miscalculation of the amount of work involved, and my pedantic insistence on uniformity of style throughout the volume. Perhaps unfortunately, I tried to referee as well as edit each paper and entered into email discussions with nearly every author about details that weren't clear to me. Although I believe this process has resolved ambiguities and brought about many small improvements, it has also caused unacceptable delays. I apologize to all authors for taking so long to get their papers into print, and I thank them for their patience and understanding.

The papers in these Proceedings are grouped into broad topics and, with minor exceptions, appear in the order in which they were presented in Antalya. Abstracts of posters appear at the end of the volume, arranged alphabetically by first author.

I thank Michael Feast and Jeffrey Linsky for sharing with SOC members the task of chairing the paper sessions. All SOC members deserve thanks for their numerous communications and ideas during the planning stages, and for their active participation in Antalya.

The discussion that followed most of the talks was recorded by Turkish student volunteers who handed sheets of paper to each discussant. I hope that my subsequent editing has not cleansed the comments of too much of their original flavor.

Although I won't mention them by name, I want to acknowledge that many authors of papers in this volume provided important assistance by email. Most of what I know about LATEX I learned while editing these proceedings, and the authors were my teachers. I also thank the undergraduate students at Ohio State University who helped at different times: Melissa Holdren mounted several of the figures that required manual intervention, and Sarah Rayburn carried out various sorting and proof-reading tasks.

The photos appearing on many of the blank pages are mostly from my own camera (as digitally remastered in black & white at the Ohio State University Cop–Ēz center), and I apologize for the perhaps too frequent appearance of family members in them. I also thank Sunetra Giridhar, Kunio Noguchi, Irene Little, Janet Mattei, and Jeff Linsky for kindly sending photos, some of which are included on these pages.

The artistic talents of astronomers may not be sufficiently appreciated. Bengt Gustafsson sent his visualization (p. 480) of the traditional Turkish

Hoca story of "Duck Soup," which he told at the end of his talk. And quite recently Pierre North kindly sent to me, as a *souvenir* of the symposium, the sketches of participants that appear on pp. 140, 206, and 302 . I thank him for allowing their inclusion in the proceedings. Pierre points out that they are not portraits, since they were done from memory in his hotel room — he calls them *impressions visuelles* — and he hopes that none of his unwitting subjects will take offense!

A grant from TÜBİTAK, the Science and Technical Research Council of Turkey, allowed us to waive the registration fees of Turkish participants. It also provided bus transportation for the excursion to Bakırlıtepe (with box lunches), as well as computer support.

We thank Akdeniz University for the use of poster boards and other materials, and for the services of computer systems manager Yavuz Kömür, who helped us set up terminals with an Internet connection in the Talya Hotel for the use of participants.

Many participants have told me how pleased they were with their stay at the Talya Hotel. I would like to thank the management and staff of the hotel for maintaining such an excellent facility and for their courteous service, which contributed greatly to the pleasant atmosphere of the Symposium.

Participants and their guests were given Turkish shoulder-bags (*heybeler*) made of kilim material, each one different. I thank Hülya Çalışkan for shopping for these in Istanbul. She was able to get a very good price, simply by agreeing to buy every *heybe* in the store!

I thank Sylvia and Muammer Önder for their help and encouragement, both before and during the symposium. Many participants enjoyed Sylvia's evening workshop on Turkish customs and language, and several of them appreciated her help with shopping (as did the local shopkeepers).

Most of all, I am grateful to Zeki Aslan for his enormous efforts on behalf of the Symposium. On his shoulders fell the tasks of securing the conference rooms, hotel accommodations, and transportation, making arrangements for tours and the banquet, making hotel reservations for everyone, dealing with agents, and making sure that everything was paid for. He also organized the LOC, enlisted the help of numerous Turkish astronomers, students, and secretaries, and obtained the financial support of TÜBİTAK. Zeki, it was a pleasure working with you — *teşekkür ederim*, and may the Turkish National Observatory have clear skies!

Columbus, Ohio
November 1999

Robert F. Wing

# TRIBUTE TO NÜZHET GÖKDOĞAN

Solar physicist Nüzhet Gökdoğan receiving a Certificate of Appreciation from SOC Chair Robert Wing, while LOC Chair Zeki Aslan shows his approval. Prof. Gökdoğan was thanked for her many contributions to Turkish astronomy, including her role as Turkey's first National Representative to the IAU. The brief ceremony took place on May 27, 1996, during the symposium's Welcoming Reception in the gardens of the Antalya Archaeological Museum.

## Biographical Notes

Nüzhet Gökdoğan was born in Istanbul on August 14, 1910. After completing her secondary education she was granted a national scholarship to study Mathematics and Physics in France in 1928. In 1934, with a B.Sc. in Mathematics from the University of Lyon and after a period of work at Paris Observatory, she returned to Turkey to a faculty position in the Astronomy Department of Istanbul University as the first female member of the University's Science Faculty and the only Turkish member of its Astronomy Department. There she prepared her dissertation with the title "Contribution aux recherches sur l'existence d'une matière obscure interstellaire homogène autour du soleil" under the supervision of Prof. E. Finley-Freundlich and obtained her Ph.D. degree in 1937.

During the years 1936-46 she held an appointment in Mathematics at Istanbul Technical University in addition to her work at Istanbul University. In 1948 she was promoted to the rank of Professor at Istanbul University. Soon thereafter she was elected by Faculty Council to the University Senate, and in 1954 she became Dean of the Faculty. Thus Prof. Gökdoğan had the honor of being the first lady senator and Dean in Turkey.

From 1958 to 1980 Prof. Gökdoğan directed the Astronomy Department, which expanded from 5 to 18 staff members, all trained by her. She supervised 11 Ph.D. dissertations and translated several textbooks from both English and French. She also developed collaborative research programs with the Observatories of Meudon and Nice in France, Basel in Switzerland, and Asiago and Padova in Italy.

Prof. Gökdoğan was the founder of the Turkish Astronomical Society and served as its President for 20 years. She was also the first Turkish national representative to the International Astronomical Union (which Turkey joined in 1961) and was a member of its Commission 12 on Radiation and Structure of the Solar Atmosphere. She was also one of the founders of the Turkish University Women's Association. She has organized several conferences.

Prof. Gökdoğan, who speaks French, English, German, and Greek fluently, was married to Prof. Mukbil Gökdoğan, who died in 1992. Their two children are both university professors: Gönül Gökdoğan is Professor of Music at Istanbul University, and Can Gökdoğan is Professor of Medicine in the Cerrahpaşa Medical Faculty of Istanbul University.

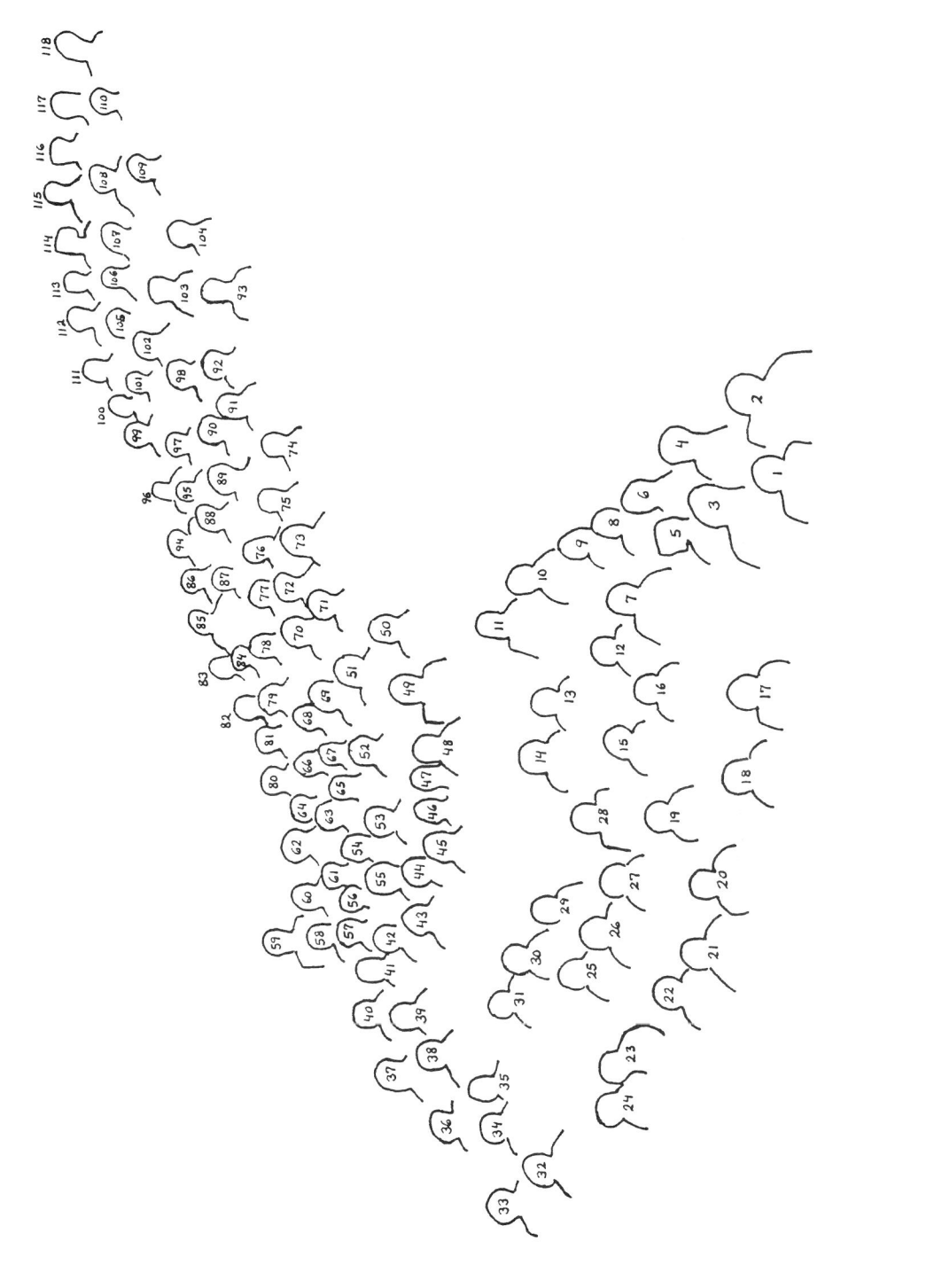

# CONFERENCE PHOTO IDENTIFICATIONS

1. Željko Ivezić 2. John Lattanzio 3. Fatih Berktaş 4. Moshe Elitzur 5. Cengiz Sezer 6. Ömer Değirmenci 7. Robert Wing 8. France Allard 9. Angela Speck 10. James Brewer 11. Jacco van Loon 12. Sandy Yorka 13. Serge Demers 14. Wako Aoki 15. Marie-Odile Mennessier 16. Anne-Marie LeSqueren 17. Andreas Gauger 18. Varol Keskin 19. Irene Little-Marenin 20. Tuncay Özdemir 21. Eric Bakker 22. Gerard Muratorio 23. Michael Feast 24. Robert McClure 25. François Querci 26. Bertrand Plez 27. Katia Cunha 28. Mário Magalhães 29. Nils Ryde 30. Martin Asplund 31. Dimitri Doikov 32. Janet Mattei 33. Talat Saygaç 34. Tsitsino Simonia 35. Irakli Simonia 36. Kenes Kuratov 37. Bernhard Aringer 38. Victor Zubko 39. Orhan Gölbaşı 40. Mutlu Yıldız 41. Keiichi Ohnaka 42. Hideyuki Izumiura 43. Michelle Creech-Eakman 44. Nicole van der Bliek 45. Michael Lindqvist 46. Silvia Lorenz Martins 47. Margit Kipper 48. Tony Kipper 49. Jay Frogel 50. Paul Green 51. Franz Kerschbaum 52. Michael Mickelson 53. Fredrik Larsen 54. Vanni Calamai 55. Michel Grenon 56. Takashi Iijima 57. Kunio Noguchi 58. Hans Olofsson 59. Hans Van Winckel 60. Isabelle Cherchneff 61. Stefano Bagnulo 62. Omar Kurtanidze 63. Bengt Gustafsson 64. Jeffrey Linsky 65. Anita Richards 66. Patricia Whitelock 67. M. Parthasarathy 68. Rita Loidl 69. Sophie Van Eck 70. Nami Mowlavi 71. Donald Luttermoser 72. Jan Martin Winters 73. Uffe Jørgensen 74. Inma Dominguez 75. Takashi Tsuji 76. Christiane Helling 77. Beate Patzer 78. Nathan Netzer 79. Josef Hron 80. Jurij Frantsman 81. Andrejs Alksnis 82. Rodrigo Alvarez 83. Agnès Lèbre 84. Peter Sarre 85. Axel Fleischer 86. Martin Groenewegen 87. Tom Lloyd Evans 88. Josef Wagenhuber 89. Anja Andersen 90. Takashi Kozasa 91. Carlos Abia 92. Laimons Začs 93. Nüzhet Gökdoğan 94. Ernst Dorfi 95. Susanne Höfner 96. Sunetra Giridhar 97. Detlef Schönberner 98. Matthias Steffen 99. Ed Totten 100. Ken Hinkle 101. Pierre North 102. Tom Ake 103. Alain Jorissen 104. Dursun Koçer 105. Irmela Bues 106. Karin Sedlmayr 107. Erwin Sedlmayr 108. Aleksandra Borysow 109. Ilfan Bikmaev 110. Don Fernie 111. Dick Joyce 112. Bruce Hrivnak 113. Jack MacConnell 114. Frank Molster 115. Russell Cannon 116. Zeki Aslan 117. Cahit Yeşilyaprak 118. Serdar Özer

# List of Participants

*(Affiliations are given as of the time of the symposium; e-mail addresses have been updated in 1999 to the best of the editor's knowledge; Registered Guests are listed under the corresponding participant.)*

**Carlos Abia**, Dpt. Física Teórica y del Cosmos, Universidad de Granada, Granada, SPAIN    cabia@goliat.ugr.es
**Hasan Ak**, Ankara Univ. Observatory, Ankara, TURKEY    ak@astro1.science.ankara.edu.tr
**Thomas B. Ake**, Goddard Space Flight Center, Greenbelt MD, USA    ake@pha.jhu.edu
   Barbara N. Ake
**Berahitdin Albayrak**, Ankara Univ. Observatory, Ankara, TURKEY    berahitdin@astro1.science.ankara.edu.tr
**Andrejs Alksnis**, Radioastrophysical Observatory, Riga, LATVIA    aalksnis@acad.latnet.lv
**France Allard**, Department of Physics, Wichita State University, Wichita KS, USA    fallard@ens-lyon.fr
**Rodrigo Alvarez**, GRAAL, Université Montpellier II, Montpellier, FRANCE    ralvarez@astro.ulb.ac.be
**Anja C. Andersen**, University Observatory, Niels Bohr Institute, Copenhagen, DENMARK    anja@nbivms.nbi.dk
**Wako Aoki**, Institute of Astronomy, The University of Tokyo, Osawa, Mitaka, Tokyo, JAPAN    waoki@mtk.ioa.s.u-tokyo.ac.jp
**Bernhard Aringer**, Institut für Astronomie, Vienna, AUSTRIA    aringer@astro.ast.univie.ac.at
**Turgut Aslan**, Astronomisches Institut der FAU, Dr. Remeis-Sternwarte, Bamberg, GERMANY    aslan@sternwarte.uni-erlangen.de
**Zeki Aslan**, Physics Dept., Akdeniz University, Antalya, TURKEY    aslan@pascal.sci.akdeniz.edu.tr
**Martin Asplund**, Astronomical Observatory, Uppsala, SWEDEN    martin@astro.uu.se
**Marc Azzopardi**, Canada-France-Hawaii Telescope, Kamuela HI, USA    azzopardi@observatoire.cnrs-mrs.fr
**Stefano Bagnulo**, Armagh Observatory, Armagh, N. IRELAND    sba@star.arm.ac.uk
**Eric J. Bakker**, Dept. of Astronomy, Univ. of Texas, Austin TX, USA    e.j.bakker@fel.tno.nl

**Ilfan Bikmaev**, Engelhardt Astronomical Observatory, Kazan State University, Tatarstan, RUSSIA    ilfan.bikmaev@ksu.ru

**Çetin Bolcal**, Istanbul University Observatory, İstanbul, TURKEY    ebolcal@yahoo.com

**Aleksandra Borysow**, University Observatory, Niels Bohr Institute, Copenhagen, DENMARK    aborysow@stella.nbi.dk

**James P. Brewer**, European Southern Observatory, Santiago, CHILE    jbrewer@eso.org

**Irmela Bues**, Dr. Remeis-Sternwarte, Bamberg, GERMANY    bues@sternwarte.uni-erlangen.d400.de

**Giovanni Calamai**, Osservatori di Arcetri, Firenze, ITALY    vanni@arcetri.astro.it

**Russell D. Cannon**, Anglo-Australian Observatory, Epping NSW, AUSTRALIA    rdc@aaoepp.aao.gov.au

**Isabelle Cherchneff**, Physics Dept., UMIST, Manchester, UK    imc@ast.ma.umist.ac.uk

**Michelle J. Creech-Eakman**, Dept. of Physics & Astronomy, Univ. of Denver, Denver CO, USA    mce@huey.jpl.nasa.gov

**Katia Cunha**, Departamento de Astrofísica, Observatorio Nacional-CNPq, Rio de Janeiro, BRAZIL    katia@baade.physics.utcp.edu

**Hülya Çalışkan**, Istanbul University Observatory, İstanbul, TURKEY    caliskan@istanbul.edu.tr

**Taşkın Çay**, Istanbul University Observatory, İstanbul, TURKEY    ipek@istanbul.edu.tr

**Ömer L. Değirmenci**, Ege Univ. Observatory, İzmir, TURKEY    omerd@astronomy.sci.ege.edu.tr

**Serge Demers**, Département de physique, Université de Montréal, Montréal QC, CANADA    demers@astro.umontreal.ca
    Helene Forest

**Dimitri N. Doikov**, Astronomy Observatory, Odessa State University, Odessa, UKRAINE    jvi@dtp.odessa.ua (for D.N.Doikov)

**Ekaterina M. Doikova**, Astronomy Observatory, Odessa State University, Odessa, UKRAINE    jvi@dtp.odessa.ua (for E.M.Doikova)

**Inma Dominguez**, Dpt. Física Teórica y del Cosmos, Universidad de Granada, Granada, SPAIN    inma@goliat.ugr.es

**Ernst A. Dorfi**, Institut für Astronomie, Vienna, AUSTRIA    ead@astro.ast.univie.ac.at

**Moshe Elitzur**, Dept. of Physics & Astronomy, University of Kentucky, Lexington KY, USA    moshe@pa.uky.edu

Michael W. Feast, Astronomy Dept., University of Cape Town,
    Rondebosch, SOUTH AFRICA    mwf@uctvms.uct.ac.za
  Con Feast
J. D. Fernie, David Dunlap Observatory, Richmond Hill ON, CANADA
    fernie@astro.utoronto.ca
  Yvonne Fernie
Axel J. Fleischer, Institut für Astronomie & Astrophysik, Technische
    Universität Berlin, Berlin, GERMANY
    fleischer@physik.TU-Berlin.de
Jurij Frantsman, Radioastrophysical Observatory, Riga, LATVIA
    frants@acad.latnet.lv
Jay A. Frogel, Department of Astronomy, Ohio State University,
    Columbus OH, USA    frogel@astronomy.ohio-state.edu
Andreas H. Gauger, Los Alamos National Laboratory, Theoretical
    Division, Los Alamos NM, USA
    gauger@bock.physik.tu-berlin.de
Sunetra Giridhar, Indian Institute of Astrophysics, Bangalore, INDIA
    giridhar@iiap.ernet.in
Zhigang Gong, Yunnan Observatory, Kunming, PEOPLE'S REPUBLIC
    OF CHINA    zgong@physics1.usc.edu
Ersin Göğüş, Physics Dept., Middle East Technical University,
    Ankara, TURKEY
Fatma Gök, Physics Dept., Akdeniz University, Antalya, TURKEY
    gok@pascal.sci.akdeniz.edu.tr
Orhan Gölbaşı, Physics Dept., Akdeniz University, Antalya, TURKEY
    ogolbasi@pascal.sci.akdeniz.edu.tr
Paul J. Green, Harvard-Smithsonian Center for Astrophysics,
    Cambridge MA, USA    pgreen@cfa.harvard.edu
  Avivah Goldman
Michel Grenon, Observatoire de Genève, Sauverny, SWITZERLAND
    michel.grenon@obs.unige.ch
Martin Groenewegen, Max-Planck Institut für Astrophysik, Garching,
    GERMANY    groen@mpa-garching.mpg.de
Bengt Gustafsson, Astronomical Observatory, Uppsala, SWEDEN
    bg@astro.uu.se
Mehmet Ali Gülver, Physics Dept., Middle East Technical University,
    Ankara, TURKEY
İpek Hamami, Istanbul University Observatory, İstanbul, TURKEY
    ipek@istanbul.edu.tr
Christiane Helling, Astronomisk Observatorium, Niels Bohr Institutet,
    Copenhagen, DENMARK    helling@physik.tu-berlin.de

**Kenneth Hinkle**, NOAO, Tucson AZ, USA
  hinkle@noao.edu
**Susanne Höfner**, Institut für Astronomie, Vienna, AUSTRIA
  hoefner@astro.ast.univie.ac.at
**Bruce J. Hrivnak**, Department of Physics & Astronomy, Valparaiso
  University, Valparaiso IN, USA   bhrivnak@exodus.valpo.edu
**Josef Hron**, Institut für Astronomie, Universität Wien, Vienna,
  AUSTRIA      hron@astro.ast.univie.ac.at
**Takashi Iijima**, Osservatorio Astrofisico, Università di Padova, Asiago,
  ITALY       iijima@astras.pd.astro.it
**Željko Ivezić**, Dept. of Physics & Astronomy, University of Kentucky,
  Lexington KY, USA      ivezic@astro.princeton.edu
**Hideyuki Izumiura**, Dept. of Astronomy & Earth Sciences,
  Tokyo Gakugei University, Koganei, Tokyo, JAPAN
  izumiura@oao.nao.ac.jp
**Alain Jorissen**, Institut d'Astronomie et d'Astrophysique,
  Université Libre de Bruxelles, Bruxelles, BELGIUM
  ajorisse@astro.ulb.ac.be
**Richard R. Joyce**, NOAO, Tucson AZ, USA
  rjoyce@noao.edu
**Uffe Gråe Jørgensen**, University Observatory, Niels Bohr Institute,
  Copenhagen, DENMARK      uffegj@nbivms.nbi.dk
**Gülçin Kandemir**, Physics Dept., Istanbul Technical University,
  İstanbul, TURKEY      kandemir@itu.edu.tr
**Franz Kerschbaum**, Institut für Astronomie, Vienna, AUSTRIA
  kerschbaum@astro.ast.univie.ac.at
**Varol Keskin**, Ege Univ. Observatory, İzmir, TURKEY
  keskinv@astronomy.sci.ege.edu.tr
**Hüseyin Kılıç**, Physics Dept., Akdeniz University, Antalya, TURKEY
  hkilic@pascal.sci.akdeniz.edu.tr
**Tõnu Kipper**, Tartu Observatory, Tartumaa, Tõravere, ESTONIA
  tk@aai.ee
  Margit Kipper
**Dursun Koçer**, Istanbul University Observatory, İstanbul, TURKEY
  kocer@istanbul.edu.tr
**Tuba Koktay**, Istanbul University Observatory, İstanbul, TURKEY
  tuba@subaru.naoj.org
**Takashi Kozasa**, Dept. of Earth and Planetary Sciences, Kobe University, Kobe, JAPAN   kozasa@icluna.kobe-u.ac.jp
**Kenesken S. Kuratov**, Astrophysical Institute, Almaty, KAZAKHSTAN
  kurt@tsao.academ.alma-ata.su
  Nazira Kuratova

**Omar M. Kurtanidze**, Abastumani Astrophysical Observatory, Abastumani, REPUBLIC OF GEORGIA    okur@abao.kheta.ge
**Kaan K. Kutluata**, Ankara Univ. Observatory, Ankara, TURKEY
   kaan@astro1.science.ankara.edu.tr
**Vladimir P. Kuz'kov**, Main Astronomical Observatory, Kiev, UKRAINE
   kuzkov@mao.gluk.apc.org
**İbrahim Küçük**, Physics Dept., Erciyes University, Kayseri, TURKEY
   kucuk@vm.cc.erciyes.edu.tr
**Fredrik Larsen**, Stockholm Observatory, Saltsjöbaden, SWEDEN
   fredrik@astro.su.se
**John Lattanzio**, Department of Mathematics, Monash University, Clayton, Victoria, AUSTRALIA
   johnl@flash.maths.monash.edu.au
**Agnès Lèbre**, GRAAL, Université Montpellier II, Montpellier, FRANCE
   lebre@graal.univ-montp2.fr
**Anne-Marie LeSqueren**, Arpeges, Observatoire de Meudon, Meudon, FRANCE    lesqueren@mesiob.obspm.fr
   M LeSqueren
**Michael Lindqvist**, Onsala Space Observatory, Onsala, SWEDEN
   michael@oso.chalmers.se
   Maret Sooaru
   Max Lindqvist (2.5 years)
**Jeffrey L. Linsky**, JILA, University of Colorado, Boulder CO, USA
   jlinsky@jila.colorado.edu
   Lois F. Linsky
**Irene R. Little–Marenin**, Whitin Observatory, Wellesley College, Wellesley MA, USA    ilittle@casa.colorado.edu
   Stephen J. Little
**Thomas H. H. Lloyd Evans**, South African Astronomical Observatory, Observatory, SOUTH AFRICA    tle@saao.ac.za
   Marlene Lloyd Evans
**Rita Loidl**, Institut für Astronomie, Vienna, AUSTRIA
   loidl@astro.ast.univie.ac.at
**Silvia Lorenz Martins**, Observatorio Nacional, Rio de Janeiro, BRAZIL
   silvia1@belatrix.on.br
**Donald G. Luttermoser**, Applied Research Corporation, Landover MD, USA    lutter@access.etsu.edu
**D. Jack MacConnell**, Space Telescope Science Institute / CSC, Baltimore MD, USA    macconnell@stsci.edu
**Antonio Mário Magalhães**, Instituto Astronômico e Geofísico, Universidade de São Paulo, São Paulo, BRAZIL
   mario@argus.iagusp.usp.br

**Janet Akyüz Mattei**, AAVSO, Cambridge MA, USA
jmattei@aavso.org
**Robert D. McClure**, Dominion Astrophysical Observatory,
Victoria BC, CANADA     mcclure@dao.nrc.ca
**Marie-Odile Mennessier**, GRAAL, Univ. Montpellier II,
Montpellier, FRANCE     menes@graal.univ-montp2.fr
**Michael E. Mickelson**, Dept. of Physics & Astronomy, Denison
University, Granville OH, USA     mickelson@cc.denison.edu
Diana Mickelson
**Frank Molster**, Astronomical Institute "Anton Pannekoek,"
Amsterdam, THE NETHERLANDS     frankm@astro.uva.nl
**Nami Mowlavi**, Observatoire de Genève, Sauverny, SWITZERLAND
nami.mowlavi@obs.unige.ch
**Gerard Muratorio**, Observatoire de Marseille, Marseille, FRANCE
muratorio@obmara.cnrs-mrs.fr
**Nathan Netzer**, ORT Braude College, Karmiel, ISRAEL
nanetzer@ort.org.il
**Maria G. Nikolashvili**, Abastumani Astrophysical Observatory,
Abastumani, REPUBLIC OF GEORGIA
mgnik@abao.kheta.ge
**Kunio Noguchi**, National Astronomical Observatory, Osawa, Mitaka,
Tokyo, JAPAN     knoguchi@optik.mtk.nao.ac.jp
**Pierre North**, Institut d'Astronomie, Université de Lausanne,
Chavannes-des-Bois, SWITZERLAND
pierre.north@obs.unige.ch
**Keiichi Ohnaka**, Institute of Astronomy, The University of Tokyo, Osawa,
Mitaka, Tokyo, JAPAN     ohnaka@mtk.ioa.s.u-tokyo.ac.jp
**Hans Olofsson**, Stockholm Observatory, Saltsjöbaden, SWEDEN
hans@astro.su.se
**Sacit Özdemir**, Ankara Univ. Observatory, Ankara, TURKEY
sacit@astro1.science.ankara.edu.tr
**Tuncay Özdemir**, Physics Dept., Inonu University, Malatya, TURKEY
tuncay@inonu.edu.tr
**Serdar U. Özer**, Physics Dept., Akdeniz University, Antalya, TURKEY
serdar@pascal.sci.akdeniz.edu.tr
**M. Parthasarathy**, Indian Institute of Astrophysics, Koramangala,
Bangalore, INDIA     partha@iiap.ernet.in
**Beate Patzer**, Institut für Astronomie & Astrophysik,
Technische Universität Berlin, Berlin, GERMANY
patzer@physik.tu-berlin.de

**Bertrand Plez**, Niels Bohr Institute, University Observatory,
  Copenhagen, DENMARK    plez@graal.univ-montp2.fr
**François R. Querci**, Observatoire Midi-Pyrénées, Toulouse, FRANCE
  querci@obs-mip.fr
**Anita M. S. Richards**, Jodrell Bank, Macclesfield, Cheshire, UK
  amsr@jb.man.ac.uk
**Nils Ryde**, Astronomical Observatory, Uppsala, SWEDEN
  ryde@astro.uu.se
**Peter J. Sarre**, Department of Chemistry, The University of Nottingham,
  Nottingham, UK    psa@star.le.ac.uk
**Talat Saygaç**, Istanbul University Observatory, İstanbul, TURKEY
  saygac@istanbul.edu.tr
**Detlef Schönberner**, Astrophysikalisches Institut Potsdam,
  Potsdam, GERMANY    DeSchoenberner@aip.de
**Erwin Sedlmayr**, Institut für Astronomie & Astrophysik, Technische
  Universität Berlin, Berlin, GERMANY
  sedlmayr@physik.tu-berlin.de
**Karin Sedlmayr**, Institut für Astronomie & Astrophysik, Technische
  Universität Berlin, Berlin, GERMANY
  sedlmayr@physik.tu-berlin.de
**Cengiz Sezer**, Ege Univ. Observatory, İzmir, TURKEY
  sezerc@astronomy.sci.ege.edu.tr
**Ian Shelton**, David Dunlap Observatory, Richmond Hill ON, CANADA
  shelton@subaru.naoj.org
**Irakli Simonia**, Tbilisi Laboratory of the Abastumani Astrophysical
  Observatory, Tbilisi, REPUBLIC OF GEORGIA
  root@aod.ge
**Tsitsino Simonia**, Tbilisi Laboratory of the Abastumani Astrophysical
  Observatory, Tbilisi, REPUBLIC OF GEORGIA
  root@aod.ge
**Verne V. Smith**, Department of Physics, Univ. of Texas at El Paso,
  El Paso TX, USA    verne@balmer.physics.utep.edu
**Angela K. Speck**, Physics & Astronomy Department,
  University College London, London, UK    aks@star.ucl.ac.uk
**Matthias Steffen**, Astrophysikalisches Institut Potsdam,
  Potsdam, GERMANY    msteffen@aip.de
**Mehmet Tanrıver**, Ankara Univ. Observatory, Ankara, TURKEY
  tanriver@astro1.science.ankara.edu.tr
**Ed J. Totten**, Dept. Pure and Applied Physics, Queen's
  University Belfast, Belfast, NORTHERN IRELAND
  e.totten@queens-belfast.ac.uk

**Takashi Tsuji**, Institute of Astronomy, The University of Tokyo, Osawa, Mitaka, Tokyo, JAPAN    ttsuji@mtk.ioa.s.u-tokyo.ac.jp
**Nicole S. van der Bliek**, Sterrewacht Leiden, Leiden, THE NETHERLANDS    nvdbliek@not.iac.es
**Sophie Van Eck**, Institut d'Astronomie et d'Astrophysique, Bruxelles, BELGIUM    svaneck@astro.ulb.ac.be
**Jacco Th. van Loon**, European Southern Observatory, Garching bei München, GERMANY    jacco@ast.cam.ac.uk
**Hans Van Winckel**, Instituut voor Sterrenkunde, K.U. Leuven, Heverlee, BELGIUM    hans@ster.kuleuven.ac.be
**Miroslav Vetešnik**, Dept. of Astrophysics, Masaryk University, Brno, CZECH REPUBLIC    vetesnik@astro.sci.muni.cz
   Jitka Vetešnikova
**Josef Wagenhuber**, Max-Planck-Institut für Astrophysik, Garching, GERMANY    jow@mpa-garching.mpg.de
**Patricia A. Whitelock**, South African Astronomical Observatory, Observatory, SOUTH AFRICA    paw@saao.ac.za
**Robert F. Wing**, Astronomy Department, Ohio State University, Columbus OH, USA    wing@astronomy.ohio-state.edu
   Ingrid M. Wing
   Sylvia Wing Önder
   Muammer Önder
   Timur A. Önder (3 months)
**Jan Martin Winters**, Institut für Astronomie & Astrophysik, Technische Universität Berlin, Berlin, GERMANY    winters@physik.TU-Berlin.de
**Cahit Yeşilyaprak**, Physics Dept., Akdeniz University, Antalya, TURKEY    cahit@pascal.sci.akdeniz.edu.tr
**Mutlu Yıldız**, Physics Dept., Middle East Technical University, Ankara, TURKEY    yildiz@astronomy.sci.ege.edu.tr
**Sandra B. Yorka**, Dept. of Physics & Astronomy, Denison University, Granville OH, USA    yorka@cc.denison.edu
   Barbara Bruce
**Kutluay Yüce**, Ankara Univ. Observatory, Ankara, TURKEY    kyuce@astro1.science.ankara.edu.tr
**Laimons Začs**, Radioastrophysical Observatory, Baldone, LATVIA    zacs@acad.latnet.lv
**Victor Zubko**, Institute of Astronomy, N. Copernicus University, Toruń, POLAND    zubko@astri.uni.torun.pl

# INTRODUCTION

Hülya Çalışkan and Sylvia Önder paying homage to Ulugh Beg (1394-1449) outside the Physics Department of Istanbul University (from symposium "scouting expedition" of August 1994).

# INTRODUCTORY REMARKS: WATCH THE DETAILS

*How to Deal with Unexpected Scientific Results*

WILLIAM P. BIDELMAN
*Warner & Swasey Observatory*
*Case Western Reserve University*
*Cleveland, OH 44106, U.S.A.*

I have always been fascinated by red stars, partly because they were so easy to find in a telescope field, and partly also because they seemed to warm things up a bit on a cold winter night. Thus I early became acquainted with the celebrated carbon star 280 Schjellerup, better known as WZ Cassiopeiae, which was recognized about 100 years ago as an unusual member of a group of red stars then known as the stars of Secchi's fourth type. The Upsala astronomer Dunér (1899) stated that "so far as my experience goes the great strength in the spectrum of this star of band 4 (= sodium 'D'), combined with the remarkable faintness of band 6 (= the $\lambda 5635$ sequence of $C_2$), is met with in the same degree in no other spectrum." Remarkably good photographic spectrograms, reproduced by Hale, Ellerman & Parkhurst (1903), confirmed Dunér's visual observation that the carbon bands in WZ Cas were "relatively very feeble." The Yerkes investigators also published a wavelength table for this star extending from $\lambda 4430$ to $\lambda 5786$ which to my knowledge is the only one yet published, though Dr. Keenan is planning to remedy this to some extent. We now know, of course, that WZ Cas is the brightest object in which the atmospheric abundance ratio of carbon to oxygen is very close to one, though this situation is no doubt rather temporary. The star has many other interesting properties, including strong Li I, $H_2$, CO, etc.

My purpose here, however, is not to discuss this stellar oddball in any detail, but rather to make the point that sometimes observers note things that don't seem to make much sense, but which later are realized to have been very significant indeed. A good example is the statement in Hale et al. (1903), referring to the $\lambda 4606$ blue CN sequence, that "for some reason the maximum intensity of these flutings seems to have been attained in so slightly developed a fourth-type star as 280 Schjellerup." Some 12 years later the aptly named Michigan red star investigator W. C. Rufus (1915), in his study of the spectra of the R and N stars, noted that "the cyanogen bands appear to be somewhat stronger in the spectra of class N stars (than in class R), especially $\lambda 4553.6$ and $\lambda 4606$." And finally, in 1928, C. D. Shane emphasized that "the behavior of the $\lambda 4606$ CN group in the carbon stars is not typical of the entire (blue) cyanogen spectrum." Since the same lower electronic state is involved this was rather surprising.

The mystery was solved, as far as I was concerned, with the discovery at the Dearborn Observatory of the very weak-banded star BD $+15°726 =$ GP Orionis (Bidelman 1950[1]). This object exhibited an extremely strong feature near $\lambda 4606$ that could not possibly be due to cyanogen, but was in fact due to the resonance line of neutral strontium. In my innocence I assumed at the time that the great strength of the low-lying lines in GP Ori, WZ Cas, and other similar objects was solely a consequence of low temperature, which is certainly only part of the story. My student Courtney Gordon (1968) was far ahead of me in recognizing the importance of atomic abundance variations among carbon stars.

However, I must here record the embarrassing fact that the Russian astronomer G. A. Shajn had already pointed out (in English!) in 1942 that the $\lambda 4606$ cyanogen sequence heads were likely to be severely blended with the resonance lines of Sr I and Ba II. I was a graduate student at the time, concerned with more important matters.

I hope you get my point: that if something seems a bit strange it is worth doing some serious thinking to try to make sense of it. At the very least do tell others about it; though perhaps hard to believe, they may be smarter than you! This policy may not make you popular with the establishment but the risk is well worth taking.

### References

Bidelman, W. P. 1950, *ApJ*, 112, 219
Dunér, N. C. 1899, *ApJ*, 9, 119
Gordon, C. P. 1968, *ApJ*, 153, 915
Hale, G. E., Ellerman, F. & Parkhurst, J. A. 1903, *Publ. Yerkes Obs.*, Vol. 2, p. 251
Rufus, W. C. 1915, *Publ. Michigan Obs.*, Vol. 2, p. 139
Shajn, G. A. 1942, *Bull. Abastumani Astr. Obs.*, No. 6, p. 1
Shane, C. D. 1928, *Lick Obs. Bull.*, Vol. 13, p. 125 (No. 396)

---

[1] A somewhat more informative photograph of the spectrum appears in Vol. I of *Stellar Astronomy*, ed. H.-Y. Chiu, R. L. Warasila and J. L. Remo (New York: Gordon & Breach), p. 203, 1969.

## Discussion

**Wing:** I thank Dr. Bidelman for sending these remarks to open our conference and am sorry that he couldn't be here to present them himself.

We are here to discuss the many kinds of stars that have managed — whether by themselves or with help from a companion — to change their surface chemical compositions. This large field of study began with the recognition of various groups of spectroscopically peculiar stars. Dr. Bidelman's sharp eye for noticing subtle spectral peculiarities, and his role in recognizing and defining the properties of the Barium stars (Bidelman & Keenan 1951, *ApJ*, 114, 473), are well known. Less well known is his contribution to the idea that stars of types M, S, and C form a continuous sequence in which the C/O ratio is the significant variable. The idea itself is fairly old, going back at least to the work of Y. Fujita (1939, *Japanese J. Astron. Geophys.*, 17, 17) who showed that many of the properties of S stars could be accounted for by simply increasing the C/O ratio of an M star. Bidelman's contribution, a decade later, was to find real stars to fill out the M–S–C sequence and to identify examples of the intermediate MS and SC classes. The clincher was his observation that the highly unusual spectrum of GP Ori has exactly the characteristics expected for a star in which O and C are precisely balanced. This 9th-magnitude star was thus shown to link the SC stars that show ZrO bands to the SC stars that show $C_2$, thereby tying together the whole M–MS–S–SC–C sequence.

Papers that have great impact on individuals are not always the well-known classics. Bidelman's fullest discussion of the pivotal position of GP Ori in the M–S–C sequence was given at one of the early Liège Colloquia (1954, *Mém. 8° Soc. Roy. Sci. Liège*, 4th ser., Vol. 14 [Liège Contr. No. 357], p. 402). I have seldom seen citations to that paper, but its effect on me was profound. When I stumbled across it as a graduate student in 1964, on a cloudy night at Lick Observatory, I suddenly had the exhilarating feeling that many things were starting to make sense. That paper influenced the direction of my dissertation, and consequently of my entire career, up to and including the organization of this conference.

# WHAT THEORISTS THINK THEY KNOW ABOUT AGB STARS

JOHN C. LATTANZIO
*Department of Mathematics, Monash University*
*Clayton, VIC 3168, Australia*

**Abstract.** In this paper I discuss our basic picture of the structure of AGB stars, and how their earlier evolution led them to this situation.

## 1. Introduction

The present paper is intended as an introduction to the structure of low and intermediate mass Asymptotic Giant Branch stars. It should provide a useful pointer into the literature for non-experts, and also serve to ensure that everyone understands the basic theory so that we can use this to debug the data!

## 2. Discussion

We discuss the evolution of a low mass star $(1\,M_\odot)$ and an intermediate mass star $(5\,M_\odot)$ in a purely qualitative sense. The evolution is followed from the ZAMS through to the AGB. The emphasis is on the internal structure, with particular reference to mixing and nucleosynthesis. After discussing core hydrogen exhaustion and the First Dredge-up, we briefly discuss semiconvection (which is seen in the low mass stars during core helium burning). Following core helium exhaustion the stars ascend the AGB. The more massive models will experience the Second Dredge-Up. Both models converge to a similar qualitative structure as they ascend the AGB and begin their thermally pulsing evolution.

After discussing the detailed structure and convection changes during a thermal pulse cycle, we discuss the position and extent of the $^{13}C$ pocket, which appears necessary for the production of s-process elements in these stars. This $^{13}C$ was initially believed to be burned when it was engulfed by the convective pocket at the next thermal pulse, but recent calculations indicate that the $^{13}C$ burns radiatively (Straniero et al. 1995) during the inter-pulse phase. We also discuss how thermal pulses can produce $^{19}$F (but we defer to Mowlavi et al. 1996 for details).

We then discuss the phenomenon of "hot bottom burning," where the convective envelope penetrates the top of the hydrogen burning shell. This process appears to explain the existence of super Li-rich stars (Boothroyd & Sackmann 1992) as well as the lack of carbon stars at high luminosity (Wood, Bessell & Fox 1983). The effect on other elements, such as CNO and $^{26}$Al, is also discussed. We then emphasise the importance of ion-probe analysis of meteoritic grains as a source of highly accurate quantitative information about stellar compositions (e.g. Zinner 1995).

## 3. Further Reading

Due to space limitations, this paper is not presented in this volume, but two very similar papers can be found in Frost & Lattanzio (1996) and Lattanzio et al. (1996).

## References

Boothroyd, A. I. & Sackmann, I.-J. 1992, *ApJ*, 393, L21
Frost, C. A. & Lattanzio, J. C. 1996, in *Stellar Evolution: What Should Be Done?*, Proceedings of the 32nd Liège Colloquium, ed. A. Noels et al., p. 307
Lattanzio, J. C., Frost, C. A., Cannon, R. C. & Wood, P. R. 1996, *Mem. Astron. Soc. Italia*, 67, 729
Mowlavi, N., Jorissen, A. & Arnould, M. 1996, *A&A*, 311, 803
Straniero, O., Gallino, R., Busso, M., Chieffi, A., Raiteri, C. M., Limongi, M. & Salaris, M. 1995, *ApJ*, 440, L85
Wood, P. R., Bessell, M. S. & Fox, M. W. 1983, *ApJ*, 272, 99
Zinner, E. K. 1995, in *Nuclei in the Cosmos III*, ed. M. Busso, R. Gallino and C. M. Raiteri (AIP Press: New York), p. 567

## Discussion

**Wagenhuber**: Do you use the mixing-length theory for your HBB calculations?

**Lattanzio**: The evolution calculations are carried out with the mixing-length theory and assume instantaneous mixing. For the time-dependent mixing required in the nucleosynthesis calculations we use a phenomenological method due to Cannon (1993, *MNRAS*, 263, 817). This requires a convective velocity and a local mixing length as input. At present we take these from the standard mixing-length theory. But the method is actually a combination of a "conveyor belt" model and a diffusion equation. Although it uses some inputs from the mixing-length theory, it is quite different in construction.

**Plez**: From observations it seems we do not see the signature of the $^{22}$Ne neutron source in any AGB stars. What does the theorist say about that?

**Lattanzio**: The models of low-mass stars do not produce temperatures high enough for this source (except for a very brief period near the peak of the pulse) so this is consistent with the models. Of course, the models do not at present produce the $^{13}$C required either! But this is a problem with the details of convective boundaries, so we can be excused (at least a little).

**Little-Marenin**: Do your models give $s$-process enhancements (such as Tc) and mixing to the surface?

**Lattanzio**: At present we do not calculate any $s$-processing. We include a small iron-peak network so that we can estimate the neutron density, but we do not attempt to determine the details of the $s$-processing. This is at present impossible to do in a self-consistent manner, because it requires a small pocket of $^{13}$C, and how to obtain this is still a matter of some debate. Nevertheless, for reasonable assumptions about the size of this $^{13}$C pocket, Gallino and collaborators can obtain very good agreement with observations of $s$-process elements (see Gallino et al. 1995, in the Proceedings of the "Nuclei in the Cosmos" meeting). The agreement is further improved by the recent discovery that the $^{13}$C burns radiatively during the intershell phase rather than during ingestion into the convective pocket at the next pulse.

**Whitelock**: There are galactic stars with high Li abundances and masses not much more than one solar mass. Do we understand how these form?

**Lattanzio**: The standard picture requires HBB on the AGB, and this requires a larger stellar mass. Boothroyd and Sackmann estimate that the Li-rich AGB stars have masses between about 4 and 6 $M_\odot$, and our calculations agree with that. Nevertheless, Boothroyd (private communication) is able to produce Li enhancements in low-mass first giant branch stars by a process of deep mixing on the giant branch (also known as "cool bottom processing"), if the mixing is not too fast. There seems to be observational evidence for this effect on the first giant branch.

Some of the galactic carbon stars which are rich in $^7$Li may also form from this mechanism: their mass is too small for HBB. But if cool processing occurs on the first giant branch, then it may also occur on the AGB. Of course, if it is rotationally induced then braking on the RGB may mean there is no mixing on the AGB. We just don't know. I believe (Boothroyd, private communication) that some observations (e.g. $^{18}$O abundances) are indicative of mixing on the AGB, but are not conclusive.

# Session I

SURVEYS

Bakırlıtepe, a 2485-m (8150 ft) peak in the Toros range, is the site of the new Turkish National Observatory, headed by Zeki Aslan.

# DEEP IV–N AND YELLOW–RED SPECTRAL SURVEYS FOR CARBON STARS

O. M. KURTANIDZE AND M. G. NIKOLASHVILI
*Abastumani Astrophysical Observatory,*
*383762 Abastumani, Republic of Georgia*

**Abstract.** We present the results of the First and Second Abastumani Objective Prism Spectral Surveys along with other surveys for carbon stars carried out in recent years.

## 1. Introduction

Theoretical models for Asymptotic Giant Branch (AGB) stars indicate that the carbon-star phase may last up to $10^6$ years. It is strongly limited by rapid mass-loss rates of up to $10^{-4}$ $M_\odot$ yr$^{-1}$ (Kleinmann 1989), and it is critical in their evolution since it determines their fate as planetary nebulae and white dwarfs. This lifetime is consistent with the analysis of Claussen et al. (1987), who used the space distribution of high-luminosity carbon stars and their main-sequence progenitors to argue that the lifetime of carbon stars lies between $10^5$ and $10^6$ years (Jura 1991).

Carbon stars are also the most likely sources of carbon grains, PAHs, and any carbon-cage molecules and *s*-process species that may be found in the interstellar medium. Furthermore, a few of them are super-rich in lithium and these objects may be important in synthesizing lithium in the Galaxy. The low effective temperature of carbon stars means that their atmospheres are molecular and hence their winds are primarily in molecular form, and consequently they develop an extensive, cold, molecular circumstellar envelope due to mass loss (Knapp et al. 1990). Not all carbon stars lie on the AGB. The hottest R-type stars may have luminosities of only 100 $L_\odot$.

Carbon stars are promising standard-candle tracers of large-scale galactic structure. They are intrinsically bright and numerous, so they can be identified anywhere in the Galaxy despite dust extinction. Carbon stars with known distances and velocities are powerful indicators of disk dynamics, providing an opportunity to probe the evolutionary history of the Galaxy (Weinberg 1993). They can also play an important role in the estimation of galactic dark matter out to 5 $R_\odot$ (Schechter 1993).

Interest in obtaining an accurate census of these stars by various types of surveys is stimulated by the goals of measuring their range of lifetime in the AGB phase and the extent to which they dominate the chemical evolution of the Galaxy.

Numerous spectral surveys have been carried out in the past several decades. Some of them have been analyzed by Mavridis (1971), McCarthy & Blanco (1978), Blanco & McCarthy (1981), and MacConnell (1995). We will limit our attention to extensive spectral surveys of the galactic plane carried out by means of objective prisms.

## 2. Objective-Prism Surveys

The Henry Draper catalogue, produced at Harvard by Annie J. Cannon and her assistants, was the primary source of data on carbon stars until the 1940s. Starting with the pioneering work of Lee et al. (1940), low-dispersion objective-prism spectra have been widely used for the identification of red stars. Nassau & Velghe (1964) introduced the use of the near-infrared (6800–8800 Å) spectral range for the detection and classification of late-type stars. They showed that carbon stars could be easily identified by their pronounced cyanogen (CN) bands at 7945, 8125 and 8320 Å nearly to the limit of their plates.

### 2.1. THE I–N SPECTRAL SURVEYS

A near-infrared spectral survey of the northern part of the galactic belt was carried out in Cleveland by Nassau and his collaborators (1949, 1950, 1954; 1954a,b,c; 1957, 1964) to a limiting magnitude of about $I = 10.5$, and the distribution of M, C and S stars was studied. All these near-infrared (Kodak I–N + Wr 89 filter), low dispersion (1700 and 3400 Å mm$^{-1}$ at the A-band) spectral surveys were carried out with the 24/36-inch Schmidt telescope of the Warner and Swasey Observatory equipped with 4° or 2° prisms. The area in galactic longitude from $l = 333°$ through 0° to $l = 210°$ and $b = 0°$, ±4° was covered and more than seven hundred carbon stars were found, among them six hundred new ones. An extension to southern declinations was carried out by Blanco & Münch (1955, $200° \leq l \leq 270°$, $|b| \leq 2°$) at Tonantzintla Observatory and by Smith & Smith (1956, $180° \leq l \leq 360°$, $|b| \leq 1°.5$) at Bloemfontein. They identified 172 and 186 carbon stars, respectively. The limiting near-infrared magnitude of these surveys was $I \approx 10.2$.

To extend the galactic belt surveys to fainter magnitudes, a complete low-dispersion spectral survey (2100 Å mm$^{-1}$ at the A-band) of the southern Milky Way, covering a 10° equatorial belt from $l = 235°$ to $l = 7°$, was carried out by Westerlund (1971) with the 50-cm Schmidt telescope at the Uppsala Southern Station at Mount Stromlo Observatory to a limiting infrared magnitude $I = 12.5$. On the basis of the spectral material obtained, 1124 carbon stars were discovered including more than 950 new ones.

Statistics of these surveys are summarized in Table 1, which gives the approximate area of the sky covered in sq. deg., the limiting magnitude in $I$ or $V$, the number of C stars recorded, and the number of these that were new.

TABLE 1. Spectral Surveys of the Milky Way

| Observer | Area | $I$ | $V$ | All | New |
|---|---|---|---|---|---|
| Nassau et al. | 3,700 | 10.5 | | 700 | 600 |
| Westerlund | 1,320 | 12.5 | | 1,124 | 950 |
| Fuenmayor | 644 | 13.0 | | 283 | 123 |
| MacConnell | 1,200 | 13.0 | | | 600 |
| Maehara, Soyano | 1,325 | 13.0 | | 884 | 259 |
| Alksne, Alksnis et al. | 1,200 | | 14.5 | | 318 |
| Kurtanidze, Nikolashvili | 1,500 | | 16.5 | 1,503 | 862 |
| Kurtanidze, Nikolashvili | 350 | 15.5 | | 496 | 384 |
| Nikolashvili, Kurtanidze | 150 | 15.5 | | 257 | 196 |
| Aaronson, Blanco et al. | 400 | 15.0 | | 921 | 423 |
| Stephenson | 2,400 | 14.0 | | 2,100 | |

In the early 1970s, MacConnell (1988, 1995) began taking infrared plates of dispersion 3400 Å mm$^{-1}$ at the A-band with the 61/91-cm Curtis Schmidt telescope at Cerro Tololo Inter-American Observatory to search for cool supergiants and carbon stars. Plates were taken along the full southern galactic half-circle, from $l = 230°$ to $l = 30°$, covering a band 13° wide. His deep plates were ammonia-sensitized and reach $I = 13.5$. The discovery of more than 600 new carbon stars has been reported to date.

During nearly the same period (1978–1985), Maehara and Soyano were making a deep survey for cool carbon stars along the galactic plane with the 105/150-cm Schmidt telescope of the Kiso Station of the Tokyo Astronomical Observatory. They used hypersensitized Kodak I–N plates behind a Schott RG 695 filter and the 4° prism, giving a dispersion of about 1000 Å mm$^{-1}$ at the A-band. Direct IIa–D plates were taken to determine the positions and $V$ magnitudes of the objects discovered. The limiting magnitude of the survey is $I = 13.0$. In a series of papers (Maehara & Soyano 1987a, 1987b, 1988, 1990; Soyano & Maehara 1991, 1993) they published accurate positions, visual magnitudes and finding charts for 884 carbon stars found in the Cassiopea ($l \sim 120°$), anticenter ($l \sim 180°$), Cygnus ($l \sim 90°$), Serpens-Aquila-Scutum ($l \sim 30°$), Perseus-Camelopardalis ($l \sim 150°$) and Monoceros ($l \sim 210°$) regions. The survey covers about 1300 sq. deg. and 272 new carbon stars were identified.

To study the center-anticenter asymmetry in the surface and space distributions of carbon stars in the Galaxy, Fuenmayor (1981) carried out a spectral survey at 1700 Å mm$^{-1}$ on Kodak I–N plates hypersensitized in a bath of distilled water at 5°C for 2 minutes. This procedure resulted in a limiting infrared magnitude of $13\overset{m}{.}0$. Each survey area of 322 sq. degrees located in the galactic center and anticenter directions was observed with similar Schmidt-type telescopes located in the southern and northern hemispheres. As a result, 283 carbon stars were identified, of which 123 were new.

## 2.2. THE YELLOW–RED SPECTRAL SURVEYS

Using the 80/120-cm Schmidt telescope of the Radioastrophysical Observatory at Baldone near Riga, Alksne and collaborators (1989) carried out a green-yellow (A600 films, 600 Å mm$^{-1}$ and 1130 Å mm$^{-1}$ at H$_\gamma$) and near-infrared (Kodak I–N + RG 1, 2500 Å mm$^{-1}$ at the A-band) objective-prism spectral survey of strategically selected northern Milky Way regions ($168° < l < 200°$, $b = \pm 7°$; $84° < l < 96°$ and $172° < l < 180°$, $|b| < 2°.5$; $128° < l < 140°$, $b = +7°$; $80° < l < 96°$, $|b| < 9°.5$). As a result of these surveys they discovered 318 new cool C stars.

In the beginning of 1975 we began laboratory tests at Abastumani Astrophysical Observatory for hypersensitization of the Kodak emulsions 103a–O, IIa–O,D and IIIa–J,F by baking in dry air for 4–10 hours at 50–70°C. Applied to types IIIa–J,F this technique yielded speed gains of a factor of 3 to 5. At first these hypersensitized plates were used for direct photography of nearby clusters of galaxies ($z < 0.05$, $B = 21.5$), and later on at the end of 1977 they were used for low-dispersion spectroscopy for the identification of different kinds of stars. Carbon stars were identified on these plates to a limiting magnitude of $V = 16.5$, which is equivalent to the limit of Westerlund for objects having $V - I \approx 3$. Therefore, in the middle of 1978 we undertook an extensive low-dispersion spectral survey of the northern equatorial belt of the Milky Way for the identification of OB, B8–A3, M, and C stars (Kurtanidze & Nikolashvili 1981). Our principal objective was the extension of Westerlund's southern galactic belt survey to the northern Milky Way.

The spectra obtained for this survey have a dispersion of 1250 Å mm$^{-1}$ at H$_\gamma$. They were taken with the 2° objective prism attached to the 70/98/210-cm meniscus telescope of the Abastumani Observatory. Kodak IIIa–J,F plates hypersensitized in dry air or in nitrogen and exposed for 20 and 50 min yielded an approximate limiting magnitude of $B = 18.5$ (A0 V).

The plate material covers the region of the Milky Way from $l = 30°$ to $l = 165°$ and $l = 195°$ to $l = 210°$, $b = 0°$, $\pm 3°.6$. A survey of the region $165° < l < 196°$ was carried out by Fuenmayor (1981) in the near-infrared. On these plates the groups of OB, B8–A3, M, and C stars are easily distinguished. Carbon stars are identified by the presence of C$_2$ bands at 4737 Å, 5167 Å (IIIa–J,F), and 5635 Å (IIIa–F).

It must be noted that the survey of the region $195° < l < 210°$, $|b| < 5°$ was carried out on Kodak I–N + RG 5 but is included here because the limiting magnitudes of the I–N and J,F surveys are nearly the same for late-type stars.

As a result of the First Abastumani Survey for early and late-type stars, more than 1500 C stars were identified, including more than 860 new ones (Nikolashvili 1987a,b; Kurtanidze & Nikolashvili 1981, 1988b, 1989a,b,c; 1994, 1995), and their distribution was studied.

## 2.3. THE IV–N SPECTRAL SURVEYS

Thanks to all these low-dispersion spectral surveys carried out in the near-infrared (I–N) and yellow-red spectral regions, the Milky Way equatorial ten-degree belt is now uniformly covered to a near-infrared magnitude of $12^m.5 - 13^m.0$.

To probe the outermost parts of the galactic plane we initiated a still deeper spectral survey in the mid-1980s. It covers the region from $l = 50°$ to $l = 115°$ and $b = 0°$, $\pm 3.5°$. The area covered by $12 \times 12$ cm plates is equal to $\sim 12$ sq. degrees. All observations have been carried out with the 70-cm meniscus telescope equipped with the 2° prism giving a reciprocal dispersion of about 7000 Å mm$^{-1}$ at the $\Lambda$ band. Fine-grained Kodak IV–N plates hypersensitized in silver-nitrate solution (AgNO$_3$) in combination with a 2-mm Schott RG 8 filter yielded a passband from 6800 to 8800 Å. Carbon stars are identified on the infrared-sensitive spectral plates by the presence of an easily noticeable dip nearly in the middle of the spectra, due to unresolved CN bands at $\lambda\lambda$ 7945, 8025, 8320 Å. Despite the very low dispersion used, and consequently the low resolution of $\sim 200$ Å (R = 40), they are easily distinguished from M–type stars, the TiO bands of which are well separated in these spectra. On plates obtained with the 1° prism ($\sim 15000$ Å mm$^{-1}$ at the A-band, $\Delta\lambda = 400$) carbon stars are indistinguishable from other late-type stars. As a result of the survey of about 500 sq. degrees, 752 carbon stars were revealed to a limiting near-infrared magnitude of $15^m.5$, among them 580 new ones (Kurtanidze & Nikolashvili 1988a; Nikolashvili & Kurtanidze 1989). The limiting magnitude was estimated on spectral plates of IC 5146 (Forte & Orsatti 1984).

In an excellent and unique project to measure the rotation of the Milky Way at large galactocentric radii, Aaronson, Blanco et al. (1989, 1990) carried out a very deep spectral survey in 16 regions of low absorption. All observations were obtained using the Curtis and Burrell Schmidt telescopes equipped with objective prisms giving reciprocal dispersions of about 1800 Å mm$^{-1}$ at 8000 Å. Silver nitrate hypersensitized IV–N plates were used with an RG 695 filter. Each of the plates obtained covers an area of 25 sq. deg. They identified 498 known and 487 probable C stars. Spectroscopic observations of the latter with higher resolution show that there are 423 new discoveries and 64 misidentifications including 13 new S stars. Radial velocities and *JHK* photometry have been obtained for more than 800 and 1000 objects, respectively.

Stephenson (1989) used the Burrell Schmidt telescope of the Warner and Swasey Observatory on Kitt Peak to conduct the extensive spectral survey of the Milky Way from $l = 0°$ to $l = 240°$, $|b| < 6°$. The spectral dispersion provided by the 4° prism was close to 1700 Å mm$^{-1}$ at the telluric A-band. A near-infrared magnitude $I = 13.5$ was reached using Kodak IV–N emulsion hypersensitized in AgNO$_3$. As a result about 2100 carbon stars were identified.

The data on all the surveys described were compiled by Stephenson (1989) in *A General Catalog of Cool Galactic Carbon Stars* containing 5987 entries, the faintest of which are as faint as $15^m.5$ in the near-infrared spectral region.

## 3. Further Prospects

While carbon stars are usually identified on the basis of optical surveys, they are so cool (except for the warmer carbon stars of the R0–R3 or C0–C2 subclasses) that they emit most of their radiation at wavelengths longward of one micron, at which the Galaxy is almost transparent. Therefore, the DENIS and 2MASS surveys will play a very important role in studies of the space distribution of carbon stars and the variation of their characteristics with galactocentric distance, as well as in other galactic problems. Thanks to the very faint limiting magnitude reached ($K = 14$), practically all galactic giant carbon stars will be detected including the warmer ones, although the separation of those having $J-K < 1.5$ from other kinds of objects will not be possible (Wood 1994). Therefore, spectral surveys in the optical region will still play an important role for the identification and classification of the warmer carbon stars. These surveys could be carried out at 10 Å resolution and should be as deep as possible.

It is great pleasure to thank the Chair of the SOC, Dr. R. Wing, and the IAU for invaluable financial support which enabled us to attend the meeting.

## References

Aaronson, M., Blanco, V. M., Cook, K. H. & Schechter, P. L. 1989, *ApJ Suppl.*, 70, 637
Aaronson, M., Blanco, V. M., Cook, K. H., Olszewski, E. W. & Schechter, P. L. 1990, *ApJ Suppl.*, 73, 841
Alksne, Z., Alksnis, A., Ozolinya, V. & Platais, I. 1989, *Investigations of the Sun and Red Stars*, 30, 40
Blanco, V. M. & Münch, L. 1955, *Bol. Obs. Tonantzintla y Tacubaya*, 12, 17
Blanco, V. M. & McCarthy, M. F. 1981, in *Physical Processes in Red Giants*, ed. I. Iben Jr. and A. Renzini (Reidel), p. 147
Claussen, M. J., Kleinmann, S. G., Joyce, R. R. & Jura, M. 1987, *ApJ Suppl.*, 65, 385
Forte, J. C. & Orsatti, A. M. 1984, *ApJ Suppl.*, 56, 211
Fuenmayor, F.J. 1981, *Rev. Mex. Astron. y Astrofís.*, 6, 83
Jura, M. 1991, *Astron. Astrophys. Rev.*, 2, 227
Kleinmann, S. G. 1989, in IAU Coll. 106: *Evolution of Peculiar Red Giant Stars*, ed. H. R. Johnson and B. Zuckerman (Cambridge), p. 13
Knapp, G. R., Rauch, K. P. & Wilcots, E. M. 1990, in *The Evolution of the Interstellar Medium*, ed. L. Blitz, ASP Conf. Ser., 12, 151
Kurtanidze, O. M. & Nikolashvili, M. G. 1981, *Astrophysica*, 17, 576
Kurtanidze, O. M. & Nikolashvili, M. G. 1988a, *Reports of the Georgian Sci. Res. Institute of Sci. and Tech. Information*, no. 430-G88, 1
Kurtanidze, O. M. & Nikolashvili, M. G. 1988b, *Astrophysica*, 29, 470
Kurtanidze, O. M. & Nikolashvili, M. G. 1989a, in IAU Coll. 106: *Evolution of Peculiar Red Giant Stars*, ed. H. R. Johnson and B. Zuckerman (Cambridge), p. 63
Kurtanidze, O. M. & Nikolashvili, M. G. 1989b, *Astrophysica*, 31, 507
Kurtanidze, O. M. & Nikolashvili, M. G. 1989c, unpublished
Kurtanidze, O. M. & Nikolashvili, M. G. 1994, in IAU Symp. 161: *Astronomy from Wide-Field Imaging*, ed. H. T. MacGillivray et al. (Kluwer), p. 467
Kurtanidze, O. M. & Nikolashvili, M. G. 1995, in IAU Coll. 148: *The Future Utilisation of Schmidt Telescopes*, ed. J. Chapman, R. Cannon, S. Harrison and B. Hidayat, ASP Conf. Ser., 84, 337
Lee, O., Gore, G. & Bartlett, T. 1940, *Ann. Dearborn Observ.*, 4, 16
MacConnell, D. J. 1988, *AJ*, 96, 354
MacConnell, D. J. 1995, in IAU Coll. 148: *The Future Utilisation of Schmidt Telescopes*, ed. J. Chapman, R. Cannon, S. Harrison and B. Hidayat, ASP Conf. Ser., 84, 323
Maehara, H. & Soyano, T. 1987a, *Ann. Tokyo. Astron. Obs.*, 2nd ser., 21, 293
Maehara, H. & Soyano, T. 1987b, *Ann. Tokyo. Astron. Obs.*, 2nd ser., 21, 423
Maehara, H. & Soyano, T. 1988, *Ann. Tokyo. Astron. Obs.*, 2nd ser., 22, 59
Maehara, H. & Soyano, T. 1990, *Publ. Nat. Astron. Obs. Japan*, 1, 207
Mavridis, L. N. 1971, in *Structure and Evolution of the Galaxy*, ed. L. N. Mavridis (Springer-Verlag), p. 110
McCarthy, M. F. & Blanco, V. M. 1978, *Mem. Soc. Astr. It.*, 49, 281
Nassau, J. J. & Blanco, V. M. 1954a, *ApJ*, 120, 118
Nassau, J. J. & Blanco, V. M. 1954b, *ApJ*, 120, 129
Nassau, J. J. & Blanco, V. M. 1954c, *ApJ*, 120, 464
Nassau, J. J. & Blanco, V. M. 1957, *ApJ*, 125, 195

Nassau, J. J., Blanco, V. M. & Morgan, W. W. 1954, *ApJ*, 120, 478
Nassau, J. J. & Colasevich, A. 1950, *ApJ*, 111, 199
Nassau, J. J. & van Albada, G. B. 1949, *ApJ*, 109, 391
Nassau, J. J. & Velghe, A. G. 1964, *ApJ*, 139, 190
Nikolashvili, M. G. 1987a, *Astrophysica*, 26, 209
Nikolashvili, M. G. 1987b, *Astrophysica*, 27, 197
Nikolashvili, M. G. & Kurtanidze, O. M. 1989, unpublished
Schechter, P. L. 1993, in *Back to the Galaxy*, ed. S. S. Holt and F. Verter, AIP Conf. Proc., 278, 571
Smith, E. v. P. & Smith, H. J. 1956, *AJ*, 61, 273
Soyano, T. & Maehara, H. 1991, *Publ. Nat. Astron. Obs. Japan*, 2, 203
Soyano, T. & Maehara, H. 1993, *Publ. Nat. Astron. Obs. Japan*, 3, 259
Stephenson, C. B. 1989, *Publ. Warner and Swasey Obs.*, 3, 53
Weinberg, M. D. 1993, in *Back to the Galaxy*, ed S. S. Holt and F. Verter, AIP Conf. Proc., 278, 347
Westerlund, B. E. 1971, *A&A Supp.*, 4, 51
Wood, P. R. 1994, in *Science with Astronomical Near-Infrared Sky Surveys*, ed. N. Epchtein, A. Omont, B. Burton and P. Persi, *Astrophys. Space Sci.*, 217, 121

## Discussion

**Frogel**: Can you comment on the *relative* number of C stars you find, i.e. relative to the number of M stars or to the overall stellar density?

**Kurtanidze**: The $C/M5^+$ ratios along the galactic plane are presented by M. Nikolashvili (these proceedings). As regards the other Natural Groups that can be detected by the meniscus J,F survey, the work is still in progress, although the penetration would differ for different groups of stars. Early estimates indicate a space density of C stars of the order of $10^{-7}$ $pc^{-3}$, or about 25 C stars within 400 pc of the sun. On the other hand if the surface density of disk dwarf C stars to $V = 18$ is 1.3 $deg^{-2}$ (de Kool & Green 1995, *ApJ*, 449, 236), then at least $900 \times 1.3 / 4 = 300$ C stars of the deepest IV–N surveys (Table 1) should be dwarfs.

# SPACE DISTRIBUTION OF CARBON STARS IN OUR GALAXY

K. NOGUCHI
*National Astronomical Observatory of Japan*
*Mitaka, Tokyo 181, Japan*

AND

Z. QIAN, J. SUN, G. WANG, J. WANG AND Y. RAO
*Beijing Astronomical Observatory*
*Beijing 100080, P. R. China*

**Abstract.** We are attempting to investigate the space distribution of carbon stars in our Galaxy by evaluating the distances of individual carbon stars. We estimate the distance by evaluating the total radiant energy of each star and assuming a value for the bolometric magnitude of carbon stars. Since the flux maxima of carbon stars are mostly in the near-infrared (NIR), NIR photometric data are most useful for evaluating the total energy. We have compiled NIR photometric data for 694 carbon stars at all galactic longitudes and have made new observations of 470 carbon stars in the galactic longitude region between 20° and 160°. We discuss the space distribution of 1164 galactic carbon stars.

## 1. Introduction

The asymptotic giant branch (AGB) stars are classified into three main types on the basis of their atmospheric composition: carbon-rich, oxygen-rich (O-rich) and S-type. It is interesting to compare the space distributions of these AGB stars in order to understand what kind of AGB stars evolve as carbon stars.

Jura (1990) reported that there is a measurable galactocentric gradient in the surface density (SD) projected onto the galactic plane of the O-rich Miras, but that there is no decrease with galactocentric radius in the SD of carbon stars for at least 3 kpc beyond the solar circle. The space distribution of carbon stars is different from that of O-rich AGB stars in the solar neighborhood. However, these investigations have been limited to the brighter carbon stars (Claussen et al. 1987; Jura & Kleinmann 1989, 1990; Groenewegen et al. 1992).

*A General Catalog of Cool Galactic Carbon Stars*, 2d ed. (GCCS2), compiled by Stephenson (1989), includes 5987 carbon stars. The majority of these carbon stars are distributed near the galactic plane on the $l - b$ diagram. It is interesting to note that the stellar density of carbon stars is lower in the longitude range between about −50° and +50°, i.e. in the direction towards the galactic center.

Guglielmo et al. (1993) reported near-infrared (NIR) photometric data for 1332 IRAS sources. Combining the measured $K$–$L$ color with the IRAS [12–25] color, they classified these sources into IRCS (Infrared Carbon Stars) and O-rich stars. On the basis of their data, they reported that the SD of IRCS is constant in the galactocentric distance range 5–11 kpc but might present a cut-off at about 4 kpc from the Sun in the galactic center direction. However, further observations are required to clarify the global space distribution of carbon stars in our Galaxy, because the observations have been carried out only in limited regions of the Galaxy and their stellar classification is based only on infrared photometric data.

We have attempted a more extensive investigation of the space distribution of the carbon stars in GCCS2, which mostly are classified on the basis of optical spectra. How to estimate distances for individual stars is an important point in investigating the SD distribution. In this work, distances are determined by evaluating the total radiant energies of stars under an assumption of a certain bolometric luminosity for carbon stars. The space distribution of 694 relatively bright carbon stars is discussed in § 3. The distribution of carbon stars, including 470 newly observed stars, is discussed in § 4.

## 2. Determination of Distances to Carbon Stars

Distances to carbon stars are determined by comparing the observed radiant energy with the assumed intrinsic luminosity. All the carbon stars are assumed to be radiating with $10^4 L_\odot$. Observed total radiant energies from individual stars are evaluated by integrating the spectral energy distributions (SEDs). In order to get SEDs of stars and to evaluate total radiant energies accurately, NIR photometric data are very important for carbon stars because the SEDs of the majority of carbon stars show their flux maxima in the NIR. Therefore, we estimated the distance only for stars with NIR photometric data. NIR photometric data and IRAS photometric data are used to get SEDs on $\lambda \cdot F_\lambda$ vs. $\lambda$ diagrams. Total radiant energies are evaluated by integrating the model spectra which are fitted to the observed SEDs.

Individual SEDs of carbon stars are fitted by either of two types of model spectra of carbon stars. Since the SEDs of relatively blue carbon stars are usually well fitted by blackbody spectra, 'blue' carbon stars are approximated by blackbody spectra. But red carbon stars, in most cases, have SEDs that are flatter than blackbody spectra. Therefore we made a model spectrum which fits better to the SEDs of red carbon stars on the average. In order to classify stars into blue or red classes, the $K$–$L$ color (or the wavelength where $\lambda \cdot F_\lambda$ shows its maximum when there is no $K$–$L$ data) is used.

## 3. Distribution of Relatively Bright Carbon Stars

We have compiled NIR ($J,H,K,L$) photometric data for carbon stars in the GCCS2 from the observations reported by Catchpole et al. (1979), Epchtein et al. (1987, 1990), Guglielmo et al. (1993), and Noguchi et al. (1981, 1991). We added our unpublished NIR photometric data for 138 relatively bright carbon stars.

In total, we have compiled NIR photometric data for 694 relatively bright carbon stars. Total radiant energies were evaluated for these stars and distances of individual stars were determined. The derived SD distribution, projected onto the galactic plane, is shown in Figure 1. Since the observed NIR data were collected from several papers reported by different groups, the observations do not cover the Galaxy uniformly. However, the SD distribution obtained looks rather uniform within a circle with a radius of about 4 kpc centered on the Sun.

## 4. Space Distribution of Carbon Stars

In order to investigate the space distribution of carbon stars in our entire Galaxy, more extensive NIR photometric observations including fainter stars are required. We carried out NIR photometry of carbon stars in the GCCS2 in October 1995. Since our observations were made from the northern hemisphere (from the Xinglong Station of the Beijing Astronomical Observatory United Laboratory of Optical Astronomy, Chinese Academy of Sciences), the area observed is limited to the longitude region between about 20° and 160°. NIR data were obtained in the $J$, $H$, and $K$ bands for 470 stars. Distances for these 470 stars were estimated and the SD distribution has been derived. The total number of carbon stars for which distances have been estimated is 1164 including the relatively bright carbon stars. The SD distribution of these stars is shown in Figure 2. The paucity of stars in the range $60° \leq l \leq 70°$ and the higher stellar density in the longitude range $120° \leq l \leq 160°$ is due to an observational bias. Although

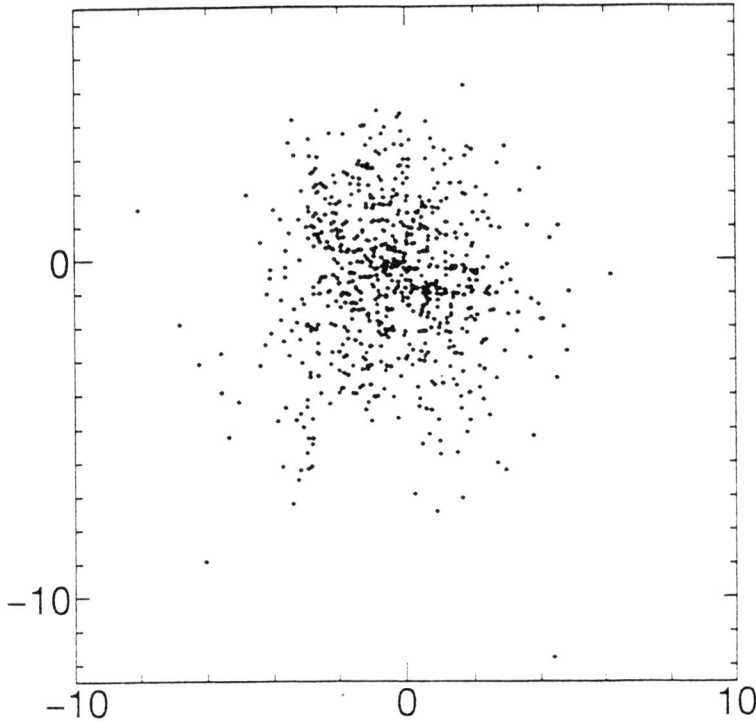

*Figure 1.* Space distribution of 694 relatively bright carbon stars projected onto the galactic plane. The Sun is located at the (0,0) position and the Galactic Center is to the right. Distances derived by assuming luminosities of $10^4 L_\odot$ are plotted in kpc.

the observed region does not cover sufficient area, the SD distribution of the carbon stars observed does not show a clear dependence on galactocentric distance. The reason for the paucity of GCCS2 stars in the longitude range $-50° \leq l \leq 50°$ in the $l - b$ plane is still not clearly understood. More observations of carbon stars, especially in the galactic center direction, are required.

Aaronson et al. (1989, 1990) reported $J$, $H$, and $K$-band photometric data for carbon stars in the southern Milky Way (an additional 373 stars in the interval $238° \leq l \leq 315°$) and in the northern Milky Way (an additional 308 stars with $54° \leq l \leq 185°$). Although our present analysis does not include their data, they will be very valuable for future analysis.

The space distribution of R stars is different from that of N stars. However, the distribution examined in our present analysis is for the R and N stars combined. It will be important to examine the space distribution of N stars separately from R stars in future work.

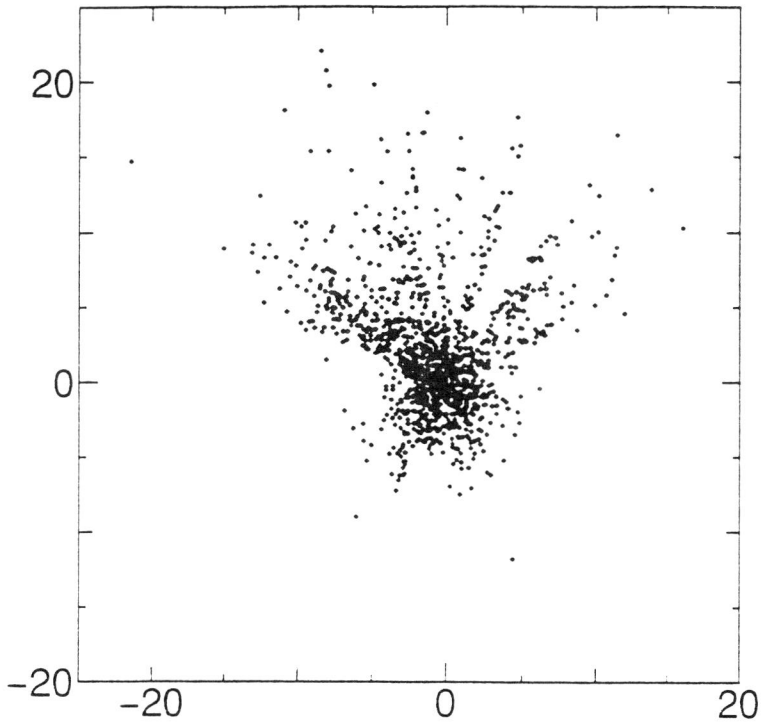

*Figure 2.* Space distribution of 1164 carbon stars in the GCCS2 catalog.

## References

Aaronson, M., Blanco, V. M., Cook, K. H. & Schechter, P. L. 1989, *ApJ Supp.*, 70, 637
Aaronson, M., Blanco, V. M., Cook, K. H., Olszewski, E. W. & Schechter, P. L. 1990, *ApJ Supp.*, 73, 841
Catchpole, R. M., Robertson, B. S. C., Lloyd Evans, T. H. H., Feast, M. W., Glass, I. S. & Carter, B. S. 1979, *SAAO Circulars*, 1, 61
Claussen, M. J., Kleinmann, S. G., Joyce, R. R. & Jura, M. 1987, *ApJ Supp.*, 65, 385
Epchtein, N., Le Bertre, T. & Lépine, J. R. D. 1990, *A&A*, 227, 82
Epchtein, N., Le Bertre, T., Lépine, J. R. D., Marques dos Santos, P., Matsuura, O. T. & Picazzio, E. 1987, *A&A Supp.*, 71, 39
Groenewegen, M. A. T., de Jong, T., van der Bliek, N. S., Slijkhuis, S. & Willems, F. J. 1992, *A&A*, 253, 150
Guglielmo, F., Epchtein, N., Le Bertre, T., Fouqué, P., Hron, J., Kerschbaum, F. & Lépine, J. R. D. 1993, *A&A Supp.*, 99, 31
Jura, M. 1990, in *From Miras to Planetary Nebulae: Which Path for Stellar Evolution?*, ed. M. O. Mennessier and A. Omont (Editions Frontières), p. 41
Jura, M. & Kleinmann, S. G. 1989, *ApJ*, 341, 359
Jura, M. & Kleinmann, S. G. 1990, *ApJ*, 364, 663
Noguchi, K., Kawara, K., Kobayashi, Y., Okuda, H., Sato, S. & Oishi, M. 1981, *Publ. Astron. Soc. Japan*, 33, 373
Noguchi, K., Sun, J. & Wang, G. 1991, *Publ. Astron. Soc. Japan*, 43, 275
Stephenson, C. B. 1989, *Publ. Warner & Swasey Obs.*, 3, No. 2

## Discussion

**Hron:** I would like to draw your attention to the ongoing near-infrared sky survey (DENIS) and a planned survey (2MASS) which will give us all the carbon stars in the Galaxy – but it will take time.

**Noguchi:** Thanks for your comment.

# UNDERSTANDING THE ELUSIVE DWARF CARBON STAR

*Ender Bulunan Cüce Karbon Yıldızını Anlama*

PAUL J. GREEN

*Smithsonian Astrophysical Observatory*
*Harvard-Smithsonian Center for Astrophysics*
*60 Garden St., Cambridge, MA 02140, U.S.A.*

**Abstract.** Most stars in our Galaxy with photospheric $C/O > 1$ (carbon stars) are not giants but dwarfs. The newly-recognized class of dwarf carbon stars joins the growing family of stars with peculiar abundances that are now recognized as products of mass-transfer binary (MTB) evolution. The dozen examples now known span a wide range of evolutionary histories, ages, and abundances. These stars can already provide some much-needed constraints on the formation of AGB C stars in the disk and spheroid populations, and on the parameters characterizing binary evolution there. A larger sample, with some bright members, would hasten our progress.

## 1. Introduction

The 'C' and the star on the Turkish flag make this conference a good place to set the stage for a new perspective on carbon stars. Just so we get it straight, the cool, luminous AGB stars that dominate C-star theory and observation (and the rest of this conference) are in all likelihood atypical. If a carbon star is defined as a star showing carbon molecular features ($C_2$ bands, with CH, CN and/or $s$-process enhancements often associated), then the vast majority of C stars are not giants at all but dwarfs. Recognition of this paradigm shift is at first glance troublesome, both for observers, who can no longer assume that a C-star spectrum implies a giant luminosity, and for theorists, who must explain how a dwarf can show $C/O > 1$ when no carbon is produced by main-sequence hydrogen burning. I'll explain briefly how we came to recognize that dwarf carbon (dC) stars are not some rare freak of nature, but constitute a whole class of stars that fall naturally into the family of post mass-transfer binary systems.

## 2. Finding Faint High-Latitude Carbon Stars

Our initial search for Faint High Latitude Carbon (FHLC) stars was motivated by the impression that models of the chemical and dynamical properties of the Galactic spheroid (the 'halo') are still rather weakly constrained. In the grand scheme, did a monolithic protogalaxy undergo rapid collapse and enrichment (Eggen, Lynden-Bell & Sandage 1962), or did many smaller dwarf galaxies merge together (Searle & Zinn 1978)? Both processes may contribute, with mergings ongoing today. More modestly, how can we best characterize the mass-to-light ratio, the velocity ellipsoid and systemic rotation of the outer halo? Intrinsically bright stars visible to large galactocentric distances ($10 - 100$ kpc) provide the best opportunity. Faint C stars have been sought as excellent tracers of the outer halo because they were thought to be distant giants, and because they are readily recognizable from their strong $C_2$ and CN absorption bands. As tracers, they do not suffer as do globular clusters from selection effects complicated by tidal interactions with the disk.

Objective-prism photography with wide-field Schmidt telescopes has yielded low-dispersion spectra for thousands of objects over substantial portions of the sky. At high galactic latitudes, we find mostly CH stars, and possibly some R stars. (Unless stated explicitly, it is to these warmer types that I refer here.) Fewer than 1% of the 6000 stars in Stephenson's (1989) catalogue are the faint, high-latitude carbon (FHLC) stars ($V > 13$, $|b| > 40°$) most useful as dynamical probes of the outer halo. The two most prolific sources of published FHLC stars, the Case low-dispersion survey (CLS; Sanduleak & Pesch 1988) and the University of Michigan – Cerro Tololo survey (UM; MacAlpine & Williams 1981) appear to be complete to about $V = 16$ and have provided about 30 FHLC stars. Emission-line objects, not FHLC stars, were the primary goal of these photographic surveys, and known FHLC stars were not examined to help predefine selection criteria or estimate completeness. The surface density of FHLC stars from objective-prism surveys is low, about one per 50 deg$^2$ to $V \approx 16$. The challenge was then to go deeper than the objective-prism surveys, while still covering a wide area with CCD imaging. We sought to expand shallower objective-prism survey samples to provide a significant-sized sample of distant halo C stars. The combined sample could be used for tests of the spheroid density law, and for dynamical analysis. We used a 3-filter, two-color photometric technique to distinguish C stars from other late-type stars via intermediate band ($\sim 200$ Å FWHM) filters (e.g. Cook & Aaronson 1989; Palmer & Wing 1982). One ("77") is centered on a region of TiO absorption near $\lambda 7752$ Å, and the other ("81") is on a CN absorption band near $\lambda 8104$ Å. The $77 - 81$ color thus separates C stars from other stars of similar effective temperature (and similar $V-I$) because C stars appear particularly faint in the 81 filter relative to the 77 filter, while the converse prevails in M stars. By the end, our CCD survey covered 52 deg$^2$ of high galactic latitude sky to a depth of about $V = 18$. Only one highly ranked $V = 17$ candidate was found to have strong carbon and CN bands. That star, ($\alpha_{1950} = 03^h 11^m 44\overset{s}{.}08$, $\delta = +07°33'38\overset{''}{.}0$), is of course my favorite in the whole sky. So perhaps the hubris of using one star to estimate a surface density will be pardoned. To a depth of $V = 18$, the surface density of FHLC stars in our CCD survey is about 0.02 deg$^{-2}$ (Green et al. 1994), the same as the surface density from objective-prism surveys. We went deeper, so why didn't we find more? First, there might well be more FHLC stars to $V = 18$ than to $V = 16$, but we can't tell within the large uncertainties. If the apparent difference is real, it could be that we begin to "run out of Galaxy" at these magnitudes. Also remember that most photographic surveys pick up Swan bands of $C_2$, while our survey picks up CN, so that slightly different trends or normalization may obtain with metallicity and/or galactocentric radius.

## 3. Stumbling over Dwarfs

While taking follow-up optical spectra of candidate FHLC stars from our survey at the KPNO 2.1-m telescope, we also obtained numerous spectra of previously known FHLC stars (Green & Margon 1990). One of these objects, CLS 96, also bore the comment "LP 328-57?" in the Sanduleak & Pesch (1988) list, indicating a possible identification with a high-proper-motion (p.m.) star. Bidelman and MacConnell had independently noticed this association of CLS 96 with the Luyten p.m. star at around the same time. When Peter Pesch alerted us to this, we immediately checked the p.m. catalogues (via SIMBAD) for other faint C stars and found another probable association, that of LHS 1075 and C*22 from the UM survey. Astrometry using the 30-year baseline between the POSS and the HST Quick V Survey plates yielded yet another high p.m. C star, CLS 31 (Green, Margon & MacConnell 1991). Large proper motions could only occur at small distances for a star bound to the Galaxy, thus mandating that these faint C stars must be dwarfs.

Up until then, the only known flaw in the otherwise promising technique of using FHLC stars as halo tracers had been the existence of one lone star, G77–61, a $V = 13.9$ dC also with a high proper motion (Dahn et al. 1977). The object was subsequently shown to be a single-lined spectroscopic binary of period 245 d and parallax $\pi_{abs} = 0\overset{''}{.}017 \pm 0\overset{''}{.}003$ (Dearborn et al. 1986). These authors argued that, since main-sequence stars can't dredge up carbon, the most reasonable explanation for the prominent $C_2$ bands in the dwarf's spectrum is photospheric deposition of mass from a now unseen companion during the companion's second ascent of the giant branch. Even given such different origins, we found that the optical spectra of dCs at resolution $\geq 1$ Å are strikingly similar to those of Pop II carbon giants, from their wide distribution of $C^{13}/C^{12}$ ratios to their enhanced $s$-process abundances (Green & Margon 1994). 'Faint' doesn't guarantee 'distant' anymore; C stars are nearly on the same footing as other late-type stars.

How could we know if there remained dCs of lower p.m. in the FHLC star sample? Since IR colors are commonly used to determine C giant luminosities, I took the time to plot the published IR colors of FHLC stars (e.g. Bothun et al. 1991; Mould et al. 1985). The three dwarfs were redder in $H-K$ than most other C stars, but so were two or three other stars. A more thorough p.m. survey of all known FHLC stars (Green et al. 1992) showed that the other odd-colored stars were also moving! This brought the total up to five and strongly suggested that dCs may have $JHK$ colors distinct enough to identify them as dwarfs.

## 4. Reign of the Dwarfs

Our Monte Carlo simulations indicated that the proper motion survey, sensitive to p.m. $> 0.''1$/year, could detect Pop II dCs brighter than $V = 18$ 98% of the time. Therefore, since we detected proper motions for 5 of 39 FHLC stars in our survey, we know that at least 13% of FHLC stars to that magnitude are dCs. Presumably, deeper surveys will find higher fractions, because they will begin to probe beyond the Galaxy for giants. But as it stands, how does the space density of dCs locally compare with other types of C stars? Simply taking 13% of FHLC stars, with their a surface density of 0.02 deg$^{-1}$, yields about 100 dCs. All dCs with absolute magnitude estimates to date have $M_V \approx 10$. Assuming this holds for all dCs, a survey limit of $V = 18$ corresponds to a sphere of 400 pc radius. Within this volume, the number of dCs easily surpasses the sum of *all other types of carbon stars combined*, including N, R, CH, Ba and sgCH (or dBa) stars (Green et al. 1992). My simplistic calculation is nevertheless quite conservative since all these latter types will be brighter by several magnitudes than dCs within the same volume. Furthermore the estimated C dwarf to C giant fraction must be a lower limit, because it concerns only high-proper-motion dCs. The binary mass transfer explanation for photospheric carbon in dCs predicts that they should exist in the disk as well. Disk dCs would tend to have small proper motions. If, for example, 2 disk dCs in our sample were counted incorrectly as giants, then the true fraction is closer to 20%.

Soon after our p.m. survey, Warren et al. (1993) found two faint dCs in the south by means of their high proper motions. Heber et al. (1993) found a composite spectrum DA/dC binary system, PG 0824+289, with a 60,000 K white dwarf. The prototype dC G77–61 was also known to have a faint companion ($T_{\text{eff}} < 6000$ K from an IUE upper limit), but PG 0824+289 is truly the smoking gun, white hot evidence for the mass-transfer hypothesis. In addition, since PG 0824+289 had no detectable proper motion and disk kinematics, it may represent the first known disk dC. Only a year passed before a similar DA/dC composite, CBS 311, was found (Liebert et al. 1994). Neither could have been found via proper motion selection. How many such dCs are out there?

## 5. Predicting the Space Density of Dwarf Carbon Stars

If the space density of disk dCs scales with the ratio of disk/halo space densities (at least $\sim 500$: Bahcall & Soneira 1984; Morrison 1993), then we would expect there to be about 1.3 deg$^{-2}$ to our nominal limiting magnitude of $V = 18$. But this exceeds the *total* surface density of FHLC stars by about a factor of 6. This back-of-the-napkin calculation shows that the dC fraction must actually be much lower in the disk than in the halo.

Age and metallicity differences offer a likely explanation for the larger fraction of dwarfs that are dCs in the spheroid compared to the disk. Other effects such as the means of accretion (e.g. Roche lobe overflow, or wind) and mixing efficiency in the accreting star currently preclude simple analytic predictions. De Kool & Green (1995) have constructed simulated samples of dC stars to determine whether reasonable assumptions lead to dC space densities compatible with observations, and to investigate how these assumptions affect the expected properties of dCs. A simulated population of dCs is constructed by following the evolution of a large number of binaries using simple analytic fits to detailed evolutionary calculations, and determining which ones would presently contain a dC star. The zero-age parameters of the sample are chosen randomly from observed distributions of unevolved binaries. The space density of halo dC stars that we predict ($\sim 2-4 \times 10^{-7}$ pc$^{-3}$) is in agreement with current observational constraints. The predicted local space density of disk dC stars ($\sim 1 \times 10^{-6}$ pc$^{-3}$) may be a bit high, since it still predicts nearly as many disk dCs as there are FHLC stars observed. The fraction of binaries that produces dCs depends strongly on initial metallicity, and virtually no dCs are formed in systems with an initial metallicity of more than half solar. Thus all disk dCs are predicted to be in binaries that formed in the very early phases of disk star formation, and their number depends strongly on assumptions about the age-metallicity relation during this epoch. The predictions for the halo are much less model-dependent. In either population, we may expect that dCs on average will exhibit lower than average metallicities. We also predict that dCs in the disk, for instance, will eventually be shown to have a scale-height consistent with old or thick-disk ages. The simulated dC orbital period distributions are bimodal, with one peak between $10^3$ and $10^5$ days and another peak between $10^2$ and $10^3$ days. The shorter-period component is caused by systems that have gone through a common envelope phase, while most have accreted from an AGB wind. The simulated period distributions bear a strong resemblance to the observed orbital period distribution of barium and CH giants, which may be the evolved descendants of the disk and halo dC populations we modeled.

## 6. Joining the Mass Transfer Binary (MTB) Family

Because they are considerably more luminous, barium (Ba), CH, and S stars have been better studied and characterized than dCs. These giants show peculiar abundances, with $C/O \geq 1$ and a strong overabundance of $s$-process elements, thought to be produced during shell burning on the AGB. Some S stars are indeed on the AGB. Those suspected of being MTB products (the 'extrinsic' S stars) show no sign of the unstable element technetium (Tc) in their spectra, unlike their AGB (or 'intrinsic') S analogs. Ba, CH and extrinsic (non-Tc) S stars are red giants that have not undergone the thermal AGB pulsations necessary to produce their observed peculiar photospheric abundances. Observations are consistent with the MTB hypothesis as an explanation for all of them. McClure & Woodsworth (1990) present orbital periods for CH stars; Jorissen & Boffin (1992) have collected orbital parameters and abundances for Ba stars, and Jorissen & Mayor (1992) for S stars. The mass functions derived for systems with detected periods are consistent with white dwarf companions. A fair fraction have no period yet detected, which almost certainly means that it is very long. This implies wind accretion, as does the non-zero eccentricity of many orbits. Diagrams of $e$ vs. $\log P$ can reveal the typical Roche lobe radius of the former AGB primaries in a sample, since tidal effects or Roche lobe overflow will circularize an orbit with separation near that radius.

CH stars are halo giants whose unevolved precursors could be halo dCs. Similarly, the Ba giants, with old-disk kinematics and near-solar metallicities, are likely to represent the high-mass end of the population of disk stars that have experienced mass transfer from an AGB companion. The CH subgiant stars (Smith, Coleman & Lambert 1993), perhaps better described as Ba dwarfs (dBa), are main-sequence (MS) counterparts, or perhaps precursors of Ba giants (North, Jorissen & Mayor 2000). Extrinsic S stars probably represent somewhat cooler, lower mass MTB disk giants (Jorissen & Mayor 1992) whose precursors have yet to be postulated. The spectrum of any post-MTB giant (CH, Ba, or extrinsic S star) may differ substantially from that of its unevolved dwarf precursor, particularly if the mass of the convective zone changes greatly during evolution.

Lower limits to ages of white dwarf companions (e.g. Johnson et al. 1993; Smith et al. 1993; Böhm-Vitense et al. 1984) generally exceed the lifetime of the giant phase, so that the mass transfer episode *must* have occurred while the contemporary giant was still on the main sequence. Since MS lifetimes are much longer than giant lifetimes, the MS precursors to Ba, CH, and extrinsic S giants should abound, but are clearly more difficult to recognize or detect. We may explain the unique existence of dCs in both disk and halo as a consequence of their mass range, the lowest of any post-mass-transfer objects so far discussed (mostly near $0.5 M_\odot$). Still, the undetected companion of G77–61 in contrast to the very hot DA companions to the dCs PG 0824+289 and CBS 311 reveals a truly wide range of ages since mass transfer among dCs.

## 7. Finding More about dC Stars

I'll just summarize four areas that can be clarified by observations in the near future: (1) Long-term radial velocity monitoring is needed to prove duplicity, and to determine the mass function and eccentricity. These must be correlated with measured abundances. A trend predicted in our simulations, and observed in Ba giants, is a small but significant anti-correlation between $s$-process overabundance and orbital period. (2) Good abundance determinations are needed, particularly of Tc, heavy to light $s$-process element ratios, metallicity, C/O and carbon isotope ratios. More rapid progress in our understanding will also be made when model atmospheres for dCs have been matched to UV/optical/IR spectrophotometry and to luminosities derived from trigonometric parallax. (3) Does enriched accreted material mix beyond the shallow convective zones of higher-mass dwarfs? Proffitt & Michaud (1989) argued that the higher mean molecular weight of the accreted material leads to instability and mixing even into radiative layers. If there is no such mixing, the predicted number of disk dCs increases by an order of magnitude, and extends to higher-mass dwarfs. (4) More FHLC and more dC stars must be found. There may still be low p.m. dCs lurking even in the sample of currently known FHLC stars. We need to know whether *JHK* colors are a good luminosity discriminant for disk dCs, and *why* they are a good discriminant for halo dCs. Ancient generations of AGB stars, the physics of mass transfer, and important parameters describing binary mass ratios and separations may be probed by such measurements.

Some of these suggestions for dCs mimic hard work already done on peculiar red giants, which have led to enormous leaps in our understanding of binary evolution. I hate to point out that the work on dCs may be even harder, because to date all are fainter than about 14th visual magnitude. There are few surveys specifically designed to net new C stars, but we now know that deeper surveys should reveal more dCs locally. As an example, I have initiated a deep CCD multicolor Schmidt survey, and a survey using the CCD grism transit scans of Schmidt, Schneider & Gunn (1995). Multicolor searches for CH and dC stars in globular clusters have also begun, from which relative distance and age uncertainties are largely removed. We hope that the Sloan Digital Sky Survey will be a major source of new discoveries of FHLC and dC stars in the field. A large, well-quantified sample of dC stars will go a long way toward a better understanding of how evolution in mass transfer binary systems is affected by age, metallicity and other factors. Let's just say there's a lot of physics packed into these dwarfs, at once the most elusive and most common type of carbon star in the Galaxy.

## References

Bahcall, J. N. & Soneira, R. M. 1984, *ApJ Supp.*, 55, 67
Böhm-Vitense, E., Nemec, J. & Proffitt, C. 1984, *ApJ*, 278, 726
Bothun, G., Elias, J. H., MacApine, G., Matthews, K., Mould, J. R., Neugebauer, G. & Reid, I. N. 1991, *AJ*, 101, 2220
Cook, K. H. & Aaronson, M. 1989, *AJ*, 97, 923
Dahn, C. C., Liebert, J., Kron, R. G., Spinrad, H. & Hintzen, P. M. 1977, *ApJ*, 216, 757
Dearborn, D. S. P., Liebert, J., Aaronson, M., Dahn, C. C., Harrington, R., Mould, J. & Greenstein, J. L. 1986, *ApJ*, 300, 314
deKool, M. & Green, P. J. 1995, *ApJ*, 449, 236
Eggen, O. J., Lynden-Bell, D. & Sandage, A. R. 1962, *ApJ*, 136, 748
Green, P. J. & Margon, B. 1990, *PASP*, 102, 1372
Green, P. J. & Margon, B. 1994, *ApJ*, 423, 723
Green, P. J., Margon, B., Anderson, S. F. & Cook, K. H. 1994, *ApJ*, 434, 319
Green, P. J., Margon, B., Anderson, S. F. & MacConnell, D. J. 1992, *ApJ*, 400, 659
Green, P. J., Margon, B. & MacConnell, D. J. 1991, *ApJ*, 380, L31
Heber, U., Bade, N., Jordan, S. & Voges, W. 1993, *A&A*, 267, L31
Johnson, H. R., Ake, T. B. & Ameen, M. 1993, *ApJ*, 402, 667
Jorissen, A. & Boffin, H. M. J. 1992, in *Binaries as Tracers of Stellar Formation*, ed. A. Duquennoy and M. Mayor (Cambridge Univ. Press), p. 110
Jorissen, A. & Mayor, M. 1992, *A&A*, 260, 115
Liebert, J., Schmidt, G. D., Lesser, M., Stepanian, J. A., Lipovetsky, V. A., Chaffee, F. H., Foltz, C. B. & Bergeron, P. 1994, *ApJ*, 421, 733
MacAlpine, G. M. & Williams, G. A. 1981, *ApJ Supp.*, 45, 113
McClure, R. D. & Woodsworth, A. W. 1990, *ApJ*, 352, 709
Morrison, H. L. 1993, *AJ*, 106, 578
Mould, J. R., Schneider, D. P., Gordon, G. A., Aaronson, M. & Liebert, J. W. 1985, *PASP*, 97, 130
North, P., Jorissen, A. & Mayor, M. 2000, in IAU Symp. 177: *The Carbon Star Phenomenon*, ed. R. F. Wing (Kluwer), p. 269
Palmer, L. G. & Wing, R. F. 1982, *AJ*, 87, 1739
Proffitt, C. R. & Michaud, G. 1989, *ApJ*, 345, 998
Sanduleak, N. & Pesch, P. 1988, *ApJ Supp.*, 66, 387
Searle, L. & Zinn, R. 1978, *ApJ*, 225, 357
Schmidt, M., Schneider, D. P. & Gunn, J. E. 1995, *AJ*, 110, 68

Smith, V. V., Coleman, H. & Lambert, D. L. 1993, *ApJ*, 417, 287
Stephenson, C. B. 1989, *Pub. Warner and Swasey Obs.*, 3, No. 2
Warren, S. J., Irwin, M. J., Evans, D. W., Liebert, J., Osmer, P. S. & Hewett, P. C. 1993, *MNRAS*, 261, 185

## Discussion

**Frogel**: How does proper-motion selection bias your results, and can you tell the difference between dC stars and CH stars like those in $\omega$ Cen?

**Green**: Proper-motion selection more or less guarantees we find only halo dCs. I believe the CH stars in globular clusters fall among giants in the *JHK* color-color diagram. If I'm wrong, then I suppose we have to rely on radial-velocity membership.

**Frogel**: Have radial-velocity surveys revealed the binary nature of dC stars?

**Green**: No survey has been done to date that would be sensitive to the periods and radial-velocity amplitudes common to, say, CH or Ba star systems.

**Little-Marenin**: What is the $^{12}C/^{13}C$ ratio in dC stars?

**Green**: We see a range similar to CH stars, from $\sim 3$ to 20 or more.

**Giridhar**: What is Fe/H for dwarf C stars?

**Green**: [Fe/H] $\approx -4$ for the prototype G77–61, but no recent metallicity measurements have been published for other dCs.

**McClure**: There was an argument, by Luck & Bond I think, that the large amount of mass transferred in a binary system, in order to form a Barium star, should cause a dwarf to move up the main sequence. Thus you might not expect to find many late-type dwarf C stars. Is this not a proper argument?

**Green**: The amount of mass that needs to be transferred to achieve C/O $> 1$ is less in lower-metallicity systems. Also, even if a large amount of mass is transferred to a brown dwarf, it may end up as a late-type dC. So if the initial mass function increases at the low end, late-type dCs could be quite common.

# A PROPER–MOTION SEARCH FOR DWARF CARBON AND S STARS

D. J. MacCONNELL
*Computer Sci. Corp./Space Telescope Science Institite*
*Baltimore, MD 21218, U.S.A.*

R. L. WILLIAMSON II
*Space Telescope Science Institute*
*Baltimore, MD 21218, U.S.A.*

AND

W. J. ROBERTS
*Box 16329, Baltimore, MD 21210, U.S.A.*

**Abstract.** We report on an effort to identify further members of the dwarf carbon class and new members of a putative dwarf S group in the general field through determination of proper motions of catalogued stars. Examination of nearly 1500 C stars and over 300 S stars reveals some interesting false alarms but no new dwarf members of these classes.

## 1. Introduction

The majority of galactic carbon stars have been found on objective-prism photographic plates of low resolution / dispersion taken with wide-angle cameras in any of several spectral bands from the blue to the near-infrared. In the blue-visual region, the Swan bands of $C_2$ are the defining criteria for membership in the class, while in the red and near-IR, the great strength of the CN bands is taken as sufficient for inclusion in the class. Having such evolved atmospheres, all C stars are expected to be on the AGB, and their high luminosities and distinctive optical spectra make them excellent probes of the kinematics and structure of the Galaxy to large distances. A challenge to this scenario, however, came in 1977 with the realization that the high-proper-motion star G77–61 was a normal-appearing C star but of $M_V \sim +10$ (Dahn et al. 1977), and it became the prototype of the dwarf carbon (dC) class. Since stars of dwarf mass cannot produce and dredge up carbon to the photosphere, G77-61 was explained as the product of mass transfer in a close binary of large mass ratio — the dC has received its carbon from an originally more massive AGB companion which is now a white dwarf. Indeed, Dearborn et al. (1986) found that G77-61 is a single-lined spectroscopic binary of $P = 245$ days and that the undetected, degenerate companion must have $T < 6000$ K. It remained the only dC until several others were identified early in this decade via their proper motions (Green et al. 1991, 1992). To date, 11 stars have been presented as dCs, and two of these, PG 0824+289 (Heber et al. 1993) and CBS 311 = SBS 1517+5017 (Liebert et al. 1994), are spectrum binaries in which a DA spectrum is seen in the blue while that of a carbon star is evident in the red, thus providing strong support to the mass-exchange hypothesis. These two have small proper motions, $< 0.''035$/yr, whereas most dCs have been identified by motions $> 0.''1$/yr which place an upper limit on their luminosities several magnitudes below the AGB.

De Kool & Green (1995) and Green (2000) have estimated that the dCs may outnumber the AGB C stars per unit volume with $\sim 110$ halo dCs detectable to $V \sim 18$; however, only about 10% of this number are known and none has been found in the last few years. Since those known have been found without the benefit of a systematic search, it is difficult to draw firm conclusions when discussing their origin, frequency, and kinematics. One way to attempt to close this gap is to measure known C stars for appreciable proper motion. The relation between a star's absolute and apparent magnitudes, its tangential velocity and proper motion is

$$M = m - 5(\log V_t - 1.68 - \log \mu) \ .$$

Thus, for a given apparent magnitude and measured proper motion, an assumed maximum tangential velocity leads to a limit on the star's luminosity if the star is to be bound to the Galaxy. For example, if a 15th magnitude C star were found to have a proper motion of $0.''1$/yr, it could be no brighter than $M_V \approx +4.9$ unless it were escaping from the Galaxy. Such a star would be a dC.

## 2. The Proper Motion Search

We have begun a systematic program to search for motions among a large number of catalogued C stars subject to certain conditions. We use the archive of digitized scans made from the 1950–era POSS red (E) plates and the 1980–era V and SERC-J plates used for the HST *Guide Star Catalogue* which are available on 10×–compressed CD-ROMs accessible on a jukebox at the Space Telescope Science Institute. The early-epoch scans are not available on CD south of $\delta$ –10°, so all of the stars examined for proper motion are north of that. In the magnitude range $12 < V < 18$, proper motions can be determined down to 35 milliarcseconds/yr (mas/yr) from these scans with a formal error of about ±10 mas/yr over the typical 30-year epoch difference. A more complete description of the procedures and software is given in MacConnell et al. (1995).

At the outset, we confined our program to stars in Stephenson's (1989) *A General Catalog of Cool Galactic Carbon Stars, Second Edition* which are at $|b| \geq 7°$, not of types R or CH, and within the declination and brightness limits specified above. We extracted $7\rlap{.}'3 \times 7\rlap{.}'3$ frames for both epochs on the resulting $\sim 300$ stars and ran Roberts' nearest-neighbor matching algorithm and proper-motion program; the average number of reference stars per field was about two dozen. We measured a further 30 high-latitude C stars resulting from the First Byurakan Survey and published by Abramyan and Gigoyan in *Astrofizika* in the years 1989–1995.

From analysis of this material, we thought we had found one new dC. No. 3635 in Stephenson's catalogue has a proper motion of $0\rlap{.}''06$/yr; if its tangential velocity were of order 100 km s$^{-1}$, its $M_V$ would be $\sim 5.3$. At $V \sim 13$, it would have been the brightest member of the dC class. However, at P. Green's request, J. Huchra obtained a spectrum of CCS 2-3635 with the FAST spectrograph at the 1.5-m telescope on Mt. Hopkins. The spectrum shows no characteristics of a C star but is a good match with M0 V. W. P. Bidelman kindly examined the Burrell Schmidt objective-prism plate on which the star was found and agrees that it should not have been published as a C star. We have continued with extractions on northern C stars within 7° of the galactic plane. We report that among 1400 C stars in the Stephenson catalogue meeting our criteria and of R.A. $< 21^h$, as well as among the Byurakan stars, there are no new dCs, i.e. no stars having $\mu > 0\rlap{.}''035$/yr.

## 3. Dwarf S Stars?

There are sound reasons to believe that dwarf S stars in mass-transfer binaries should also exist, but none is known. Subject to the same constraints as above, we searched for motions among 322 stars in Stephenson's (1984) *A General Catalogue of Galactic S Stars, Second Edition* and additional ones published by him in 1990 and found one with sizeable motion: no. 1237 has $\mu = 0\rlap{.}''16$/yr, which is about 10% below the limit for Luyten's survey for the general field. W. P. Bidelman has also examined the spectrum of this star on the discovery plate at our request and finds that it does not have ZrO absorption at 6474 Å but instead has strong CaH at 6385 Å; the two bands are separated by only about one resolution unit at the dispersion of the plate. Thus, no. 1237 joins the other ex-S stars, nos. 25, 780, and 875, which Stephenson (1986) found to be early M dwarfs with strong CaH, making them possibly similar to Kapteyn's star (see Wing et al. 1976).

**References**

Dahn, C. C., Liebert, J., Kron, R. G., Spinrad, H. & Hintzen, P. M. 1977, *ApJ*, 216, 757
Dearborn, D. S. P., Liebert, J., Aaronson, M., Dahn, C. C., Harrington, R., Mould, J. & Greenstein, J. L. 1986, *ApJ*, 300, 314
de Kool, M. & Green, P. J. 1995, *ApJ*, 449, 236
Green, P. J. 2000, in IAU Symp. 177: *The Carbon Star Phenomenon*, ed. R. F. Wing (Kluwer), p. 27
Green, P. J., Margon, B., Anderson, S. F. & MacConnell, D. J. 1992, *ApJ*, 400, 659
Green, P. J., Margon, B. & MacConnell, D. J. 1991, *ApJ*, 380, L31
Heber, U., Bade, N., Jordan, S. & Voges, W. 1993, *A & A*, 267, L31
Liebert, J., Schmidt, G. D., Lesser, M., Stepanian, J. A., Lipovetsky, V. A., Chaffee, F. H., Foltz, C. B. & Bergeron, P. 1994, *ApJ*, 421, 733
MacConnell, D. J., Roberts, W. J. & Williamson, R. L. II 1995, in IAU Colloquium 148: *The Future Utilisation of Schmidt Telescopes*, ed. J. Chapman, R. Cannon, S. Harrison and B. Hidayat, ASP Conf. Ser., 84, 224
Stephenson, C. B. 1984, *Publ. Warner & Swasey Obs.*, 3, 1
Stephenson, C. B. 1986, *PASP*, 98, 467
Stephenson, C. B. 1989, *Publ. Warner & Swasey Obs.*, 3, 53
Wing, R. F., Dean, C. A. & MacConnell, D. J. 1976, *ApJ*, 205, 186

# LARGE MAGELLANIC CLOUD CARBON STARS: READING THE ROSETTA STONE OF STELLAR EVOLUTION

JAY A. FROGEL
*Department of Astronomy, The Ohio State University*
*Columbus, OH 43210, U.S.A.; and*

*Physics Department, University of Durham*
*Durham, England*

AND

EDGARDO COSTA
*Departamento de Astronomia, Universidad de Chile*
*Santiago, Chile*

**Abstract.** We discuss new results based on *RI* and *JHK* photometry for 888 and 204 carbon stars, respectively, of the 1035 C stars found by Blanco and his collaborators in the Large Magellanic Cloud (LMC). Bolometric magnitudes and effective temperatures for these stars are calculated and compared with theoretical predictions. We find a spatial gradient in the transition luminosity between M and C type stars. This has implications for the age of the most recent major epoch of star formation in the LMC.

## 1. Introduction and Rationale

Just as the Rosetta Stone provided the key to understanding a previously incomprehensible symbolic language (Uhlemann 1853), so too the Magellanic Clouds have been the key to the interpretation of the visible signs of stellar evolution. This has been particularly true for the final stages of a star's existence. A spectacular recent example has of course been SN 1987A. Much less glamorous but perhaps of equal importance has been the extensive study of asymptotic giant branch (AGB) stars in the Clouds with a continual interplay between theory and observation. In this review we will concentrate on the critical role the Magellanic Clouds have played in the development of a coherent picture for the formation and evolution of carbon (C) stars on the AGB.

First, consider several reasons why understanding the evolution of C stars should be of interest to other than the specialist: C stars are the final stage of AGB evolution for most if not all stars with masses between 1 and 7 $M_\odot$. They are the major source of carbon-rich molecules and grains for the interstellar medium. Carbon stars dominate the bolometric luminosity and the near-IR light of intermediate-age stellar systems which, in turn, implies that they can be an excellent age diagnostic for integrated-light observations of stellar clusters and galaxies.

The importance of the Magellanic Clouds in C-star research comes about because we can get complete samples of AGB C stars in the Clouds, i.e. samples which are not magnitude-limited (e.g. Blanco et al. 1980, hereafter BMB; Blanco & McCarthy 1983, hereafter BM; Rebeirot et al. 1993). The distance to the Clouds is known to a relatively high degree of accuracy and their distension along the line of sight is small relative to their distance. These facts combined with a low and relatively uniform reddening to each Cloud imply that accurate absolute luminosity functions can be derived for C stars. Finally, the LMC and SMC have significant numbers of populous star clusters which contain large numbers of AGB stars (of both C and M type) and whose ages and chemical compositions can be well estimated and are closely related to one another (e.g. Cohen 1982; Bica et al. 1986).

Blanco and his collaborators (BM & BMB) identified more than 1000 C stars in the course of their low dispersion red spectroscopic surveys for C and M giants in 52 fields in the LMC. We (Costa & Frogel 1996) have obtained *RI* photometry for an unbiased sample of about 900 of these C stars and *JHK* photometry for about 200 of the stars with *RI* data. Our objectives in undertaking such an extensive program of single-channel photometry fall into two general categories: first, to empirically derive physical properties of LMC C stars with a sample that is nearly an order of magnitude larger than any previous one and to compare these properties with predictions of stellar evolution theory; and second, to see what can be learned about the history of star formation in the LMC by searching for spatial dependences of these properties.

## 2. The Luminosity Function for LMC Field C Stars

We (Costa & Frogel 1996) have calculated accurate bolometric magnitudes for the 197 C stars observed at *RIJHK*. With these calculations and the tight relation between $BC_K$ and $J-K$ that exists for both M and C stars (Frogel et al. 1980), it is possible to derive an equation which relates the difference between the $I$ magnitude and $m_{\rm bol}$ to the $R-I$ color with an

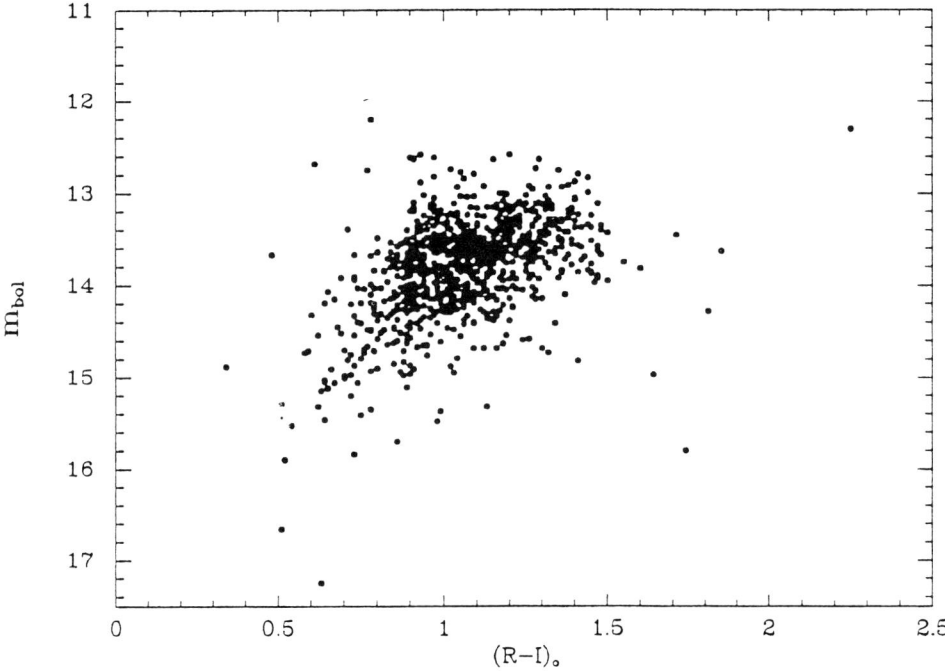

*Figure 1.* The $m_{bol}$ vs. $(R-I)_o$ color-magnitude diagram (Costa & Frogel 1996) for 888 C stars from the surveys of BMB and BM. The upper and lower magnitude limits to the distribution of stars is intrinsic and not due to observational selection effects.

uncertainty of ±0.34 mag. Figure 1 is a color-magnitude diagram for the unbiased sample of 888 C stars from the Blanco et al. surveys for which we have obtained $RI$ photometry and calculated bolometric magnitudes. Figure 2 is the luminosity function for these stars.

The cluster C star luminosity function, as well as the field C star function of Cohen et al. (1981, hereafter CFPE), both of which were based on samples of fewer than 100 stars, are statistically indistinguishable from the new LMC field C star function based on nearly 1000 stars. Thus, the comparisons of the first two functions with stellar models by, *inter alia*, Iben (1981), Iben & Renzini (1983), Lattanzio (1986), and Groenewegen & de Jong (1993) remain valid.

Our new results (Costa & Frogel 1996) emphasize the nearly complete absence of C stars with $m_{bol} \leq 12.5$ — there are only 2 in our entire sample, or ~0.2%. Since the Blanco et al. surveys may have missed a small number of C stars at the two extremes of the color distribution, the actual number of luminous stars may be slightly greater than 0.2%. This limit to the presence of any *significant* population of luminous C stars in the field

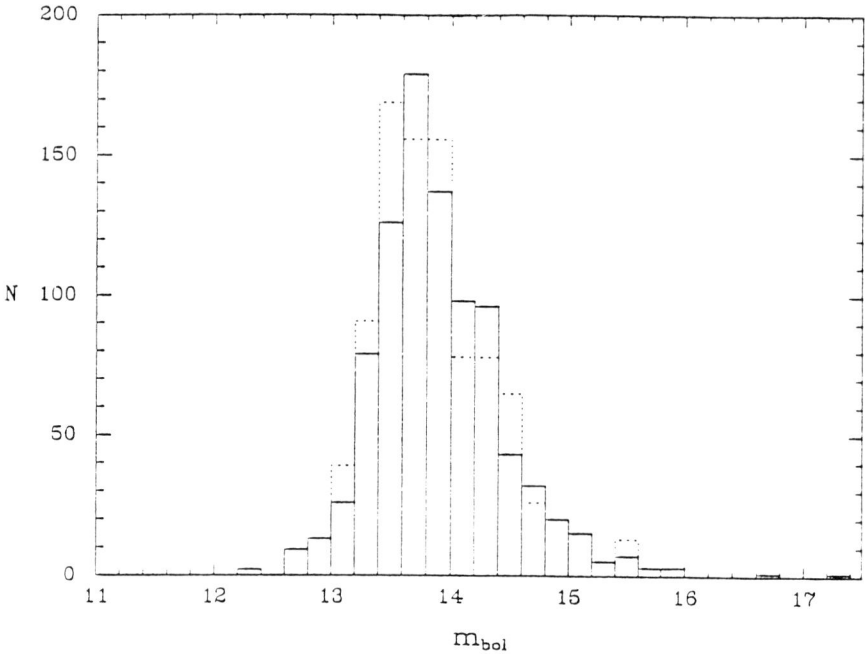

*Figure 2.* The apparent bolometric luminosity function for the 895 LMC field C stars from Figure 1 is shown by the solid line. The dashed line is the same function for the 69 C stars identified in SWB 3.5–6.5 clusters of the Magellanic Clouds by FMB. This later function has been scaled by 895/69 to facilitate comparison.

of the LMC, when combined with results from Frogel & Blanco (1990) on field M stars and from Frogel et al. (1990, hereafter FMB) for clusters, leads us to conclude that one or more processes such as a superwind phase or convective overshooting strongly inhibits AGB stars in the LMC from becoming brighter than $M_{bol} \approx -6$. There is no evidence that the missing bright C stars have reverted back to M stars. Similar results come from the surveys of Reid et al. (1990) for luminous AGB stars in the LMC, of Reid & Mould (1990) in the SMC, and of Wood et al. (1992) for OH/IR stars.

Since the Blanco et al. surveys are not magnitude-limited, we can also draw rather definitive conclusions regarding the faint end of the luminosity function. From our sample of ~900 field C stars, only 3.8% are fainter than $m_{bol} = 15$. Examination of the color-magnitude diagram in Figure 1 shows that most of these bolometrically faint stars are also quite blue and will tend to be among the faintest stars in $I$ as well. Thus, since effects due to misidentification (some could be M stars) and crowding could become important for these stars, 15.0 or 15.2 are good estimates for the faint limit to the C star luminosity function in the LMC.

## 3. Colors and Effective Temperatures of LMC Field C Stars

We have just shown that the field and cluster C stars have indistinguishable distributions in bolometric magnitude. Now consider their colors. Figure 3 compares the $J$–$H$, $H$–$K$ colors of the field stars from Costa & Frogel with the colors of cluster stars from FMB. The extent of the cluster star distribution in Figure 3 is well covered by that of the field C stars. But although they have similar blue limits to their distributions, the field stars extend to considerably redder colors than do the clusters stars: there are only two cluster C stars, both of the LMC, redward of $H$–$K = 0.75$. This difference is not simply due to a difference in sample size. Fully 11% of the field C star sample have $J$–$K \geq 1.9$ while only 3% of the cluster sample is this red. FMB have also noted that the LMC Bar West field has a higher percentage of red, luminous M stars than their cluster M star sample.

An excess of red M and C stars in the field could arise from a metallicity distribution skewed to higher values in the field than in the cluster sample. Higher metallicity produces redder C stars since the evolutionary track of such a star will be shifted to cooler temperatures than that of a lower metallicity star of the same luminosity (as will the tracks of M stars). Also, the formation of molecules and grains will be enhanced in the higher metallicity star due to both lower temperatures and larger numbers of heavy atoms. These enhancements could, in turn, result in increased blanketing by molecular absorption bands, higher mass-loss rates (cf. Groenewegen et al. 1995), and increased circumstellar thermal emission. For the same age, a higher metallicity star will have a greater mass than a lower metallicity star; thus, its AGB evolution would terminate at a brighter magnitude. Also, the luminosity which marks the transition between M and C stars would be brighter in the more metal-rich population, thus accounting for the elevated luminosity of the red field M stars.

Except for their bluer colors, the luminous CH stars found by Hartwick & Cowley (1988) in the LMC (for which infrared data are given by Suntzeff et al. 1993 and Feast & Whitelock 1992) appear to be indistinguishable in terms of their measureable properties from the redder C stars studied by Costa & Frogel (1996). They constitute only a small percentage of the total C star population of the LMC and we propose that these CH stars are just the blue tail of the distribution of redder C stars rather than a distinct class of stars.

Costa & Frogel (1996) derived effective temperatures for the C stars from their $R$–$I$ and $J$–$K$ colors. Three different estimates yield results with a mean dispersion less than 150 K for an individual star. A comparison of the absolute magnitude, $T_{\text{eff}}$ diagram with theoretical tracks indicates that a 1 $M_\odot$ model does indeed represent the lower mass limit for a moderately

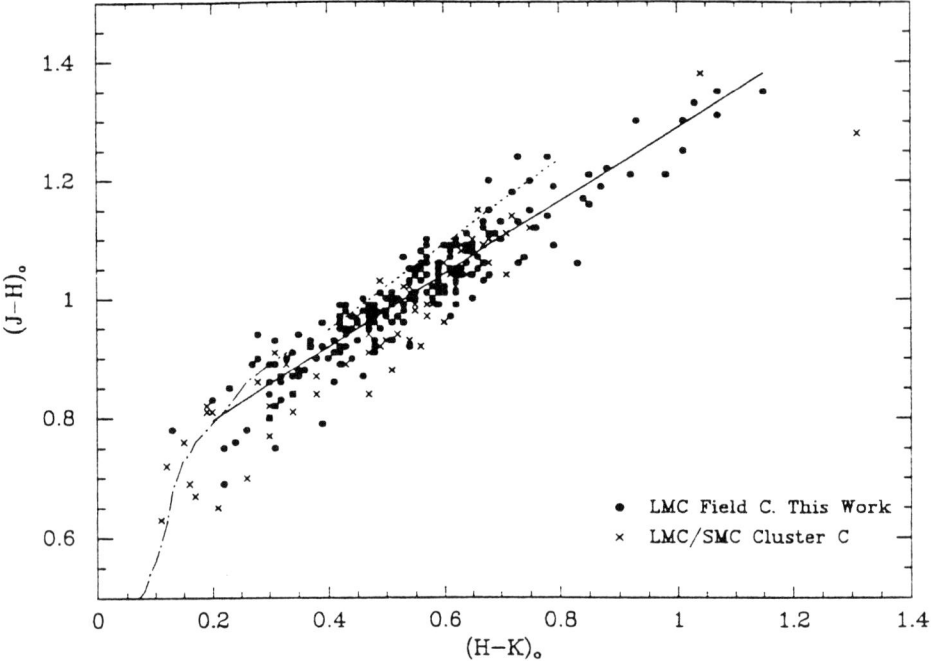

*Figure 3.* The near-IR color-color diagram for the 204 LMC field C stars with *JHK* photometry from Costa & Frogel (1996, filled circles). C stars in clusters of the Magellanic Clouds (FMB) are also plotted. The least-squares fit to the field data is the solid line. The mean relation for galactic C stars is the dotted straight line (CFPE); for galactic M stars it is the dot-dashed curved line (Frogel et al. 1978).

metal-poor C star (cf. Lattanzio 1989; FMB; see also Fig. 4 in Westerlund et al. 1995). In lower mass stars, dredge-up of processed material is not sufficient to bring about C/O > 1 in the envelope. To turn a more metal-rich or younger (i.e. more massive) M star into a C star requires a combination of higher luminosity and lower temperature.

## 4. The Transition Luminosity between M and C Stars: Implications for the Star Formation History of the LMC

FMB show that for Magellanic Cloud clusters that contain both C and M stars, there exists a transition luminosity $m_{bol}(t)$ between the faintest C stars and the brightest M stars, and that this transition luminosity is correlated with the cluster's SWB type (Searle et al. 1980) and, hence, with its age and metallicity in the sense that the transition luminosity increases with increasing Z and decreasing age. Costa & Frogel estimated $m_{bol}(t)$ values for each field in which at least 8 C stars were observed by averaging the magnitudes of the 4 faintest C stars. These estimates are almost all between $m_{\rm bol}(t) = 14.0$ and 15.0, comparable to the fainter transition luminosities for SWB V–VI type Cloud clusters (see FMB's Table 3 and Fig. 14) and significantly fainter than the transition luminosities of the earlier, III–IV, SWB types which are between 13.1 and 13.5. Since the technique used by FMB to calculate transition luminosities involved averaging the $m_{\rm bol}$ of the faintest C stars with the $m_{\rm bol}$ of the brightest M stars, and since the brightest M stars are typically fainter than the faintest C stars in any given cluster by a few tenths of a magnitude, the $m_{\rm bol}(t)$ values calculated by Costa & Frogel (1996) would need to be made yet fainter for a proper comparison with the FMB values. Given the close linkage between age and metallicity for the clusters (e.g. Cohen 1982; Bica et al. 1986) this result implies that throughout the LMC there is a significant population of stars similar to those found in the older, metal-poor end of the distribution of intermediate age LMC clusters. In other words, none of the fields surveyed by BM has an *exclusively* younger stellar population such as is found in clusters of SWB types II–IV.

Figure 4 shows a significant brightening of $m_{\rm bol}(t)$ with increasing distance from the center of the LMC. The change in $m_{\rm bol}(t)$ is from $\sim 14.7$ near the Bar to $\sim 14.1$ in the outermost fields. If we compare this with the relation between $m_{\rm bol}(t)$ and turnoff mass in Fig. 19 of FMB based on clusters of known ages and metallicities, we can estimate the oldest epoch of C star formation as a function of position. To effect this comparision the present values of $m_{\rm bol}(t)$ must be made fainter by a few tenths of a magnitude to take into account the fact that our estimates for $m_{\rm bol}(t)$ were made without the help of any M stars. We conclude that if the field C stars follow the same [Fe/H] – age relation as the clusters, then the earliest epoch of C-star formation near the Bar of the LMC corresponds to clusters of SWB type 6.5, while near the periphery it corresponds to clusters of type 5.5, or an age difference of a few Gyr according to Table 3 of FMB (a younger "earliest epoch" on the periphery than near the Bar) and an [Fe/H] difference of 0.2 to 0.4 dex (a higher [Fe/H] near the periphery). If the mean [Fe/H] at the periphery were forced to be the same as that near the Bar, the implied age difference would be greater, i.e. the periphery would be younger still. For reasons we do not fully understand, the dispersion in the relationship between $m_{\rm bol}(t)$ and distance is considerably less if fields containing populous clusters are excluded from the analysis. This may have to do with C stars from the clusters contaminating the field C stars.

Carbon stars are just the tip of the iceberg that represents a major epoch of star formation occurring a few Gyr ago in the LMC and first identified by Butcher (1977). Our finding of a systematic spatial variation in $m_{\rm bol}(t)$ suggests that this epoch occurred more recently in the periphery

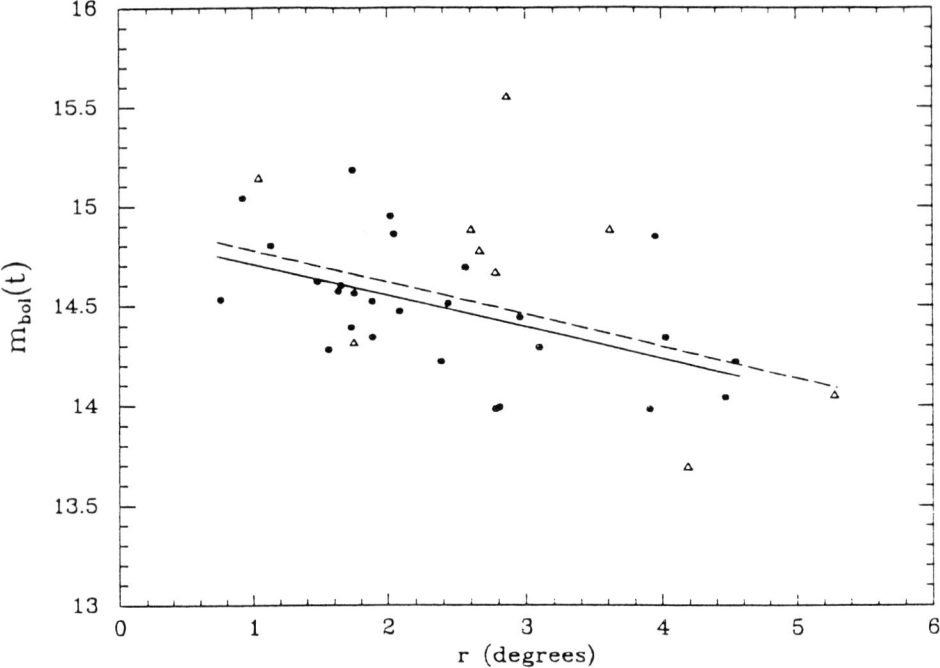

*Figure 4.* The transition luminosity $m_{bol}(t)$ for the 36 fields which have 8 or more C stars observed by us, shown as a function of the field's radial distance from the center of the LMC. Solid dots represent fields without clusters; open triangles fields centered on populous clusters. The dashed line is the linear regression of $m_{bol}(t)$ on r for all 36 fields. The solid line is the same regression for only the 27 fields without clusters.

than in the central regions of the LMC. This result could be tested by obtaining optical CMDs that include the main sequence turnoff. Given the variation in $m_{bol}(t)$, one would expect to find a variation in the C to M star ratio as well. However, in view of the small size of the change in $m_{bol}(t)$, examination of the cluster data in FMB indicates that the absence of a variation in the C to M star ratio (BM) is not unexpected and can be attributed to statistical fluctuations in star numbers. BM did find large spatial gradients in the C to M star ratio in the SMC. Thus it would be useful to carry out a study similar to ours of the C stars they identified in the SMC.

JAF was a staff member and EC a visiting astronomer at CTIO when these observations were made. CTIO is operated by AURA, Inc. under contract with the NSF. JAF also thanks Roger Davies and PPARC for a Visiting Senior Research Fellowship at the Univeristy of Durham where this review was prepared.

## References

Bica, E., Dottori, H. & Pastoriza, M. 1986, *A&A*, 156, 261
Blanco, V. M. & McCarthy, M. F. 1983, *AJ*, 88, 1442 (BM)
Blanco, V. M., McCarthy, M. F. & Blanco, B. M. 1980, *ApJ*, 242, 938 (BMB)
Butcher, H. 1977, *ApJ*, 216, 372
Cohen, J. G. 1982, *ApJ*, 258, 143
Cohen, J. G., Frogel, J. A., Persson, S. E. & Elias, J. H. 1981, *ApJ*, 249, 481 (CFPE)
Costa, E. & Frogel, J. A. 1996, *AJ*, 112, 2607
Feast, M. W. & Whitelock, P. A. 1992, *MNRAS*, 259, 6
Frogel, J. A. & Blanco, V. M. 1990, *ApJ*, 365, 168
Frogel, J. A., Mould, J. & Blanco, V. M. 1990, *ApJ*, 352, 96 (FMB)
Frogel, J. A., Persson, S. E., Aaronson, M. & Matthews, K. 1978, *ApJ*, 220, 75
Frogel, J. A., Persson, S. E. & Cohen, J. G. 1980, *ApJ*, 239, 495
Groenewegen, M. A. T. & de Jong, T. 1993, *A&A*, 267, 410
Groenewegen, M. A. T., van den Hoek, L. B. & de Jong, T. 1995, *A&A*, 293, 381
Hartwick, F. D. A. & Cowley, A. P. 1988, *ApJ*, 334, 135
Iben, I., Jr. 1981, *ApJ*, 246, 278
Iben, I., Jr. & Renzini, A. 1983, *Ann. Rev. Astron. Astrophys.*, 21, 271
Lattanzio, J. C. 1986, *ApJ*, 311, 708
Lattanzio, J. C. 1989, *ApJ*, 344, L25
Rebeirot, E., Azzopardi, M. & Westerlund, B. E. 1993, *A&A Supp.*, 97, 603
Reid, N. & Mould, J. 1990, *ApJ*, 360, 490
Reid, N., Tinney, C. & Mould, J. 1990, *ApJ*, 348, 98
Searle, L., Wilkinson, A. & Bagnuolo, W. G. 1980, *ApJ*, 239, 803 (SWB)
Suntzeff, N. B., Phillips, M. M., Elias, J. H., Cowley, A. P., Hartwick, F. D. A. & Bouchet, P. 1993, *PASP*, 105, 350
Uhlemann, M. 1853, "Inscriptionis Rosettanae Hieroglyphicae Decretum Sacerdotale," (Lipsiae: Dykiana)
Westerlund, B. E., Azzopardi, M., Breysacher, J. & Rebeirot, E. 1995, *A&A*, 303, 107
Wood, P. R., Whiteoak, J. B., Hughes, S. M. G., Bessell, M. S., Gardner, F. F. & Hyland, A. R. 1992, *ApJ*, 397, 552

## Discussion

**Gustafsson:** What does the tendency of $m_{bol}$ to vary with distance from the LMC Bar correspond to in terms of age differences?

**Frogel:** The difference is $\sim 30\%$ in turnoff mass. If SWB III clusters have an age of a few hundred Myr, then the age difference from Bar to edge is a couple of Gyr or 30–50%.

**Gustafsson:** Is it possible to separate age effects from metallicity effects for the LMC clusters and from that deduce the significance of metallicity alone for the C star phenomenon? I suppose that would be difficult?

**Frogel:** We have to *assume* that the age – [Fe/H] relation in clusters is one-to-one since the scatter is comparable to the uncertainties. This means we assume the same correlation for C stars. The dependence noted in $m_{bol}$(transition) with distance *may* change this.

**Kerschbaum**: With your near-IR sample, how can you be sure not to miss the high-mass-loss objects? Also, with only near-IR data the estimation of $m_{bol}$ should be more uncertain for higher-mass-loss objects. To avoid these problems, two ISO projects, namely by Trams and by Loup et al., are studying AGB stars in the LMC in the mid- and far-IR.

**Frogel**: An $L$-band survey that Harvey Richer and I did a long time ago did not reveal any significant number of "hidden" C stars. Also, as part of our cluster work we scanned many of the clusters in the near-IR on the CTIO 4-m and found only one star that is too faint in the $I$ band to have been detected by the Blanco et al. surveys. Third, Reid & Mould concluded, on the basis of a deep $I$ band imaging survey, that the numbers of dust-enshrouded C stars in the LMC and SMC have to be very small.

# SURVEYS FOR CARBON STARS IN EXTERNAL GALAXIES

MARC AZZOPARDI
*Observatoire de Marseille, F-13248 Marseille, France*
and
*Canada-France-Hawaii Telescope, Kamuela, HI 96743, U.S.A.*

**Abstract.** Prominent molecular absorption bands in the spectra of carbon stars make their detection possible, even in relatively distant external galaxies. Although extensive surveys for carbon stars have been carried out mainly in the Magellanic Clouds and the dwarf spheroidals in the Galactic halo, more distant galaxies in the Local Group and beyond have been successfully searched for this kind of object.

## 1. Introduction

AGB star populations are of special interest for the study of the morphology, stellar evolution, and kinematics of galaxies. For instance, it has been shown that the ratio of carbon stars (C stars) to late-type M stars is very sensitive to metallicity and age. In addition, the AGB star populations are of special interest for the study of the stellar evolution, structure and kinematics of nearby systems. Consequently, a number of extensive surveys for C stars have been carried out in the Magellanic Clouds and the dwarf spheroidals in the Galactic halo as well as in selected regions of other nearby galaxies beyond those systems. The presence of prominent molecular bands of $C_2$, CN and TiO make AGB star detection possible as far away as NGC 2403 in the M81/NGC 2403 group of galaxies.

## 2. Late-Type Star Surveys

Spectroscopic surveys for late-type stars have been successfully carried out in the Magellanic Clouds and the Galactic halo dwarf spheroidal galaxies. Photographic plates obtained either with Schmidt telescopes equipped with low-dispersion objective prisms or at the prime focus of 4-m class telescopes equipped with wide-field correctors and very low dispersion transmission gratings (grism or grens) have been employed in extensive surveys of these systems. Occasionally, suitable intermediate to broad bandpass filters have been used to restrict the instrumental spectral range and keep the number of overlaps as low as possible. At present, the same observing technique using CCD cameras behind focal reducers at the Cassegrain focus of 2-m to 4-m class telescopes is discovering fainter C stars in some of the Galactic halo dwarf spheroidals (see section 4) that escaped detection in previous photographic surveys.

Slitless transmission grating detections cannot reach AGB stars effectively beyond the more distant Galactic halo dwarf spheroidal galaxies. Hence, the only possible way to probe these objects in other nearby galaxies is the imaging technique. Based on the Palmer & Wing (1982) survey technique, intermediate-band photometric color systems measuring CN and TiO molecular bands have been defined (Cook & Aaronson 1989; Richer et al. 1985) to identify more distant C and late-M stars. However, subsequent spectroscopic observations are mandatory to confirm the nature of the AGB-star candidates detected in this manner.

Carbon stars can be identified mainly through their near-infrared CN bands at 7945, 8125 and 8320 Å. However, they have also been successfully found by means of their Swan $C_2$ bands at 4737, 5165 and 5636 Å in the blue-green spectral domain. The two survey techniques are basically complementary (see McCarthy 1987). Note that the most luminous C stars, with strong stellar winds responsible for high mass-loss rates, can only be detected at infrared or radio wavelengths.

Further information on C star surveys in external galaxies is provided by Azzopardi (1994) in the proceedings of the third CTIO/ESO workshop.

## 3. Magellanic Clouds Surveys

Early spectroscopic surveys for field C stars were carried out in the Large Magellanic Cloud (LMC) by Sanduleak & Philip (1977) using, in the blue-green spectral domain, the thin prism attached to the Curtis Schmidt at Cerro Tololo Inter-American Observatory (CTIO), and by Westerlund at Mount Stromlo Observatory using, in the near infrared, the Uppsala Southern Station Schmidt telescope (Westerlund et al. 1978). These led to the detection of 474 and 302 candidates, respectively, leading to a total number of about 700 probable C stars found in this system. These surveys were limited to stars brighter than $V \approx 16.5$, close to the typical magnitude of the fainter carbon stars in the LMC, so that completeness was not assured. Note that Westerlund's attempt to identify C stars in the Small Magellanic Cloud (SMC) using the same instrumentation and survey technique did not succeed.

The near-infrared photographic transmitting – grating survey for red giant stars in selected areas in the Magellanic Clouds (52 circular LMC fields of 0.12 deg$^2$ each, and 28 circular SMC fields of 0.12 deg$^2$ each plus 9 square fields of 0.38 deg$^2$ each) carried out at the prime focus of the CTIO 4-m telescope by Blanco et al. (1980) and Blanco & McCarthy (1983, 1990) resulted in the identification of 1045 C star candidates in the LMC and 860 such candidates in the SMC. From the integration of the C star surface densities of the selected sample fields, Blanco & McCarthy (1983) inferred the overall surface distribution of these objects in the two Clouds and estimated at 11,000 and 2,900 the total numbers of C stars in the LMC and SMC, respectively.

A photographic blue-green grism survey for field C stars in the SMC (13 partially overlapping circular fields of 0.78 deg$^2$ each) was carried out by our team at the prime focus of the European Southern Observatory (ESO) 3.6-m telescope (Westerlund et al. 1986; Azzopardi 1993). This search led to the detection of 1707 probable C stars in the central regions of this galaxy, of which fewer than 12% were already known (Rebeirot et al. 1993). Using a color–magnitude diagram obtained from the photometry of the slitless grism spectra, Rebeirot and associates identified a sample of about 80 faint and relatively blue C-star candidates. Subsequent medium-resolution slit spectroscopy and $JHK$ photometry for 50 of them resulted in the discovery of 38 low-luminosity C stars with $M_{bol}$ ranging from –1.4 to –3.0 mag (Westerlund et al. 1992, 1995: see also Fig. 2 in Azzopardi 1994). The faintest objects of this sample are among the least luminous C stars ever found in an external galaxy and are in many ways similar to those found in the Galactic bulge (Westerlund et al. 1991).

More recent AGB star surveys are moving toward the detection of C– and M–type stars in the outer regions of the Magellanic Clouds and the intercloud region. These intermediate-age and cool giant stars, which are representative of the age of the majority of the stellar population of the Clouds and can be found in their outer halos, are suitable for the study of the interaction dynamics of the LMC–SMC bulk system. The two-color $(R, B_J)$ UK Schmidt Telescope (UKST) survey in the outer halos and intercloud region of the Magellanic Clouds by Demers et al. (1993) led to the identification of 57 very red stars ($14 < R < 17$; $B$–$V > 1.75$). Subsequent spectroscopy by the author and $JHK$ infrared photometry by Feast & Whitelock (1994) resulted in the discovery of 30 proven C stars. Since then, more very red stars have been found in an extended survey of the outer regions of the Clouds using the same imaging detection technique (Kunkel et al. 1997; Demers et al. 2000), and the C–star nature of more than 500 objects has been confirmed, the spectra being of sufficient quality to provide accurate radial velocities ($\pm 5$ km s$^{-1}$).

The blue-green medium-dispersion objective-prism UKST survey of the outer parts of the SMC by Morgan & Hatzidimitriou (1995) resulted in the detection of 1185 C star candidates. From this sample these authors came to the conclusion that the total number of C stars in the SMC would be about 3060, in good agreement with the estimates 2900 and 3100 by Blanco & McCarthy (1983) and Azzopardi & Rebeirot (1991), respectively. The surface distribution of the C stars found by Rebeirot et al. (1993), together with the new ones found by Morgan & Hatzidimitriou, shows a low-eccentricity elliptical pattern with major axis parallel to the Bar, and this supports the assertion by Hardy et al. (1989) that SMC C stars, as well as planetary nebulae, belong to a spheroidal-like system.

## 4. Galactic Halo Dwarf Spheroidal Surveys

Following the pioneering work of Demers & Kunkel (1979), who discovered 66 very red giants in the Fornax dwarf spheroidal galaxy and suspected that some of them could be carbon stars, early photographic surveys for this kind of object were carried out in seven Galactic halo dwarf spheroidal galaxies in a manner similar to the ones carried out in the Magellanic Clouds. Carbon stars were searched for in:

- Fornax and Sculptor with the CTIO 4-m reflector by Blanco & McCarthy (Frogel et al. 1982), as well as by Richer & Westerlund (1983) and Westerlund et al. (1987) with the ESO 3.6-m telescope, all using transmission gratings in the near infrared;
- Draco, Leo I, Leo II and Ursa Minor, through their near-infrared CN bands, by Aaronson et al. (1982, 1983);
- Carina, thanks to a color-magnitude diagram derived from a pair of plates taken with the 3.9-m Anglo-Australian Telescope (AAT) by Cannon et al. (1981) and two objective-prism plates obtained, in the blue-green spectral domain, with the UKST (Mould et al. 1982).

Those seven dwarf spheroidals were revisited by us (Azzopardi et al. 1985, 1986), using our slitless spectroscopy technique based on the photographic detection of the blue-green $C_2$ band, at the Canada-France-Hawaii (CFHT) and ESO (Carina, only) 3.6-m telescopes. We discovered new C star candidates, whose nature has been confirmed by subsequent medium-resolution slit spectroscopy. They had escaped previous surveys in all systems except Ursa Minor. The more outstanding results were obtained for Leo I (19 C stars found, only one previously known) and Fornax (29 new C stars found, where 45 were previously known). We (Azzopardi, Breysacher, Muratorio & Westerlund) are now searching for fainter C stars in the Galactic halo dwarf spheroidals with the same detection technique using focal reducers and CCD cameras:

TABLE 1. Census of field carbon stars in Galactic halo galaxies

| (1) System | (2) No. C Stars n(C) | (3) Total Number N(C) | (4) References |
|---|---|---|---|
| LMC | 849 | 11000 | (2)(8)(14) |
| SMC | 1718 | 3060 | (8)(12)(18)(19) |
| Fornax | 104 | ~115 | (1)(5)(13)(16)(21) |
| Leo I | 16 | | (7)(10)(11) |
| Carina | 9 | 9 | (3)(6)(10)(11)(21) |
| Leo II | 8 | 8 | (7)(10)(21) |
| Sculptor | 8 | | (5)(9)(10)(11) |
| Sagittarius | 4 | | (20) |
| Draco | 3 | | (4)(11) |
| Ursa Minor | 1 | | (7)(11) |
| Sextans | 0 | | (15) |
| Galactic bulge | 34 | | (17) |

Column 1: Name of the system.
Column 2: Number of confirmed C stars now identified in the system.
Column 3: Estimated total number of C stars in the system.
Column 4: References in chronological order as follows:
    (1) Aaronson & Mould (1980)
    (2) Blanco, McCarthy & Blanco (1980)
    (3) Cannon, Niss & Norgaard-Nielsen (1981)
    (4) Aaronson, Liebert & Stocke (1982)
    (5) Frogel, Blanco, McCarthy & Cohen (1982)
    (6) Mould, Cannon, Aaronson & Frogel (1982)
    (7) Aaronson, Olszewski & Hodge (1983)
    (8) Blanco & McCarthy (1983)
    (9) Richer & Westerlund (1983)
    (10) Azzopardi, Lequeux & Westerlund (1985)
    (11) Azzopardi, Lequeux & Westerlund (1986)
    (12) Westerlund, Azzopardi & Breysacher (1986)
    (13) Westerlund, Edvardsson & Lundgren (1987)
    (14) Blanco & McCarthy (1990)
    (15) Irwin, Bunclark, Bridgeland & McMahon (1990)
    (16) Lundgren (1990)
    (17) Azzopardi, Lequeux, Rebeirot & Westerlund (1991)
    (18) Rebeirot, Azzopardi & Westerlund (1993)
    (19) Morgan & Hatzidimitriou (1995)
    (20) Ibata, Gilmore & Irwin (1995)
    (21) Azzopardi (present paper)

TABLE 2. Carbon star surveys in M 31

| System | Field n | Field arcmin | C/late M | C/M | References |
|---|---|---|---|---|---|
| M 31 | 1 | 1.8×1.1 | 5C/41M5$^+$ | 0.12 | Richer & Crabtree, 1985 |
| | 2 | 2.5×1.5 | C/M5$^+$ | 0.08 | Cook et al. 1986 |
| | 1 | 3.5×2.1 | 6C/39M5$^+$ | 0.15 | Richer et al. 1990 |
| | 5 | 7.0×7.0 | 55C/2884M | 0.019 | Brewer et al. 1995 |
| | | | 26C/740M | 0.035 | |
| | | | 75C/828M | 0.091 | |
| | | | 82C/727M | 0.113 | |
| | | | 5C/56M | 0.089 | |

TABLE 3. Carbon star surveys in other nearby galaxies

| System | Field n | Field arcmin | C/late M | C/M | References |
|---|---|---|---|---|---|
| M 33 | 1 | 2.5×1.5 | C/M5$^+$ | 1.0 | Cook et al. 1986 |
| NGC 6822 | 2 | 2.5×1.5 | C/M5$^+$ | 33 | Cook et al. 1986 |
| NGC 205 | 1 | 1.8×1.1 | 7C/17M2$^+$ | 0.41 | Richer et al. 1984 |
| IC 1613 | 2 | 2.5×1.5 | C/M5$^+$ | >14 | Cook et al. 1986 |
| WLM | 1 | 2.5×1.5 | C/M5$^+$ | >14 | Cook et al. 1986 |
| NGC 300 | 1 | 5.0×3.0 | 16C/23M5$^+$ | 0.41 | Richer et al. 1985 |
| NGC 55 | 1 | 5.0×3.0 | 14C/7M5$^+$ | 2.0 | Pritchet et al. 1987 |
| NGC 2403 | 1 | 1.8×1.1 | some C stars/ several M stars | ??? | Hudon et al. 1989 |

- Carina was observed at the ESO 3.5-m NTT with EMMI, but no further C stars were found;
- Leo II was observed at the ESO 3.6-m reflector with EFOSC and at the CFHT with MOS, and two new C stars were found and their nature confirmed afterwards;
- Fornax was observed at the ESO 3.5-m NTT with EMMI as well as at the CFHT with MOS, and several new faint C star candidates were found. Up to now, the C–star nature of 30 of them has been confirmed with MEFOS attached to the ESO 3.6-m telescope. Most are probably low-luminosity C stars like those we discovered in the SMC. With 104 C stars presently known, Fornax contains by far the largest number of these objects in the Galactic halo dwarf spheroidals.

In Table 1, these results are collected together with those secured for two more dwarf galaxies, namely Sextans (no C star found) and Sagittarius (4 C stars found), that have been detected more recently in the Galactic halo by Irwin et al. (1990) and Ibata et al. (1995), respectively.

## 5. Other Nearby Galaxy Surveys

Two teams have carried out photometric surveys for C and late-M stars in selected fields of external galaxies beyond the most distant dwarf spheroidals in the Galactic halo. M 31, M 33, NGC 6822, IC 1613 and the Wolf-Lundmark-Melotte galaxy (WLM) were observed by Aaronson and collaborators (Aaronson et al. 1984; Cook et al. 1986) using the Steward Observatory 2.3-m telescope (see Cook & Aaronson 1989 for more information on the photometric color system used). Meanwhile, using similar intermediate-band filters, Richer and associates (see for instance Richer, Crabtree & Pritchet 1984) were observing M 31, NGC 205 and NGC 2403 with the CFHT, as well as NGC 300 and NGC 55 at the CTIO 4-m reflector. Tables 2 and 3 give the number and size of the fields surveyed, the C to late-M star ratio, and the reference paper for each system. Among those galaxies, M 31 forms the subject of several late-type star surveys, and the recent work by Brewer et al. (1995, 2000) is called to the reader's attention.

## References

Aaronson, M., Da Costa, G. S., Hartigan, P., Mould, J. R., Norris, J. & Stockman, H. S. 1984, *ApJ*, 277, L9
Aaronson, M., Liebert, J. & Stocke, J. 1982, *ApJ*, 254, 507
Aaronson, M. & Mould, J. 1980, *ApJ*, 240, 804
Aaronson, M., Olszewski, E. W. & Hodge, P. W. 1983, *ApJ*, 267, 271
Azzopardi, M. 1993, *The Messenger*, No. 71, 29
Azzopardi, M. 1994, in $3^{rd}$ CTIO/ESO Workshop: *The Local Group: Comparative and Global Properties*, ed. A. Layden, R. C. Smith and J. Storm, ESO Conference and

Workshop No. 51, p. 129
Azzopardi, M., Lequeux, J., Rebeirot, E. & Westerlund, B. E. 1991, *A&A Supp.*, 88, 265
Azzopardi, M., Lequeux, J. & Westerlund, B. E. 1985, *A&A*, 144, 388
Azzopardi, M., Lequeux, J. & Westerlund, B. E. 1986, *A&A*, 161, 232
Azzopardi, M. & Rebeirot, E. 1991, in IAU Symp. 148: *The Magellanic Clouds*, ed. R. Haynes and D. Milne (Kluwer), p. 71
Blanco, V. M. & McCarthy, M. F. 1983, *AJ*, 88, 1442
Blanco, V. M. & McCarthy, M. F. 1990, *AJ*, 100, 674
Blanco, V. M., McCarthy, M. F. & Blanco, B. M. 1980, *ApJ*, 242, 938
Brewer, J. P., Richer, H. B. & Crabtree, D. R. 1995, *AJ*, 109, 2480
Brewer, J. P., Richer, H. B. & Crabtree, D. R. 2000, in IAU Symp. 177: *The Carbon Star Phenomenon*, ed. R. F. Wing (Kluwer), p. 59
Cannon, R. D., Niss, B. & Nørgaard-Nielsen, H. U. 1981, *MNRAS*, 196, 1P
Cook, K. H. & Aaronson, M. 1989, *AJ*, 97, 923
Cook, K. H., Aaronson, M. & Norris, J. 1986, *ApJ*, 305, 634
Demers, S. & Kunkel, W. E. 1979, *PASP*, 91, 761
Demers, S., Irwin, M. J. & Kunkel, W. E. 1993, *MNRAS*, 260, 103
Demers, S., Kunkel, W. E. & Irwin, M. J. 2000, in IAU Symp. 177: *The Carbon Star Phenomenon*, ed. R. F. Wing (Kluwer), p. 528
Feast, M. & Whitelock, P. 1994, *MNRAS*, 269, 737
Frogel, J. A., Blanco, V. M., McCarthy, M. F. & Cohen, J. G. 1982, *ApJ*, 252, 133
Hardy, E., Suntzeff, N. B. & Azzopardi, M. 1989, *ApJ*, 344, 210
Hudon, J. D., Richer, H. B., Pritchet, C. J., Crabtree, D., Christian, C. A. & Jones, J. 1989, *AJ*, 98, 1265
Ibata, R. A., Gilmore, G. & Irwin, M. J. 1995, *MNRAS*, 277, 781
Irwin, M. J., Bunclark, P. S., Bridgeland, M. T. & McMahon, R. G. 1990, *MNRAS*, 244, 16P
Kunkel, W. E., Irwin, M. J. & Demers, S. 1997, *A&A Supp.*, 122, 463
Lundgren, K. 1990, *A&A*, 233, 21
McCarthy, M. F. 1987, in ESO Conference and Workshop No. 27: *Stellar Evolution and Dynamics of the Outer Halo of the Galaxy*, ed. M. Azzopardi and F. Matteucci, p. 203
Morgan, D. H. & Hatzidimitriou, D. 1995, *A&A Supp.*, 113, 539
Mould, J. R., Cannon, R. D., Aaronson, M. & Frogel, J. A. 1982, *ApJ*, 254, 500
Palmer, L. G. & Wing, R. F. 1982, *AJ*, 87, 1739
Pritchet, C. J., Richer, H. B., Schade, D., Crabtree, D. & Yee, H. K. C. 1987, *ApJ*, 323, 79
Rebeirot, E., Azzopardi, M. & Westerlund, B. E. 1993, *A&A Supp.*, 97, 603
Richer, H. B. & Crabtree, D. R. 1985, *ApJ*, 298, L13
Richer, H. B., Crabtree, D. R. & Pritchet, C. J. 1984, *ApJ*, 287, 138
Richer, H. B., Crabtree, D. R. & Pritchet, C. J. 1990, *ApJ*, 355, 448
Richer, H. B., Pritchet, C. J. & Crabtree, D. R. 1985, *ApJ*, 298, 240
Richer, H. B. & Westerlund, B. E. 1983, *ApJ*, 264, 114
Sanduleak, N. & Philip, A. G. D. 1977, Publ. Warner & Swasey Obs., v. 2, No. 5
Westerlund, B. E., Azzopardi, M. & Breysacher, J. 1986, *A&A Supp.*, 65, 79
Westerlund, B. E., Azzopardi, M., Breysacher, J. & Rebeirot, E. 1992, *A&A*, 260, L4
Westerlund, B. E., Azzopardi, M., Breysacher, J. & Rebeirot, E. 1995, *A&A*, 303, 107
Westerlund, B. E., Edvardsson, B. & Lundgren, K. 1987, *A&A*, 178, 41
Westerlund, B. E., Lequeux, J., Azzopardi, M. & Rebeirot, E. 1991, *A&A*, 244, 367
Westerlund, B. E., Olander, N., Richer, H. B. & Crabtree, D. R. 1978, *A&A Supp.*, 31, 61

# A PHOTOMETRIC AND SPECTROSCOPIC SURVEY OF AGB STARS IN M31

JAMES P. BREWER
*European Southern Observatory, Santiago 19, Chile*

HARVEY B. RICHER
*Dept. of Physics & Astronomy, Univ. British Columbia
Vancouver, B.C., V6T 1Z4 Canada*

AND

DENNIS R. CRABTREE
*CFHT Corporation, Kamuela, HI 96743, U.S.A.*

**Abstract.** We have used a four-band photometric system capable of distinguishing C, S, and M stars to undertake a survey of AGB stars in M31. We discuss the results from this survey and from follow-up spectroscopy.

## 1. Introduction

The Andromeda Galaxy (M31) is ideal for the study of AGB stars; it is sufficiently close that its AGB population is easily resolved with a four-meter class telescope, while being sufficiently distant that many stars can be observed in a single CCD field. Observations of AGB stars in a galaxy such as M31 allow the interplay between AGB stars and star-forming history, metallicity, and the ISM to be better understood. Additionally, as the distance to M31 is well established, M31's AGB stars make an ideal testbed for models of AGB evolution. With the above in mind, we undertook a survey of AGB stars in M31.

## 2. A Photometric System for Identifying C Stars

To identify C stars in the crowded fields of Local Group (LG) galaxies, groups led by Richer and Aaronson followed a suggestion by Palmer & Wing (1982) and developed a four-band photometric system (FBPS). The FBPS uses two narrowband filters to provide low-resolution spectral infor-

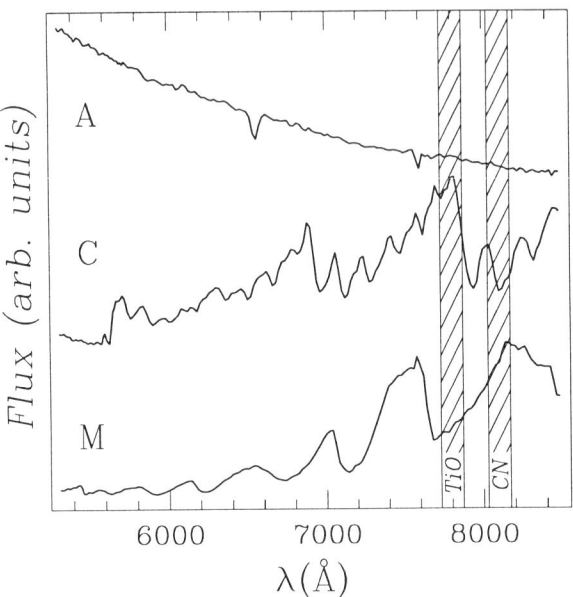

*Figure 1.* Spectra of stars of types A, M, and C. Superimposed on the spectra are the bandpasses of the *CN* and *TiO* filters.

mation and two broadband filters, typically $V$ and $I$, to provide a colour temperature. In Figure 1 we plot spectra of A, M, and C stars and superimpose the bandpasses of Richer's narrowband filters. The bluer of the two filters, the *TiO* filter, lies on a strong TiO absorption band in the M–star spectrum and on pseudo-continuum in the C-star spectrum. The other filter, the *CN* filter, lies on a strong CN band in the C-star spectrum and on pseudo-continuum in the M-star spectrum. Fig. 1 shows that the A star will have $(CN-TiO) \approx 0$, while the C star will have a positive $(CN-TiO)$ index and the M star a negative $(CN-TiO)$ index. In Figure 2 we plot a selection of Galactic and LMC stars in a $(CN-TiO, V-I)$ diagram, along with their spectral types. This figure shows the bifurcation in $(CN-TiO)$ which splits the stars into C and M types. In the $(CN-TiO, V-I)$ diagram, S stars lie between the C and M stars.

## 3. Photometric Observations

We used the CFHT to image five $7' \times 7'$ fields in M31 at galactocentric distances (henceforth $R_{M31}$) of between 4 and 32 kpc. The fields were imaged through $CN$, $TiO$, $V$ and $I$ filters, and calibrated photometry was obtained for the stars in each of these fields (see Brewer et al. 1995, henceforth BRC95).

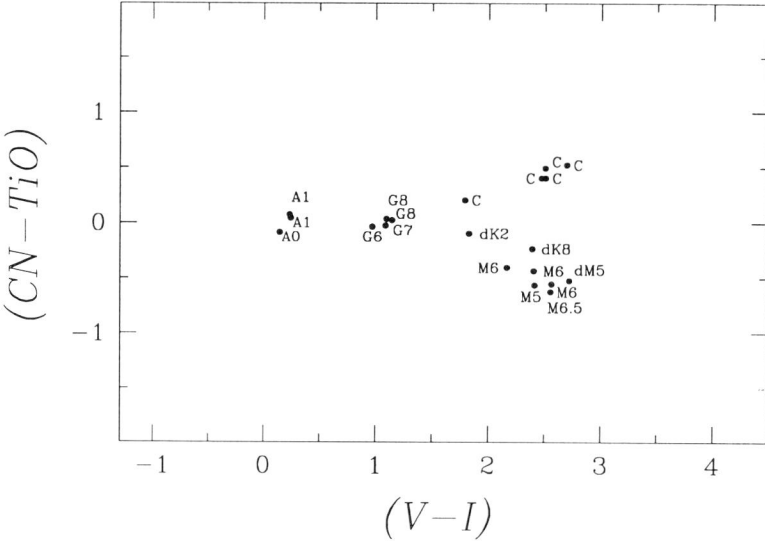

*Figure 2.* The ($CN$–$TiO$, $V$–$I$) diagram for a sample of Galactic and LMC stars. The data for this plot are from Richer et al. (1985).

3.1. TWO-COLOUR DIAGRAMS

In Figure 3 we plot the ($CN$–$TiO$, $V$–$I$) diagrams of our five M31 fields. The lower-right panel of Fig. 3 shows the bounds used to define C and M stars (see BRC95 and Brewer et al. 1996, henceforth BRC96).

3.1.1. *C Stars: Smoky Standard Candles?*
The high luminosity, red colour, and strong spectral characteristics of C stars make them potentially good distance indicators *if* there is a universal C–star luminosity function (LF). In certain types of systems, there is strong observational evidence for a common C–star LF: Richer et al. (1985) concluded that galaxies with [Fe/H] $> -1.8$ and $M_V < -12.9$ will have similar C-star LFs to those of Fornax, the Magellanic Clouds, the Milky Way, M31 and NGC 205.

We used the well-defined samples of C stars in Fields 2, 3 and 4 to construct a C–star LF whose mean was $I_\circ = 19.61 \pm 0.03$. Taking the mean absolute $I$ magnitude of C stars to be $M_I = -4.75$ (see BRC95) gives $(m-M)_\circ = 24.36 \pm 0.03$ for M31, in agreement with, e.g., Freedman & Madore (1990). Additionally, the mean luminosities of the Field 2, 3 and 4 C stars are consistent with each other. This is remarkable given that: (1) Field 3 has undergone much recent star formation; (2) the AGB LF of Field 2 is dissimilar to those of Fields 3 and 4; and (3) the observations of Blair

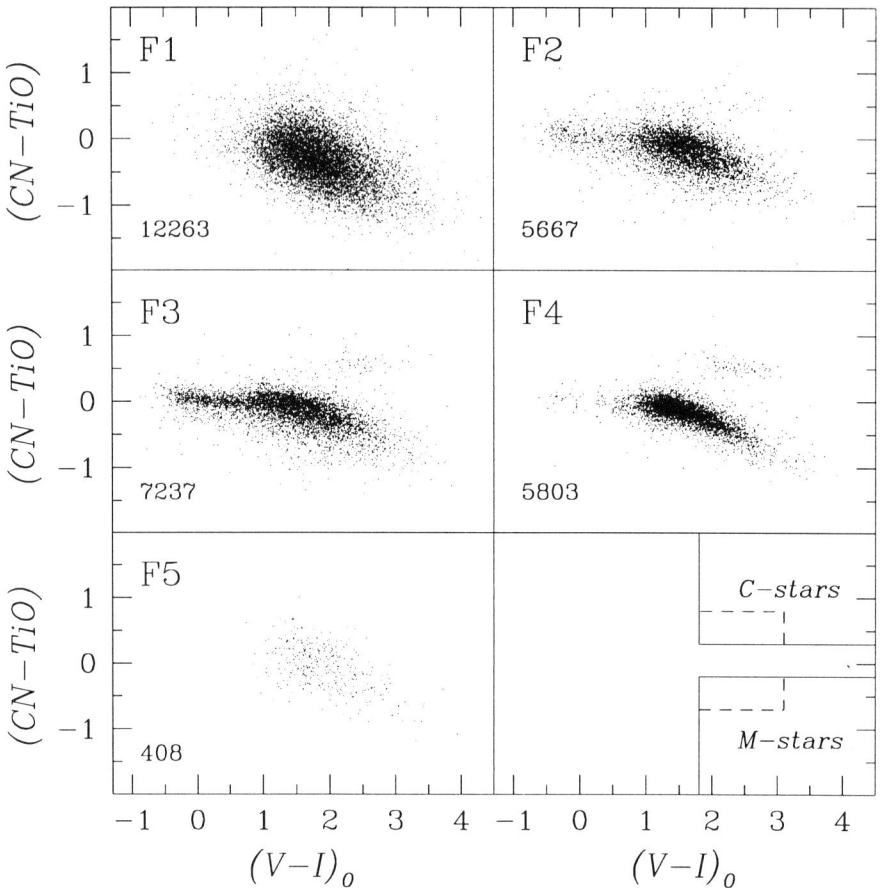

*Figure 3.* The labeled panels show the ($CN-TiO$, $V-I$) diagrams of the five M31 fields. Field 1, at $R_{M31} = 4.6$ kpc, is the innermost field, while Field 5, at $R_{M31} = 31.5$ kpc, is the outermost. The lower right panel shows the colour criteria used to identify the C and M stars.

et al. (1982, henceforth BKC82) indicate a $\sim 0.3$ dex metallicity difference between Fields 2 and 4. This suggests that *the C-star LF provides a robust standard candle* and is relatively insensitive to differences in star formation history and metallicity.

### 3.1.2. *The C/M Ratio*

Blanco et al. (1978) noted that the ratio of C stars to late M stars (hereafter referred to as the C/M ratio) was much greater in the metal-poor Magellanic Clouds than it was in the metal-rich Galactic Nuclear Bulge, a difference they attributed to metallicity and/or age. The C/M ratio is presently believed to be determined mainly by the metallicity of the gas

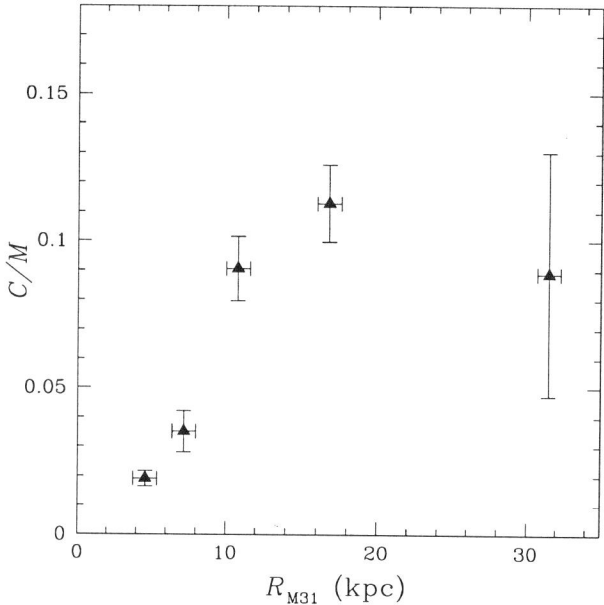

*Figure 4.* The measured C/M ratios in the five M31 fields as a function of $R_{M31}$.

from which the stars condensed, though the effects of star formation history and age remain unclear. Although models predict an anticorrelation between the C/M ratio and metallicity, the predictions are *two orders of magnitude* too small (see Scalo 1981).

In Figure 4 we show the dramatic variation of the C/M ratio as a function of $R_{M31}$. What is the cause of this? The AGB LFs of the five fields were found to have significant differences (see BRC95), though a comparison of the AGB LFs with C/M ratios showed no correlation. From this we concluded that star-forming history, and hence age and mass, is not the dominant factor driving the C/M ratio in M31.

### 3.1.3. *C/M: An Alternative Abundance Indicator?*

As seen in Fig. 4, the C/M ratio increases with $R_{M31}$. This is expected as in M31 metallicity is inversely correlated with $R_{M31}$. The metallicity gradient in M31 has been most recently investigated by BKC82. How well do estimates of the metallicity in our fields (derived via C/M ratios) compare with BKC82's measurements? We obtain abundance estimates in our fields by using measurements from other LG galaxies to derive a relationship between the C/M ratio and abundance (see BRC95). In Figure 5 we plot M31 abundance estimates derived by BKC82 and by us. Reasonable agreement is seen between the abundance estimates up to the limit of the BKC82 data,

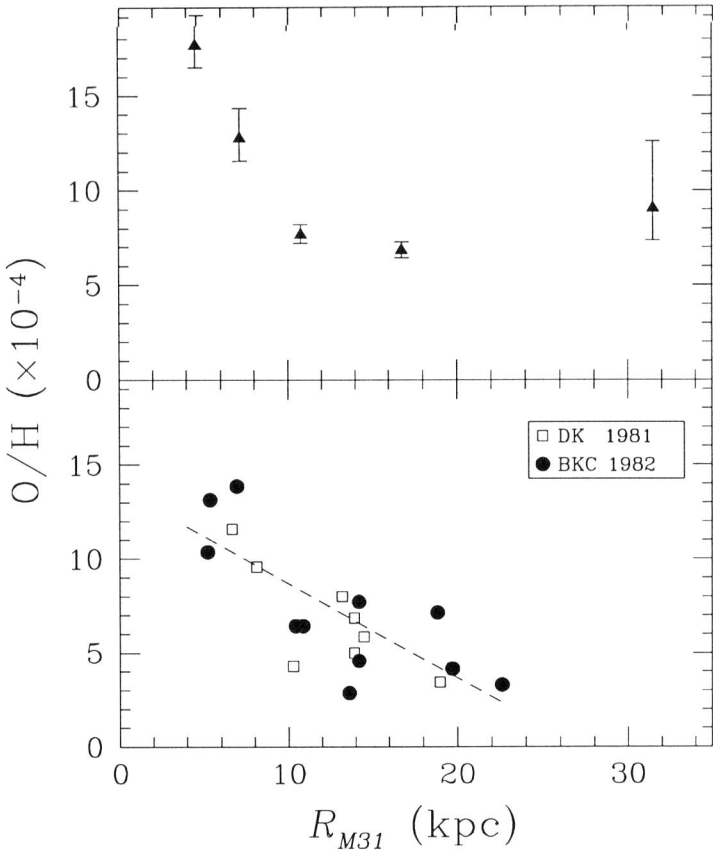

*Figure 5.* The O/H ratio *vs.* distance $R_{M31}$ from the centre of M31. Data in the lower panel are from Fig. 6 of BKC82. The dashed line is the fit of BKC82 to their data (filled circles), while open squares are data from a previous study (see BKC82). In the upper panel we show the O/H number ratio derived from our measurements of the C/M ratio.

while beyond this the Field 5 datum hints at a flattening of the metallicity gradient.

Why is the C/M ratio anticorrelated with metallicity? The C/M ratios can be explained if either: (1) the number of late M stars increases with metallicity; (2) the number of C stars decreases with metallicity; or (3) a combination of (1) and (2). A physical reason for (1) is the redward shift of the AGB with increasing metallicity, and the classification of more AGB stars as M type stars. If it is this which determines the C/M ratios, then it would be expected that the C–star density should scale with field luminosity. This is clearly not so. Although the colour of the AGB may play a small role in determining the C/M ratio, we look to the second possibility for an explanation.

Why might it be easier to make metal-poor C stars? Possible reasons are: (1) a metal-poor star will have a less extended envelope, and so convection does not need to reach so deep for dredge-up to occur; and (2) in a metal-poor star, less dredge-up is needed to drive C > O. A competing explanation for the decreasing number of C stars with metallicity is that metal-rich C stars have higher mass-loss rates and are either comparatively short-lived or evolve into infrared C stars. Clearly more investigation (by both observers and theoreticians) is needed if the cause of this *very pronounced* effect is to be fully understood.

### 3.1.4. *The C/M Ratio and the Interstellar Medium*

Iben & Renzini (1983) note that the different proportions of C stars in the SMC, LMC and Milky Way should lead to different relative proportions of carbonaceous and siliceous grains in the ISM of these galaxies. They mention that the differences between the ultraviolet extinction curves of these three galaxies correlate with their C–star abundances. If extinction laws are affected by grain composition then M31's varying C/M ratio might lead to a radial dependence of the extinction law. Searle (1982) found evidence for M31's reddening law varying systematically with $R_{M31}$. Searle & Thompson (1985) speculate that the radial variation in the dust properties is due to a change in M31's C/M ratio. We have shown that such a variation does exist, and consequently have added weight to the hypothesis that the C/M star ratio has a significant impact on the ISM.

## 4. Spectroscopic Observations

Follow-up multi-object spectroscopy of C, S, and M star candidates was made at the CFHT. The spectra showed that the FBPS did an excellent job in identifying the C stars; all the stars within our chosen C–star region were found to be C stars, while only a handful lay just outside this region. Furthermore, the spectroscopy confirmed the identification of an S star, making it the first known S–type star in M31 and also the most distant S star known. Of the 48 C stars for which we obtained spectra, we found 7 with strongly enhanced $^{13}$C bands (J stars), 2 with strong H$\alpha$ emission, and 3 which exhibited enhanced Li absorption. In Fig. 6 we show spectra of M, S, and C stars in M31.

### 4.1. COMPARISON WITH MODELS

To compare AGB models with our observed stars, we applied a bolometric correction (BC) to our stars' $M_I$ magnitudes. Our BCs are provisional as they were derived from a fit to 4 C stars with known BCs (see BRC96).

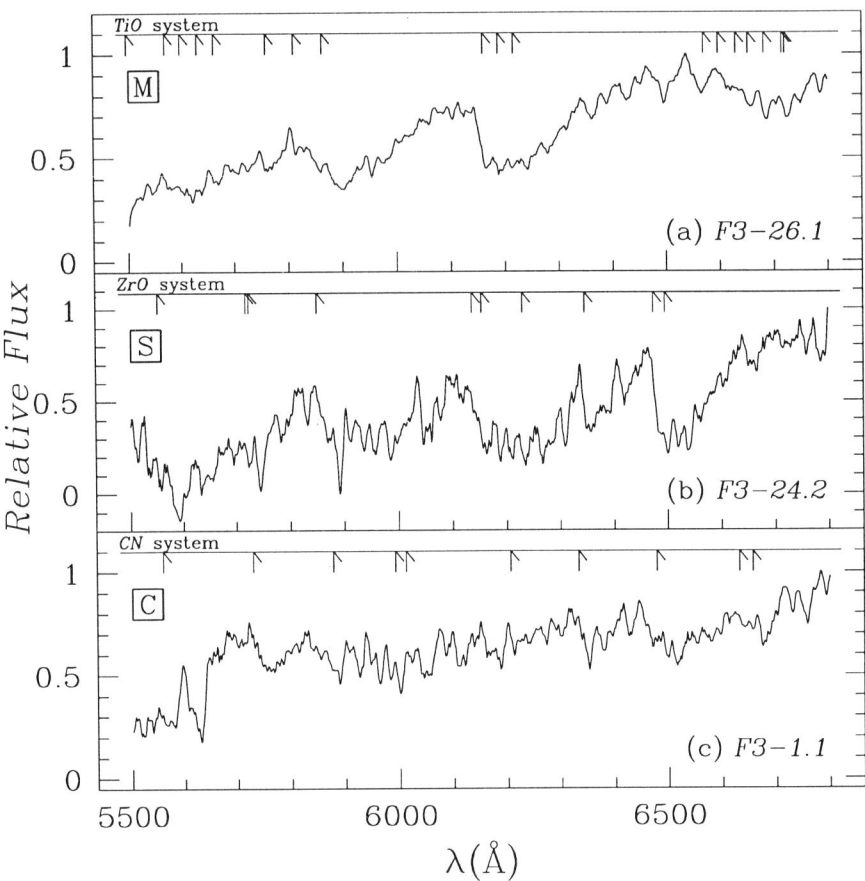

*Figure 6.* Spectra of three M31 AGB stars whose initial spectral classification was made using the FBPS.

### 4.1.1. *J Stars*

The third dredge-up mechanism can lead to an increase in the $^{12}C/^{13}C$ abundance ratio of a star by bringing to the surface $^4He$ that has been burnt to $^{12}C$ by the triple-$\alpha$ process. However, J stars exhibit a $^{12}C/^{13}C$ ratio which is lower than the ratio at the onset of the thermally pulsing AGB! Hot bottom convective envelope (HBCE) burning, in which the CN cycle converts $^{12}C$ to $^{13}C$ and $^{14}N$, is able to explain this unusual ratio.

The models of Boothroyd et al. (1993) predict the existence of two C–star luminosity boundaries. The first boundary, at $M_{bol} = -6.4$, is the maximum magnitude at which C stars can exist, while the second boundary, at $M_{bol} = -6.3$, divides the C stars from J stars. Our observations are inconsistent with these models; we found our J stars to be around 1 to 2 magnitudes fainter than model predictions. Indeed, we found the J stars to be amongst the fainter C stars.

### 4.1.2. *Stars with Enhanced Lithium*

Sackmann & Boothroyd (1992) modeled the surface abundance of Li produced by the Cameron-Fowler mechanism (Cameron & Fowler 1971) and predicted that super-rich Li stars should lie in the range $M_{bol} \approx -6.2$ to $-6.8$. Our C stars which show enhanced Li are around 2 magnitudes fainter than predicted by these models.

The observations of the J stars and the Li stars strongly suggest that HBCE burning *occurs at fainter magnitudes than AGB models predict.* There is now evidence for this in both the LMC (Richer et al. 1979; Smith et al. 1995) and M31 (this study). It appears that modifications to existing theory must be made. Another possible explanation for the disagreement lies in the BC applied. Could it be that some of the C stars observed are evolving into infrared C stars and require larger BCs than applied? We urge future observers undertaking similar spectroscopic surveys to obtain infrared photometry.

## 5. Conclusions

Unlike observations of Galactic C stars, observations of C stars in LG galaxies allow us to easily explore the interplay between C–star populations and environment. The distances to LG galaxies are generally well known, making possible comparison between observations and models. Further investigation of AGB stars in LG galaxies can only increase our understanding of the carbon star phenomenon.

This work was supported by grants to HBR from the Natural Sciences and Engineering Research Council of Canada. JPB is indebted to UBC for providing a University Graduate Fellowship for the years 1991–1994.

## References

Blair, W. P., Kirshner, R. P. & Chevalier, R. A. 1982, *ApJ*, 254, 50 (BKC82)
Blanco, B. M., Blanco, V. M. & McCarthy, M. F. 1978, *Nature*, 271, 638
Boothroyd, A. I., Sackmann, I.-J. & Ahern, S. C. 1993, *ApJ*, 416, 762
Brewer, J. P., Richer, H. B. & Crabtree, D. R. 1995, *AJ*, 109, 2480 (BRC95)
Brewer, J. P., Richer, H. B. & Crabtree, D. R. 1996, *AJ*, 112, 491 (BRC96)
Cameron, A. G. W. & Fowler, W. A. 1971, *ApJ*, 164, 111
Freedman, W. L. & Madore, B. F. 1990, *ApJ*, 365, 186
Iben, I. Jr. & Renzini, A. 1983, *Ann. Rev. Astron. Astrophys.*, 21, 271
Palmer, L. G. & Wing, R. F. 1982, *AJ*, 87, 1739
Richer, H. B., Olander, N. & Westerlund, B. E. 1979, *ApJ*, 230, 724
Richer, H. B., Pritchet, C. J. & Crabtree, D. R. 1985, *ApJ*, 298, 240
Sackmann, I.-J. & Boothroyd, A. I. 1992, *ApJ*, 392, L71

Scalo, J. M. 1981, in *Physical Processes in Red Giants*, ed. I. Iben Jr. and A. Renzini (Reidel), p. 77
Searle, L. 1982, in Carnegie Year Book, pp. 622–624
Searle, L. & Thompson, I. 1985, in Carnegie Year Book, pp. 68–69
Smith, V. V., Plez, B., Lambert, D. L. & Lubowich, D.A. 1995, *ApJ*, 441, 735

## Discussion

**Frogel**: The ratio of M stars to C stars is a strong function of distance from the center of the Milky Way, but there is no evidence for variation in the reddening law as you claim for M31.

**Brewer**: Our location in the galactic plane means that it is hard to observe at large heliocentric distances in the galatic plane, and also that we are uncertain of where the material responsible for any reddening lies. Obviously external galaxies are the best place to test whether the C/M star ratio has an impact on the interstellar medium.

**Gustafsson**: Could you trace any tendencies for the fraction of J stars to change with distance from the center of M31?

**Brewer**: Unfortunately our statistics are insufficient to explore this question.

# Session II

## Models and Abundance Determinations

Exterior view of part of the Talya Hotel, the site of the symposium, looking West across the harbor.

Pool area of the Talya Hotel.

# MODEL ATMOSPHERES FOR NORMAL AND PECULIAR RED GIANT STARS

BERTRAND PLEZ
*Astronomiska observatoriet, S-751 20 Uppsala, Sweden*
and
*Atomspektroskopi, Fysiska Institutionen*
*S-223 62 Lund, Sweden*

**Abstract.** I review the current status of model atmospheres for red giants, with special emphasis on recent progress and newer grids. I draw attention to some specific problems regarding opacity sources and present current and forthcoming efforts in cool-star atmospheric modeling.

## 1. Background

This paper follows up the excellent review articles of Johnson (1986), Gustafsson & Jørgensen (1994), and Gustafsson (1995). I will emphasize specific problems, recent achievements in producing and using newer grids, and upcoming models.

I shall review models for all types of red giants, the O-rich K and M stars, the C-rich R, N, and J–type stars, the intermediate MS, S and SC stars, as well as hydrogen-deficient (HdC) carbon stars and R CrB stars. At this point it is useful to recall that the chemistry in the atmosphere of these stars is dominated mostly by the C/O ratio. In O-rich objects, with C/O < 1, most C goes into CO, the most stable molecule, and the remaining O binds into oxides like TiO, VO, and $H_2O$. In a C-rich mixture, most O is bound in CO, and the remaining C forms CN, $C_2$, CH, HCN, $C_2H_2$, etc. When C/O $\approx$ 1, most C and O is in CO, most N in $N_2$, and metals with lower abundances start building compounds in significant amounts: ZrO, LaO, YO, and so on. This effect is often enhanced in S stars by the overabundance of *s*-elements dredged up on the AGB. Note also that at lower metallicity, molecules like CH are privileged over species such as $C_2$, the abundance of which drops faster with decreasing C abundance. This explains the dominant spectral features of CH stars.

## 2. Model Atmospheres: What For?

Model atmospheres are a crucial ingredient in our study of stars. Most observations lack any sense or cannot be fully exploited without the help of a model atmosphere. Obvious examples include the interpretation of spectra and the derivation of elemental abundances, and the understanding of interferometric observations in terms of stellar diameters or center-to-limb variations. To close the discussion of interferometric observations, let us mention that synthetic visibilities from static or hydrodynamic models are so far unsuccessful in matching high-quality observations of even normal red giants (see Quirrenbach et al. 1993). A general feature is that the amplitude of variation of the radius with wavelength is too small in the models. This is certainly a very strong constraint for the models, although not yet exploited.

Models are necessary for the generation of synthetic spectra at high (for abundance analysis) or low resolution (for example, for population synthesis or construction of color–$T_{\text{eff}}$ diagrams).

The use of (good) model atmospheres has proved to be of particular importance in the modeling of red and brown dwarf evolution (Chabrier, Baraffe & Plez 1996). The outer boundary condition influences the calculated internal structure to a point which cannot be neglected. There are speculations that the atmospheric boundary condition is also of significance in red giants, especially for hot-bottom burning. The inclusion of model atmospheres is however more difficult in giant stars, mostly because of the lack of an easily defined radius.

Finally, the construction of model atmospheres might help us to understand the physics of real stars' outer layers. It is worth stressing again that this is not the primary purpose of models. Rather we want models that are good enough to understand and interpret our observations, yet still simple enough that we can easily use them and assess their behavior.

In short, model atmospheres are a link between our theories and observations, and a key to our understanding of observations.

## 3. Ingredients for a Good Model Atmosphere

In the following I will mostly discuss standard or classical models which are static, one-dimensional (i.e. consisting of homogeneous plane-parallel (PP) or spherically-symmetric (Sph) layers), in LTE, and in radiative and convective equilibrium (the convection treated with the mixing-length formalism). All these hypotheses are of course masking a much more complex reality with pulsation, 3–D convective motions, shock waves, and non-equilibrium chemistry and radiative processes. Some of these assumptions have been relaxed in more ambitious modeling (see §9). A great advantage of standard models is that they are relatively easy and inexpensive to compute. Great effort has thus been invested in including very detailed and complete line opacity in these models. In contrast, hydrodynamical models of pulsating red giants include only very crude radiative transfer (see §9). Only classical models are currently used in routine analysis of, e.g., chemical composition.

I will consider only two specific aspects of good modeling. Other, more detailed discussions can be found in e.g. Gustafsson & Jørgensen (1994). Atomic and molecular data are the main ingredient of any realistic models. Partition functions and dissociation energies are of course needed for the computation of partial pressures of species and the equation of state. Continuous and line opacities have a strong effect on the spectrum and thermal structure (see Carbon 1979 for an excellent discussion of blanketing). In O-rich objects TiO and $H_2O$ dominate the spectrum, while CN, HCN and other species characterize the spectra of C-rich objects. Some molecules affect both the spectrum and the thermal structure (TiO, $H_2O$, HCN), while others affect mainly the spectrum (LaO, ...). This involves a large quantity of factors like the position of the absorption band in the spectrum, the depth in the atmosphere where it occurs, with what strength, how many lines, etc. Note also that if lines mostly scatter instead of absorbing (as suggested for electronic transitions of diatomic molecules by Hinkle & Lambert 1975), their effect on the thermal structure is much weaker. A huge effort has therefore been put into collecting and evaluating molecular line lists for a variety of species supposed to dominate the blanketing in cool stars. Gustafsson (1995) discusses current knowledge (or absence thereof) on opacities in detail.

The computation of models involves a discretization on frequencies. Various methods have been devised, dictated by the then-available computer resources. The line opacities are sampled or averaged in various ways. Jørgensen (1992) provides a review of these approximations. Let us just say that with current computer facilities only the best method should be used: namely, opacity sampling (OS). It consists, in a Monte Carlo spirit, of selecting a number of frequencies (in most cases 1000 to 10 000 seems to be enough) in the spectrum. The opacity is then calculated exactly at these particular points by summing up contributions from all lines and continua. The solution of the radiative transfer at all these frequencies then allows the evaluation of the integrals on the radiative field (specific intensity, flux, etc) needed for the solution of the model-atmosphere problem. All other methods — straight mean (SM), opacity distribution function (ODF), etc — perform some average of the opacity prior to any solution of the radiative transfer. This is of course not equivalent to the OS approximation, but in some cases is not so bad (ODF only). The ODF method offers an alternative to OS (but with less flexibility and no extremely significant gain in computing time), with the notable exception of carbon stars with $T_{\text{eff}}$ lower than about 3000 K (Ekberg, Eriksson & Gustafsson 1986). It will therefore most likely be abandoned in the near future.

## 4. Problems

We need line positions (not so critical for the calculation of model structure or photometry), line strengths, and broadening parameters. The data is too often of low-quality, incomplete or non-existent. Even such basic data as dissociation constants are sometimes not accurate enough (see the debate around $D_0^{\circ}(\text{CN})$). The situation is now considerably better than at the end of the last decade. I will here only review recent progress for a few specific species.

The electronic transition systems of TiO give rise to most of the features in the optical spectra of M giants. The situation has been unsatisfactory for many years, especially regarding the transition probabilities of the bands. The lifetimes of most states have recently been measured accurately by Hedgecock, Naulin & Costes (1995). This should provide good line-strengths when combined with the extensive line lists of Plez, Brett & Nordlund (1992) or Jørgensen (1994). One persistent problem, however, is the lack of a definite lifetime for the $E^3\Pi$ state which gives rise to the $\epsilon$ band system, of importance around 1 $\mu$m. Laboratory measurements only provide a lower limit. Also the $\phi$ and $\delta$ system strengths still rely on the absorption measurements by Davis, Littleton & Phillips (1986). Before better laboratory determinations become available, there is the possibility of semi-empirically calibrating these bands by careful use of narrow-band colors (see §5).

Absorption bands of $H_2O$ fully dominate the infrared spectra of the coolest O-rich stars. Real *ab initio* line lists (Jørgensen & Jensen 1994, Miller et al. 1994) are now starting to be used instead of the mean opacities of Auman (1967) and Ludwig et al. (1973) or the statistically generated lists of Alexander, Augason & Johnson (1989) and Plez, Brett & Nordlund (1992). The new lists are being tested in M dwarf models and seem to provide better models. They should be applied to giants very soon.

VO appears in the near IR spectra of the coolest M giants. There is unfortunately no laboratory measurement of the strength of the A–X or B–X bands. Brett (1990) performed an astrophysical calibration of these bands by comparing model spectra with observations. This is a delicate exercise, as errors in the model thermal structures or spectra, caused by missing or poorly calibrated opacities, may induce systematic errors in the calibration. We hope to perform a better calibration of these bands with newer models and better data (see §5).

## 5. Illustration of Progress

The computation of synthetic spectra, especially at high resolution, requires more complete and higher quality line data than model construction. Despite the problems mentioned above, and other missing species such as LaO and YO, progress has been steady over the past few years. Plez, Smith & Lambert (1993) carried out an abundance analysis of bright AGB stars in the SMC (Li, $^{12}$C, $^{13}$C, metals, $s$-elements) using extensive line lists for TiO. The simultaneous fit of various spectral regions was possible in a self-consistent way, without the need for invoking any extra "fudge" opacity as in previous studies. Fluks et al. (1994) derived a new $T_{\rm eff}$ scale for M giants by fitting synthetic spectra with carefully calibrated observations in the optical region. This scale is in excellent agreement with the Ridgway et al. (1980) scale, based on angular diameter measurements. We are reaching good agreement over most of the spectrum. Comparisons of synthetic spectra and colors with observations yield consistent values of $T_{\rm eff}$. There seem to remain some problems for stars cooler than about 3100 K (later than M6–M7), probably due to a mismatch in $H_2O$ opacity. It is not fully clear yet if the new *ab initio* line lists will resolve this discrepancy.

For carbon stars, the extensive chemical analysis of Lambert et al. (1986), using ODF models with polyatomic opacities, was able to explain the $H_2$ line strength as a natural consequence of the blanketing effect of the polyatomic molecules. Earlier attempts had predicted $H_2$ lines that were too strong. However, the CH and HCN bands appear too strong in the models. This situation is still unresolved and awaits further work in the modeling of carbon-rich atmospheres (OS with newer opacities, in progress; see §8). There may be a need for inclusion of more opacity-contributing molecules as well as dust in the cooler models. It is also worth mentioning that Ohnaka & Tsuji (1996) find carbon isotope ratios much lower than Lambert et al. and that the reasons for this difference still need to be explained.

Alvarez & Plez (1998) are carrying out a study of Mira variables using the observations of Lockwood (1972) with 5 narrow-band filters in the near-infrared. These filters, based on the Wing system (see Wing 2000), isolate (more or less, the region is crowded!) specific TiO and VO absorption bands. The synthetic colors match pretty well the observations of reference non-variable M giants, thanks to the use of high-quality model atmospheres and very extensive line lists. It is possible to tune the strength of some of these bands (the ones which have no accurate lifetime measurement) to obtain a better fit in color-color diagrams. This is only allowed because the models are of high quality (they have been well tested in other respects: $T_{\rm eff}$–color relations, spectroscopy, ...) and they are not so much affected by these particular changes. Narrow-band colors appear quite powerful in assessing the quality of, or even for calibrating, model spectra.

## 6. Existing Grids

It is now time to describe the existing models and to discuss them briefly. The following table is largely inspired by Table 1 in Gustafsson & Jørgensen (1994). In general, models using the OS approximation (or possibly ODF for O-rich atmospheres) should be preferred. For carbon stars, the most reliable models are the ones that include opacities from polyatomic molecules in the OS approximation. The uncertainties in opacity sources and strengths are larger for C-rich mixtures. This is made more critical by the generally low $T_{\rm eff}$ of carbon stars. In M-type star models, there is still the uncertainty

TABLE 1. Grids of model atmospheres for red giant stars

| Sp. Type | $T_{\rm eff}$ | log $g$ | [M/H], C/O | Notes | Ref. |
| --- | --- | --- | --- | --- | --- |
| O–K | 3500 –50000 | 0.0 – 5.0 | −5.0 – 1.0 | OS | (1) |
| G–K III | 3750 – 6000 | 0.75 – 3.0 | −3.0 – 0.0 | ODF, no TiO | (2) |
| M I–III | 2600 – 4200 | −2.0 – 2.5 | 0.0 | VAEBM | (3) |
| M III | 2500 – 4000 | 0.0 – 2.0 | 0.0 | OS+SM | (4) |
| M III | 3000 – 4000 | 0.0 – 2.0 | 0.0 | OS | (5) |
| M I–III | 3000 – 4000 | −0.5 – 1.5 | 0.0 | OS, Sph | (6) |
| N, R | 2600 – 4500 | −1.0 – 1.0 | C/O=3.2 etc | ODF | (7,8) |
| N, S | 2500 – 3500 | 0.0 | C/O=0.6 – 2.0 | OS+SM | (9) |
| N | 2500 – 3500 | −1.0 – 0.0 | −1.0 – 0.0 C/O=1.01 – 1.5 | ODF, poly | (10, 11) |
| N | 2500 – 3400 | −1.0 – 0.5 | −1.0 – 0.0 C/O=1.02 – 2.0 | OS, poly, Sph | (12) |
| R | 3800 – 4800 | 2.0 – 3.0 | 0.0 C/O=1.02 – 3.5 | ODF | (13) |
| R | 4200 – 5400 | 2.0 – 3.0 | 0.0 C/O=1.74 | OS | (14) |
| Miras | 2300 – 3350 | $L \approx 10^4 L_\odot$ | 0.0 | SM, Sph, dyn | (15) |
| R CrB, HdC | 5500 – 9500 | −0.5 – 2.0 | C/He=0.1 – 10% | OS+ODF | (16) |

Notes: OS: opacity sampling, ODF: opacity distribution function, VAEBM: Voigt-analog Elsasser band-model, SM: straight mean, Sph: spherical, poly: including polyatomic opacity sources, dyn: hydrodynamical models

References: (1) Kurucz (1992); (2) Bell et al. (1976); (3) Tsuji (1978); (4) Johnson et al. (1980); (5) Brown et al. (1989); (6) Plez et al. (1992); (7) Querci et al. (1974); (8) Querci & Querci (1975); (9) Johnson (1982); (10) Eriksson et al. (1986); (11) Lambert et al. (1986); (12) Jørgensen et al. (1992); (13) Olander (1981); (14) Johnson & Yorka (1986); (15) Bessell et al. (1989, 1996); (16) Asplund et al. (1996)

due to $H_2O$ at $T_{\text{eff}} < 3100\,\text{K}$.

## 7. Latest News

A new grid of models for R CrB and HdC stars has just been completed by Asplund et al. (1997). Previous grids by Schönberner (1975) and Jones (1991) included older continuous opacities and no line-blanketing in the former case. The Asplund et al. models are based on the MARCS code (Gustafsson et al. 1975), with improved and expanded continuous opacities (dominated by C I) and line-blanketing in a mixture of OS and ODF. The grid covers $5000\,\text{K} \leq T_{\text{eff}} \leq 9500\,\text{K}$, $-0.5 \leq \log g \leq 2.0$, $0.001 \leq \text{C/He} \leq 0.1$. The inclusion of blanketing steepens the temperature gradient and provides an explanation for the presence of $C_2$ and CO lines up to $T_{\text{eff}} = 7000\,\text{K}$, and for C II and He I lines down to the same temperature. The model for R CrB itself matches the observed spectrum very well from the UV to the IR. The authors also discuss the density inversion present in their models in the He ionization region and its possible link with an instability and episodic mass-loss. They conclude that non-LTE effects on C are not very large, and they announce upcoming models including spherical symmetry.

## 8. Coming Soon

The above grid is a step toward a more ambitious project in progress in the Uppsala group. The goal is the completion of an "ultimate" grid of standard models for stars of spectral types later than about F. After all the progress of the past years, especially in opacity quality and completeness, the time is now ripe for the production of a consistent grid of models including all the best data and covering a large area of the HR diagram, extending Kurucz' work to cooler temperatures and more exotic chemical compositions. This involves the careful (and painful) compilation, computation and evaluation of line and continuous opacity data. Only the best possible data will be included, with the least possible ad-hoc or astrophysical calibration, and with laboratory data whenever possible. Partition functions and molecular equilibrium constants will also be reviewed. The models will be PP or Sph, in LTE, static, with a range of [Fe/H], CNO, $\alpha$–elements, ... covering all foreseeable needs. The emphasis will be placed on obtaining good synthetic spectra and colors.

## 9. ... And Later, beyond Classical Models

Some attempts have been made to relax some of the classical model hypotheses. Hydrodynamical models have been constructed, and a first attempt at including non-LTE (for Ti, Ti$^+$ and TiO) in the model calculation has recently been reported (Hauschildt et al. 1996).

Radially pulsating atmospheric models have been produced by Bowen (1988) and Bessell, Scholz & Wood (1996) for O-rich stars and by the Berlin (e.g. Winters et al. 1995) and Vienna (Höfner, Feuchtinger & Dorfi 1995) groups for C-rich objects, with various degree of sophistication in the description of dust formation, cooling/heating, etc. Many exciting and sometimes surprising results have come from these studies, but some aspects still need improvement. Most dynamical modeling has so far relied on very crude descriptions of the radiative field. A better treatment, with about 100 frequency points in LTE, will probably be included in the course of the next few years. The driving of the pulsation is included as a piston at the bottom of the envelope. The atmosphere acts as an active filter which results sometimes in multiple periodicity, or semi-periodic behavior (Winters et al. 1994). The inclusion of the driving zone in a self-consistent way will be necessary to understand why and which stars show multi-periodicity or only semi-periodicity. The coupling between convection and pulsation may play a role too, but cannot be studied before a more realistic description of convection is included in the models.

Another direction of research is offered by the extension to cool stars of the 3-D modeling of convection (as done by Nordlund and collaborators for solar type stars: see Nordlund & Dravins 1990, and references in Gustafsson & Jørgensen 1994).

Extensive non-LTE included at the modeling stage is the third development beyond standard models. Especially important is the coupling between electron donors, H$^-$, and molecules. The main, and very serious, problem is the lack of collisional cross-sections for most relevant species, and for virtually all molecules. Large model atoms are not sufficient. Good quality data are also needed here if meaningful results are to be expected.

I shall conclude here by saying that there is a strong motivation to construct better model atmospheres; many reasons may be found in this volume. We will soon access regions of the spectrum that are only poorly known (ISO). Large telescopes will allow us to extend our studies to other galaxies. With the advent of "modern" classical models and the systematic exploration of non-classical directions, the next decade should provide for a lot of excitement and a rich harvest of new results.

## References

Alexander, D. R., Augason, G. C. & Johnson, H. R. 1989, *ApJ*, 345, 1014
Alvarez, R. & Plez, B. 1998, *A&A*, 330, 1109
Asplund, M., Gustafsson, B., Kiselman, D. & Eriksson, K. 1997, *A&A*, 318, 521
Auman, J. Jr. 1967, *ApJ Supp.*, 14, 171

Bell, R. A., Eriksson, K., Gustafsson, B. & Nordlund, Å. 1976, *A&A Supp.*, 23, 37
Bessell, M. S., Brett, J. M., Scholz, M. & Wood, P. R. 1989, *A&A*, 213, 209
Bessell, M. S., Scholz, M. & Wood, P. R. 1996, *A&A*, 307, 481
Bowen, G. H. 1988, *ApJ*, 329, 299
Brett, J. M. 1990, *A&A*, 231, 440
Brown, J. A., Johnson, H. R., Alexander, D. R., Cutright, L. C. & Sharp, C. M. 1989, *ApJ Supp.*, 71, 623
Carbon, D. F. 1979, *Ann. Rev. Astron. Astrophys.*, 17, 513
Chabrier, G., Baraffe, I. & Plez, B. 1996, *ApJ*, 459, L91
Davis, S. P., Littleton, J. E. & Phillips, J. G. 1986, *ApJ*, 309, 449
Ekberg, U., Eriksson, K. & Gustafsson, B. 1986, *A&A*, 167, 304
Eriksson, K. et al. 1986, unpublished
Fluks, M. A., Plez, B., Thé, P. S., de Winter, D., Westerlund, B. E. & Steenman, H. C. 1994, *A&A Supp.*, 105, 311
Gustafsson, B. 1995, in *Astrophysical Applications of Powerful New Databases*, ed. S. J. Adelman and W. L. Wiese, ASP Conf. Series, 78, 347
Gustafsson, B., Bell, R. A., Eriksson, K. & Nordlund, Å. 1975, *A&A*, 42, 407
Gustafsson, B. & Jørgensen, U. G. 1994, *A&A Rev.*, 6, 19
Hauschildt, P. H., Allard, F., Alexander, D. R., Schweitzer, A. & Baron, E. 1996, in IAU Symp. 176: *Stellar Surface Structure*, ed. K. G. Strassmeier and J. L. Linsky (Kluwer), p. 539
Hedgecock, I. M., Naulin, C. & Costes, M. 1995, *A&A*, 304, 667
Hinkle, K. H. & Lambert, D. L. 1975, *MNRAS*, 170, 447
Höfner, S., Feuchtinger, M. U. & Dorfi, E. A. 1995, *A&A*, 297, 815
Johnson, H. R. 1982, *ApJ*, 260, 254
Johnson, H. R. 1986, in *The M-Type Stars*, ed. H. R. Johnson and F. R. Querci, NASA SP–492, p. 323
Johnson, H. R., Bernat, A. P. & Krupp, B. M. 1980, *ApJ Supp.*, 42, 501
Johnson, H. R. & Yorka, S. B. 1986, *ApJ*, 311, 299
Jones, K. 1991, Ph.D. thesis, St Andrews University
Jørgensen, U. G. 1992, *Rev. Mex. Astron. Astrof.*, 23, 195
Jørgensen, U. G. 1994, *A&A*, 284, 179
Jørgensen, U. G. & Jensen, P. 1994, *J. Mol. Spec.*, 161, 219
Jørgensen, U. G., Johnson, H. R. & Nordlund, Å. 1992, *A&A*, 261, 263
Kurucz, R. L. 1992, in IAU Symp. 149: *The Stellar Populations of Galaxies*, eds. B. Barbuy and A. Renzini (Kluwer), p. 225
Lambert, D. L., Gustafsson, B., Eriksson, K. & Hinkle, K. H. 1986, *ApJ Supp.*, 62, 373
Lockwood, G. W. 1972, *ApJ Supp.*, 24, 375
Ludwig, C. B., Malkmus, W., Reardon, J. E. & Thomas, J. A. L. 1973, *Handbook of Infrared Radiation from Combustion Gases*, NASA SP–3080
Miller, S., Tennyson, J., Jones, H. R. & Longmore, A. J. 1994, in IAU Coll. 146: *Molecules in the Stellar Environment*, ed. U. G. Jørgensen (Springer-Verlag), p. 296
Nordlund, Å. & Dravins, D. 1990, *A&A*, 228, 155
Ohnaka, K. & Tsuji, T. 1996, *A&A*, 310, 933
Olander, N. 1981, Uppsala Astr. Obs. Report No. 21
Plez, B., Brett, J. M. & Nordlund, Å. 1992, *A&A*, 256, 551
Plez, B., Smith, V. V. & Lambert, D. L. 1993, *ApJ*, 418, 812
Querci, F. & Querci, M. 1975, *A&A*, 39, 113
Querci, F., Querci, M. & Tsuji, T. 1974, *A&A*, 31, 265
Quirrenbach, A., Mozurkewich, D., Armstrong, J. T., Buscher, D. F. & Hummel, C. A. 1993, *ApJ*, 406, 215
Ridgway, S. T., Joyce, R. R., White, N. M. & Wing, R. F. 1980, *ApJ*, 235, 126
Schönberner, D. 1975, *A&A*, 44, 383
Tsuji, T. 1978, *A&A*, 62, 29

Wing, R. F. 2000, in IAU Symp. 177: *The Carbon Star Phenomenon*, ed. R. F. Wing (Kluwer), p. 127
Winters, J. M., Fleischer, A. J., Gauger, A. & Sedlmayr, E. 1994, *A&A*, 290, 623
Winters, J. M., Fleischer, A. J., Gauger, A. & Sedlmayr, E. 1995, *A&A*, 302, 483

## Discussion

**Luttermoser**: A comment about the NLTE photospheric modeling that is currently underway: it is important to include the chromospheric radiation field flowing backward down onto the photosphere. It has an important impact on the excitation and ionization of neutral metals in the photosphere, especially in the upper photosphere (i.e. $\tau_{\mathrm{Ross}} < 1.0$).

**Plez**: Yes, you are perfectly right!

**Steffen**: What is your choice for the mixing length parameter of convection, and how sensitive are your models to the assumed convective efficiency? What about microturbulence?

**Plez**: We use $\alpha = 1.5$. This only affects the structure of the red giant models at large optical depth ($> 10$). The impact on spectra or colors is negligible. For microturbulence we use about $2 \ \mathrm{km\,s^{-1}}$. This enters only in the broadening of spectral lines. It is not included in the turbulent pressure term, but could be.

**Mowlavi**: Could you comment on the importance of using good model atmospheres for the calculation of stellar structure evolution?

**Plez**: The atmospheric boundary condition most probably influences the Hot Bottom Burning. Good atmospheres are also of course necessary for relating the model parameters to observations ($T_{\mathrm{eff}}$, ...).

# QUANTITATIVE ANALYSIS OF CARBON ISOTOPE RATIOS IN N–, SC–, AND J–TYPE CARBON STARS

K. OHNAKA* AND T. TSUJI

*Institute of Astronomy, The University of Tokyo*
*Mitaka, Tokyo, 181 Japan*

* Now at Technische Universität Berlin, Germany

**Abstract.** We present the results of a quantitative analysis of $^{12}C/^{13}C$ ratios in 62 N–type, 15 SC–type, and 26 J–type carbon stars. The $^{12}C/^{13}C$ ratios are determined from lines of the CN Red System around 8000 Å, based on the iso-intensity method and line-blanketed model atmospheres. The resulting $^{12}C/^{13}C$ ratios in N– and SC–type carbon stars are consistent with a scenario in which M giants evolve through SC– to N–type carbon stars, as $^{12}C$ produced during thermal pulses is added to the envelope.

## 1. Introduction

The carbon isotope ratio brings us important information about the nucleosynthesis and mixing in the stellar interior, because carbon is one of the elements involved in the nucleosynthesis during thermal pulses and mixing to the surface. If the $^{12}C$ synthesized in thermal pulses is responsible for the formation of carbon stars, then $^{12}C/^{13}C$ ratios are expected to be enhanced as compared with those of their progenitors, and to have a correlation with C/O ratios.

To demonstrate this correlation, we have carried out a quantitative analysis of carbon isotope ratios from lines of CN located around 8000 Å for a large sample of carbon stars, including N, SC, and J stars. Our main purpose is to determine the distribution of $^{12}C/^{13}C$ ratios of each type of carbon star based on a large sample, and to find some connections between carbon stars and K and M giants, which are considered to be their progenitors.

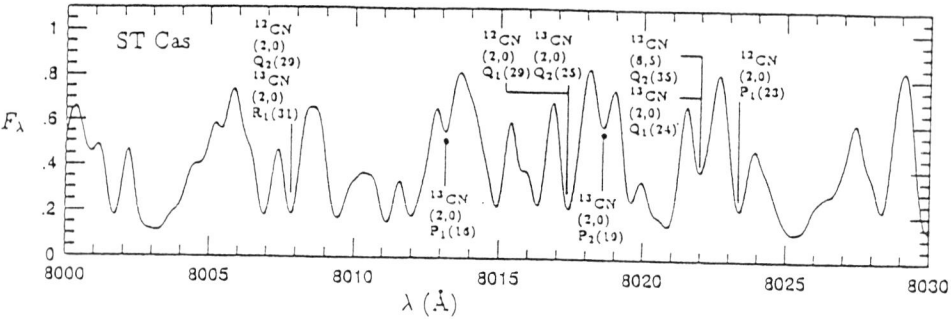

*Figure 1.* Observed spectrum of an N-type carbon star, ST Cas.

## 2. Observations

All observations were carried out using the 74-inch telescope of Okayama Astrophysical Observatory (OAO[1]), during the period between October 1987 and March 1994. Our sample consists of 66 N-type, 19 SC-type, and 26 J-type carbon stars. Lists of our program stars are given in Ohnaka & Tsuji (1996, 1998). All spectra consist of two adjacent exposures, covering from 7800 to 8030 Å. Each exposure covers 120 Å with an overlap of 10 Å. The spectral resolution is about 20,000. The signal-to-noise ratios are about 100, although somewhat lower for faint stars and much higher for some of the brightest stars.

## 3. Analysis

The observed region is very crowded with lines due to the Red System of $^{12}$CN and $^{13}$CN and the Phillips System of $C_2$, as can be seen in Figure 1. We selected about 10 lines of $^{12}$CN and 2 lines of $^{13}$CN for each star. Some lines are blended with other lines at almost the same wavelength. We used them in the analysis only if the blending line is weak enough compared to the line in question, considering its $gf$-value and lower excitation potential. Central depths are normalized by the fictitious continuum level, which is drawn to go through the highest point in the observed region and is assumed to be constant over the region.

We adopted the iso-intensity method for the analysis. It is similar to the curve-of-growth method, but we use the central depths of lines as the ordinate, instead of equivalent widths. For lines of $^{12}$CN and $^{13}$CN, logarithms of the observed central depths normalized by the fictitious continuum level are plotted against $\log(gf\Gamma)$, which is the line intensity predicted by the weighting-function method (e.g. Cayrel & Jugaku 1963). The horizontal

---

[1] OAO is a branch of the National Astronomical Observatory of Japan (NAOJ). This work was carried out under the common use program of OAO.

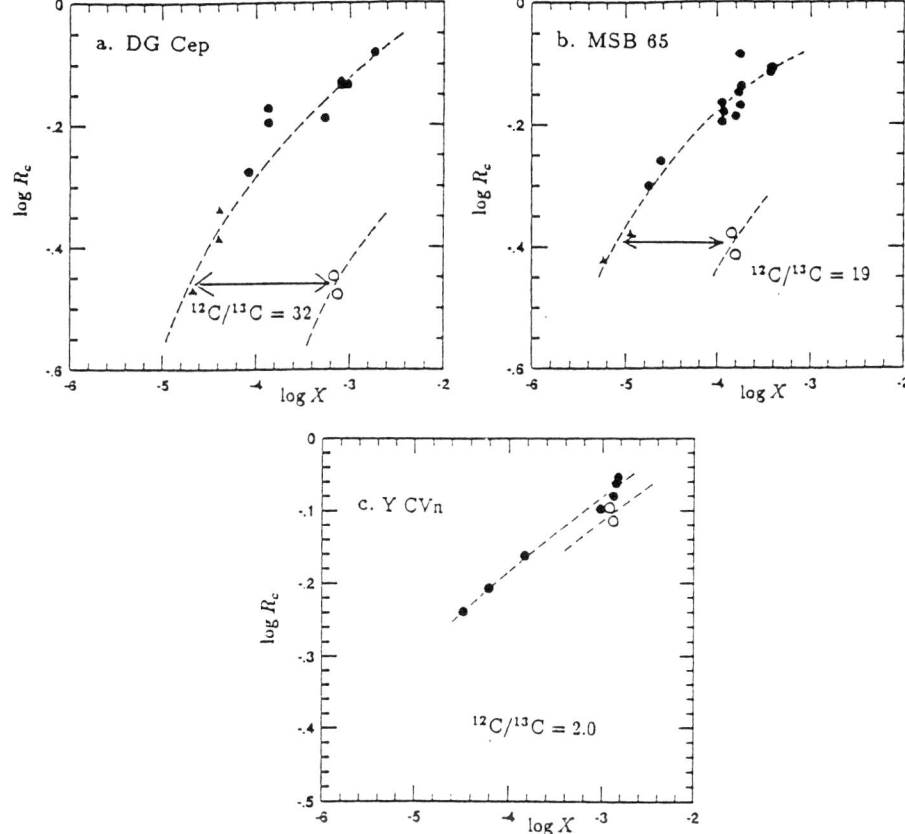

*Figure 2.* Examples of curves-of-growth: (a) DG Cep; (b) MSB 65; (c) Y CVn.

shift between the two curves gives the isotopic ratio. Examples are shown in Figure 2.

It should be noted that the horizontal shift between the curves for $^{12}$CN and $^{13}$CN is measured at the same depth; hence, the continuum location does not affect the resulting isotopic ratio as long as the continuum is constant in the observed region. Since the carbon isotope ratios are determined by lines of $^{12}$CN and $^{13}$CN of the *same* intensity, what really matters is that the central depths normalized by the fictitious continuum are a kind of measure of line intensity which serves to recognize that the lines of $^{12}$CN and $^{13}$CN are *iso*-intensity.

## 4. Results

The distributions of the derived $^{12}$C/$^{13}$C ratios of each type of carbon star are shown in Figure 3. The average $^{12}$C/$^{13}$C ratio is 27 ± 11 (standard deviation) for N–type carbon stars and $22 \pm 14$ for SC–type carbon stars. The average $^{12}$C/$^{13}$C ratio in the 26 J–type carbon stars is $4.7 \pm 2.8$. The $^{12}$C/$^{13}$C ratios found for each star are given in Ohnaka & Tsuji (1996, 1998).

Figure 3a clearly shows that the most frequent $^{12}$C/$^{13}$C ratios among 62 N stars are between 20 and 30. Nearly half of our program stars have ratios in this range. This is in marked contrast with the result of Lambert et al. (1986) who reported that the most frequent ratios among the 24 N stars they studied are from 50 to 60. As one possible reason for this disagreement, we noticed that their model atmospheres may be cooler by more than 500 K in the line-forming region than our models of the same effective temperature. However, this difference in the model atmospheres may be due to the inclusion of turbulent pressure in our model atmospheres, since the comparison of the models was done on the $\log P_g - T$ plane. Thorough tests of the model atmospheres should be done in order to resolve this discrepancy.

## 5. Discussion

### 5.1. N AND SC STARS

We now discuss how an oxygen-rich star evolves into a carbon star, in terms of chemical compositions. We consider the scenario in which an M giant evolves to an N type carbon star as $^{12}$C produced in the helium shell flash is dredged up to the surface, and SC stars are transitional objects between oxygen-rich stars and carbon stars. First, analyses of chemical compositions of M giants were done by Smith & Lambert (1985, 1990), who showed that C/O = 1/2 – 1/3 and $^{12}$C/$^{13}$C = 10 – 20 in M giants. On the other hand, C/O ratios in N stars have been analyzed by Lambert et al. (1986), who found that the average C/O ratio in 24 N stars is 1.1. In other words, if we assume that the oxygen abundance is not modified by the third dredge-up, the carbon abundance is doubled or tripled by the addition of $^{12}$C in the progression from an M giant to an N star. Therefore, the $^{12}$C/$^{13}$C ratio should also be doubled or tripled from 10 up to $\sim 20$ or 30, which is consistent with our result for N stars. Besides, C/O ratios of SC stars are considered to be nearer to unity, that is, less than the value 1.1 of N stars. This means that the amount of $^{12}$C added to the surface of SC stars is almost equal to that added to the surface of N stars. This implies that the $^{12}$C/$^{13}$C ratios of SC stars should have a distribution similar to that of N stars. Our results (Figs. 3a and 3b) show that the range of the $^{12}$C/$^{13}$C ratios of SC stars is not so different from that of the $^{12}$C/$^{13}$C ratios of N stars.

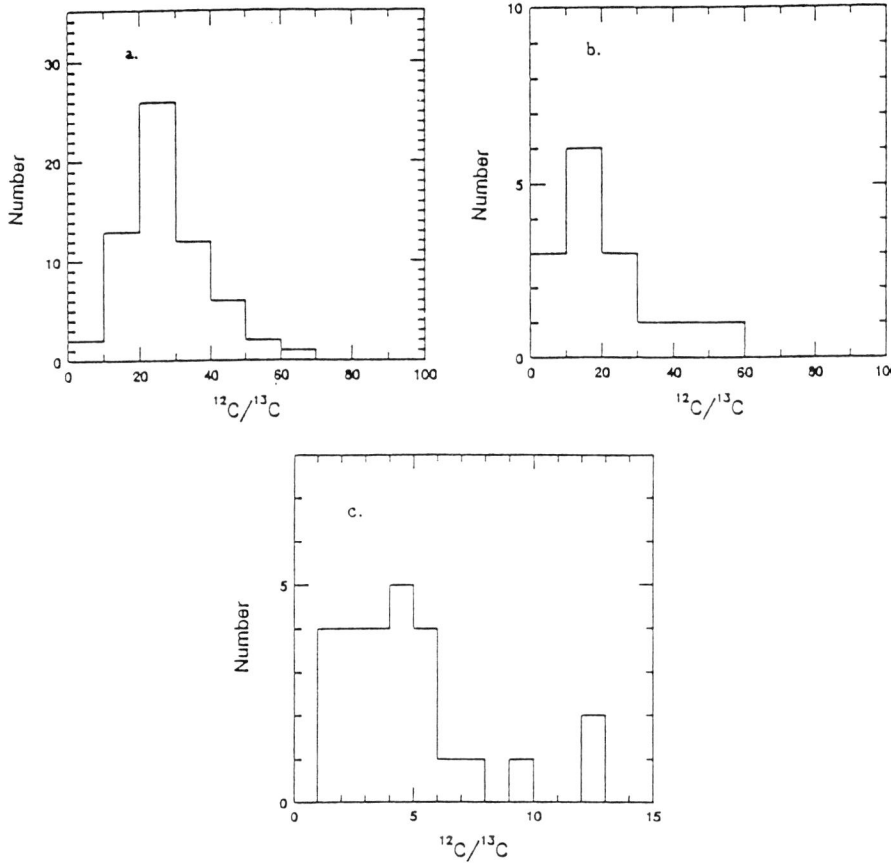

*Figure 3.* Histograms of the $^{12}C/^{13}C$ ratios obtained for (a) N-type, (b) SC-type, and (c) J-type stars.

## 5.2. J STARS

In Figure 4, we plot the galactic latitude $|b|$ of each J star against the derived $^{12}C/^{13}C$ ratio. Though the plot appears to be rather scattered, it seems that the J stars with relatively large $^{12}C/^{13}C$ ratios tend to be at low galactic latitudes whereas the extremely $^{13}C$-rich J stars are distributed at all galactic latitudes. However, since the number of high-latitude stars observed on our program is relatively small, it would be desirable to analyze more stars at high galactic latitudes to confirm the above correlation.

We now turn to the silicate carbon stars in our sample. The $^{12}C/^{13}C$ ratios are $5.3 \pm 0.9$ for EU And, $4.8 \pm 0.8$ for BM Gem, $4.8 \pm 0.8$ for V 778 Cyg, $3.0 \pm 1.0$ for NC 83, and $5.6 \pm 1.4$ for GCCCS 447. These results are perfectly consistent with the previous identification of these stars as J-type. And it is worth noting that the five silicate carbon stars have $^{12}C/^{13}C$ ratios

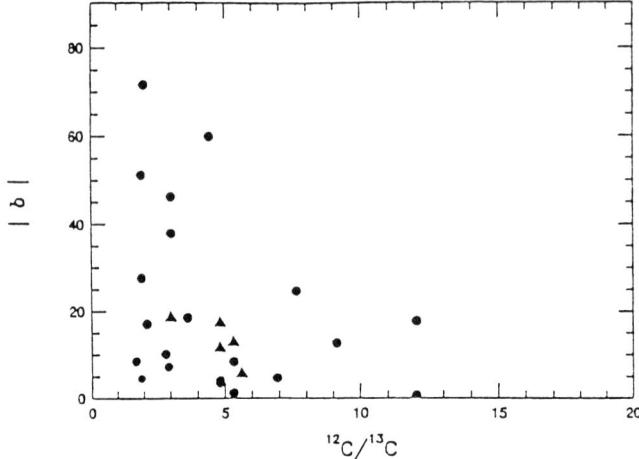

*Figure 4.* Galactic latitude $|b|$ of each J star is plotted against its $^{12}C/^{13}C$ ratio. Filled triangles represent the silicate carbon stars.

which are among the most common values in our sample. In other words, these stars exhibit no peculiar $^{12}C/^{13}C$ ratios which would be associated with the presence of the silicate emission feature.

## 6. Future Work

To clarify the complicated evolution of carbon stars and related objects, further analyses of the elemental abundances of C, N, O, and their isotopes are indispensable. In addition, determinations of the abundances of the $s$-process elements are highly desirable. We should also investigate the mechanism of mass loss, which no one doubts plays an important role in stellar evolution at the final stage. The Infrared Space Observatory (ISO) will bring us spectra of objects at the final stage of stellar evolution. These spectra will help to clarify the physics and chemistry of circumstellar shells. This kind of approach is expected to show how mass loss during the evolution from the AGB to planetary nebulae is related to the formation of carbon stars and related objects.

## References

Cayrel, R. & Jugaku, J. 1963, *Ann d'Astrophys*, 26, 495
Lambert, D. L., Gustafsson, B., Eriksson, K. & Hinkle, K. H. 1986, *ApJS*, 62, 373
Ohnaka, K. & Tsuji, T. 1996, *A&A*, 310, 933
Ohnaka, K. & Tsuji, T. 1998, *A&A*, submitted
Smith, V. V. & Lambert, D. L. 1985, *ApJ*, 294, 326
Smith, V. V. & Lambert, D. L. 1990, *ApJS*, 72, 387

## Discussion

**Little-Marenin**: Some J–type stars are relatively close to us (e.g. Y CVn), and even though they are at high galactic latitude, their distance above the galactic plane is fairly small. Will this affect the $^{12}C/^{13}C$ – latitude effect that you found? Did you consider this?

**Ohnaka**: Distances for C stars are difficult to determine; I will consider this in more detail.

**Gustafsson**: The difference in $^{12}C/^{13}C$ between Ohnaka & Tsuji and Lambert et al. (1986) is not easy to explain. The methods are different — Lambert et al. used $\Delta v = 2$ CO lines and CN Red System lines in the 2.2 $\mu$m band, and there we claim that the continuum can be traced, while you rely on comparing lines with similar depths without knowing the level of the continuum. Your models have lower gas pressures than ours (basically because you include a turbulent pressure) but I doubt whether that matters. Our temperature scale is about 200 K lower than yours, but that would only – at the most – explain half of the effect. The reason for the discrepancy and the model sensitivity of the two methods should be further explored.

**Ohnaka**: First, the models should be compared in the $\tau - T$ plane. That comparison will tell us more exactly how different the two models are. Second, a re-analysis of the spectra that Lambert et al. (1986) analyzed should be done with our model atmospheres. This should serve to explore the reasons for the discrepancies in the $^{12}C/^{13}C$ ratios.

# $^{12}$C/$^{13}$C RATIOS AND Li ABUNDANCES IN LOW-MASS C STARS

CARLOS ABIA
*Dpto. Física Teórica y del Cosmos*
*Universidad de Granada, 18071 Granada, Spain*

AND

JORDI ISERN
*Centre d'Estudis Avançats de Blanes (CSIC)*
*18032 Blanes, Spain*

**Abstract.** High-resolution and high signal-to-noise-ratio spectroscopy is used to derive $^{12}$C/$^{13}$C ratios from the Red System of the CN molecule at 8000 Å in several galactic low-mass C stars previously analyzed for lithium. It is found that Li-rich C stars usually have low isotopic ratios ($< 15$). This suggests that the Li and $^{13}$C enrichments in C stars might be produced by a similar mechanism. However, due to the large uncertainties in the abundance derivations for both Li and carbon isotope ratios, this result must be confirmed with further studies.

## 1. Introduction

Currently the study of the origin of lithium is one of the most active fields of research in astrophysics. During the past few years new observational data and theoretical work have been added to the study of this elusive element. From the observational point of view, recent observations of a lithium abundance dispersion in the stars of M92 (Deliyannis et al. 1995) constitute the most compelling evidence for some (perhaps a severe) depletion in the *Spite plateau* stars. Although this result should be confirmed with further observations, it casts doubt on the widely accepted primordial lithium abundance (Li/H $\approx 10^{-10}$). In any case, whatever the primordial abundance of lithium might be, there is a wide consensus that in order to produce the current abundance of this element (Li/H $\approx 10^{-9}$), an extra source is needed, perhaps of stellar origin. Concerning this, the detection of an intense gamma-ray emission in the Orion region (Bloemen et al. 1994) has been linked with the existence (until now hypothetical) of a low-energy component in the galactic cosmic-ray spectrum. This low-energy component might have important consequences on the production of the light elements (Li, Be, B) by spallation reactions in the insterstellar medium. In addition, two new sites of primary lithium production have been proposed recently: nucleosynthesis induced by neutrinos in supernova explosions (Woosley et al. 1990), and the thermonuclear runaway during a nova event (Hernanz et al. 1996). However, apart from the large uncertainties involved in these calculations, these two mechanisms still lack observational confirmation (if they are possible at all). Hence, at present the only compelling evidence for the stellar production of lithium are the AGB stars (S and C stars).

The production of Li in AGB stars was initially proposed by Cameron (1955) and expanded by Cameron & Fowler (1971). Basically, it requires a hot convective envelope where the series of reactions $^3\text{He}(\alpha,\gamma)^7\text{Be}(e^-,\nu)^7\text{Li}$ might take place. $^7$Be produced in the first reaction is carried out by convection to the outer and cooler layers of the star where it decays to $^7$Li. At the same time $(p,\alpha)$ reactions destroy some of the lithium produced, but a kind of *equilibrium* abundance is established in the outer envelope and, eventually, this lithium may be detectable spectroscopically. Indeed, for a long time it has been known that some AGB stars in the Galaxy (and now also in the Magellanic Clouds) show huge features at the $\lambda$ 6708 Å Li I resonance line (Sanford 1955; Torres-Peimbert & Wallerstein 1966). These *super-Li-rich* stars probably have abundances of lithium one or two orders of magnitude higher than the current galactic abundance (Li/H $\approx 10^{-8}$ – $10^{-7}$) (Abia et al. 1991).

A couple of years ago, we started a theoretical and observational project to gain a better understanding of the lithium production in AGB stars. The main goals of this project were threefold: (*a*) to derive accurate lithium abundances in as large a number of C stars as possible in order to obtain good statistics on the super-Li-rich phenomenon; (*b*) as a second step, to calculate observationally the Li yield from C stars; and (*c*) to study what exactly are the physical conditions which should occur in an AGB star to produce lithium. Here we summarize the most important results achieved concerning points (*a*) and (*b*) and, due to the great utility of the $^{12}\text{C}/^{13}\text{C}$ ratio for tracing stellar interiors, we present some new results about this isotopic ratio in C stars, trying to give some light on point (*c*).

## 2. Results

### 2.1. LITHIUM

The stars studied here constitute a flux-limited and homogeneous sample of galactic C stars taken from the *Two-Micron Sky Survey* (Neugebauer

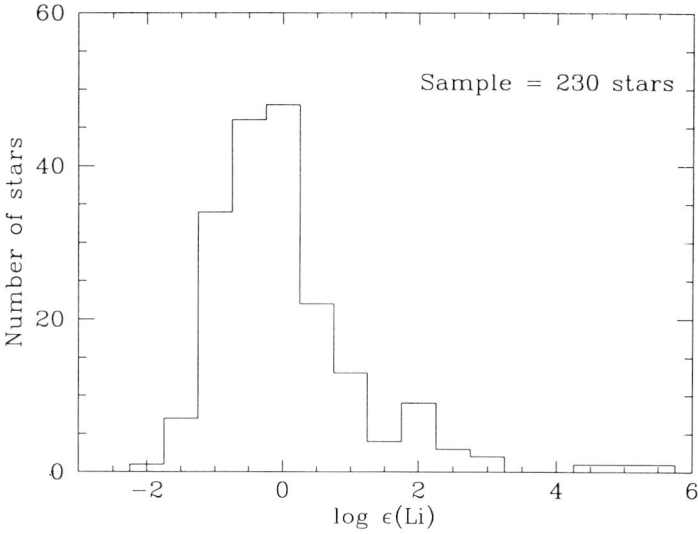

*Figure 1.* Li abundance histogram for the C stars studied.

& Leighton 1969). This survey covers almost 75% of the sky. The main characteristics of this sample are reviewed in Claussen et al. (1987); we only summarize the most important points. It is assumed that these stars have absolute 2.2 μm magnitudes similar to those of C stars in the Magellanic Clouds; thus, the sample should penetrate a volume of radius $\sim 1.5$ kpc about the Sun covering a range in bolometric magnitudes of $-3.5 \leq M_{bol} \leq -6$. From the scale height from the galactic plane ($z \sim 200$ pc), the sample should have stars with progenitor masses between 1.2 and 1.6 $M_\odot$. This is an important point (problem?) because current models for AGB stars do not produce lithium in stars with initial masses $M < 4 M_\odot$ (see Lattanzio et al. 2000).

Between 1990 and 1993, we observed about 230 galactic C stars around the λ6708 Å Li I feature using high-resolution spectroscopy (for details about observations, reduction procedures and analysis, see Boffin et al. 1993; Abia et al. 1993a). The final statistics concerning Li abundances can be seen in Figure 1. It is apparent from this figure that most of the stars are distributed around abundances of log $\epsilon$(Li) $\approx 0.0$ (on the scale of log N(H) $\equiv 12$), but it is also evident that there is a nice long tail toward higher lithium abundances ($\sim 15\%$ of the stars). Some of the stars in this tail (2–3%) have Li abundances log $\epsilon$(Li) $> 4.0$. It is straightforward to estimate the Li yield from C stars with the statistics shown in Figure 1 and the mass-loss rates derived by Claussen et al. (1987) (see their figure 10). This yield is found to be $\dot{M}_{Li} \approx 4.5 \times 10^{-14}$ $M_\odot$/yr/C-star. However, note that this yield is extremely dependent on the mass-loss rate during the

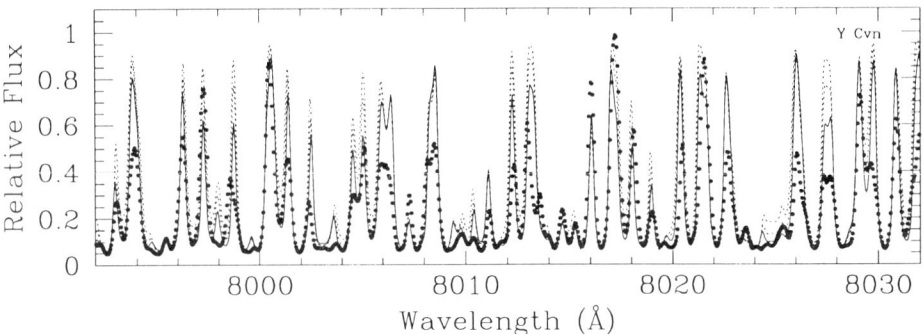

*Figure 2.* Observed (black dots) and theoretical (continuous and dotted lines) spectrum of Y CVn for three values of the $^{12}C/^{13}C$ ratio: 15, 6 and 2. The features where the synthetic spectra coincide are $^{12}CN$ lines.

super-Li-rich stage, on the actual stellar lithium abundances, and on the assumed star-formation rate during the life of the Galaxy (for details, see Abia et al. 1993b). All these parameters are rather uncertain to the extent that, for example, more than 80% of the yield given above might be due to a single star, IY Hya, which has $\log \epsilon(\text{Li}) \approx 5.0$ and $\dot{M} \geq 10^{-5}\ M_\odot/\text{yr}$. In any case, when this yield is introduced in a simple model of galactic chemical evolution for the solar neighborhood, it is possible to reproduce the observed evolution of the Li/H vs. [Fe/H] relationship (see Abia et al. 1995). This shows that AGB stars might well be the most important source of Li in the Galaxy. Further studies should be done to confirm or refute this statement. Effort should be made to improve the accuracy of the Li abundances in AGB stars. Current uncertainties in the analysis and in the model atmospheres tell us that Li abundances (and those of any other chemical element) cannot be determined with a precision higher than 0.4–0.5 dex in C stars.

## 2.2. THE $^{12}C/^{13}C$ RATIOS

The carbon isotopic ratio $^{12}C/^{13}C$ is a very useful tracer of H–burning in stellar interiors. In order to understand the conditions in which Li might be produced in AGB stars we have derived this isotopic ratio in some of the stars previously analyzed for lithium. However, the determination of this ratio in C stars is not an easy task as shown by previous studies on this topic (see e.g. Fujita & Tsuji 1977; Lambert et al. 1986). This is mainly due to saturation effects, the difficulty of locating the continuum, and the uncertainties mainly in the upper layers of the current model atmospheres, where most of the useful molecular lines for this kind of analysis are formed.

The $A^2\Pi - X^2\Sigma^+$ electronic transition of the CN molecule around 8000 Å offers an opportunity in the near infrared for this analysis in C stars. This spectral range is not free from the problems metioned above, but using synthetic spectrum analysis with high-resolution and high signal-to-noise spectra permits an analysis as accurate as other methods (see Ohnaka & Tsuji 2000). We derived the $^{12}C/^{13}C$ ratio in 40 C stars focusing our attention mainly on the stars in the tail of the distribution of Figure 1. Details about observations and analysis can be found in Abia & Isern (1996), where previous results are presented. Figure 2 shows an example of fitting to the star Y CVn, a well known J–type star. We derive for this star a mean value of $^{12}C/^{13}C = 3.5$. Note however that there are still important discrepances between the synthetic and observed spectra. Most of the stars studied are distributed around $^{12}C/^{13}C = 20$–$35$, which is in excellent agreement with the results by Ohnaka & Tsuji (2000) for a larger example of AGB stars. We estimate a total uncertainty in the isotopic ratio of $\pm\, 8 - 12$, the larger the isotopic ratio the larger the uncertainty. However, this result contrasts with that of Lambert et al. (1986) who found $^{12}C/^{13}C$ ratios mostly around 50–60 (see the Discussion section following the Ohnaka & Tsuji paper in this volume for a possible explanation for that). Most of the stars with low carbon isotopic ratio are Li-rich stars ($\log \epsilon(\text{Li}) > 1.0$). This is shown in Figure 3. From this figure it is clear that there is a correlation between lithium abundance and $^{12}C/^{13}C$ ratio (note however the error bars). We might conclude that as lithium is produced by the Cameron & Fowler mechanism, some $^{12}C$ burning occurs. In fact, despite the important uncertainties in the analysis we found that our Li-rich and $^{13}C$-rich stars have $^{12}C/^{16}O$ ratios slightly lower than 1 (however C/O $> 1$ as C $= {}^{12}C+{}^{13}C$), showing that indeed $^{12}C$ is burned in the envelope. The difficulty is that current stellar models for AGB stars cannot obtain both high Li and high $^{13}C$ abundances for initial masses lower than 4 $M_\odot$. At masses as low as 1–2 $M_\odot$ the bottom of the convective envelope in these models is too cool to develop any kind of burning. However, there is a possibility that was pointed out by Wasserburg et al. (1995): the existence of deep circulation currents below the bottom of the standard convective envelope could transport matter from the non-burning bottom of the convective envelope down to regions where some CNO processing takes place in low-mass RGB and AGB stars. This cool-bottom processing, with the simultaneous operation of the third dredge-up, may produce a C star with a low $^{12}C/^{13}C$ ratio. No calculation has been done yet concerning lithium in this situation, but the temperatures achieved in this cool-bottom burning allow us to guess that it is quite possible that the Cameron & Fowler mechanism works in such conditions. In fact Wasserburg et al. (1995) show that this mechanism works during the RGB phase.

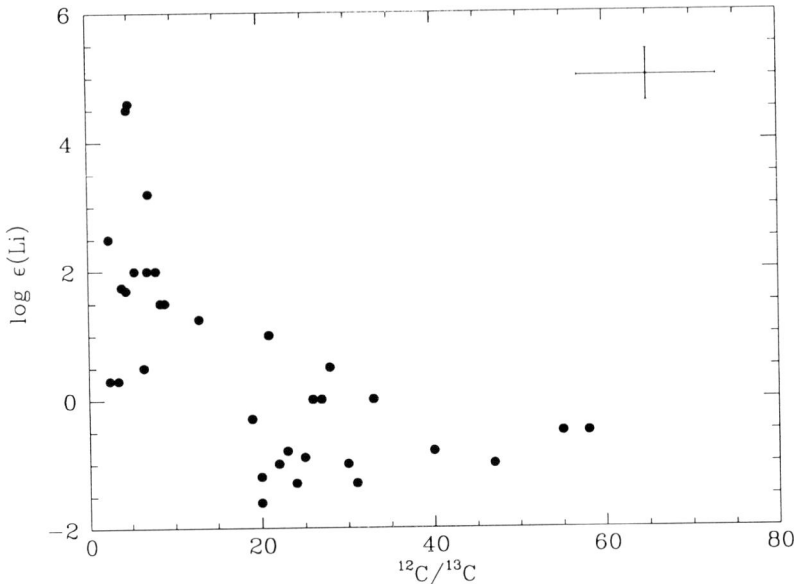

*Figure 3.* Li abundances vs. $^{12}C/^{13}C$ ratios. A correlation is clearly present.

Finally, we would like to emphasize that the derivation of abundances in AGB stars still involves large uncertainties. Note for example that varying the C/O ratio used in the analysis by a few hundredths changes the abundance derived for Li and/or $^{13}C$ by a factor 2 or even 3. At present, we cannot determine the C/O ratio in AGB stars as accurately as a few hundredths!

## References

Abia, C., Boffin, H. M. J., Isern, J. & Rebolo, R. 1991, *A&A*, 245, L1
Abia, C., Boffin, H. M. J., Isern, J. & Rebolo, R. 1993a, *A&A*, 272, 455
Abia, C. & Isern, J. 1996, *ApJ*, 460, 443
Abia, C., Isern, J. & Canal, R. 1993b, *A&A*, 275, 96
Abia, C., Isern, J. & Canal, R. 1995, *A&A*, 298, 465
Bloemen, H. et al. 1994, *A&A*, 281, L5
Boffin, H. M. J., Abia, C., Isern, J. & Rebolo, R. 1993, *A&A Supp.*, 102, 361
Cameron, A. G. W. 1955, *ApJ*, 121, 144
Cameron, A. G. W. & Fowler, W. A. 1971, *ApJ*, 164, 111
Claussen, M. J., Kleinmann, S. G., Joyce, R. R. & Jura, M. 1987, *ApJ Supp.*, 65, 385
Deliyannis, C. P., Boesgaard, A. M. & King, J. R. 1995, *ApJ*, 452, L13
Fujita, Y. & Tsuji, T. 1977, *PASJ*, 29, 711
Hernanz, M., José, J., Coc, A. & Isern, J. 1996, *ApJ*, 465, L27
Lambert, D. L., Gustafsson, B., Eriksson, K. & Hinkle, K. H. 1986, *ApJ Supp.*, 62, 373
Lattanzio, J. C., Frost, C. A., Cannon, R. C. & Wood, P. R. 2000, in IAU Symp. 177: *The Carbon Star Phenomenon*, ed. R. F. Wing (Kluwer), p. 449

Neugebauer, G. & Leighton, R. B. 1969, *Two-Micron Sky Survey: A Preliminary Catalog* (NASA SP-3047)
Ohnaka, K. & Tsuji, T. 2000, in IAU Symp. 177: *The Carbon Star Phenomenon*, ed. R. F. Wing (Kluwer), p. 81
Sanford, R. F. 1950, *ApJ*, 111, 262
Torres-Peimbert, S. & Wallerstein, G. 1966, *ApJ*, 146, 724
Wasserburg, G. J., Boothroyd, A. I. & Sackmann, I.-J. 1995, *ApJ*, 447, L37
Woosley, S. E., Hartmann, D. H., Hoffman, R. D. & Haxton, W. C. 1990, *ApJ*, 356, 272

## Discussion

**Plez**: A word of caution: The disappearance of most opacity-contributing molecules in stellar atmospheres with C/O ratios very close to 1 leads to a strong cooling of the outer layers of these atmospheres with a resulting strengthening of resonance lines like the Li I 6707 Å line. It is therefore critical to know the C/O ratio accurately. There is also the possibility of circumstellar contamination of the Li I line (observed by us for some S type stars). This can be checked at higher resolution and by looking at the K I line at 7699 Å.

**Abia**: Yes, certainly the derivation of Li abundances in AGB stars is a difficult task full of uncertainties. However, we have determined the C/O and $^{12}C/^{13}C$ ratios from other spectral ranges before determining the Li abundance. Anyway, we are currently studying the Li I lines at 6103 Å and 8126 Å, less sensitive to uncertainties in the upper atmosphere, to check the validity of the Li abundances that we have derived from the Li I resonance line at 6707 Å. Concerning the possibility of circumstellar contamination, it is not clear at all whether the 7699 Å K I line would be a useful test or not. There is an *AJ* paper by C. Barnbaum showing that this line forms in C stars at a very different optical depth than the 6707 Å Li I line.

# CARBON AND HYDROCARBON MOLECULES IN VERY COOL WHITE DWARFS

TURGUT ASLAN AND IRMELA BUES
*Dr. Remeis Sternwarte*
*Astronomisches Institut der Universität Erlangen-Nürnberg*
*96049 Bamberg, Germany*

**Abstract.** We report on our investigations in progress on the importance of polyatomic molecules — especially of carbon and hydrocarbon — in the atmospheres of very cool helium-rich white dwarfs. For two sets of abundance ratios, model atmospheres with $T_{\rm eff} = 5000\,{\rm K}$, $\log g = 8.0$ have been computed and the dependence of $C_3$, $C_2H$ and $C_2H_2$ on abundance variations of H/He, C/He, O/He and N/He has been calculated in detail, with the absorption of carbon molecules included quantitatively. We discuss the possibility that $C_2H$ features can explain the observed IR flux deficiency for LHS 1126 at $1.6\,\mu{\rm m}$ and $2.0\text{--}2.3\,\mu{\rm m}$.

## 1. Introduction

At the very cool end of the white dwarf cooling sequence with $T_{\rm eff} \leq 5500\,{\rm K}$, the objects are intrinsically faint. Even if they belong to the solar neighborhood within 25 pc, they cannot be observed with high resolution. Spectra in the blue show only weak features, and some have a pure continuum.

Even observations with a high signal/noise ratio allow only a classification (Greenstein & Liebert 1990). Strong line features of Ca (Ca I $\lambda\,4227$ Å, Ca II H & K) and Fe can be seen in LHS 69 and strong band features of CH and $C_2$ in G99-37, thus indicating helium-rich compositions *with* and *without* carbon.

New model atmospheres for G99-37 ($T_{\rm eff} = 6000\,{\rm K}$, $\log g = 8.0$, He/H = 1000, C/H = 35.5) with $C_3$ included have shown that one feature in the blue can be attributed to this molecule (Bues & Karl-Dietze 1995). G99-37 has a magnetic field of $2 \times 10^3$ Tesla which might be responsible for the intermediate ratio of H/He compared to other white dwarfs with carbon features in their spectra. The former conclusion that cooler objects with shifted $C_2$ bands have the same ratios of C/He and the shifts of the features are due to a magnetic field is no longer valid for all observed objects, according to the polarization measurements by Schmidt et al. (1995), which showed no polarization at all for ESO 439-162 and LHS 1126. These authors suggest the possibiliy that the observed features are not due to $C_2$ at all but belong to polyatomic molecules. Our current investigation is aimed at very cool helium-rich white dwarfs where the flux in the infrared region must be affected severely by the presence of molecular features, even for reduced abundances and the corresponding increase of gas pressure, which could cause shifts in wavelength of the same order as a magnetic field.

## 2. Model Atmospheres

New flux-constant model atmospheres have been calculated for extremely helium-rich compositions in the range $5500\,\text{K} > T_{\text{eff}} > 4800\,\text{K}$, $\log g = 8.0$, with varied relative abundances of H/He, C/He, O/He and N/He, where oxygen is most important for gas pressure.

The possible H,C,O reaction processes were critically reviewed. Because of the high gas pressure and a ratio $C/H \geq 1$ in agreement with the presence of $C_2$ and $C_3$ features, some formation processes have been preferred compared to those in normal carbon stars, namely:

$C_2 + C \rightarrow C_3$
$C_2 + H \rightarrow C_2H$
$C_2 + H_2 \rightarrow C_2H_2$

while other processes for the formation of $C_2H$ and $C_2H_2$ such as

$C_2 + H_2 \rightarrow C_2H + H$
$C_2H + H \rightarrow C_2H_2$
$C_2H + H_2 \rightarrow C_2H_2 + H$

have been omitted.

The constants of dissociation for $C_3$ and $C_2H_2$ were taken from Tsuji (1964), and for $C_2H$ from Perić et al. (1990). The computation of dissociative equilibria included all diatomic molecules and neutral and ionized species of H, He, C, N, and O and was solved with an iterative scheme. For absorption, the molecules were included in a smeared-line approximation, appropriate for white dwarfs due to the high pressure. For $C_2H$, transitions at 9500 cm$^{-1}$ (= 1.05 $\mu$m) and 5500 cm$^{-1}$ (= 1.82 $\mu$m) calculated by Reimers et al. (1985), and transitions at 1.78 $\mu$m, 1.81 $\mu$m and 2.61 $\mu$m calculated by Perić et al. (1992) with *ab initio* methods, have been included in our calculations.

Figure 1 shows relative partial pressures of the molecules $H_2$, $C_2$, CN, CO, CH, $C_3$, $C_2H$ and $C_2H_2$ compared to the total gas pressure versus the optical depth for a model atmosphere with the indicated parameters

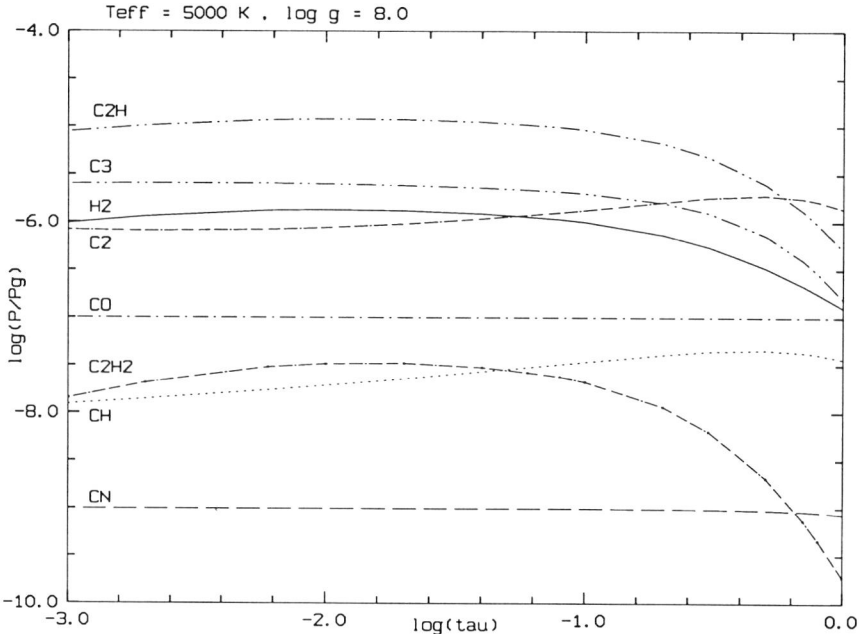

*Figure 1.* Pressure fractions of various diatomic and polyatomic molecules versus the optical depth for $T_{\text{eff}} = 5000$ K, $\log g = 8.0$, $\log(\text{H/He}) = -5$, $\log(\text{C/He}) = -5$, $\log(\text{O/He}) = -7$, and $\log(\text{N/He}) = -9$.

relevant for LHS 1126. The optical depth scale in the line-forming region is $\tau$-Rosseland.

Obviously the $C_2H$ molecule is the most dominant one in the composition given in Fig. 1, followed by $C_3$, $H_2$ and $C_2$. Diatomic CO and CH as well as $C_2H_2$ play important minor roles, while CN can be neglected.

A further reduction in H and C as well as in the relative abundance ratio yields the results shown in Figure 2. Both figures can be compared to Fig. 4 of Schmidt et al. (1995) where, for a fixed value of $P_g$, the relative abundances of hydrocarbons, dependent on temperature only, are shown for a mixture of just H, He and C. Our model-atmosphere stratification includes a steep gradient of temperature and gas pressure as well, where the O abundance is responsible for the formation of polyatomic molecules due to the formation of CO and the corresponding high pressure.

In Fig. 2 this effect is well demonstrated. Although the ratio C/H is changed by a factor of 3, CO is still the most abundant molecule; next, with smaller values than for the first mixture, we find $C_2$, $C_3$ and $C_2H$. As for depressions in flux caused by observable features, this means that $C_2$ features should be visible, at least at this $T_{\text{eff}}$. The molecules CN, $H_2$ and

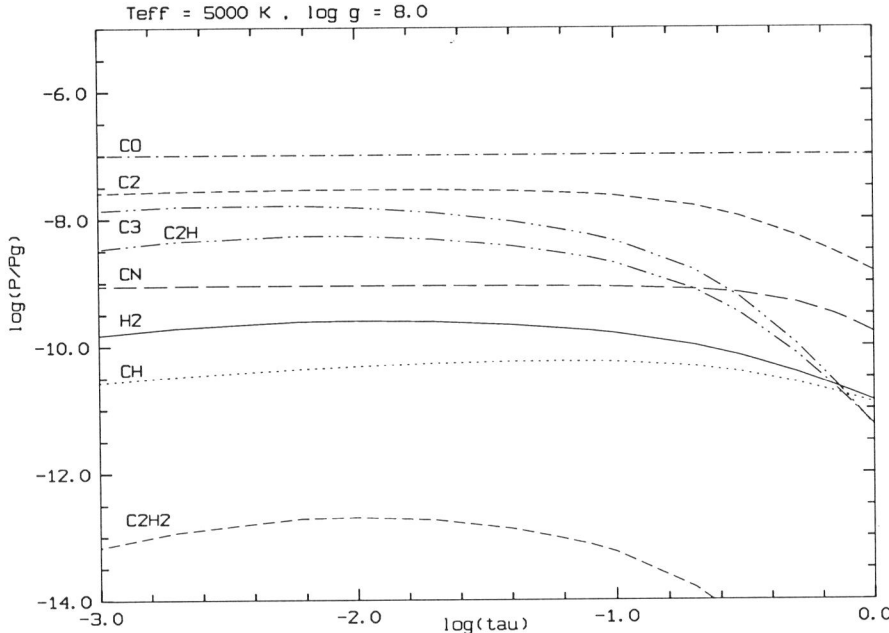

*Figure 2.* Pressure fractions for reduced abundances: $T_{\text{eff}} = 5000\,\text{K}$, $\log g = 8.0$, $\log(\text{H/He}) = -7$, $\log(\text{C/He}) = -6.5$, $\log(\text{O/He}) = -7$, $\log(\text{N/He}) = -9$.

CH play a minor role, and $C_2H_2$ can be neglected.

## 3. LHS 1126

One object to which these computations can be applied is LHS 1126. The optical spectrum shows a 12 % molecular absorption band at 4990 Å ± 100 Å. Broad absorption bands at 5450 Å ± 200 Å, 6050 Å ± 200 Å and 6680 Å ± 100 Å are present in addition to $C_3$ λ4050 Å. Our infrared spectrum, taken in May 1995, shows a weak depression near 9200 Å.

The star has already been analyzed by Bergeron et al. (1994) with blue and red spectra and model atmospheres consisting of hydrogen and helium only. The observed features at 4590 Å and 4990 Å were attributed to $C_2$, yet the shift in wavelength cannot, in their opinion, be due to pressure effects alone as their ratio H/He = 0.8 gives more electrons than our computations. To account for the infrared flux distribution, which does not correspond to a black body as shown by Lebofsky & Liebert (1984), they propose collision-induced absorption by molecular hydrogen due to collisions with helium, yet they do not discuss absorption by CH or other carbon molecules. They obtain $T_{\text{eff}} = 5400\,\text{K} \pm 200\,\text{K}$, $\log g = 7.85 \pm 0.17$, and $\log(\text{He/H}) = 0.8$

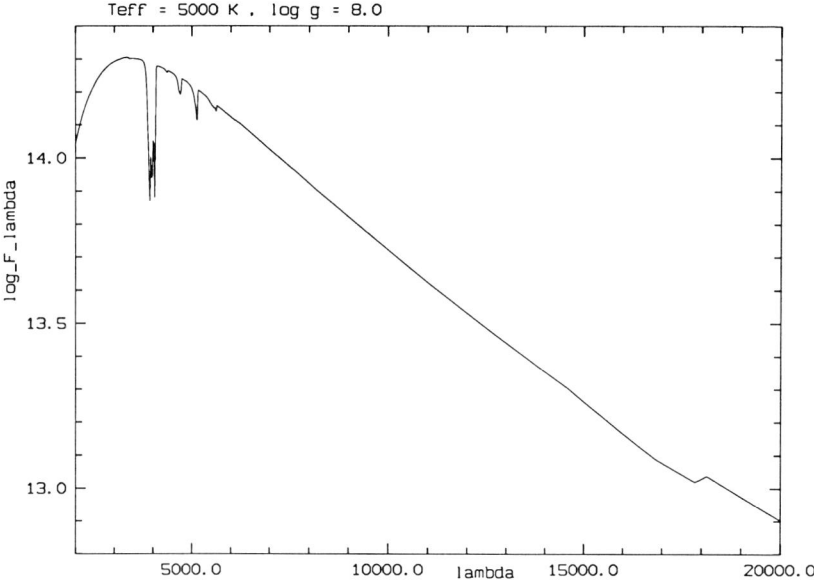

*Figure 3.* Calculated flux for the parameters given in Fig. 2. In addition to the $C_2$ and $C_3$ bands in the visual, a broad absorption feature at $\lambda = 1.78$ $\mu$m is present.

$\pm$ 0.2 for this object.

Schmidt et al. (1995) discussed the observed features in LHS 1126 and concluded that the absence of the G–band of CH is an indication of a lower hydrogen abundance than the value determined by Bergeron et al. (1994). They investigated the relative abundances of $C_n$ and $C_nH_n$ compounds for a fixed value of gas pressure, and C/He as a function of temperature (H/He = 0.1 and C/He = $10^{-2}$, $10^{-3}$ and $10^{-5}$). $C_2H$ is one of their preferably formed molecules in the photosphere of white dwarfs, a result confirmed by our computations.

With $T_{\rm eff}$ = 5000 K, log $g$ = 8.0, log (H/He) = –5, log (C/He) = –5, log (O/He) = –5, and log (N/He) = –9 (Aslan et al. 1996), agreement with the observed flux in the visible was possible. Further analysis showed that an even lower relative abundance of carbon and hydrogen was sufficient, and therefore two sets of parameters were used for the calculations (see captions to Figs. 1 and 2).

Information about the infrared region of LHS 1126 has been obtained by photometry with six filters in the *JHK* region by Lebofsky & Liebert (1984), and by spectroscopy around 2.0 $\mu$m by Bergeron et al. (1994). The observed flux deficiency between 1.6 $\mu$m and 2.1 $\mu$m could not be matched completely by the model spectra.

For the abundance ratios of Fig. 2, the corresponding flux of our model atmosphere is shown in Figure 3 from the blue to the infrared region of the spectrum. Features in the blue are mainly due to $C_3$ and $C_2$, and in the infrared to $C_2$ and $C_2H$. Our results can explain the $1.78\,\mu m$ region but not the $2.1\,\mu m$ feature. The consistency with the observed spectrum in the $1.6\,\mu m$ region is better than in the computations by Bergeron et al. (1994). Further steps of iteration in the details of abundance ratios and absorption processes in these extremely high-pressure model atmospheres are necessary.

## References

Aslan, T., Bues, I. & Karl-Dietze, L. 1996, in *Hydrogen-Deficient Stars*, ed. C. S. Jeffery and U. Heber, ASP Conf. Ser., 96, 325

Bergeron, P., Ruiz, M.-T., Leggett, S. K., Saumon, D. & Wesemael, F. 1994, *ApJ*, 423, 456

Bues, I. & Karl-Dietze, L. 1995, in *White Dwarfs*, ed. D. Koester and K. Werner (Springer-Verlag), Lecture Notes in Physics, No. 443, p. 201

Greenstein, J. L. & Liebert, J. W. 1990, *ApJ*, 360, 662

Lebofsky, M. J. & Liebert, J. W. 1984, *ApJ*, 278, L111

Perić, M., Peyerimhoff, S. D. & Buenker, R. J. 1990, *Mol. Phys.*, 71, 693

Perić, M., Peyerimhoff, S. D. & Buenker, R. J. 1992, *Z. Phys. D*, 24, 177

Reimers, J. R., Wilson, K. R., Heller, E. J. & Langhoff, S. R. 1985, *J. Chem. Phys.*, 82, 5064

Schmidt, G. D., Bergeron, P. & Fegley, B. Jr. 1995, *ApJ*, 443, 274

Tsuji, T. 1964, *Ann. Tokyo Astron. Obs.*, 2nd ser., 9, 1

## Discussion

**Jørgensen**: Is $C_2H$ stable under the conditions of your models? Where did you take your data for the $C_2H$ molecule from?

**Aslan**: Our calculations show that $C_2H$ is stable and is responsible for the IR flux deficiency at $1.6\,\mu m$ ($H$ band). There is a strong transition at $1.78\,\mu m$. We took the data for $C_2H$ from an article by Perić et al. (1992).

**Bues**: These data are *ab initio* calculations.

# THE s-PROCESS IN THE YELLOW SYMBIOTIC AG DRACONIS

KATIA CUNHA
*Observatório Nacional, Rio de Janeiro, Brazil*

VERNE V. SMITH
*University of Texas at El Paso, TX 79968, U.S.A.*
*and*
*McDonald Observatory*
*University of Texas at Austin, TX 78712, U.S.A.*

AND

ALAIN JORISSEN
*Institut d'Astronomie, Université Libre de Bruxelles, Belgium*

**Abstract.** An abundance analysis of the yellow symbiotic system AG Draconis reveals it to be a metal-poor K giant ([Fe/H] = −1.3) which is enriched in the heavy s-process elements. This star thus provides a link between the symbiotic stars and the binary barium and CH stars which are also s-process enriched. These binary systems, which exhibit overabundances of the heavy elements, owe their abundance peculiarities to mass transfer from thermally-pulsing asymptotic giant branch stars, which have since evolved to become white-dwarf companions of the cool stars we now view as the chemically peculiar primaries. A comparison of the heavy-element abundance distribution in AG Dra with theoretical nucleosynthesis calculations shows that the s-process is defined by a relatively large neutron exposure ($\tau = 1.3$ mb$^{-1}$), while an analysis of the rubidium abundance suggests that the s-process occurred at a neutron density of about $2 \times 10^8$ cm$^{-3}$. The derived spectroscopic orbit of AG Dra is similar to the orbits of barium and CH stars. Because the luminosity function of low-metallicity K giants is skewed towards higher luminosities by about 2 magnitudes relative to solar-metallicity giants, it is argued that the lower metallicity K giants have larger mass-loss rates. It is this larger mass-loss rate that drives the symbiotic phenomena in AG Dra and we suggest that the other yellow symbiotic stars are probably low-metallicity objects as well.

*No text received*

**Discussion**

**North**: What is the orbital period of this system?

**Cunha**: A CORAVEL orbit yields $P = 549 \pm 7$ d, in agreement with the value obtained by Mikołajewska et al. (1995, *AJ*, 109, 1289).

# THE CHROMOSPHERES OF CARBON STARS

DONALD G. LUTTERMOSER
*Department of Physics, East Tennessee State University*
*Johnson City, TN 37614, U.S.A.*

**Abstract.** Most oxygen-rich late-type giant stars show evidence for chromospheres in their visual spectra (e.g. Ca II H & K emission features). Cool (i.e. N–type) *non-Mira* carbon stars, however, have never been observed to have chromospheric emission in their Ca II H & K lines. However, faint Mg II h & k lines were detected in emission in low-dispersion IUE spectra of the brightest cool carbon stars in the early 1980s. May 1984 saw the first (and only) successful high-dispersion IUE spectrum taken of a cool carbon star, TX Psc (N0; C6,2). Armed with this high-dispersion spectrum, as well as low-dispersion IUE and ground-based spectra, Luttermoser et al. (1989) made the first detailed attempt to *semiempirically* model the chromosphere of a cool carbon star. This model was successful in reproducing the Mg II lines, but it was not well constrained due to the lack of other observed high-resolution chromospheric profiles for comparison. Modeling carbon star chromospheres can now be addressed more accurately with HST/GHRS high-resolution spectra. New fluoresced emission features have been discovered in the GHRS spectra of carbon stars that are not present in their oxygen-rich counterparts.

## 1. Introduction

There is often debate as to the meaning of the term *chromosphere*. We are all familiar with the Sun's *classical* chromospheric structure (i.e. the VAL model: Vernazza, Avrett & Loeser 1981) — a temperature reversal and rise to $\sim$10,000 K just above the solar photosphere. Observationally, chromospheres present themselves by emission features of singly ionized metals caused (presumably) by an enhanced-temperature region in the outer atmosphere of the star. Theoretically, the chromosphere has been described by Linsky (1980) as an enhanced-temperature region above the stellar photosphere, mechanically heated to temperatures in the range from $T_{\text{eff}}$ to $\sim$10,000 K. However, what is the structure of this enhanced-temperature region for a given star? Is it similar to the semiempirical *solar-like* VAL

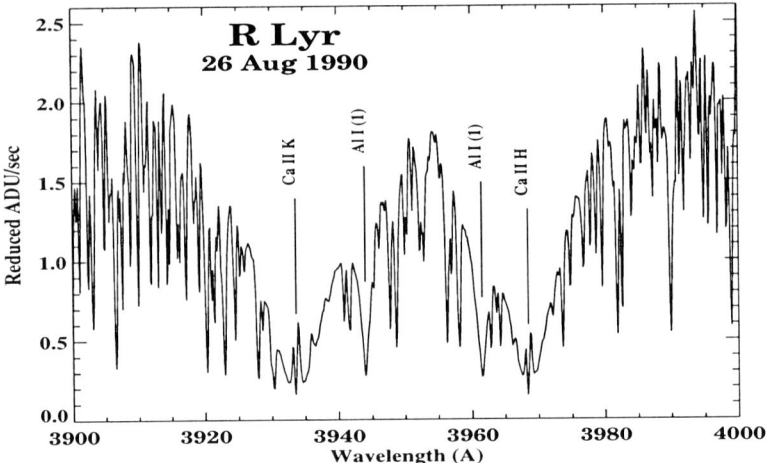

*Figure 1.* McMath–Pierce spectrum of R Lyr (M5 III) in the Ca II H & K region. Note the strong emission peaks near the cores of these lines which indicates the existence of a chromosphere in this star.

model? Or is it similar to theoretical, *dynamic* shock models that have been calculated for pulsationally unstable stars (e.g. Bowen 1988)?

The first evidence that chromospheres may exist in carbon stars was reported by Bidelman & Pyper (1963), when they identified emission features in the 3250–3300 Å region of TX Psc as Fe II (1), (6), and (7) from ground-based observations. However, comparisons of synthetic spectra to the IUE spectra of TX Psc suggest that these emission features may be *peaks* in the flux between the many strong absorption lines in this region of the spectrum (Luttermoser 1988). Surprisingly, no chromospheric features have been noted in the *non-Mira* N-type carbon stars at visual wavelengths ($\lambda > 4000$ Å). Also, the optically bright, semiregular (SR) N-type stars lack H$\alpha$ in either emission or *strong* absorption (Yamashita 1972, 1975), as is demonstrated quite nicely in Figure 6 of Johnson et al. (1995). Also, unlike their oxygen-rich counterparts (e.g. R Lyr in Figure 1), the Ca II H & K lines display no chromospheric emission cores as demonstrated in the spectrum of TX Psc in Figure 2. With the advent of ultraviolet (UV) astronomy in the early 1980s, the question of chromospheres in carbon stars could be addressed again. Querci et al. (1982) detected no IUE flux for two SR N-type carbon stars (Y CVn and WZ Cas) and two Mira-type cool carbon stars (U Cyg and SS Vir). From this they deduced that either cool carbon stars have no permanent chromospheres, or substantial overlying opacity from carbon condensates *hides* this chromospheric emission.

However, Johnson & O'Brien (1983) were successful in detecting the UV spectra of 3 other N-type carbon stars (BL Ori, TX Psc, and T Ind). Later,

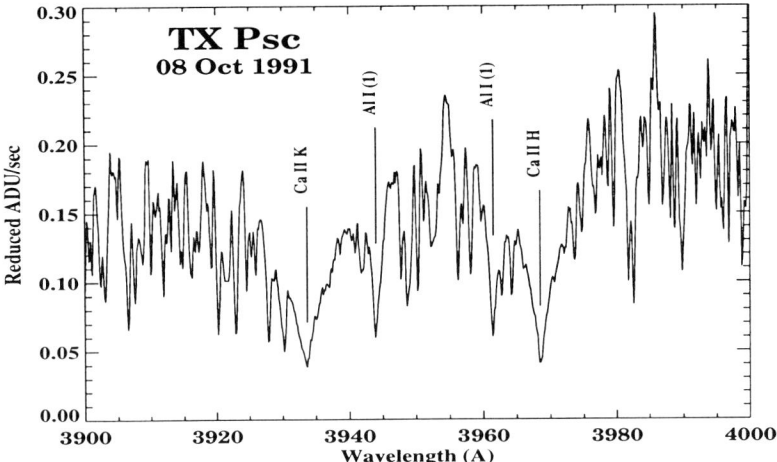

*Figure 2.* McMath–Pierce spectrum of TX Psc (N0) in the Ca II H & K region. Note the lack of emission in the cores of these lines.

Eaton et al. (1985) reported on the UV spectra of R–type carbon stars, with the late R stars (R5–R8) having UV chromospheric emission similar to late K and M stars. The strongest features seen in the N–type stars are the Mg II h & k lines near 2800 Å and the C II] (UV 0.01) intersystem multiplet near 2325 Å, as can be clearly seen in Figure 4 of Johnson & Luttermoser (1987). The strength of the Mg II lines in carbon stars is similar to that of the coolest non-Mira M–type giant stars.

Querci & Querci (1985) first reported on carbon-star UV emission line variability in the IUE spectra of TW Hor. At least a factor of 3 variation in the Mg II flux was observed, and this was used to empirically deduce a short-period acoustic wave model for the atmospheric structure of this star. They summarize from this work that the chromosphere of TW Hor is extended with a gradual temperature rise. Shortly thereafter, Johnson et al. (1986) found a factor of 8 variability in the Mg II line strength for TX Psc and a factor of 5 for the C II] (UV 0.01) multiplet. This sets the SR variable carbon stars apart from their oxygen-rich counterparts which typically display no more than a 10% variation in emission-line strength (Judge et al. 1993). Indeed, this amount of variability is consistent with that observed for oxygen-rich Mira variables (e.g. Brugel, Willson & Cadmus 1986).

## 2. Semiempirical Model Chromospheres

To determine the temperature structure of the UV emission-line regions, past studies often employed the technique of semiempirical chromospheric modeling (Linsky 1980). In this technique, one attaches a temperature rise (as a function of column mass or height) to a radiative equilibrium photospheric model representative of the star in question. Adjustments then are made to this temperature rise until the calculated synthetic spectrum matches the observed spectrum. Figure 2 of Luttermoser, Johnson & Eaton (1994) demonstrates this technique, which was used to semiempirically deduce the chromospheric structure of the M6 giant star g Her.

Early attempts at chromospheric modeling of carbon stars, based on low-dispersion IUE spectra of TX Psc, were made by Avrett & Johnson (1984) and de la Reza (1986). Avrett & Johnson found that the Mg II k line can be produced to within a factor of 2 – 5 of its observed strength while producing no H$\alpha$ emission. However, no C II] (UV 0.01) emission was produced from this model.

TX Psc is perhaps the most important carbon star in the sky since it is one of the brightest and also lies in the ecliptic. Because of this, an angular diameter is known from lunar occultation, which, in turn, allows us to make an absolute flux comparison between synthetic and observed spectra. Prior to HST, only one high-dispersion IUE spectrum existed for an N–type carbon star, that of TX Psc (Eriksson et al. 1986). As such, TX Psc was chosen to be the prototypical N–type carbon star for the chromospheric structure study of Luttermoser et al. (1989). This model was to become known as the LJAL model.

From this work, Luttermoser (1988) and Luttermoser et al. (1989) noted that NLTE must be used to model the outer atmospheres of these stars, and the effects of partial redistribution is important in the formation of the Ly$\alpha$ and the Mg II resonance lines. Also, getting background opacities correct is of the utmost importance when trying to model the chromospheric features, particularly for Mg II, where the Ca I bound-free opacity from the metastable $4p\ ^3P^\circ$ state (edge at 2940 Å) and the Mg I (UV1) line wing dominate *all* background opacities. The LJAL model was able to reproduce (excluding the circumstellar absorption) the Mg II h & k profiles (Figure 3), match the integrated flux of the C II] (UV 0.01) multiplet, and yet produce no Balmer line emission. To achieve this fit, the chromosphere must have low enough densities and the temperature gradient great enough to prevent the strong neutral metal lines from going into emission, yet produce the proper Mg II and C II] flux (Luttermoser et al. 1989).

The Mg II and C II] emission arises in a region of the atmosphere that is close to the photosphere ($\sim$ 2–3 % of $R_\star$). Therefore, the plane-parallel assumption in hydrostatic equilibrium (HSE) was sufficient to reproduce the Mg II profiles. Much was learned concerning the NLTE radiative transport in cool, low-density atmospheres. The radiative *detailed balance* approximation is *not* valid for the Ly$\alpha$ line despite its enormous optical depth

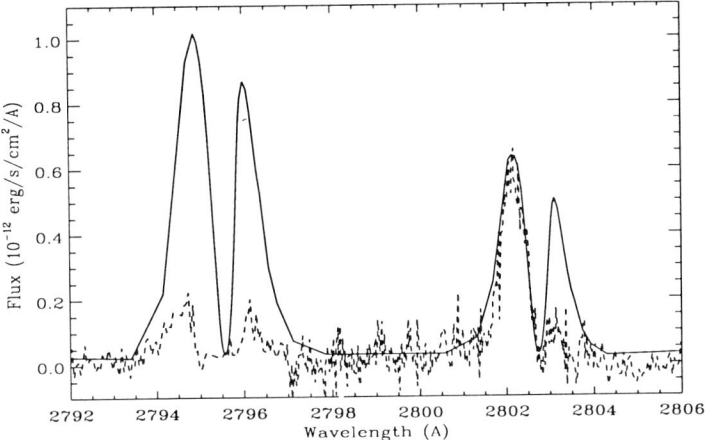

*Figure 3.* NLTE synthetic profiles from the LJAL chromospheric model of TX Psc (*solid line*) as compared to the high-resolution IUE spectrum (*dashed line*) of this star. Note that circumstellar absorption from Mn I (UV1) and Fe I (UV3) *hides* much of the emission of the k line.

(Luttermoser & Johnson 1992). This is due to the fact that cool temperatures produce collisional rates smaller than the net radiative rates for strong resonance lines. This had a large impact on the excited and ion states of hydrogen and on the strength of the H$\alpha$ line (see Figure 12 in Luttermoser & Johnson 1992). Also, the ionization and excitation of neutral metals in the upper photosphere are strongly influenced by UV chromospheric photons flowing back down. Photospheric lines must be carefully selected when carrying out abundance analyses for these stars. Strong lines should be avoided since they will form in the upper photosphere or chromosphere.

## 3. Future Modeling with HST Data

Johnson and collaborators (Johnson et al. 1995; Carpenter et al. 1997) have obtained several FOS (see Figure 1 of Johnson et al. 1995) and GHRS (Figure 4) spectra of N–type carbon stars. The *red light leak* compromises the usefulness of the HST/FOS spectra for these stars. However, the GHRS spectra, with its superior resolution, signal-to-noise, and dynamic range as compared to IUE, have presented additional chromospheric indicators to refine the NLTE semiempirical models. One of the new discoveries was a new emission line, Fe I (UV45) $[z\,^5H° \ (J=5) \rightarrow a\,^5F \ (J=4)]$ at 2807 Å, which was never previously seen in a cool giant star spectrum. Actually, this feature appeared in the original IUE spectrum but was partially hidden by a cosmic ray hit. As part of the Johnson et al. team, D. G. Luttermoser identified this feature as Fe I (UV45) and suggested that it is a fluoresced

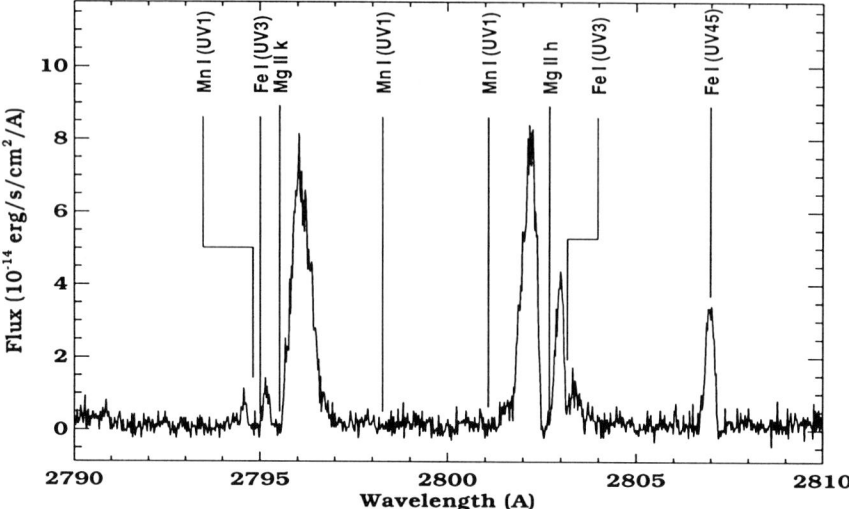

*Figure 4.* HST/GHRS spectrum of TX Psc. Note that the effects of the circumstellar absorption overlying the k line now can be clearly seen in this spectrum. The fluorescent Fe I line at 2807 Å is discussed in the text.

line since other lines in this multiplet were not seen in emission. However, no suitable pump was found in Moore's UV Multiplet tables. Another member of the team, R. F. Wing, then carried out calculations that showed that an Fe I transition in the UV13 multiplet should have a line at 2325.32 Å [$a\,^5D$ (J=4) $\rightarrow z\,^5H°$ (J=5)] which does not appear in the multiplet tables. This lies virtually on top of the strongest line in the C II] (UV 0.01) multiplet. Later, when the TX Psc spectrum in the C II] region was obtained, the effect of this pump was seen in the C II] line at 2325.40 Å (Figure 5).

Also note the substantial changes seen in the Mg II profiles between the IUE spectrum (31 May 1984) and the GHRS spectrum (4 Dec 1994). Does this result from changes in overlying absorption or from changes in the actual Mg II strength? Although changes in the opacity of the overlying circumstellar shell certainly plays a part in the variability of the Mg II lines, comparing the GHRS spectrum to the IUE spectrum certainly makes a strong case for actual changes in the chromospheric emission region of the atmosphere. The differences in these spectra provide probable evidence for a non-static chromosphere for carbon stars.

Is the LJAL model a reasonable one to describe the chromosphere of TX Psc and carbon stars in general? Luttermoser, Johnson & Eaton (1994) showed that the Mg II and C II] features of g Her could not be simultaneously modeled with an HSE, 1–D, plane-parallel atmosphere. The final g Her chromosphere model was based on fitting the Mg II profiles. However, this model produced C II] line ratios that were inconsistent with the IUE

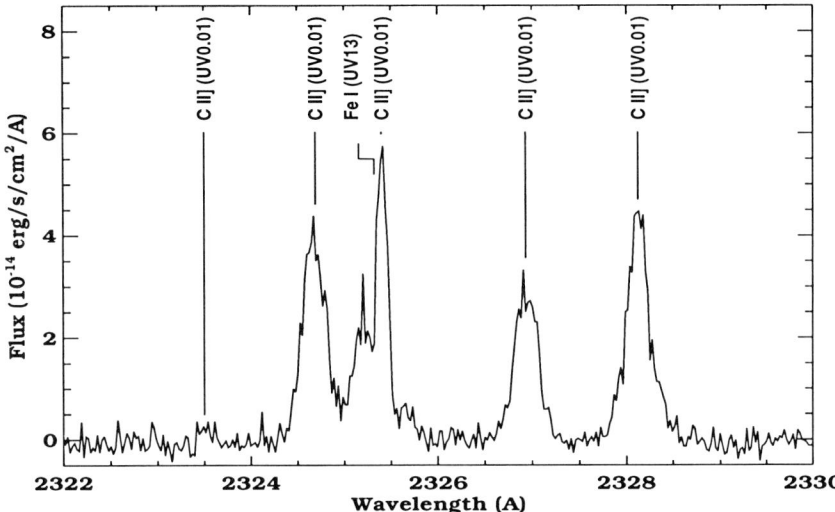

*Figure 5.* C II] (UV 0.01) profiles in the GHRS spectrum of TX Psc. Note the overlying absorption from Fe I (UV13) near the C II] λ2325.40 line that gives rises to the fluoresced emission feature at 2807 Å.

observations. The C II] lines indicated that the electron density in the line-emitting region had to be increased by two orders of magnitude to achieve the proper fit. Since these C II] lines form at the same depths in the model as the Mg II peak flux (i.e. the inner chromosphere), this would require either moving the chromosphere inward or enhancing the temperature of the inner chromosphere. Neither of these techniques would work, since moving the chromosphere inward would produce strong emission in the Mg I (UV1) and optical 'b' lines, and in Hα. Meanwhile, enhancing the temperature in the inner chromosphere would produce much too broad Mg II emission. Hence the assumptions used in the model cannot be valid. With the HST data, we can now compare the synthetic C II] profiles from the LJAL model to the observed lines. As was the case with g Her, the C II] line ratios from the model for TX Psc are inconsistent with the observations. The C II] lines indicate that the electron density must be higher by a factor of 100 with respect to the final HSE model. Essentially, we need to force C II] to form in a different atmospheric region than the Mg II lines. Luttermoser et al. (1998) show that this can be done with *dynamic* models.

Finally, Jørgensen & Johnson (1991) have also shown that the LJAL model should produce emission cores in the infrared CO and HCN lines — but none are observed! Are the chromospheres of carbon stars inhomogeneous? Jørgensen & Johnson suggest that they are, having a chromospheric filling factor less than 10%.

## 4. Conclusion

The era of semiempirically modeling the chromospheric structure of carbon stars and late-type M giant stars is over — the *classical* semiempirical method (i.e. a 1–D, HSE, homogeneous atmosphere) produces inconsistent results! So what must be done to achieve more realistic models for the outer atmospheres of these stars? First, NLTE techniques must be incorporated. Second, dynamical processes must be included, both long-period and short-period waves. Finally, inhomogeneities may also have to be included should the CO and HCN emission problems still exist in the radiative-hydrodynamic calculations. Can we answer the question: *What mechanism(s) heats the chromospheres of these stars?* The answer is NO, not yet! But with the HST data as a guide and more sophisticated modeling efforts, we can start to approach the correct answer.

I wish to thank the IAU for financial assistance to attend this meeting and the AAS for an NSF-funded International Travel Grant. Further support was granted under NASA contract NAS 5–32863 to Applied Research Corporation. I wish to thank K. G. Carpenter for the HST spectra of carbon stars discussed here, and E. H. Avrett and R. Loeser for use of the PANDORA radiative transfer code. Finally, my greatest thanks go to one of the true pioneers in this field, Hollis R. Johnson, who got me interested enough in these *cool* stars to devote a Ph.D. dissertation to this research.

## References

Avrett, E. H. & Johnson, H. R. 1984, in *Cool Stars, Stellar Systems, and the Sun*, ed. S. L. Baliunas & L. Hartmann (Springer-Verlag), p. 330
Bidelman, W. P. & Pyper, D. M. 1963, *PASP*, 75, 389
Bowen, G. H. 1988, *ApJ*, 329, 299
Brugel, E. W., Willson, L. A. & Cadmus, R. 1986, in *New Insights in Astrophysics: 8 Years of UV Astronomy with IUE*, ed. E. J. Rolfe, ESA, SP–263, p. 213
Carpenter, K. G., Robinson, R. D., Johnson, H. R., Eriksson, K., Gustafsson, B., Pijpers, F. P., Querci, F. & Querci, M. 1997, *ApJ*, 486, 457
de la Reza, R. 1986, in *The M–Type Stars*, ed. H. R. Johnson & F. R. Querci, NASA, SP–492, p. 373
Eaton, J. A., Johnson, H. R., O'Brien, G. T. & Baumert, J. H. 1985, *ApJ*, 290, 276
Eriksson, K., Gustafsson, B., Johnson, H. R., Querci, F., Querci, M., Baumert, J. H., Carlsson, M. & Olofsson, H. 1986, *A&A*, 161, 305
Johnson, H. R., Baumert, J. H., Querci, F. & Querci, M. 1986, *ApJ*, 311, 960
Johnson, H. R., Ensman, L. M., Alexander, D. R., Avrett, E. H., Brown, A., Carpenter, K. G., Eriksson, K., Gustafsson, B., Jørgensen, U. G., Judge, P. D., Linsky, J. L., Luttermoser, D. G., Querci, F., Querci, M., Robinson, R. D. & Wing, R. F. 1995, *ApJ*, 443, 281
Johnson, H. R. & Luttermoser, D. G. 1987, *ApJ*, 314, 329
Johnson, H. R. & O'Brien, G. T. 1983, *ApJ*, 265, 952
Jørgensen, U. G. & Johnson, H. R. 1991, *A&A*, 244, 462
Judge, P. G., Luttermoser, D. G., Neff, D. H., Cuntz, M. & Stencel, R. E. 1993, *AJ*, 105, 1973

Linsky, J. L. 1980, *Ann. Rev. Astron. Astrophys.*, 18, 439
Luttermoser, D. G. 1988, Ph.D. Thesis, Indiana University
Luttermoser, D. G., Bowen, G. H., Willson, L. A. & Brugel, E. W. 1998, in preparation
Luttermoser, D. G. & Johnson, H. R. 1992, *ApJ*, 388, 579
Luttermoser, D. G., Johnson, H. R., Avrett, E. H. & Loeser, R. 1989, *ApJ*, 345, 543
Luttermoser, D. G., Johnson, H. R. & Eaton, J. A. 1994, *ApJ*, 422, 351
Querci, F., Querci, M., Wing, R. F., Cassatella, A. & Heck, A. 1982, *A&A*, 111, 120
Querci, M. & Querci, F. 1985, *A&A*, 147, 121
Vernazza, J. E., Avrett, E. H. & Loeser, R. 1981, *ApJ Suppl.*, 45, 635
Yamashita, Y. 1972, *Ann. Tokyo Astr. Obs.*, 13, 169
Yamashita, Y. 1975, *Ann. Tokyo Astr. Obs.*, 15, 47

## Discussion

**Dorfi**: In the case of chromospheres, it must be important to include a pulsational model to have an appropriate inner boundary condition for your calculations.

**Luttermoser**: Yes, very much so since such boundary conditions will be important for the strength (i.e. temperature) of the outward-propagating shocks, which of course has a major impact on the emergent spectrum and *global* structure of the atmospheric model.

**Little–Marenin**: Are your models affected by the fact that TX Psc has no dust emission? Is the spectrum of UU Aur, which does have dust emission, significantly different from that of TX Psc?

**Luttermoser**: It won't affect the NLTE model of TX Psc at all, since we neglected dust in those calculations. Does dust affect the UV spectra of UU Aur (and other "dustier" carbon stars)? The UV spectra of many of the optically bright carbon stars (both GHRS and IUE low-dispersion) look very similar (see Johnson & Luttermoser 1987). However, some of the optically bright N stars have no observed UV flux (Y CVn is a good example). The presence of dust probably has a major impact on the UV spectra of these stars.

**Totten**: I find it surprising that you say no N–type C stars have been found with emission lines, because Sanduleak & Pesch reported an N–type star with H$\alpha$ and H$\beta$ emission in 1984 or 1986. Also, the APM high galactic latitude survey has observed several N–type stars which have distinctive H$\alpha$ and H$\beta$ emission lines. All of these are "classical" cool N–type stars with no known variability.

**Luttermoser**: Thank you for that information. Actually, I had tremendous difficulty keeping H$\alpha$ in absorption in my modeling attempts. Perhaps I should have qualified that comment by saying, "no *optically bright*, non-Mira N–type carbon stars have been observed with H$\alpha$ emission." The fact that some semi-regular variable (by the way, "classical" N–type carbon stars, i.e. the SR and Lb variables, vary in brightness by at least a few tenths of a magnitude) N–type carbon stars are observed as hydrogen Balmer emission carbon stars (like Miras) further supports my claim that the atmospheric structure of these stars is similar to the Mira variables.

**Whitelock**: In support of what you say about SRs and Miras, the IR and optical observations suggest that the difference between SRs and Miras is quite ambiguous in the case of C-rich stars.

**Luttermoser**: I agree 100%! The UV spectra of these stars also point in that direction.

**Elitzur**: SiO maser observations indicate the presence of magnetic fields of order 2–10 gauss. Leaving out the effect of magnetic fields may involve the neglect of a leading dynamic factor.

**Luttermoser**: This may be the case. I ignored magnetic fields mainly due to the fact that it is already difficult to carry out NLTE radiative transfer calculations in these low-density atmospheres, especially when one relaxes the plane-parallel assumption and includes macroscopic velocity fields. It has been argued in the past that due to their large sizes and slow rotation rates, these cool giants should have negligible magnetic fields since the dynamo effect should be very weak. However, recent observations of "hot spots" on the surfaces of these stars (e.g. HST observations of $\alpha$ Ori by A. Dupree) suggest that this might not be a correct assumption. Perhaps rising granules give rise to a surface magnetic field. I, myself, won't touch such radiative-magneto-hydrodynamic calculations with a "ten-foot pole."

**Wing**: Some people have expressed skepticism that the Fe I fluorescence mechanism you mentioned can be effective if the pump line at 2325 Å is only a predicted transition that is not listed in the Multiplet Tables. I would like to add that this is a rather special transition, namely the one that raises neutral iron from its ground state (in fact, from the lowest sublevel of the ground state) to the upper state of the observed emission line at 2807 Å. And if you ask why this pump line is not included in the Multiplet Tables when the other lines of the same multiplet are listed, I propose that this is due to the same coincidence with C II $\lambda 2325$ that causes the fluorescence in carbon stars — there may have been some soot on the electrodes used in producing the laboratory Fe I spectrum!

**Luttermoser**: Thank you for that explanation. I have no doubt that this Fe I (UV 13) transition exists. The fact that you predicted it from the appearance of the fluoresced 2807 Å line before the TX Psc C II (UV 0.01) spectrum was obtained is more than enough proof for me.

# Session III

## Observed Spectra
## and
## Energy Distributions

Sunetra Giridhar relaxing with Ingrid Wing, Sylvia Önder, and baby Timur.

Jack MacConnell, Mário Magalhães, Jay Frogel, Muammer Önder, and Sylvia Önder enjoying a Turkish lunch, complete with *rakı*.

# SPECTRAL CHARACTERISTICS OF RV TAURI STARS

SUNETRA GIRIDHAR
*Indian Institute of Astrophysics*
*Bangalore, India*

**Abstract.** Spectroscopic properties of RV Tau variables are summarized. We report on our detailed spectroscopic investigation of a sample of RV Tau stars covering all three RV Tau subgroups. Though the observed abundance pattern is similar to those of a few post–AGB stars, we find considerable variation in the observed chemical compositions indicating that there is some non-homogeneity among subgroups of RV Tau variables. Possible scenarios of dust–gas separation in the material ejected by the dusty wind are discussed to explain the observed abundances. This mechanism would work better if the star happens to possess a binary companion. Interestingly, one of our sample stars, EP Lyr, shows long-term variations in radial velocity possibly caused by the presence of a companion.

## 1. Introduction

The RV Tau stars have been classified as pulsating variables of Population II, but their light variations are not as regular as those of Type II Cepheids. They are post–AGB objects evolving off the AGB after mass loss has reduced the envelope mass to such an extent that the thermal flashes in the helium-burning region cannot be sustained, and they later end up as central stars of planetary nebulae. Figure 1 shows the color-magnitude diagram for the globular cluster M5 (Buonnano et al. 1981). The figure also shows the positions of several RV Tau stars and evolutionary tracks for 0.605 $M_\odot$ and 0.546 $M_\odot$ post–AGB models by Schönberner (1993). The positions of these well-known RV Tau stars in Fig. 1 substantiate the view that these are post–AGB stars.

The important features of RV Tau stars can be summarized as follows. The light curves are very distinctive with alternating deep primary and shallow secondary minima. The period is defined as the time elapsed between the two consecutive deep minima. The observed periods are in the

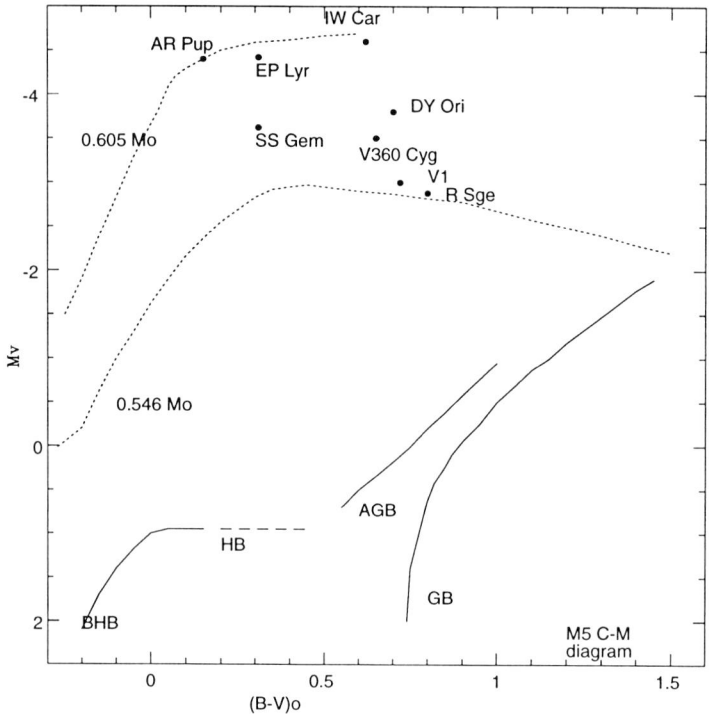

*Figure 1.* C–M diagram for M5 and positions of well-known RV Tau stars. The values of $M_V$ for the RV Tauri stars were calculated by Gonzalez using the $P$–$L$ relation for RV Tau stars described in Gonzalez (1994). The color excesses needed for calculating $(B-V)_o$ are taken from a number of sources that are listed in Gonzalez, Lambert & Giridhar (1997).

range of 30 to 150 days. Irregularities in the light period are noticed for many, but hotter members of the class generally have more stable periods. In the *General Catalogue of Variable Stars* (Kukarkin et al. 1969), RV Tau stars with constant mean brightness levels are classified RVa. RV Tauris for which the mean brightness varies with a longer period and with a larger amplitude of variation are classified RVb. An infrared excess is observed for most RV Tauris (Gehrz & Wolf 1970; Gehrz 1972). The infrared fluxes of RV Tauris show a peak near 11 $\mu$m for all RV Tau subgroups whereas a peak near 2–3 $\mu$m is seen only for RV Tauris with variable mean magnitude. These infrared excesses indicate the presence of circumstellar shells.

Preston et al. (1963) subdivided the RV Tauri stars into three spectroscopic groups (A, B, C) based on low-dispersion spectra. The RV A stars have spectra indicating spectral types G–K but abnormal strength of the CN bands or the G band of CH. Near light minimum, TiO bands of abnormal strength are seen. RV B stars show spectra of type F; the most interesting feature of these stars is the appearance of CH and CN bands that become very strong at light minimum. Though these stars generally seem to be metal-poor, numerous C I lines are seen at all phases. RV C stars are very similar to RV B stars but the CN and CH bands (the most remarkable features of group B stars) are either weak or absent at all phases. A few RV Tau stars belonging to globular clusters are found to be RV Tau stars of group C. An example is V1 in $\omega$ Cen, observed by Gonzalez & Wallerstein (1994).

The radial velocity varies in phase with the light variation, indicating a pulsational variation. However, it is difficult to draw a smooth radial-velocity curve as lines become double between phase 0.1 to 0.2. The line doubling of several strong lines and emission components in hydrogen lines are caused by the passage of a shock wave through the atmospheric layers.

The galactic latitudes of these stars are generally within $\sim 30°$ of the galactic plane. Estimates of Z (their distance from the galactic plane) range from 100 pc to 2 kpc. They appear likely to be a mixed population with a larger fraction being of the thick disk population. This view is also supported by the chemical composition studies as we shall see shortly.

1.1. CHEMICAL COMPOSITION

The chemical composition is an important datum for any kind of star. This is particularly so in the case of post–AGB stars where many processes may have affected the surface composition of the star, such as deep convective mixing or processes that might blow away a large fraction of the outer envelope exposing deeper layers. A complete abundance analysis covering elements formed by different nucleosynthesis chains and isotopic abundance ratios is mandatory for understanding the evolutionary history of the star and its current status.

In the case of RV Tau stars such attempts in the past have been few and none has really been complete. Besides, the older workers used outdated atomic data and analysis techniques and hence did not attain the desired accuracy. The study of AC Her by Baird (1979) is relatively better. The studies of R Sct by Luck (1981) and of U Mon and RU Cen by Luck & Bond (1989) use good-quality spectra and modern techniques, but they do not cover the light elements. Our study of the RV B star IW Car (Giridhar et al. 1994, hereafter Paper I) was perhaps the first comprehensive work on a field star using high-resolution and high S/N CCD spectra covering a large number of elements, and we found many intriguing features. The Fe-peak elements are deficient generally by an order of magnitude, but the elements S and Zn are almost solar. This pattern is also shown by a few high-galactic-latitude post–AGB stars. The abundance pattern of IW Car is not the one expected of a normal metal-poor star but it correlates well with the elemental depletions observed in the ISM, i.e. the elements with high condensation temperatures are strongly depleted. On the other hand Gonzalez & Wallerstein (1994) found evidence of CNO processing and enhancement of $s$-process elements for variable V1 in $\omega$ Cen though the star is generally metal-poor. We decided to explore this non-uniformity of chemical composition of RV Tau stars by studying a sample containing members of different light curve types, Preston spectral groups, periods, galactic latitudes, and infrared fluxes. Our initial sample is presented in Table 1.

TABLE 1. Basic parameters of the RV Tau program stars

| Parameter | IW Car | V360 Cyg | SS Gem | EP Lyr | DY Ori | AR Pup | R Sge | V1 in $\omega$ Cen |
|---|---|---|---|---|---|---|---|---|
| $<V>$ | 8.5 | 11.3 | 8.9 | 10.4 | 11.7 | 9.6 | 9.3 | 11.0 |
| $<B-V>$ | 0.83 | 0.94 | 1.07 | 0.76 | (0.92) | 0.76 | 1.07 | 0.6 |
| Period (days) | 67.5 | 70.4 | 89.3 | 83.4 | 60.3 | 38.9 | 70.6 | 58.7 |
| Gal. lat. (°) | −9.2 | −11.8 | 1.3 | 6.7 | −3.4 | −3.0 | −9.8 | 15 |
| IR excess | yes | no | yes | no | yes | yes | yes | no |
| Preston group | B | C | B | B | B | B | A | B–C |
| Var. type | RVb | RVa | RVa | RVa | ? | RVb | RVb | RVa |

The RV Tau star SS Gem deserves special mention. It was considered RV A by Preston et al. (1973), and subsequent papers have referred to it by the same spectroscopic designation. However, high resolution spectra of this star show numerous C I lines and also CH lines of moderate strength near 4290 Å. It is obviously an RV B star.

## 2. Observations and Abundance Determination

High-resolution spectra were obtained for over 1.5 years at the McDonald Observatory 2.10-m telescope with a Cassegrain echelle spectrograph and a Reticon 1200×400 pixel CCD. Generally a S/N ratio of 100 and two-pixel resolution of 50,000 was attained. Each star was observed at several epochs and care was taken in selecting phases when the atmosphere was not perturbed by the passage of a shock wave that manifests itself through distortions in Balmer lines and line doubling.

We have used an updated version of the LTE line-analysis program MOOG (Sneden 1973) that calculates the abundance for each spectral line individually or produces a synthesized spectrum of the required region.

We have used the new grid of model atmospheres by Kurucz (1992). The atmospheric parameters ($T_{\text{eff}}$, $\log g$, $v_t$) were determined from a set of Fe I and Fe II lines. The details of our method and error analysis can be found in Paper I; the results are summarized in Table 2.

TABLE 2. Derived abundances for the sample RV Tau stars

| Element | $T_c$ | EP Lyr | DY Ori | AR Pup | R Sge | IW Car | V1 $\omega$ Cen | SS Gem | V360 Cyg |
|---|---|---|---|---|---|---|---|---|---|
| O  | 200  | −0.04 | —     | 0.14  | −0.64 | −0.33 | −0.68 | −0.49 | −0.22 |
| C  | 90   | −0.37 | −0.18 | 0.05  | −0.41 | 0.32  | −1.07 | −0.44 | —     |
| S  | 648  | −0.61 | 0.16  | 0.44  | 0.37  | 0.36  | −0.89 | −0.22 | −0.65 |
| Zn | 660  | −0.70 | 0.21  | —     | −0.19 | −0.04 | −1.40 | −0.36 | −0.97 |
| Na | 970  | −1.14 | 0.04  | −0.15 | 0.10  | 0.34  | −1.58 | −0.11 | −0.95 |
| Mn | 1190 | −1.95 | —     | —     | −0.16 | —     | —     | −0.91 | —     |
| Cr | 1277 | −1.83 | —     | —     | −0.28 | −0.98 | −1.43 | −1.16 | —     |
| Si | 1311 | −1.26 | −1.41 | −0.39 | 0.08  | −0.57 | −1.12 | −0.22 | −1.08 |
| Fe | 1336 | −1.80 | −2.30 | −0.87 | −0.50 | −1.06 | −1.77 | −1.00 | −1.27 |
| Mg | 1340 | −1.85 | −2.11 | −1.03 | —     | −0.98 | −1.18 | −0.55 | —     |
| Ca | 1518 | −1.82 | −1.70 | −1.37 | −0.95 | −1.97 | −1.36 | −1.24 | −1.20 |
| Ni | 1354 | −1.59 | —     | −1.30 | −0.50 | −0.96 | −1.51 | −1.03 | −1.10 |
| Ti | 1549 | −2.01 | —     | —     | −1.34 | —     | −1.52 | −2.11 | —     |
| Y  | 1592 | −1.87 | —     | —     | −1.68 | —     | −1.16 | −1.46 | —     |
| Sc | 1644 | −2.11 | —     | −2.16 | −1.48 | −2.13 | −1.91 | −2.10 | —     |

## 3. Results and Discussion

In Paper I we found that the abundance pattern for the RV Tau star IW Car is similar to that of post–AGB stars and that the abundances correlate very strongly with the elemental depletions observed in the ISM. The elements that condense more easily (i.e. with higher condensation temperature) are more strongly depleted. We have plotted in Figures 2 and 3 the abundances of our sample stars against the condensation temperature $T_c$, which is defined as the temperature at which half of a particular element in a gaseous environment condenses onto grains. We have adopted the values of $T_c$ given by Wasson (1985) for the solar mix at a pressure of $10^4$ atm, the pressure at which the data are most complete. As one can see the plots fall into two categories: (a) those with a single slope like EP Lyr, and (b) those like AR Pup showing a plateau up to $T_c = 1000$ K, followed by a negative slope.

The abundances in field RV Tauri stars show a strong correlation with $T_c$, particularly for the elements with $T_c \geq 1000$ K. These stars have nearly solar S and Zn but very low Ca, Sc, and Ti abundances. The depletions

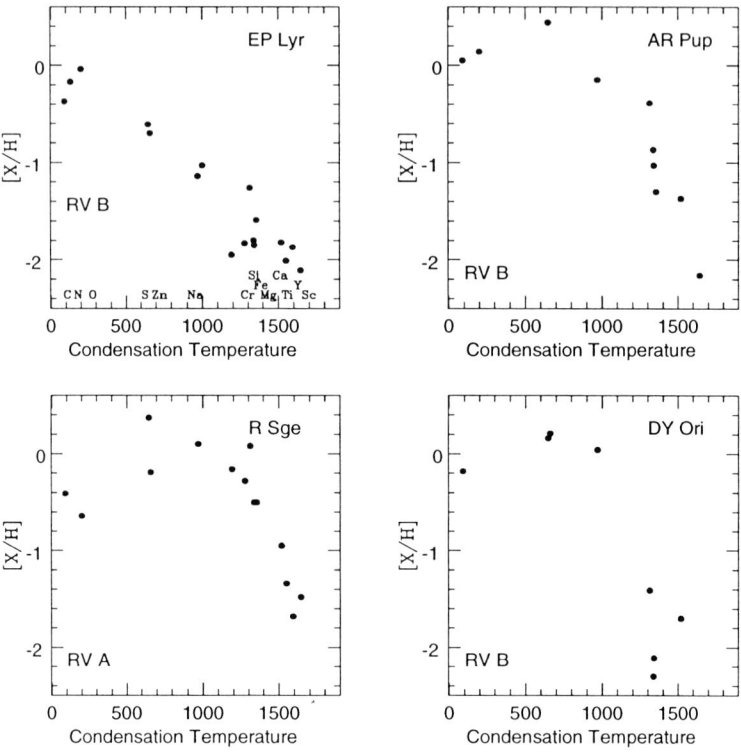

*Figure 2.* Relative abundances [X/H] *vs.* the condensation temperature $T_c$ for EP Lyr, AR Pup, R Sge, and DY Ori.

show a very shallow correlation with $T_c$ in the case of V1 in $\omega$ Cen. Even two moderately Fe-poor stars, AR Pup and R Sge, show better correlations of [X/H] with $T_c$. The abundance patterns of our sample stars are quite different from that found for unevolved or less evolved objects. Past abundance studies of large samples of such objects would have suggested for Fe-poor stars like EP Lyr and DY Ori ([Fe/H] $\approx$ –2) a value of [S/Fe] around +0.4 and of [Zn/Fe] around 0.0 instead of the values (1.2, 1.0) and (1.0, 2.5), respectively, observed for these stars. The abundance pattern of the sample RV Tau stars is also quite different from that of solar-metallicity supergiants.

3.1. INTERPRETATION IN TERMS OF DUST–GAS SEPARATION

As we mentioned earlier, an infrared excess is found for most RV Tau stars. Jura (1986) estimated the mass-loss rate from the 12 $\mu$m flux, an adopted emissivity of dust, an assumed outflow velocity of 10 km s$^{-1}$, and a distance

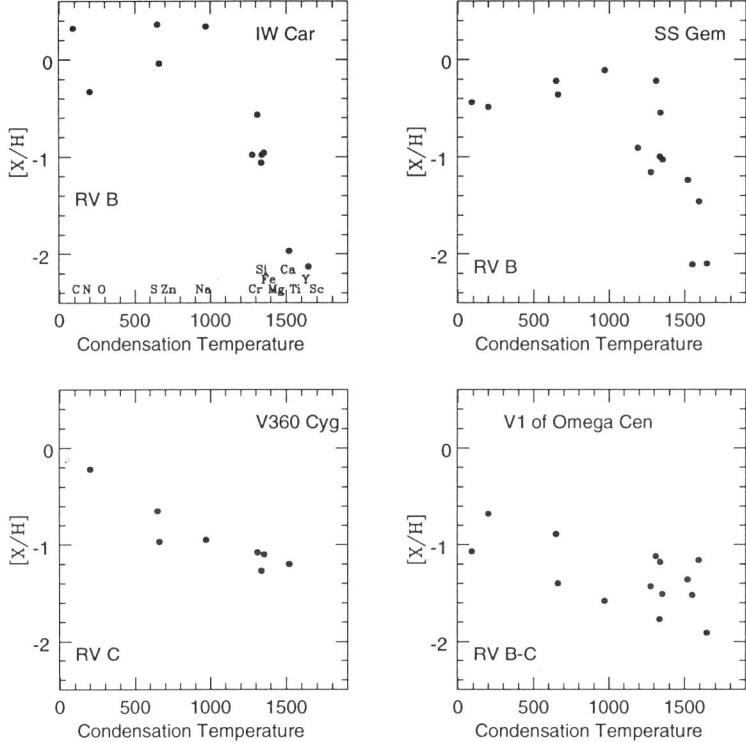

*Figure 3.* Relative abundances [X/H] *vs.* $T_c$ for IW Car, V360 Cyg, SS Gem, and V1 in $\omega$ Cen.

calculated from a period-luminosity law. A typical value for the mass-loss rate for the dust is $\dot{m}(d) \approx 10^{-8}\,M_\odot/\mathrm{yr}$. One expects the dust to drag along some gas. If the dust to gas ratio is 1% by mass, the total mass-loss rate $\dot{m}(d)+\dot{m}(g)$ would be $\approx 10^{-6}\,M_\odot/\mathrm{yr}$. This mass loss has to be fed by a stellar envelope. A star evolving from the AGB to the planetary nebula phase and the white dwarf cooling track has an envelope mass $M_e$ of around $10^{-3}\,M_\odot$ (Schönberner 1981). Other RV Tau scenarios put them on blue-loops from the AGB to lower luminosities. Gingold (1974) modeled the post–AGB evolution of a $0.6\,M_\odot$ metal-poor star and found the envelope mass to be as small as $0.005\,M_\odot$ for the star to leave the AGB. Hence, the envelope masses are expected to be in the 0.005 to 0.001 $M_\odot$ range. A photosphere containing a mass $M_\mathrm{ph} = 10^{-6}\,M_\odot$ cannot feed the wind for long even if one considers only the mass-loss rate for dust alone. Consequently, it would progressively eat into the envelope. A change of chemical composition is probable since some elements condense more easily onto grains, and at lower gas densities the coupling of the gas to the grain will be weak. It is likely that a dusty wind will effect a chemical fractionation which, if gas falls back to the photosphere, will be reflected in the chemical composition of the photosphere. One can argue that the photospheres of RV Tauri stars are too warm to initiate dust formation, which in the outer and cooler layers the gas density in hydrostatic equilibrium may be too low to sustain a dusty wind. We propose that the pulsation characteristics of RV Tau variables periodically enhance the gas and dust densities at high altitudes. Between these episodes of replenishment of the gas, it is most likely that some gas may fall back into the photosphere. The impact of the fractionated gas on the chemical composition will be larger if the efficiency with which the base of the photosphere is mixed with the entire envelope is lower.

In the scenario of dust–gas separation, the photospheric abundance anomalies should correlate with conditions more favorable for elements to condense onto grains. A strong correlation between photospheric abundances and $T_c$, first found for IW Car and later for EP Lyr and DY Ori, gives strong support to this mechanism. Then, it would appear that elements exhibiting little depletion in the ISM are a better indicator of the star's original metallicity. In this regard, S and Zn would be better metallicity indicators than even C, N, or O whose surface abundances get altered by nuclear reactions and dredge-up. Two stars, V360 Cyg and V1 of $\omega$ Cen, appear to be metal-poor as well as metal-depleted as indicated by their abundances of S and Zn relative to other elements more affected by condensation. Generally, the [X/H] vs. $T_c$ diagram has a nearly flat portion and a sudden break at $T_c$ around 1000 K (an exception is EP Lyr with a single slope). We might suggest that the distance at which the condensation took place will vary from star to star depending upon the existence of circumstellar matter (maintained far away due to the existence of a binary companion or via sporadic strong ejection of dusty wind). If condensation took place far away where the elements with lower $T_c$ could condense, then depletion would be seen for all elements with similar or higher $T_c$. Perhaps we are witnessing that effect in IW Car and EP Lyr. But if there is no way of maintaining circumstellar matter far from the star, then only the elements with $T_c > 1000$ K or so will be depleted. Van Winckel et al. (1995) have proposed an attractive way of attaining dust–gas separation far from the star in a circumbinary disk with the main star accreting gas but not dust. That makes the search for stellar companions to RV Tau stars very important.

Spectroscopic observations of Gonzalez for EP Lyr spanning 1.5 years show an indication of velocity perturbation that is different from pulsation. The light curve of EP Lyr is very stable but still the spectra taken at nearly identical pulsation phases show a variation in heliocentric velocity ranging up to 30 km s$^{-1}$. Both photometric and spectroscopic observations are being used to determine if the star indeed has a binary companion. But a similar search would be of great importance for other RV Tau stars showing a single slope in depletion diagrams, and long-term radial-velocity monitoring is advocated. The study of a larger sample of field and cluster members is essential for a better understanding of these fascinating objects.

The work presented here is part of an ongoing RV Tau program in collaboration with D. L. Lambert, N. Kameswara Rao and G. Gonzalez). I am grateful to Prof. Lambert for hospitality at the University of Texas at Austin as well as for a number of very important suggestions. It is a pleasure to thank Dr. Gonzalez for providing important data that went into the preparation of Figure 1. I also thank the IAU and INSA for supporting my visit to Turkey to attend IAU Symposium 177 on "The Carbon Star Phenomenon."

## References

Baird, S. R. 1979, Ph.D. thesis, Univ. of Washington
Buonanno, R., Corsi, C. E. & Fusi Pecci, F. 1981, *MNRAS*, 196, 435
Gehrz, R. D. 1972, *ApJ*, 178, 715
Gehrz, R. D. & Woolf, N. J. 1970, *ApJ*, 161, L213
Gingold, R. A. 1974, *ApJ*, 193, 177
Giridhar, S., Kameswara Rao, N. & Lambert, D. L. 1994, *ApJ*, 437, 476 (Paper I)
Gonzalez, G. 1994, *AJ*, 108, 1312
Gonzalez, G., Lambert, D. L. & Giridhar, S. 1997, *ApJ*, 479, 427
Gonzalez, G. & Wallerstein, G. 1994, *AJ*, 108, 1325
Jura, M. 1986, *ApJ*, 309, 732
Kukarkin, B. V., Kholopov, P. N., Efremov, Yu. N., Kukarkina, N. P., Kurochkin, N. E., Medvedeva, G. I., Perova, N. B., Fedorovich, V. P. & Frolov, M. S. 1969, *General Catalogue of Variable stars*, Moscow.
Kurucz, R. L. 1992, private communication
Luck, R. E. 1981, *PASP*, 93, 211
Luck, R. E. & Bond, H. E. 1989, *ApJ*, 342, 476
Preston, G. W., Krzeminski, W., Smak, J. & Williams, J. A. 1963, *ApJ*, 137, 401
Schönberner, D. 1981, *A&A*, 103, 119
Schönberner, D. 1993, in IAU Symp. 155: *Planetary Nebulae*, ed. R. Weinberger and A. Acker (Kluwer), p. 415
Sneden, C. 1973, Ph.D. thesis, Univ. of Texas at Austin
Van Winckel, H., Waelkens, C. & Waters, L.B.F.M. 1995, *A&A*, 293, L25
Wasson, J. T. 1985, *Meteorites: Their Record of Early Solar-System History* (New York: W. H. Freeman)

# APPLICATIONS OF NARROW-BAND PHOTOMETRY IN THE STUDY OF PECULIAR RED GIANTS

ROBERT F. WING
*Ohio State University, Columbus, OH 43210, U.S.A.*

**Abstract.** Narrow-band photometry, carried out with filters or spectrum scanners, is useful for measuring molecular bandstrengths and continuum energy distributions in late-type stars. This review emphasizes observations by the writer on three different multicolor photometric systems in the near infrared ($0.75 - 4.0$ $\mu$m); a summary of available data is given. While applications to date have been primarily qualitative (classification, recognition of peculiarities, relative temperatures), future applications are expected to be quantitative (determinations of effective temperatures, luminosities, and abundances) and based upon comparison with synthetic spectra.

## 1. Introduction

Narrow-band photometry is a technique that provides both photometric information (magnitudes and colors) and spectroscopic data (the strengths of spectral features) in a single observation. This simultaneity of measurement of colors and feature strengths is an important advantage in work on variable stars, as well as a great convenience. The technique is especially well suited to studies of cool stars since spectral regions that are heavily depressed by molecular band absorption — and the relatively clear intervals between the bands — are often on the order of 50–100 Å wide and can be measured conveniently by a number of methods. Applications to late-type stars thus include both measurements of molecular band strengths and measurements of continuum colors.

In fact, applications of narrow-band photometry have been somewhat miscellaneous, often serving as supplemental to other data. I will not attempt a complete listing of relevant observations; rather, I will use this opportunity to discuss the special advantages of the technique, to give examples of interesting applications, to summarize the extensive but largely unpublished sets of data that are now available on certain well-defined systems of narrow-band photometry, and to suggest and encourage applications of the available data. Because of major advances in atmospheric models and line lists relevant to late-type stars (see, for example, Plez 2000), many of the older data sets are now finding important new applications.

In the sections that follow, I discuss data obtained by the writer and his colleagues with three kinds of equipment: (1) scanner observations with an S-1 photomultiplier at selected wavelengths in the near infrared out to 1.1 $\mu$m; (2) filter photometry in the same spectral region; and (3) scanner observations with an InSb detector extending the wavelength range longward to about 4 $\mu$m. All three are single-channel instruments, with which the different wavelengths must be measured sequentially. To a large extent these instruments are no longer supported and have been replaced by spectrographs with CCDs and infrared array detectors. When I started these programs, there was no alternative to the single-channel instruments, and frankly I was not concerned by their "inefficiency" since many thousands of stars are so bright that good photometric data could be acquired at a large number of wavelengths in reasonable amounts of time on a small telescope. That, of course, is still true, and since so much work on bright stars remains to be done, it is unfortunate that it is becoming difficult or impossible to continue observing in the same manner, thereby staying on precisely the same photometric systems.

Many of my remarks, however, apply equally well to spectra obtained with either single-channel or array detectors. The difference is basically in what you take home with you from the telescope. With a single-channel instrument, you must decide in advance which wavelengths are to be measured, and the time at the telescope is spent integrating at those wavelengths only. With an array detector, you record the whole spectrum (over some wide range) at the telescope, and do the photometry at home on the computer. The end result is often the same. By recording the complete spectrum, you can analyze it in different ways and change your mind as often as you please about which wavelengths to use in the analysis, but at the cost of dealing with much larger data sets. For bright-star work, I still prefer the single-channel approach, not only because the data sets are so much smaller but also because it helps to ensure that the results will be expressed on a stable, well-defined photometric system.

Well, it is not my purpose here to discuss the relative merits of different observing techniques, and it would be pointless to advocate observing by methods that are no longer supported at the major observatories. Rather, my purpose is to describe data sets that already exist but have not yet been fully exploited. Most of the applications that I've pursued to date have involved extracting information directly from the data, without input from other sources. Examples include spectral classifications based on molecular band strengths, comparisons of the continuum slopes (i.e. temperatures) of M, S, and C stars, discussions of chemical composition *differences* between stars and recognition of peculiarities based on bandstrength–color diagrams, and determinations of the reddening and distances of M supergiants. On the other hand, quantitative interpretations of the photometry had to wait for the appearance of model atmospheres and line lists capable of reproducing the spectra of cool stars. Applications based on comparisons of observed and synthetic photometry — including, for example, determinations of effective temperatures and certain abundance ratios — are only beginning, and some specific suggestions along these lines are outlined in the final section.

## 2. Scanner Observations in the 1 $\mu$m Region

My long-term interest in studying late-type stars by means of narrow-band photometry in the near infrared began abruptly in July 1965, when an automated single-channel spectrum scanner designed by E. J. Wampler became available for use at Lick Observatory's Crossley telescope. The timing was perfect for me, because I was a Berkeley student looking for a thesis topic. I had been studying the blue/visual spectra of very cool stars (especially Miras) on the basis of high-dispersion spectra and various kinds of photometry but had become discouraged by the difficulty of extracting useful information from the jumble of atomic and molecular features that occur at short wavelengths. Wampler's scanner, however, when used with an S-1 photocell, gave access to the far-red/near-infrared region out to 1.1 $\mu$m. This region is nearly devoid of strong atomic lines but contains strong bands of a number of different molecules, usually with stretches of relatively clean continuum between them. A decision was quickly made to explore this region, measuring molecular band strengths and continuum slopes in as many kinds of late-type stars as possible (Wing 1967).

The observed colors of cool stars on almost any photometric system are strongly affected by molecular absorption, which is due to almost completely different sets of molecules in stars of types M, S, and C. Thus one of the major observational problems in studying late-type stars is to find color indices that can serve as reliable temperature indicators, unaffected by composition differences.

Using Wampler's scanner first in the continuous-scan mode (with integrations every 10 Å through a 20 Å exit slot), I explored the spectra of representative cool stars from 0.75 to 1.10 $\mu$m. It was then possible to select the best continuum points for each type of spectrum, as well as the points most sensitive to the presence of various molecules. Unfortunately, the most useful wavelength points in stars of types M, S, and C have very little overlap. Even among the normal M stars, one would chose different continuum points in early M and late M stars. Also, for the purpose of measuring molecular band strengths, it is desirable to include both the intrinsically strongest bands of each molecule (to be able to measure them when weak) and also some weaker bands (to avoid saturation when the absorption is stronger). Thus a general program intended to be useful for all kinds of late-type stars necessarily involves a large number of wavelengths. The scans given in my thesis (Wing 1967) consist of integrations first at 26, and later at 27, wavelengths. The bandpass used throughout was 30 Å.

The observations were reduced as 27-color all-sky photometry, with extinction and transformation coefficients determined in the usual manner. The data were transformed to a system of absolute fluxes by assuming a model-atmosphere energy distribution for $\alpha$ Lyr.

When the calibrated flux measurements are plotted against wavelength, the upper envelopes of points are seen to follow blackbody curves quite closely in most kinds of cool stars (exceptions being the very late M and S stars, later than about M7). That is, there are enough data points falling close enough to the same blackbody curve that one can believe that this blackbody curve is a reasonable approximation to the star's actual line-free continuum, and that the temperature of this blackbody is related to the effective temperature of the star. However, this blackbody continuum is defined by different points in different stars: no simple color index (magnitude difference) can be defined from the 27-color data that would provide a blanketing-free temperature index in stars of all kinds. I therefore developed a numerical scheme for fitting blackbody curves to the scans as part of the standard reduction procedure. The continua so obtained provide both a color temperature and a reference continuum with respect to which the strengths of absorption features could be measured. An important advantage of measuring absorption strengths vertically, with respect to the local continuum, is that such indices are reddening-free, so that spectral classifications and other quantities derived from them are also reddening-free. The color temperatures from blackbody fits are, of course, affected by reddening, but they are obtained in a consistent manner in stars of all compositions.

In practice, I have used the "inverse color temperature" $\theta_c \equiv 5040/T_c$ as my continuum color index. This quantity is proportional to conventional color indices (magnitude differences) and hence can be used in the same way.

Approximately 1300 scans on the 27-color system were obtained over a period of 2.5 years. Program stars included examples of essentially all kinds of stars cooler than the Sun — i.e. all spectral types, luminosity classes, variability types, and chemical peculiarities. The tables in Wing

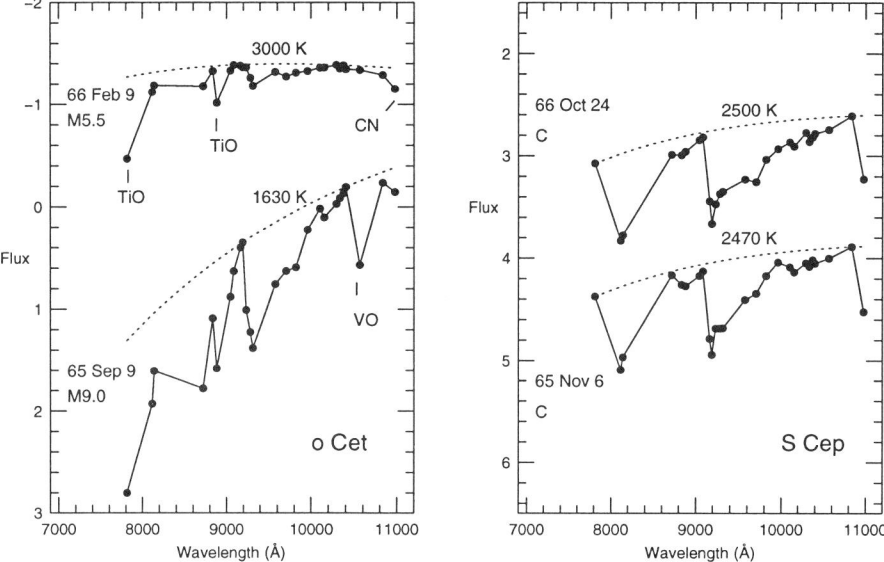

*Figure 1.* Scanner observations of Mira variables. The M–type variable *o* Cet (*left*) and the C–type variable S Cep (*right*) are each shown near the times of maximum and minimum light. Measurements taken through a 30 Å bandpass are given on an absolute flux scale and fitted with blackbody curves. Molecules responsible for absorptions in the M star are labeled; nearly all the absorption in the C star is due to CN.

(1967) give averaged data for 155 non-variable stars observed an average of ~3 times each and 808 individual observations of 160 variable stars. In particular, several Mira variables were observed 20 or more times over 2 or 3 cycles, and these observations showed that Miras execute enormous loops — both clockwise and counter-clockwise — in diagrams of bandstrength *vs.* continuum color (Spinrad & Wing 1969; Wing & Yuan 1998).

Four examples of scanner "spectra" are shown in Figure 1. At the left are scans of *o* Cet obtained at maximum and minimum light, when the observed spectral types were M5.5 and M9.0, respectively. They illustrate the well-known fact that as an M–type Mira gets fainter, it also gets cooler, as indicated not only by the growth of the TiO and VO bands but also by the considerable change in continuum slope. At the right is shown a similar pair of scans of S Cep, a C–type Mira, also observed close to the times of maximum and minimum visual light. In this case nearly all the absorption is due to the red system of CN. Note that the maximum-light and minimum-light spectra are nearly identical. It is well known that CN is not very sensitive to temperature and that the spectra of C–type Miras change relatively little with phase; but what is surprising is that the observed continuum slope hardly changed either, suggesting that the temperature itself remained constant.

It is hard to imagine how a Mira can change in infrared brightness by a factor of two or more without an accompanying change in temperature. It seems to me that the calibrated scans of Fig. 1 are telling us that we have a problem in the measurement of color temperatures for either M–type or C–type stars (or both), and consequently that the color temperatures obtained for M and C stars by blackbody continuum fitting do not successfully place them on the same scale. Possibly the residual CN absorption in the continuum points varies with temperature in just such a way as to cancel the effect of a changing blackbody continuum slope. This question can, I hope, be settled by synthetic spectrum calculations.

The reason for discussing these old observations here is that they are, I believe, still useful. Or, I should say, they have recently become useful again, after a long lapse. During the 1970s all 75 nights of scanner observations were re-reduced with the help of improved standard fluxes and a new absolute flux calibration. Then, with the changes in technology that occurred in the late 1970s, the punched cards containing these data became unreadable, and for 15 years I did nothing further with them, concentrating instead on the filter photometry described in the next section. Recently, however, my university's computer center briefly acquired a refurbished card-reader so that people like me could have "one last chance" to transfer their data to magnetic media. At the same time I became aware of the great advances being made in the atmospheric modeling of cool stars, making possible many interesting applications of the data that had not been feasible before. I therefore went to considerable pains to coax several tens of thousands of 30-year-old punched cards through the card reader, and I am happy to report that my scanner observations have been fully restored. All scans reported in Wing (1967), as well as those acquired subsequently, are now accessible at the terminal — no more hauling dollyfuls of cards to the computer center across campus — and can be sent to interested colleagues by e-mail. Now at last it should be possible to carry out some of the objectives for which these observations were originally intended — in particular, that of placing the color temperatures of M and C stars on a single scale that is independent of molecular band strength.

## 3. Eight-Color Photometry

Since 1969, without ready access to a spectrum scanner, I have used narrow-band filter photometry to pursue similar objectives. My eight-color photometric system (Wing 1971), which uses interference filters of $\sim 50$ Å bandwidth, is largely based on the 27-color scanner program and is intended to measure continuum colors and bandstrengths of TiO, VO, and CN in

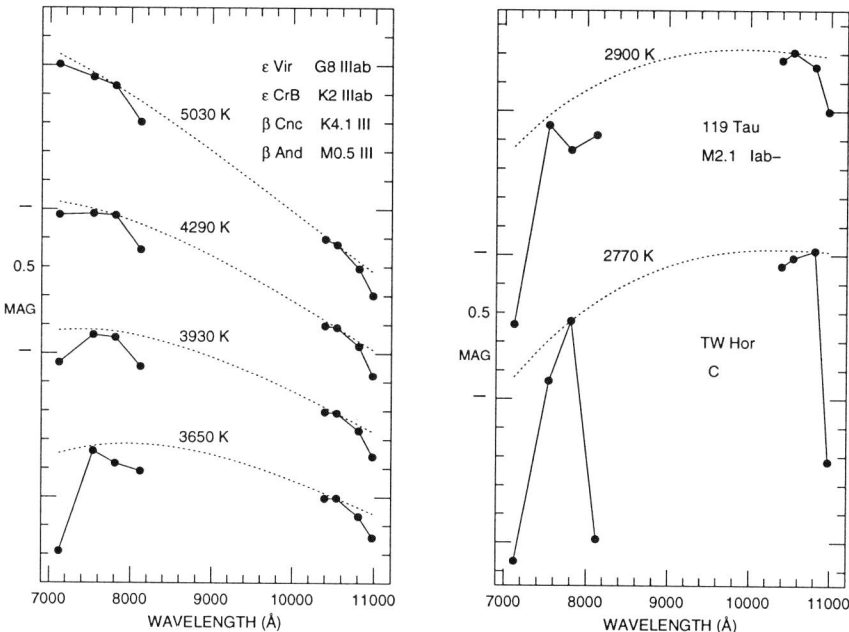

*Figure 2.* Eight-color observations of giants of types G, K, and M, an M supergiant, and a carbon star. The C-type spectrum is essentially a map of CN absorption; TiO, when present, depresses filters 1 and 3. Data are reduced to an absolute flux system and plotted with arbitrary vertical shifts. Each spectrum is compared to a blackbody curve of the indicated temperature.

the 0.71 − 1.10 $\mu$m spectra of M and C stars. In reducing the number of wavelengths from 27 to 8, it did not seem possible to do a good job of studying the more varied spectra of stars of type S; the idea was to include the best continuum points and the strongest molecular bands in both M and C stars, and this can indeed be done with 8 filters in all but the most extreme cases. The collection and use of data on this system has been my main observational activity for nearly 30 years. I don't have an up-to-date count, but an estimate made several years ago indicated that approximately 30 000 observations of about 15 000 different stars had been made on this system — mostly by the writer, but with significant contributions from 8 or 9 collaborators, most notably N. M. White, J. H. Baumert, S. B. Yorka, and E. Costa. Details of the photometric system and its associated calibrations are given in MacConnell, Wing & Costa (1992).

The information content of the eight-color photometry is best described with the help of some plotted spectra (Figure 2). It is not customary for photometrists to refer to their data as "spectra," but since this photometry is expressed on an absolutely calibrated system, a plot of the observed magnitudes *vs.* wavelength is indeed a simple spectrum, and spectroscopic (reddening-free) quantities can be derived by measuring the depression of any point, in magnitudes, with respect to the continuum at that wavelength.

In G and K stars, through type K3, the only molecule affecting the eight-color spectrum is CN. This is also nearly true of most carbon stars, so that, for example, the K2 and C stars in Fig. 2 have the same characteristic shape apart from overall differences in band strength and temperature. The K4 and M0 giants and the M2 supergiant, however, are affected by TiO as well as CN; in these cases the continuum does not pass through any of the first four filters because it has been set by an iterative procedure which allows for residual CN absorption in filter 2 to obtain a better continuum point than the TiO–contaminated filter 3.

For normal M stars of all luminosities, the eight-color system is well-suited to providing high-quality, two-dimensional spectral classifications, due to the fortunate circumstance that the two strongest molecular band systems in the 1 $\mu$m region, those of TiO and CN, are sensitive to temperature and luminosity, respectively. Classifications on the eight-color system are considerably improved over what could be obtained from the 27-color scans because the filter system extends to a shorter wavelength to include the strong $\gamma(0,0)$ band of TiO near 0.71 $\mu$m. Wing and Yorka (1979) have shown that the eight-color classifications of M–type giants are in excellent agreement with modern MK classifications. For the M dwarfs a new scale of spectral types, consistent with the giant scale, was established by means of the eight-color photometry (Wing 1973) and subsequently incorporated into the MK system by Boeshaar (1976). For M–type supergiants, the eight-color photometry has proved particularly useful: since the CN-based luminosity class can be calibrated in terms of absolute magnitude, while the reddening (and hence absorption) can be inferred from comparison of color and spectral type, the distance can be calculated from the photometric data alone. MacConnell et al. (1992) are taking advantage of this possibility to estimate distances to more than 1000 newly-discovered M–type supergiants in the southern Milky Way, including some which evidently lie in distant, highly obscured, spiral arms.

In the special case of stars that can be assumed to be unreddened, the eight-color photometry permits a different, and rather interesting, kind of two-dimensional classification. For M giants near the South Galactic Pole, the writer showed that the eight-color measures of TiO, CN, and continuum color can be used to classify the stars according to metallicity and O/C ratio (Wing 1989a,b). The actual calibration of photometric indices in terms of metallicity and O/C has not been attempted, however, but could be done with the help of synthetic spectra.

Eight-color observations of approximately 350 carbon stars were used

TABLE 1. Data Available on Eight-Color System

| Approx. No. of Objects | Program |
| --- | --- |
| 2000 | Bright Stars (including the 1000 brightest at 1 $\mu$m) |
| 1000 | IRC stars (mostly M giants) |
| 364 | C stars and related objects (1) |
| 150 | Ba II stars, S stars |
| 150 | MK–classified M supergiants (2) |
| 400 | K–M supergiants in LMC |
| >1000 | new galactic M supergiants (3) |
| 183 | M giants near the South Galactic Pole (4) |
| 150 | dM stars within 10 pc |
| 300 | LHS stars (mostly dM's), searching for dC's (5) |

References: (1) Baumert (1972); (2) White & Wing (1978); (3) MacConnell et al. (1992); (4) Wing (1989a); (5) Wing & MacConnell (2000)

by Baumert (1972, 1974) to provide color temperatures and near-infrared magnitudes for a new determination of their absolute magnitudes from statistical parallax. Other applications of this large data set have not been exploited, however. Comparisons with synthetic spectra based on modern model atmospheres should allow determinations of effective temperatures and molecular abundances.

Stars of type S are awkward to study by narrow-band photometry because of the large number of molecules affecting the near-infrared spectrum. In fact, I had decided not to consider the S stars at all when choosing filters for the eight-color system. Consequently, the system in its normal form has no explicit measure of either ZrO or LaO, and several of the filters are affected by bands which were long unidentified but are now known to be due to ZrS (Hinkle, Lambert & Wing 1989). Piccirillo (1977), however, decided to "take the bull by the horns:" noting that two of the filters of the normal eight-color system (filters 4 and 6) are not particularly useful for S stars, he replaced them with new filters designed to measure ZrO at 6510 Å and LaO at 7945 Å. In favorable cases, this modified eight-color system produces an infrared magnitude, a continuum color, and column densities for 5 molecules: TiO, ZrO, LaO, CN, and ZrS (Piccirillo 1980).

Table 1 is a summary of the observing programs that have been carried out with the (normal) eight-color photometric system. It does not include studies of individual objects or members of open or globular clusters, but only the programs which involved fairly extensive observing lists. In most cases the numbers of objects are not actual counts but estimates intended to indicate the scope of the project. Most of these data are unpublished but can be made available for collaborative studies.

A noteworthy characteristic is this observational effort is that nearly all of it has been done with small telescopes. During the first decade of this work, observations were made with an S-1 detector and were limited to rather bright stars. Although notoriously slow, good S-1 cells are extremely stable and, when used at good sites, produce excellent photometry. Early data were obtained primarily at Lowell Observatory and KPNO in the northern hemisphere and at CTIO in the southern; a considerable fraction of the observations of bright stars were made with telescopes of only 0.4-m aperture. Since about 1980, most of this work has been done at CTIO with the much more sensitive Varian LSE photocell. Recent programs, mostly using the 1.0-m telescope during bright time, have included a study of the red supergiants in the LMC, a survey for galactic red supergiants, and a search for dC stars among Luyten proper-motion stars. With the widespread closure of small telescopes in recent years and the greatly decreased availability of aperture photometers, such data are becoming increasingly difficult to obtain, and existing data are becoming correspondingly more valuable.

## 4. Continuum Measurements in the 1–4 $\mu$m region

The 1–4 $\mu$m spectra of late-type stars contain absorption bands from numerous molecules, and several of these have been studied by means of narrow-band photometry. Frogel and colleagues (Frogel et al. 1978; Aaronson et al. 1978; Cohen et al. 1978) showed that narrow-band indices of CO and $H_2O$ in the 2 $\mu$m region, when combined with wide-band $JHK$ photometry, produce useful discriminants for population synthesis studies. Rinsland & Wing (1982) used a spectrum scanner to make narrow-band measures of the first-overtone SiO bands near 4 $\mu$m, and Noguchi et al. (1981, 1986) have used spectrophotometry to measure the 3.1 $\mu$m complex due to HCN and $C_2H_2$ in cool carbon and SC stars. On the other hand, the lines of some other important molecules such as $H_2$ and OH are too scattered to be easily measured by this technique.

Measurements of the 1–4 $\mu$m continuum, when properly calibrated, are also of great interest. Since $H^-$, the dominant opacity source throughout the 1–4 $\mu$m region in nearly all cool stars, has its opacity minimum at 1.65 $\mu$m, its effect is to distort the shape of the continuum, making it measurably different from a blackbody curve. Observationally, however, these distortions have proved difficult to measure quantitatively because the energy distributions need to be absolutely calibrated and molecular line absorption must be avoided. Wing & Rinsland (1981) used a cooled grating spectrometer with an InSb detector at Kitt Peak to isolate 13 line-free

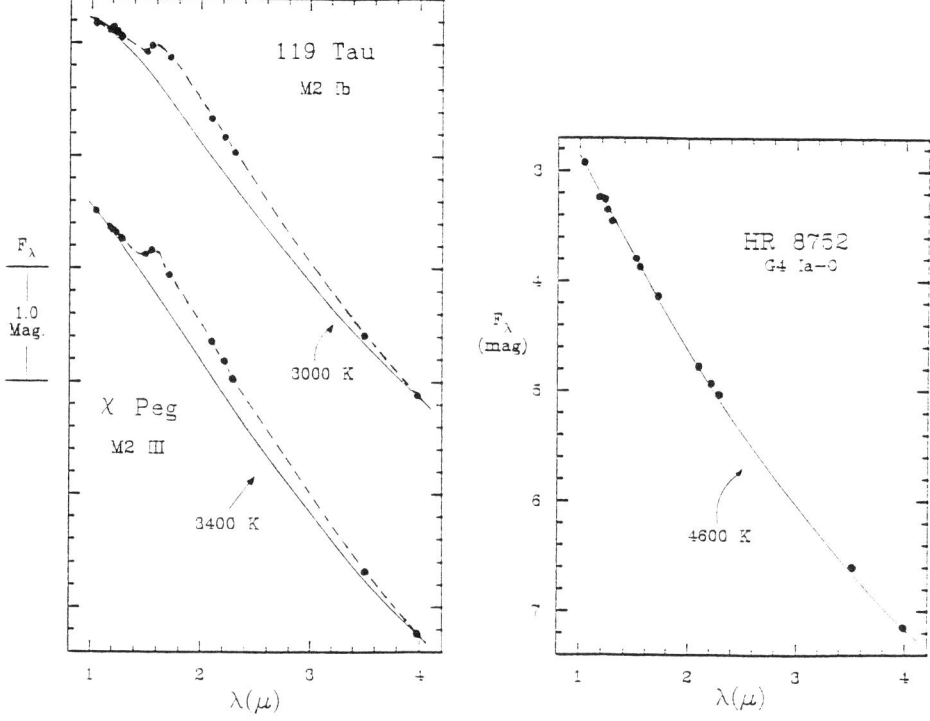

*Figure 3.* Energy distributions in the 1–4 μm region, as measured with a spectrum scanner at 13 clean continuum points. Dashed curves are hand-drawn through the observed data; solid curves are blackbody curves fitted to the endpoints at 1.04 and 4.00 μm. The differences are due almost entirely to the effects of $H^-$ opacity. Note that HR 8752 shows no sign of $H^-$ and is evidently H-poor.

continuum points between 1.04 and 4.00 μm and found that their calibrated continuum spectra agreed well with model continua, clearly showing the effects of $H^-$. Figure 3 shows three spectra obtained on this 13-color narrow-band continuum program. The two stars in the left panel, a supergiant and a giant of type M2, both show clear deviations from a blackbody curve caused by $H^-$. Similar deviations were found in several chemically-peculiar stars, including C stars, S stars, and Ba II stars. However, as pointed out by Wing & Saizar (1989) and Wing (1991), some cool stars do not show any effect of $H^-$, probably because of hydrogen deficiency. These include the HdC star HD 182040, the RV Tauri variables R Sct and U Mon, the RCB stars R CrB and XX Cam, and the luminous supergiant HR 8752 (Fig. 3, right panel). Although HR 8752 has often been cited as one of the most luminous stars in the Galaxy, its infrared energy distribution suggests that it is not a normal supergiant. Rather, the blackbody shape of its 1–4 μm spectrum seems appropriate for the hydrogen-poor envelope of a post–AGB star.

## 5. Suggested Applications

Narrow-band photometry can be used to compare real stars to model stellar atmospheres, and I have already alluded several times to the desirability of comparing the existing data to synthetic colors based on model atmospheres. Indeed, any quantitative interpretation of photometry must be based on modelling of some kind. At the same time, such comparisons can also be used to test the models themselves.

For many years, the state of models for cool stars with strong molecular bands was not sufficiently advanced to encourage detailed comparisons with observation. Simply stated, if the spectra computed from models don't look like real spectra, one does not feel inclined to use them to derive effective temperatures, abundances, etc. In recent years, however, the situation has changed dramatically. Improved methods for computing molecular opacities (e.g. Johnson 1986), including non-LTE effects (Johnson (1994) and greatly expanded molecular databases (Jørgensen 1995), have resulted in new grids of models for late-type stars (see, e.g. Gustafsson & Jørgensen 1994; Plez 2000) which are believed to be good representations of the atmospheric structure and which generate realistic-looking synthetic spectra. It is then a simple matter to multiply the computed spectra by the response functions of any photometric system to produce "synthetic photometry" that can be compared directly to observation.

I believe that the data generated by the programs described here are ideally suited for comparisons with synthetic spectra, not only because they are internally precise and carefully calibrated to an absolute flux system but also because the filters (or scanner bandpasses) have been carefully chosen. Absorption bands are measured at their optimum positions, and each photometric system includes the best available continuum points.

When a good match is found between the synthetic colors of a model and the observed colors of an unreddened star, the effective temperature of the model can be considered the effective temperature of that star. Such determinations of $T_{\text{eff}}$ may not be as "fundamental" as determinations based on angular diameters, but they are likely to be more accurate. In most cases, effective temperatures differ significantly from the color temperatures provided directly by the photometry because of the effects of $H^-$ and other opacity sources affecting the continuum, and these $T_{\text{eff}}$ determinations are therefore needed to calibrate the color temperatures. Eight-color observations of bright standard stars have already been used to improve the $T_{\text{eff}}$ scale for K and M giants (Bell & Gustafsson 1989), and suitable data are now available for similar determinations of the $T_{\text{eff}}$ scales for M-type dwarfs and supergiants, as well as individual values for hundreds of C stars and other chemically-peculiar late-type stars. Fitting the molecular band measurements with synthetic colors will convert them into column densities, leading to elemental abundances and possibly gravity indicators. We may even learn how a C–type Mira (Fig. 1) can manage to have the same color temperature at maximum and minimum light!

## References

Aaronson, M., Frogel, J. A. & Persson, S. E. 1978, *ApJ*, 220, 442
Baumert, J. H. 1972, Ph.D. dissertation, Ohio State University
Baumert, J. H. 1974, *ApJ*, 190, 85
Bell, R. A. & Gustafsson, B. 1989, *MNRAS*, 236, 653
Boeshaar, P. C. 1976, Ph.D. dissertation, Ohio State University
Cohen, J. G., Frogel, J. A. & Persson, S. E. 1978, *ApJ*, 222, 165
Frogel, J. A., Persson, S. E., Aaronson, M. & Matthews, K. 1978, *ApJ*, 220, 75
Gustafsson, B. & Jørgensen, U. G. 1994, *Astron. Astrophys. Rev.*, 6, 19
Hinkle, K. H., Lambert, D. L. & Wing, R. F. 1989, *MNRAS*, 238, 1365
Johnson, H. R. 1986, in *The M-Type Stars*, ed. H. R. Johnson and F. R. Querci, NASA SP–492, p. 323
Johnson, H. R. 1994, in IAU Coll. 146: *Molecules in the Stellar Environment*, ed. U. G. Jørgensen (Springer-Verlag), p. 234
Jørgensen, U. G. 1995, in *Astrophysical Applications of Powerful New Databases*, ed. S. J. Adelman and W. L. Wiese, ASP Conf. Ser., 78, 179
MacConnell, D. J., Wing, R. F. & Costa, E. 1992, *AJ*, 104, 821
Noguchi, K. & Akiba, M. 1986, *Publ. Astron. Soc. Japan*, 38, 811
Noguchi, K., Kawara, K., Kobayashi, Y., Okuda, H., Sato, S. & Oishi, M. 1981, *Publ. Astron. Soc. Japan*, 33, 373
Piccirillo, J. 1977, Ph.D. dissertation, Indiana University
Piccirillo, J. 1980, *MNRAS*, 190, 441
Plez, B. 2000, in IAU Symp. 177: *The Carbon Star Phenomenon*, ed. R. F. Wing (Kluwer), p. 71
Rinsland, C. P. & Wing, R. F. 1982, *ApJ*, 262, 201
Spinrad, H. & Wing, R. F. 1969, *Ann. Rev. Astron. Astrophys.*, 7, 249
White, N. M. & Wing, R. F. 1978, *ApJ*, 222, 209
Wing, R. F. 1967, Ph.D. dissertation, Univ. of California, Berkeley
Wing, R. F. 1971, in *Proc. Conf. on Late Type Stars*, ed. G. W. Lockwood and H. M. Dyck (KPNO Contr. No. 554), p. 145
Wing, R. F. 1973, in IAU Symp. 50: *Spectral Classification and Multicolour Photometry*, ed. Ch. Fehrenbach and B. E. Westerlund (Reidel), p. 209
Wing, R. F. 1989a, in *The Gravitational Force Perpendicular to the Galactic Plane*, ed. A. G. D. Philip and P. K. Lu (L. Davis Press), p. 167
Wing, R. F. 1989b, *Newsletter of Chemically Peculiar Red Giant Stars*, No. 6, p. 7
Wing, R. F. 1991, in *The Infrared Spectral Region of Stars*, ed. C. Jaschek and Y. Andrillat (Cambridge), p. 301
Wing, R. F. & MacConnell, D. J. 2000, in IAU Symp. 177: *The Carbon Star Phenomenon*, ed. R. F. Wing (Kluwer), p. 589
Wing, R. F. & Rinsland, C. P. 1981, *Rev. Mexicana Astron. Astrof.*, 6, 145
Wing, R. F. & Saizar, P. 1989, in IAU Coll. 106: *Evolution of Peculiar Red Giant Stars*, ed. H. R. Johnson and B. Zuckerman (Cambridge), p. 151
Wing, R. F. & Yorka, S. B. 1979, in IAU Coll. 47: *Spectral Classification of the Future*, ed. M. F. McCarthy, A. G. D. Philip and G. V. Coyne (Vatican Observatory), p. 519
Wing, R. F. & Yuan, Y. 1998, in *Pulsating Stars: Recent Developments in Theory and Observation*, ed. M. Takeuti and D. D. Sasselov (Universal Academy Press), p. 113

Four participants. Visual impressions by Pierre North.

# SPECTRAL CLASSIFICATION OF CARBON–PECULIAR G STARS

TUBA KOKTAY
*Astronomy and Space Science Department*
*Istanbul University*
*Istanbul, Türkiye*

AND

R. F. GARRISON
*David Dunlap Observatory*
*University of Toronto*
*Toronto, Canada*

**Abstract.** Spectroscopic observations at classification dispersion have been obtained for a set of stars having abnormal photometric indices on the Strömgren *uvby* system. Five of the stars observed have extreme carbon anomalies; two of these spectra are illustrated here.

## 1. Introduction

During a *uvby* photometric survey of ten thousand solar-type stars, Olsen (1993, 1994) isolated approximately 200 stars with uniquely peculiar photometric indices. We have observed $\sim 120$ of these stars spectroscopically at MK resolution (1–3 Å) in order to look for spectral anomalies which might account for their abnormal colors.

## 2. Spectroscopic Observations

The stars were observed with a CCD spectrograph on the 60-cm Helen Sawyer Hogg Telescope of the University of Toronto Southern Observatory on Cerro Las Campanas in Chile. The detector is a PM512 CCD with 20 micron pixels coated with MetaChrome II for improved blue response. The resolution can be either 1.7 Å (0.85 Å/pix) or 3.4 Å (1.7 Å/pix), with most of the spectra taken in the wavelength regions 4200–4600 Å or 3800–4900 Å, respectively.

Some of the stars were also observed at higher resolution with the 1024×1024 Thomson CCD and cassegrain spectrograph on the 74-inch (1.88-m) telescope of the David Dunlap Observatory. This combination gives approximately 0.2 Å/pix.

## 3. Results

Classification spectra have been taken for approximately 120 of the stars selected by Olsen as having very peculiar *uvby* indices. Five of these stars are found to have extreme carbon anomalies. One of the most interesting is HD 204848, which is a weak-lined, carbon-rich star with a preliminary classification of K0 III: Fe–2.5, CN–3, CH+1, Ba+1. This classification indicates that it is metal-weak, carbon-strong and nitrogen-weak. There are a few other stars in the sample such as HD 207687 which have similar characteristics, and others with quite different carbon-peculiarities.

HD 204848 and HD 207687 have also been observed with the DDO 1.88-m telescope. These DDO spectra from 4000 to 4600 Å are shown in Figure 1, along with similar spectra of the Sun (actually the asteroid Vesta) and the K0 III standard HD 197989.

## 4. Discussion

The advantage of studying G stars is that they are relatively easy to compare to the Sun. In G-type dwarfs the interiors are relatively simple and, in principle, they have not yet suffered any dredge-up to complicate the interpretation of the abundances. However, if they are giants, all is not so simple. Some of the spectra violate our usual concepts of the patterns of spectroscopic peculiarities. HD 204848, for example, exhibits a very strong CH band with a general weakening of the iron lines. At the same time, it shows virtually no CN, yet other features indicate that it is probably above the main sequence, though it is difficult to be sure (hence the colon) because normally luminosity-sensitive lines of Sr II and Ba II are affected by the same dredge-up process which probably produces the excess carbon, and thus their strength is not a reliable indicator of surface gravity. However, the luminosity class based on the 4376/4383 ratio, which should be relatively independent of abundance, indicates that the star is probably a subgiant or giant.

Photometric temperatures and CN indices for several of these stars are presented in an accompanying poster (Wing, Garrison & Koktay 2000). A spectroscopic analysis to determine the model parameters and composition of HD 204848 and HD 207687 is in progress by Koktay.

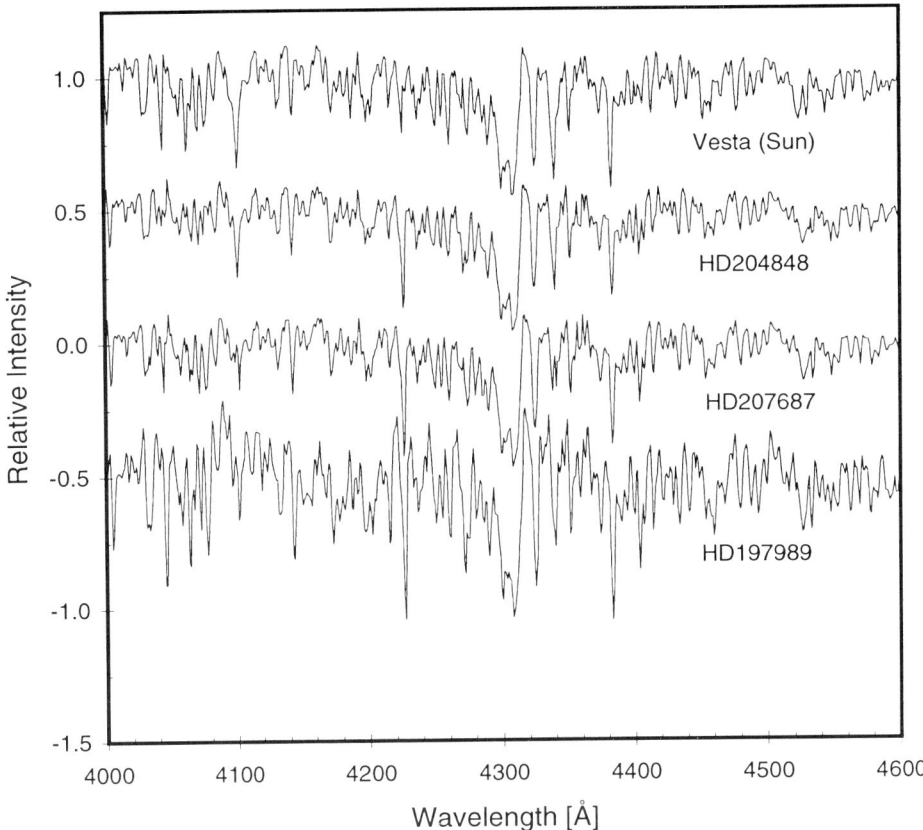

*Figure 1.* Spectra of normal and carbon-peculiar stars. From top to bottom: (a) Vesta (G2 V), (b) HD 204848 (K0 III: Fe–2.5, CN–3, CH+1, Ba+1), (c) HD 207687 (K0 III: Fe–2, CN–3, CH+2, Ba+1), and (d) the standard HD 197989 (K0 III). For clarity, the continuum levels of the spectra are offset by 0.5 units, where unity is equal to the difference between zero and pseudo-continuum.

We wish to acknowledge that the research of Koktay has been supported by a TÜBİTAK NATO A2 scholarship.

### References

Olsen, E. H. 1993, *A&A Supp.*, 102, 89
Olsen, E. H. 1994, *A&A Supp.*, 104, 429
Wing, R. F., Garrison, R. F. & Koktay, T. 2000, in IAU Symp. 177: *The Carbon Star Phenomenon*, ed. R. F. Wing (Kluwer), p. 588

# MASS-LOSING AGB STARS IN THE LMC

JACCO TH. VAN LOON AND ALBERT A. ZIJLSTRA
*European Southern Observatory*
*Garching bei München, Germany*

PATRICIA A. WHITELOCK
*South African Astronomical Observatory*

CECILE LOUP
*European Southern Observatory, Santiago, Chile*

AND

L.B.F.M. WATERS
*University of Amsterdam, The Netherlands*

**Abstract.** We show the results of an infrared study of a sample of heavily obscured AGB stars in the LMC. Both carbon-rich and oxygen-rich mass-losing AGB stars can be found at both high and low luminosities, but the percentage of carbon stars decreases with increasing luminosity. The optical depth of the circumstellar envelopes also decreases with increasing luminosity, while the mass-loss rates are (nearly) constant with luminosity. We also show tentative evidence for having found the first post-AGB stars in the LMC.

## 1. Introduction

When stars with ZAMS masses in the range $1 - 8$ $M_\odot$ evolve on the AGB, they enter a phase of thermal pulsations in the stellar interior. The convective mixing of the stellar interior that occurs during these thermal pulses can lead to a change in the carbon-to-oxygen ratio of the photosphere (the third dredge-up). In this way, the carbon-to-oxygen ratio of the photosphere can evolve from oxygen-dominated to carbon-dominated; a carbon star is formed. However, nuclear burning at the bottom of the convection mantles of the most massive AGB stars (Hot Bottom Burning) may convert carbon stars into nitrogen-overabundant S stars (see, e.g., Lattanzio in these proceedings for a discussion of the internal structure of AGB stars).

The thermal pulsing AGB is also the phase during which stars undergo strong photospheric pulsations, triggering the heaviest mass loss that these stars ever experience. With mass-loss rates typically of the order $10^{-6}$ to $10^{-4}$ $M_\odot \mathrm{yr}^{-1}$, mass loss dominates over nuclear burning, eventually terminating the star's evolution on the AGB. In the upper layers of the star's extended atmosphere dust grains form, and radiation pressure thereon drives the expelled matter away from the star. As the chemistry of the photospheres of carbon stars is carbon-dominated, so are their dusty circumstellar envelopes. This may lead to differences in the mass-loss characteristics between carbon stars and oxygen (M-type) stars (see, e.g., Fleischer et al., Sedlmayr & Winters, and Cherchneff, all in these proceedings, for a discussion on dust formation and mass loss from AGB stars).

The evolution of AGB stars can only be unravelled by studying a sample of stars with known luminosities, and covering the entire AGB. This is impossible to do in the Milky Way, but the Magellanic Clouds are perfect for it. Initially, LMC studies had been limited to optically bright AGB stars (e.g. Blanco et al. 1980; Westerlund et al. 1981), and it is only very recently that the first samples of heavily obscured AGB stars in the LMC have been compiled, based on IRAS data (Reid 1991; Wood et al. 1992; Zijlstra et al. 1996; for the SMC see the poster paper by Groenewegen & Blommaert, these proceedings). The results of ground-based near-infrared surveys, optical and near-infrared monitoring, mm-searches for maser emission, and notably mid-infrared photometry and spectroscopy by the ISO satellite are currently yielding a gigantic increase in the amount of data on mass-losing AGB stars in the LMC, expected to result in a similarly huge leap in our knowledge of mass loss and stellar evolution on the AGB.

## 2. Searching for Mass-losing AGB Stars in the LMC

Searches for near-infrared counterparts of IRAS point sources in the LMC resulted in the first samples of infrared stars in the LMC. However, besides mass-losing AGB stars, these could also be red supergiants, foreground stars, young stellar objects, or post-AGB stars. Zijlstra et al. (1996), mainly on the basis of luminosities and infrared colors, identified the mass-losing AGB stars among them. Following upon this result, we performed ground-based *JHKLN* photometry of these stars to derive accurate luminosities, and to estimate mass-loss rates and carbon-to-oxygen ratios from the infrared colors. In this analysis, all observations were reduced to a single epoch — *viz.* the epoch of the $N$-band measurement — and the sample will accordingly be referred to as the NOV94 sample.

We have attempted to extend the known sample of mass-losing AGB stars in the LMC by continuing the search for near-infrared counterparts of remaining IRAS point sources. Preliminary results of this sample extension (JAN96) are included in the present paper.

## 3. Identifying Mass-Losing Carbon and Oxygen AGB Stars

Guglielmo et al. (1993) showed that carbon and oxygen stars may be separated in a near/mid-infrared color-color diagram. Zijlstra et al. (1996) showed that in this way, a $K$–[12] versus $H$–$K$ diagram can be used to distinguish carbon from oxygen stars. A $K$–[12] versus $J$–$K$ diagram works even better, but it is more difficult to go sufficiently deep at $J$ to actually detect the most obscured AGB stars.

We show as Figure 1 the $K$–[12] versus $J$–$K$ diagram for the LMC stars of the NOV94 (squares) and JAN96 (solid circles) samples. The sequences of AGB stars in the Milky Way as used by Guglielmo et al. and Zijlstra et al. are given for carbon stars (dashed curve) and oxygen stars (dotted curve). The deviation of the oxygen sequence from a straight line is a consequence of the behavior of the 9.8 $\mu$m silicate feature in the circumstellar envelopes of these stars. It is clear that in some parts of the diagram carbon stars and oxygen stars are well separated, while in other parts there remains ambiguity. We can nevertheless take a semi-statistical approach in classifying stars as probably oxygen-rich, probably carbon-rich, or ambiguous.

## 4. Possible Post-AGB Stars in the LMC

In the JAN96 sample, a cluster of at least six stars have a location in the $K$–[12] versus $J$–$K$ diagram which is not consistent with either the carbon-star or the oxygen-star sequence. We argue that these stars may in fact be the first post-AGB stars in the LMC to have been identified as such.

Van der Veen et al. (1989) compiled a sample of presumed post-AGB stars in the Milky Way, adding (only) a few planetary nebulae and OH/IR stars. These are plotted in the $K$–[12] versus $J$–$K$ diagram as well (dots). The JAN96 cluster of deviating stars clearly falls in the region where post-AGB stars are located. Furthermore, the IRAS [12]–[25] colors for these JAN96 stars are redder than for the other stars of the JAN96 and NOV94 samples. This is expected for post-AGB stars: after departure from the AGB the mass-loss rate drops dramatically, resulting in a growing inner radius of the circumstellar dust envelope and, consequently, a decreasing temperature of the circumstellar dust.

The JAN96 post-AGB candidates could perhaps also be explained as being AGB stars that have experienced a helium shell flash but have not yet recovered from the resulting drop in luminosity and mass-loss rate. Candidates for this type of star in the Milky Way may be found in Whitelock et al. (1995), although with somewhat different infrared colors than ours.

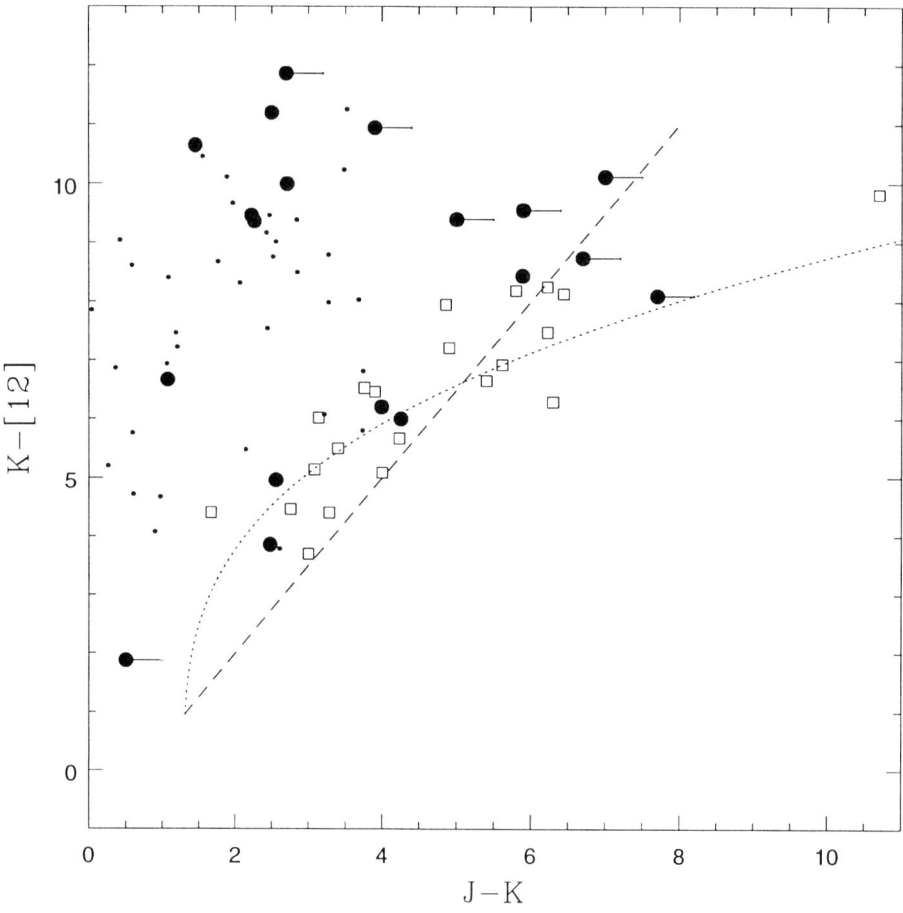

*Figure 1.* $K$–[12] versus $J$–$K$ color-color diagram for: AGB stars in the LMC, NOV94 (squares); IR stars in the LMC, JAN96 (solid circles); and post-AGB stars in the Milky Way, van der Veen et al. 1989 (dots). Mean relations for carbon stars in the Milky Way (dashed curve) and for oxygen stars in the Milky Way (dotted curve) are also shown. Bars attached to symbols indicate non-detection at $J$, i.e. lower limits to $J$–$K$.

## 5. Luminosities

Obscured AGB stars emit predominantly at near- and mid-infrared wavelengths. Obscured supergiants can be erroneously classified as normal AGB stars from optical and $I$-band data if infrared photometry is not available. The sensitivity of the IRAS satellite in the mid-infrared was near the limit needed for observing obscured AGB stars at the distance of the LMC. In the mid-infrared — and even more so in the near-infrared — these stars are highly variable, with amplitudes often exceeding one magnitude. For the NOV94 sample, the ground-based $N$-band data allow the construction of single-epoch near- and mid-infrared spectral energy distributions. From these, single-epoch luminosities can be derived. For the JAN96 sample, the spectral energy distributions are much less certain.

The luminosity distributions of the carbon and oxygen stars indicate that the ratio of carbon stars to oxygen stars decreases with luminosity. However, one of the most likely carbon stars in the sample is very luminous ($M_{\rm bol} \approx -6.8$). On the other hand, mass-losing AGB stars that are fainter (say, $M_{\rm bol} \approx -5$) are definitely not all carbon stars.

The preliminary results from the JAN96 sample confirm and improve upon the results from the NOV94 sample. Furthermore, the stars that we tentatively classify as post-AGB stars are all of relatively low luminosity ($M_{\rm bol} \approx -5$), consistent with fainter stars being both more numerous and longer living than brighter stars.

## 6. Mass-Loss Rates

Various methods exist for estimating mass-loss rates from AGB stars. However, they often disagree by an order of magnitude, and usually only an estimate for the *dust* mass-loss rate is obtained, the dust being only a minor constituent of the circumstellar matter.

We used two methods for estimating dust mass-loss rates from the NOV94 stars, using infrared colors: from $K$–[12] (Jura 1986; Whitelock et al. 1994), and from $H$–$K$ (Le Bertre 1987; Epchtein et al. 1990). When applying both methods, agreement is only obtained for the most luminous stars. For the fainter stars the $K$–[12] colors indicate greater mass-loss rates than the $H$–$K$ colors do. We explain this as a consequence of the winds from the fainter stars being optically thick, so that $K$–[12] is not a reliable measure of the mass-loss rate (Whitelock et al. 1994). The $H$–$K$ colors suggest that the mass-loss rates are consistent with no luminosity dependence, or at most a very slow increase of mass-loss rate with increasing luminosity. The total, gas+dust mass-loss rates are about $1-2 \times 10^{-5}$ $M_\odot {\rm yr}^{-1}$, similar to those of obscured AGB stars in the Milky Way (Whitelock et al. 1994).

The observed luminosity dependence of the optical depth of heavily obscured AGB stars is

$$\tau \propto L^{-k} \qquad (1)$$

with k $\approx$ 0.4–0.5. This is identical to model predictions for single, pulsating AGB stars by Elitzur & Ivezic (see their contributions in these proceedings). We show here that their formalism also applies to a sample of different stars.

## 7. Conclusions

The heavily obscured AGB stars in the LMC are shown all to have similar, high mass-loss rates. The winds of brighter stars become optically thinner as the inner radius of their dusty circumstellar envelopes increases. The higher the luminosity, the lower the percentage of carbon stars, but carbon stars can probably exist up to the very highest luminosities on the AGB, while oxygen stars remain abundant at lower luminosities on the AGB.

Our search for new near-infrared counterparts of IRAS point sources in the LMC has resulted in more mass-losing AGB stars. We have also tentatively discovered several post-AGB stars. If true, this opens exciting new possibilities for studying the transition between AGB stars and white dwarfs.

Jacco van Loon wishes to express his sincere thanks to the IAU and the Symposium organizing committee for travel and subsistence grants, which made his attendance possible. Además agradece a Dr. Montserrat Villar-Martín por las reuniones muy estimulantes, antes y después del congreso.

## References

Blanco, V. M., McCarthy, M. F. & Blanco, B. M. 1980, *ApJ*, 242, 938
Epchtein, N., Le Bertre, T. & Lépine, J. R. D. 1990, *A&A*, 227, 82
Guglielmo, F., Epchtein, N., Le Bertre, T., Fouqué, P., Hron, J., Kerschbaum, F. & Lépine, J. R. D. 1993, *A&A Supp.*, 99, 31
Jura, M. 1986, *ApJ*, 303, 327
Le Bertre, T. 1987, *A&A*, 176, 107
Reid, N. 1991, *ApJ*, 382, 143
van der Veen, W. E. C. J., Habing, H. J. & Geballe, T. R. 1989, *A&A*, 226, 108
Westerlund, B. E., Olander, N. & Hedin, B. 1981, *A&A Supp.*, 43, 267
Whitelock, P., Menzies, J., Feast, M., Marang, F., Carter, B., Roberts, G., Catchpole, R. & Chapman, J. 1994, *MNRAS*, 267, 711
Whitelock, P., Menzies, J., Feast, M., Catchpole, R., Marang, F. & Carter, B. 1995, *MNRAS*, 276, 219
Wood, P. R., Whiteoak, J. B., Hughes, S. M. G., Bessell, M. S., Gardner, F. F. & Hyland, A. R. 1992, *ApJ*, 397, 552
Zijlstra, A. A., Loup, C., Waters, L. B. F. M., Whitelock, P. A., van Loon, J. Th. & Guglielmo, F. 1996, *MNRAS*, 279, 32

## Discussion

**Linsky**: How does the wind velocity enter into the mass-loss rate equation, and what value of the velocity is assumed?

**van Loon**: The dust mass-loss rate as derived from the optical depth at 1 $\mu$m (which we estimate from the $H-K$ color) is a linear function of the expansion velocity of the dust envelope. We assume 10 km s$^{-1}$. On the

AGB we do not expect expansion velocities that differ by more than some 50 % from this assumed value.

**Feast**: Presumably it is necessary to know whether the star is C- or O-rich before one can derive a mass-loss rate from, for example, an $H-K$ color, since the opacities of carbon-rich and oxygen-rich grains are different.

**van Loon**: That is absolutely true. The mass-loss rate we obtain from the $H-K$ color is assuming that it is a carbon star, just as many of the other methods for the determination of mass-loss rates have been designed for carbon stars. We get good estimates for the optical depth of the envelope at 1 $\mu$m for our stars, but to go from these optical depths to mass-loss rates is in principle different for oxygen- and carbon-rich stars.

**Sedlmayr**: Similar theoretical relations connecting the parametric behavior of basic quantities (like $\tau$ vs. $v_{\exp}$) have been derived by Dominik some years ago. How do your empirical relations fit to these theoretical findings?

**van Loon**: Unfortunately it is very difficult to obtain all relevant stellar parameters for any sample of AGB stars. For the LMC we have the advantage of being able to obtain absolute luminosities, which are crucial for deriving many other stellar parameters from observable quantities. We are just beginning to explore the parameter space for the mass-losing AGB stars in the LMC. Therefore we can test very global and direct relationships between different stellar parameters, but to test more detailed and complex relationships is difficult at present. We are working on it, though.

**Hron**: With your multiband photometry (including 12 $\mu$m) you could try to fit the energy distributions by two blackbodies to separate C and M stars and estimate mass-loss rates. This technique has been applied successfully to field stars by Kerschbaum et al. (see their two poster abstracts in these Proceedings).

**van Loon**: In general the spectral energy distribution is not well represented by the superposition of two blackbodies. Sometimes it is not too bad a fit to the data, but then the physical parameters derived from the blackbodies still may not always be accurate, and they can even be quite different from the real physical parameters. This method should definitely be calibrated by comparison with the physical parameters found by more correct methods, before being applied. That Kerschbaum et al. show it to work for their sample might be the result of the fact that their sample is limited to AGB stars with fairly moderate mass-loss rates: they are not very much dust-enshrouded.

# Session IV

## VARIABILITY

Janet Mattei, a Turkish student, Zeki Aslan, and Tom Lloyd Evans at the construction site of the Kazan University 1.5-m telescope on Bakırlıtepe. The telescope is now in operation; the observatory itself was officially dedicated on Sept. 5, 1997.

# TREND ANALYSIS OF
# 51 CARBON LONG-PERIOD VARIABLES

JANET AKYÜZ MATTEI AND GRANT FOSTER
*American Association of Variable Star Observers*
*Cambridge, MA 02138, U.S.A.*

**Abstract.** We have performed a trend analysis of 51 long-period variables (LPVs) of spectral types C, R, N, and S using 90 years of AAVSO data. We studied the periods and amplitudes, as well as the fall time (the time from maximum light to minimum), the rise time (from minimum to maximum), and the magnitudes at maximum and minimum. We also looked for time evolution of period and amplitude in the light curves themselves. The periods are more stable than the other parameters, with longer-period stars more likely to show period fluctuations than shorter-period stars. Fall time and rise time tend to evolve oppositely (mirror evolution), keeping the period fairly constant, whereas magnitude at maximum and minimum often evolve together (parallel evolution). About half of these stars are getting fainter, especially at maximum light, showing a secular dimming of magnitude at maximum; *none* shows a secular brightening.

## 1. Introduction

The purpose of the American Association of Variable Star Observers (AAVSO) is to coordinate variable star observations made largely by amateur astronomers worldwide; to collect, evaluate, process, and archive them, and to publish and disseminate them to researchers around the world. Over 8 million observations have been compiled since the founding of AAVSO in 1911; these make up the *AAVSO International Database*. Data from 1961 to date have been digitized and processed, and are accessible. The earlier data which have been digitized are being processed and will be added into the database within one year, thus creating the longest computer-readable variable star database in the world. Annually, over 300,000 observations are submitted to the AAVSO from observers worldwide for inclusion in the database. These observations are digitized, processed, and subjected to quality control checks to ensure the highest level of reliability.

The AAVSO receives over 300 requests each year from astronomers and educators for data and services to help schedule observing runs using ground-based and satellite telescopes, to provide simultaneous optical coverage of observing targets and immediate notification of their activity during particular satellite observing programs, to correlate multi-wavelength data, and to carry out collaborative research to analyze the long-term behavior of variable stars.

## 2. Data

The AAVSO Observing Program contains 4,467 stars, of which 1,967 (44%) are *Mira* and *semiregular* type LPVs. Of these, 97 of 1361 Mira variables (7%) and 96 of 606 semiregular variables (16%) are carbon stars or S stars (spectral types C, N, R, and S).

The objective of this study is to search for trends in light curve parameters, such as period, amplitude, rise and fall time, etc. of 51 periodic carbon stars for which long-term data, i.e. over 90 years, exist in the AAVSO database. For this study we have utilized two sets of AAVSO data. The first is individual observations defining the light curve from JD 2,437,600 (October 1961) to 2,450,000 (October 1995). For a few of the more interesting stars, we extracted from the archives AAVSO data covering a longer time span, typically beginning about JD 2,420,000 (August 1913); we are preparing papers on detailed study of some of the most interesting individual stars.

For the second set of data we have utilized observed dates and magnitudes of maxima and minima, going back to 1900, that have been determined homogeneously by the AAVSO using Pogson's method. This practice was begun by Leon Campbell in 1926 (Campbell 1926) and continued by successive AAVSO directors (Campbell 1955, Mattei et al. 1975). In this data set the basic data determined are: $T_n$, time of maximum for cycle $n$; $M_n$, magnitude at maximum; $t_n$, time of minimum; and $m_n$, magnitude at minimum. These enable us to define the following *derived parameters* for each cycle: $P_n = T_{n+1} - T_n$, period; $F_n = t_n - T_n$, fall time; $R_n = T_{n+1} - t_n$, rise time; and $A_n = M_n - m_n$, amplitude (see Figure 1). For each star we studied the six parameters $P_n$, $F_n$, $R_n$, $M_n$, $m_n$, and $A_n$. The time series for these derived parameters constitute our second basic data set, and yield a very clear picture of the changes of the star's behavior over time.

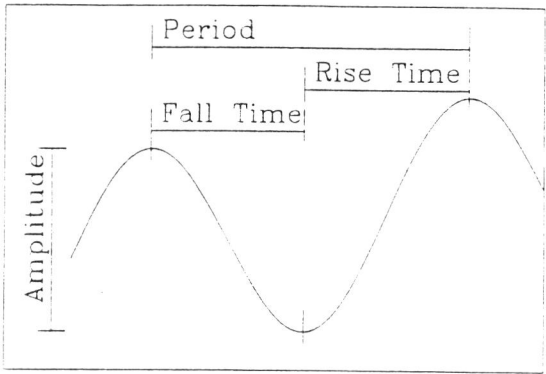

*Figure 1.* Derived parameters studied herein: Period (time from max to max), fall time (from max to min), rise time (from min to max), and amplitude.

## 3. Analysis Methods

Because we have two different types of data (derived parameters, and the individual observations themselves), we need two different approaches to search for time evolution of the fluctuations.

We searched for trends in the time series of the derived parameters by fitting low-order orthogonal polynomial functions of cycle number, from $1^{st}$ degree (linear) to $12^{th}$ degree. The lower-order polynomials detect very simple trends, while higher-order polynomials detect more complex trend patterns. Each polynomial fit generates a $\chi^2$ test statistic, and all tests were evaluated at the 95% confidence level. A statistically significant fit indicates, first and foremost, that the data are *not* merely random fluctuations about a constant mean value, *i.e.* it establishes the existence of time evolution in the data. Second, the highest polynomial degree which detects significant signal indicates the *complexity* of the trend; a linear fit is the simplest trend shape, but a $12^{th}$-order polynomial can model a very complex signal shape. Hence we recorded, for each variable, the highest significant polynomial degree. We emphasize that the lack of any statistically significant fit does *not* establish the absence of a trend; a trend might be present, but with a signal strength too low to detect. We also used the results of the $1^{st}$-degree polynomial fits (linear regression) to note the presence of secular trends.

To analyze the light curve for periodicity we applied two techniques. First, we used the CLEANEST Fourier spectrum (Foster 1995). It is designed to compensate for the difficulties associated with uneven time spacing *and* is capable of describing the time evolution of the period and amplitude. For its statistical treatment, we adopted the stricter statistical standards outlined in Foster (1996a, 1996b). Second, we used wavelet analysis. Traditional wavelet analysis is quite good for quantifying period and amplitude changes

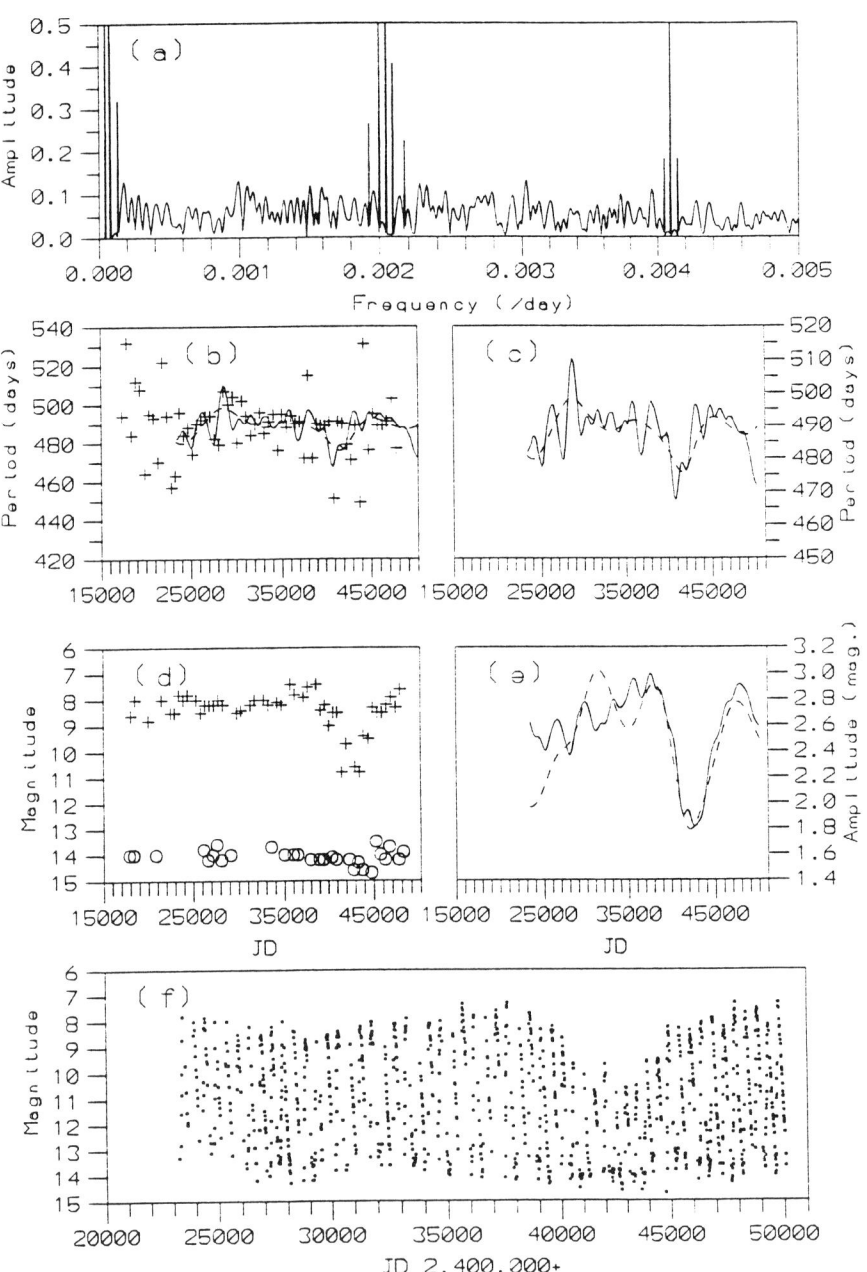

Figure 2. Results for W Aql. (a) CLEANEST Fourier amplitude spectrum. (b) Period from 90 years of AAVSO maxima and minima data (*plus signs*) compared to period analysis from CLEANEST (*dashed line*) and WWZ (*solid line*). (c) Period from CLEANEST (*dashed line*) and WWZ (*solid line*) using individual observations. (d) Magnitudes at maxima (*plus signs*) and minima (*circles*) from 90 years of AAVSO maxima and minima data. (e) Amplitude from CLEANEST (*dashed line*) and WWZ (*solid line*). (f) Light curve of 10-day averages of individual observations.

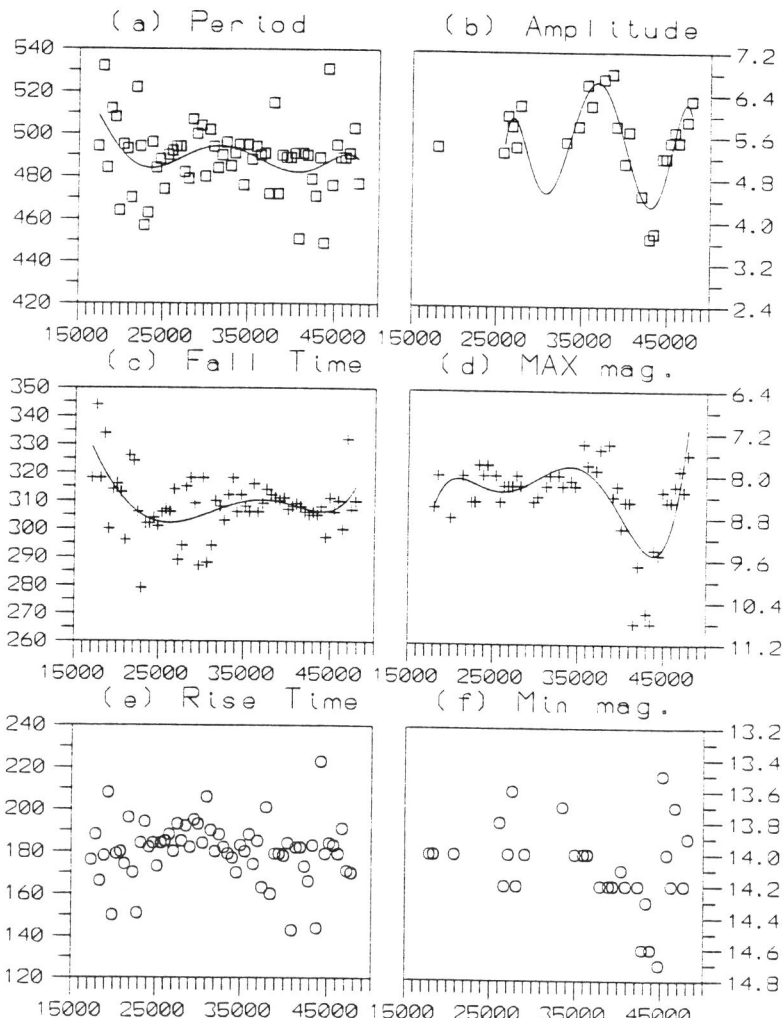

*Figure 3.* Further results for W Aql using 90 years of AAVSO maxima and minima data. (a) Period from max to max. (b) Amplitude from max to min. (c) Fall time. (d) Magnitude at maximum. (e) Rise time. (f) Magnitude at minimum.

for evenly sampled time series, but notoriously *bad* for unevenly sampled time series. Therefore we adopted the *weighted wavelet Z-transform*, or WWZ (Foster 1996c), which again is specifically designed to compensate for the difficulties of irregularly sampled data. Overall, the results of both methods were in quite good agreement with each other, and with the analysis of the derived parameters, arguing strongly for the robustness of these (relatively new) techniques. A sample of the results for each test is given in Figures 2 and 3 for the interesting S-type star W Aql.

TABLE 1. Trend results for Carbon and S–type LPVs. For the Period, Rise Time, and Fall Time, the units are *days*; for magnitudes at maximum and minimum, and amplitude, the units are *visual magnitudes*.

| Star | Period Ave. | PD | Fall Ave. | PD | Rise Ave. | PD | MAX Ave. | PD | min Ave. | PD | Amp Ave. | PD |
|---|---|---|---|---|---|---|---|---|---|---|---|---|
| R And | 409.9 | – | 259.9 | 3 | 150.0 | 5 | 7.0 | 6 | 14.4 | 3 | 7.4 | – |
| RR And | 328.4 | – | 157.4 | 5 | 171.2 | 5 | 9.2 | – | 15.0 | 8 | 6.0 | 2 |
| ST And | 334.7 | 2 | 161.4 | 6 | 173.3 | 2 | 8.9 | 10 | 11.1 | 10 | 2.2 | 9 |
| W And | 396.1 | – | 228.7 | – | 167.4 | 5 | 7.6 | 1 | 13.8 | 12 | 6.2 | 7 |
| X And | 345.7 | 7 | 219.1 | 10 | 126.6 | – | 9.1 | 4 | 14.7 | 5 | 5.7 | – |
| W Aql | 489.1 | 4 | 308.6 | 3 | 180.5 | – | 8.4 | 7 | 14.1 | – | 5.6 | 7 |
| X Aqr | 312.3 | 6 | 181.5 | 6 | 130.3 | 5 | 8.4 | – | 14.4 | – | 5.7 | – |
| V Aur | 352.4 | 3 | 163.2 | 6 | 189.3 | 11 | 9.3 | 11 | 12.2 | 3 | 2.9 | 11 |
| R Cam | 270.0 | 5 | 147.8 | 5 | 122.3 | 4 | 8.4 | 9 | 13.4 | 12 | 5.0 | 9 |
| S Cam | 327.1 | – | 158.9 | – | 168.2 | – | 8.4 | 9 | 10.4 | 5 | 1.9 | 11 |
| T Cam | 374.1 | 4 | 198.6 | 7 | 175.1 | 5 | 8.1 | 4 | 13.9 | 7 | 5.8 | 7 |
| R Cap | 346.3 | 1 | 197.1 | 6 | 148.9 | 12 | 10.8 | 6 | 13.7 | 6 | 3.1 | 4 |
| S Cas | 612.5 | 9 | 349.5 | 9 | 262.9 | – | 9.4 | 1 | 15.0 | 1 | 5.4 | 3 |
| U Cas | 276.6 | – | 155.7 | 12 | 120.9 | 5 | 8.6 | 11 | 14.9 | 3 | 6.2 | 3 |
| W Cas | 405.2 | – | 220.6 | 1 | 184.5 | 5 | 8.9 | 12 | 11.9 | 2 | 2.9 | 6 |
| X Cas | 425.1 | 12 | 189.3 | 10 | 235.8 | 4 | 10.2 | 12 | 12.3 | 6 | 2.2 | 12 |
| RV Cen | 445.2 | 11 | 197.7 | 9 | 247.2 | 12 | 7.7 | 3 | 10.2 | 6 | 2.5 | 7 |
| S Cep | 486.7 | 9 | 211.5 | 3 | 275.2 | 5 | 8.4 | 11 | 11.1 | 10 | 2.6 | 8 |
| W Cet | 351.6 | – | 170.1 | 7 | 182.3 | 9 | 8.0 | 3 | 14.4 | 4 | 6.4 | 7 |
| R CMi | 337.4 | – | 177.1 | 12 | 160.3 | 4 | 8.0 | 12 | 10.9 | 9 | 3.0 | 4 |
| T Cnc | 374.1 | – | 273.4 | 9 | 175.1 | – | 8.1 | – | 13.9 | – | 5.8 | – |
| V Cnc | 271.9 | – | 147.5 | 12 | 124.5 | – | 7.9 | 1 | 12.9 | 11 | 5.0 | 6 |
| V CrB | 357.4 | – | 211.9 | 2 | 145.5 | 10 | 7.9 | 11 | 11.2 | 10 | 3.3 | 10 |
| χ Cyg | 408.7 | 5 | 239.8 | 5 | 168.7 | 2 | 5.2 | 3 | 13.5 | 1 | 8.4 | 1 |
| R Cyg | 427.3 | – | 272.7 | – | 154.5 | – | 7.6 | 4 | 13.9 | 5 | 6.3 | – |
| RS Cyg | 416.9 | 3 | 194.3 | 9 | 222.6 | 9 | 7.3 | 4 | 8.8 | 11 | 1.5 | 6 |
| S Cyg | 322.7 | – | 161.3 | 5 | 161.3 | 3 | 10.3 | 12 | 15.8 | 10 | 5.5 | 12 |
| U Cyg | 464.5 | 6 | 241.9 | 6 | 222.6 | 2 | 7.5 | 8 | 10.8 | 5 | 3.4 | 1 |
| V Cyg | 421.0 | 12 | 227.1 | 2 | 193.8 | – | 9.3 | 11 | 12.9 | 11 | 3.6 | 11 |
| WX Cyg | 409.7 | 10 | 211.9 | 3 | 197.7 | 6 | 10.0 | 11 | 12.7 | 11 | 2.8 | 10 |
| Z Del | 304.4 | – | 158.9 | – | 145.5 | – | 9.0 | 10 | 14.6 | 1 | 5.5 | – |
| T Dra | 420.9 | 3 | 233.8 | 8 | 187.1 | 5 | 9.9 | 12 | 12.5 | 10 | 2.6 | 11 |
| R For | 388.9 | – | 186.1 | – | 202.7 | – | 9.2 | 4 | 12.3 | 4 | 3.1 | 1 |
| R Gem | 370.3 | – | 238.6 | – | 131.6 | 3 | 7.1 | 2 | 13.4 | 6 | 6.3 | 3 |
| T Gem | 287.7 | – | 150.5 | 1 | 137.1 | 6 | 8.7 | 4 | 14.2 | 7 | 5.5 | 7 |
| R Lep | 433.1 | 3 | 197.6 | 5 | 236.3 | 10 | 7.2 | 12 | 9.9 | 12 | 2.7 | 2 |
| S Lup | 342.3 | – | 165.2 | 5 | 177.0 | 6 | 8.7 | 6 | 13.1 | 5 | 4.2 | 5 |
| R Lyn | 378.5 | – | 211.9 | 3 | 166.6 | 3 | 8.0 | 4 | 13.8 | 1 | 5.8 | 7 |
| S Lyr | 438.5 | 4 | 258.8 | 8 | 179.8 | 4 | 10.9 | 5 | 15.0 | – | 4.0 | 4 |
| U Lyr | 456.6 | 3 | 216.9 | 4 | 239.2 | 8 | 9.6 | 4 | 11.9 | 12 | 2.2 | 5 |
| RR Mon | 393.2 | 9 | 234.2 | 10 | 159.4 | 10 | 9.7 | 1 | 15.0 | – | 5.3 | 3 |
| V Oph | 297.9 | – | 152.7 | 11 | 145.2 | 11 | 7.7 | 1 | 10.3 | 4 | 2.6 | 1 |
| R Ori | 379.6 | – | 219.8 | 2 | 159.8 | 7 | 9.6 | 2 | 13.1 | – | 3.6 | 2 |
| RZ Peg | 437.3 | – | 245.8 | 9 | 192.1 | 11 | 8.7 | 6 | 12.7 | – | 4.0 | 4 |
| RZ Per | 354.9 | – | 189.2 | 10 | 165.7 | 5 | 9.5 | 12 | 13.7 | 7 | 4.2 | 11 |
| Y Per | 251.5 | 2 | 126.8 | 3 | 124.8 | 3 | 8.5 | 4 | 10.4 | 11 | 1.9 | 8 |
| ST Sgr | 393.9 | – | 221.3 | 6 | 172.0 | 6 | 8.7 | – | 15.8 | – | 2.2 | – |
| T Sgr | 391.8 | 2 | 205.9 | 8 | 185.9 | 6 | 8.1 | 5 | 12.5 | 5 | 4.3 | – |
| Z Tau | 482.2 | 1 | 280.1 | – | 202.9 | 1 | 10.5 | 11 | 13.8 | 5 | 3.3 | 7 |
| S UMa | 226.0 | – | 121.2 | 10 | 104.7 | 4 | 7.9 | 5 | 11.6 | 6 | 3.7 | 6 |
| RU Vir | 436.8 | 7 | 225.7 | 11 | 210.8 | 8 | 9.7 | 11 | 13.1 | 9 | 3.3 | – |
| SS Vir | 356.2 | 4 | 167.8 | 3 | 188.4 | 4 | 6.9 | 5 | 9.0 | 8 | 2.2 | 8 |

## 4. Overall Results

Table 1 lists the results of trend analysis for each star. For each of the derived parameters, we list the average value as well as the highest significant polynomial degree ($PD$; "–" indicates *no trend* detected). The most obvious result is that for all parameters except period, $>80\%$ of the stars betray time evolution; clearly these LPVs are not constant in their fluctuations. Only about 50% of the stars reveal period evolution; this parameter is significantly more stable than the other five. It is also clear that most of the derived parameters for most of the stars show *complex* trends, as indicated by the high polynomial degree required to model the signal.

Sorting this list in order of period reveals that there is a correlation between period and period variations: longer-period stars are more likely to exhibit fluctuations in period than shorter-period stars. Only 11 of the 33 stars with $P < 407$ days (33%) show period fluctuations, while 16 of 18 (89%) with $P > 407$ do.

Another striking feature which emerges from detailed inspection of individual stars is that fall time and rise time have a *very* strong tendency to evolve oppositely ("mirror evolution"), keeping the period fairly constant. Of 40 stars showing trends in both fall time and rise time, 31 (78%) are "mirrors." In addition, both the period and the maximum magnitude of these stars show significant scatter. Faint maxima tend to be preceded by a longer cycle and followed by a shorter cycle ending in a brighter maximum. This confirms the studies of Harrington (1965) which were done on a much smaller sample of AAVSO data.

In complementary fashion, it is quite common for magnitude at maximum and magnitude at minimum to evolve together ("parallel evolution"). Several of the stars show simultaneous dimming of both maximum and minimum magnitude, which persists for many cycles, followed by recovery of both maximum and minimum magnitudes to their "normal" values.

Table 2 shows those secular trends which are revealed by linear regression. The most notable result is that the carbon LPVs are dimming at maximum magnitude; of 26 stars showing a secular trend in maximum magnitude, all 26 (100%) are getting fainter. It is also clear that for many of the stars, the amplitude is decreasing; of 17 stars showing a secular change in amplitude, 16 (94%) show a decrease. Other noteworthy results are that there is only one star (Z Tau) with significant decrease in period, and only two stars (R Lep and R Cap) show small increases in period.

## 5. Conclusion

This study establishes beyond doubt that the fluctuations of carbon LPVs tend strongly to exhibit detectable, long-term trends. We have also shown

TABLE 2. Secular Changes

| Variable | Total | trend | incr | decr | fastest | increase | fastest | decrease |
|---|---|---|---|---|---|---|---|---|
| Period    | 51 | 3  | 1  | 2  | +0.1924 | R Lep  | −0.7446 | Z Tau  |
| Fall time | 51 | 19 | 10 | 9  | +0.2175 | RV Cen | −0.2433 | SS Vir |
| Rise time | 51 | 14 | 5  | 9  | +0.2890 | SS Vir | −0.5034 | Z Tau  |
| Max magn. | 51 | 26 | 26 | 0  | +0.0500 | S Cas  | ——— | ——— |
| Min magn. | 51 | 18 | 13 | 5  | +0.0125 | S Cas  | −0.0101 | RU Vir |
| Amplitude | 49 | 17 | 1  | 16 | +0.0068 | $\chi$ Cyg | −0.0386 | S Cas |

that the trends tend to be of a complex nature, that the period is more stable than the rise time or fall time (so that the light curve *shape* shows significant evolution), and that as a whole, these stars are getting slightly fainter, and their amplitudes are decreasing. Almost certainly, still other overall behavior patterns lie undiscovered. It is also evident that individual stars show a rich variety of interesting behavior, and that many of them merit scrutiny on a case-by-case basis.

We express our sincere gratitude to the many amateur astronomers worldwide who have contributed a wealth of skill, time, and effort to variable star observing. Without their dedication, our knowledge of variable stars would be paltry at best; thanks to their observations, it is possible to study thousands of variables in great detail over nearly a century of observations. This work was partially supported by NASA grant NAGW–1493, for which we are truly grateful. J.A.M. thanks the International Astronomical Union, the American Astronomical Society, and the AAVSO for travel and accommodation grants which made it possible for her to attend this symposium.

## References

Campbell, L. 1926, *Harvard Ann.*, v. 79, no. 2
Campbell, L. 1955, *Studies of Long Period Variables*, AAVSO, Cambridge, Massachusetts
Foster, G. 1995, *AJ*, 109, 1889
Foster, G. 1996a, *AJ*, 111, 541
Foster, G. 1996b, *AJ*, 111, 555
Foster, G. 1996c, *AJ*, 112, 1709
Harrington, J. P. 1965, *AJ*, 70, 569
Mattei, J. A., Mayall, M. W. & Waagen, E. O. 1990, *Maxima and Minima of Long Period Variables 1949–1975*, AAVSO, Cambridge, Massachusetts

## Discussion

**Wing**: Can you say whether temporary decreases in amplitude, such as you illustrated in the cases of several C and S stars, also occur in M-type variables?

**Mattei**: We have not yet carried out a thorough trend analysis for the M stars, so I cannot comment about them. We will do that analysis next.

**Ake**: I'd like to point out that W Aql is a binary star. There is a short paper by Herbig in 1965 where he reports that a G–type spectrum is revealed at minimum light.

**Mattei**: I'd be very interested in that reference. [It is: Herbig, G.H. 1965, in $3^{rd}$ *Colloquium on Variable Stars*, Kleine Veröffentlichungen der Remeis-Sternwarte Bamberg, vol. 4, no. 40, p. 164 –Ed.]

**Fernie**: Where Miras show long-term trends in period, are these predominantly increasing or decreasing?

**Mattei**: Only 3 stars show secular change in period (Z Tau, R Lep, R Cap). Two others show a possible secular change. Of the 5 possible changes, 2 show a period decrease and 3 a period increase. However, only one trend (Z Tau) is a strong period decrease.

**Frogel**: For the stars that show a decrease in period and amplitude, do these quantities still lie on the mean relation between period and amplitude for all stars?

**Mattei**: Yes – the decrease in amplitude and period is episodic, not overall.

**Frogel**: Very little luminosity comes out in the $V$ bandpass for LPVs. Molecular blanketing dominates the spectrum. So small changes in blanketing could cause, in part, the temporary decreases in amplitude observed, although it isn't obvious how this would affect the period.

**Mattei**: True. One may explain shorter period and smaller amplitude if luminosity is lowered. However, the timescales for this are much, much longer than the timescales we are seeing, which are between 6 and 15 years in the four stars I mentioned.

# COMPARISON OF MEAN LIGHT CURVE PARAMETERS OF M, S AND C MIRA AND SEMI-REGULAR VARIABLE STARS USING 75 YEARS OF AAVSO DATA

MARIE-ODILE MENNESSIER AND HICHAME BOUGHALEB
*GRAAL, Université de Montpellier II*
*F-34095 Montpellier, France*

AND

JANET A. MATTEI
*American Association of Variable Star Observers*
*Cambridge, MA 02138, U.S.A.*

**Abstract.** Using 75 years of AAVSO data, mean light curve parameters of a sample of 350 long-period M, S and C Mira and semi-regular variable stars have been investigated. We compare M, S and C Mira and semi-regular stars, present a classification of the light curves of LPVs and give discriminant parameters.

## 1. Introduction

From 75 years of AAVSO data — dates and magnitudes of maxima and minima — of M, S and C Mira and semi-regular variable stars used during the preparation of the HIPPARCOS mission, we have performed a cluster analysis of the mean light curves and have examined the peculiar characteristics of each cluster.

It is well known that the differentiation between carbon- and oxygen-rich Mira stars is not obvious from photometric criteria. We show a way for identifying C Miras from our light curve classification and IRAS colors.

Another point is that long-period variable stars may show long-term variations. In the last part of this paper we give some indications of these possible long-term variations derived from the light curve parameters.

## 2. Classification of Light Curves

### 2.1. PARAMETERS

We characterize the mean light curve of each star by:

- The mean period $P$ from maximum to maximum,
- The mean amplitude $A$ of optical brightness between one maximum and the following minimum, and
- The asymmetry $f$, i.e. the mean ratio between the rising time and the period.

The corresponding values can be found in the data base ASTRID (1996).

### 2.2. CLASSIFICATION

Using the above parameters as variables we perform a dynamical clustering (Murtagh & Heck 1987) that leads to a classification into 6 clusters. The characteristics of the six best clusters according to the combinations found to describe them best are given in Table 1, and the numbers of stars of the different types in each of these clusters are given in Table 2.

TABLE 1. The characteristics of the six best clusters as achieved from an automatic classification (dynamical clustering); $n$ is the number of stars belonging to each cluster.

| Best cluster | $n$ | $P$ | $A$ | $100f - 5.9A$ | $100f - 0.022P$ |
|---|---|---|---|---|---|
| C2 | 110 | $\lesssim 275$ | $> 3.7$ | $> 10.5$ & $< 26$ | — |
| C4 | 67 | $\gtrsim 275$ | $> 3.7$ | $> 10.5$ & $< 26$ | $> 35$ |
| C1 | 54 | $\gtrsim 275$ | $< 4.7$ | $> 10.5$ & $< 26$ | $< 35$ |
| C6 | 51 | $\gtrsim 275$ | $> 4.7$ | $< 10.5$ | $< 35$ |
| C5 | 37 | $\lesssim 275$ | $< 3.7$ | $> 26$ | — |
| C3 | 35 | $\gtrsim 275$ | $< 3.7$ | $> 26$ | $> 35$ |

### 2.3. WHAT ABOUT C STARS?

We confirm the well-known characteristics:

- There are neither C stars nor semi-regular variables in the clusters corresponding to large-amplitude light curves, i.e. clusters 2, 4 and 6;

TABLE 2. Number of stars in each cluster according to different spectral and variability types.

| Type | C1 | C2 | C3 | C4 | C5 | C6 |
|---|---|---|---|---|---|---|
| O rich Miras | 48 | 106 | 12 | 59 | 25 | 40 |
| SRs | 1 |  | 3 |  | 12 |  |
| C rich Miras | 4 |  | 15 |  | 1 | 1 |
| SRs |  |  | 4 |  |  |  |
| S | 1 | 4 | 1 | 8 |  | 10 |

- The majority (80%) of C-rich variable stars and some (5%) O-rich Mira stars have a long periods, small amplitudes, and symmetric light curves, i.e. they belong to the same cluster 3.

We thus conclude that:
- The light curves of carbon stars appear to be in the continuum of semi-regular variables with longer period in $(P, A, f)$ space;
- The taxonomy from mean light curve parameters is not sufficient to differentiate C-rich from O-rich variables.

## 3. Discrimination of C Stars

The M or C spectral types for long-period variable stars depend on the abundance ratio C/O in the surface layers of these stars. Although some dust features are observed in both types, it is well known that IRAS colors alone are not discriminant parameters for C stars. Epchtein et al. (1987) propose to use the $K$ and $L$ photometric fluxes in conjunction with IRAS data. If different C/O ratios induce different properties in the circumstellar medium, we propose that the combination of light curve classification and IRAS colors may reflect the stellar and circumstellar properties and thus could distinguish carbon-rich from oxygen-rich stars.

We consider only IRAS fluxes of quality 3, and we use the ($[12] - [25]$, $[25] - [60]$) color-color diagram. We draw the following conclusions (Mennessier et al. 1997):

- IRAS colors of the oxygen-rich long-period variables belonging to cluster 3 are different from those of C stars belonging to this cluster;
- the oxygen-rich long-period variables with the same IRAS colors as the C stars belong to different clusters.

The same properties are true also when we extend the sample to all Mira stars included in the *General Catalogue of Variable Stars* with IRAS fluxes of quality 3 (Mennessier et al. 1997).

Thus the combination of light curve classification and IRAS colors allows the discrimination of carbon Miras.

## 4. Long Time Variations

We know that long-period variables exhibit variations from one cycle to another and can even present systematic long-term trends in period and amplitude (Mattei & Foster 2000). A way to detect such variations is to compute the correlation coefficients of the variables defined in § 2.1, two by two, between consecutive cycles (Boughaleb 1994). One of the most interesting significant coefficients is a positive one between magnitudes at successive maxima, i.e. a tendency for successive bright (faint) maxima. Table 3 gives the number of stars with such a significant correlation by type of variability and cluster.

TABLE 3. Number of stars with a light curve having a significant positive correlation coefficient between consecutive cycles in each cluster according to different types.

| Type | C1 | C2 | C3 | C4 | C5 | C6 | |
|---|---|---|---|---|---|---|---|
| O rich | | | | | | | |
| Miras | 3 | 1 | 7 | 4 | 7 | | 22/290 |
| SRs | 1 | | 1 | | 5 | | 7/16 |
| C rich | | | | | | | |
| Miras | 1 | | 13 | | | | 14/21 |
| SRs | | | 4 | | | | 4/4 |
| S | 1 | 1 | | 2 | | 2 | 6/24 |

It can be seen that:

- The correlation is mainly found for stars belonging to the clusters 3 (about 50% for M stars and 90% for C stars) and 5 (30%), i.e. for small-amplitude, symmetric light curves with any periods.
- The number of light curves with this correlation is much higher for C-rich (75%) than for O-rich stars (less than 10%). It could be intermediate for S stars (about 25%) but we urge caution due to the small number of stars.

This agrees with a long time variation and is consistent with the models developed by Winters et al. (2000) at Berlin.

## 5. Conclusion

The main results of our light curve studies of long-period variable stars are:

- Mean parameters of the visual light curves can be used with IRAS colors to discriminate between oxygen-rich and carbon-rich Mira stars;
- There is a similarity of some light curve parameters of C Miras and semi-regular variables;
- A tendency for a long time variability is detected in most C Miras and all C semi-regular variables. This could be linked with the influence of dust on the visual light curves and probably depends on the carbon abundance.

## References

ASTRID: "Advanced Stars: a Tool for Relating Information and Data," 1996, http://graal.univ-montp2.fr
Boughaleb, H. 1994, Ph.D. dissertation, Univ. Montpellier II
Epchtein, N., Le Bertre, T., Lépine, J. R. D., Marques dos Santos, P., Matsuura, O. T. & Picazzio, E. 1987, *A&A Supp.*, 71, 39
Mattei, J. A. & Foster, G. 2000, in IAU Symp. 177: *The Carbon Star Phenomenon*, ed. R. F. Wing (Kluwer), p. 155
Mennessier, M. O., Boughaleb, H. & Mattei, J. A. 1997, *A&A Supp.*, 124, 143
Murtagh, F. & Heck, A. 1987, *Multivariate Data Analysis*, D. Reidel Publ. Co.
Winters, J. M., Fleischer, A. J., Le Bertre, T. & Sedlmayr, E. 2000, in IAU Symp. 177: *The Carbon Star Phenomenon*, ed. R. F. Wing (Kluwer), p. 590

## Discussion

**Little-Marenin**: Could your method be used to estimate the spectral classes of stars in the AAVSO database with unknown spectral classes?

**Mennessier**: Yes.

The great theater at Aspendos, dating from 150 A.D., is said to be the best-preserved theater of the classical world.

# THE VARIABILITY OF R, N, AND C STARS FROM HIPPARCOS AND AAVSO DATA

MICHEL GRENON
*Geneva Observatory, Sauverny, Switzerland*

JANET A. MATTEI
*American Association of Variable Star Observers
Cambridge, MA 02138, U.S.A.*

LAURENT EYER
*Geneva Observatory, Sauverny, Switzerland*

AND

GRANT FOSTER
*American Association of Variable Star Observers
Cambridge, MA 02138, U.S.A.*

**Abstract.** Accurate photometry was obtained for all program stars during the 3.3-year HIPPARCOS mission. The final observing program included several hundred Mira (M), long-period semiregular (SR), and irregular (L) variables. A detailed calibration of the aging of the optics allowed the evaluation of very precise magnitudes over the whole range of star colors. Since the time coverage of the satellite observations was not sufficient to describe the behavior of M, SR, or L type variables, smooth curves were fitted statistically to the dense AAVSO observations. These curves were then transformed to the HIPPARCOS system in order to complement the HIPPARCOS photometry and thus produce precise light curves with fuller time coverage, for a set of several hundred late-type variables, including most carbon stars brighter than $V = 12.4$ at minimum luminosity. A preliminary discussion of the behavior of C stars, as observed from space in the broad $Hp$ band, is given.

## 1. LPVs in the HIPPARCOS Program

During the compilation of the HIPPARCOS Input Catalogue (HIC), special attention was given to obtaining uniform all-sky coverage of late type variables, including the poorly known ones in the Southern hemisphere. The inclusion of long-period variables (LPVs) in the HIC was possible only for those brighter than HIPPARCOS magnitude $Hp = 12.5$ (the detection threshold) during at least 80 % of their cycle. One of the prerequisites of the HIPPARCOS mission was that the brightness of the targets needed to be known in advance to allocate the appropriate observing time. However, LPVs are not strictly periodic in their amplitudes, phases, and even periods. Thus the prediction of the brightness and of the observability windows (time intervals when $Hp < 12.4$) could not be achieved without performing complementary ground-based observations before and during the mission on about 340 LPVs. The responsibility of monitoring the HIPPARCOS LPVs was taken by the AAVSO, both by continuing long-term observations and adding new variables to the AAVSO observing program (Mattei 1988). About one million long-term AAVSO observations, together with about 70 000 yearly continuing observations, were used to prepare and refine the ephemerides produced by the variable star coordinator at Montpellier, France in collaboration with the AAVSO.

## 2. Photometric Reduction of LPVs

The main-mission photometry was performed in the wide $Hp$ band extending from 380 to 900 nm. Due to irradiation by energetic solar and cosmic particles, the transmission of the detection chain suffered a severe wavelength-dependent deterioration during the mission. The standard $Hp$ system was re-defined for an epoch near mid-mission. A photometric reduction to a subset of 22 000 standard stars made it possible to fix the zero point of the $Hp$ magnitude scale to better than 0.001 mag twice per day. The very red stars were a difficult case for reduction to the standard $Hp$ system. The reddest non-variable standards have $V-I$ colors less than 1.8 whereas the mean $V-I$ of most LPVs lies in the range from 2 to 6 mag. Note that the dominant flux for late-type carbon stars is emitted in the 700 to 900 nm domain. The early reduction algorithms were polynomial relations between the instrumental magnitudes and the standard $Hp_{std}$ as a function of $B-V$. This approach failed to model the aging effects for late type stars, especially M and S type giants, inducing spurious long-term drifts and short-term flickering when the reduction relations were extrapolated to red variables.

The accurate definition of the chromatic aging, i.e. $\delta Hp$ versus $V-I$ as a function of time, was accomplished for red variables by forcing $Hp - V_{AAVSO}$ to be constant, at a given light-curve phase and $V_{AAVSO}$, throughout the mission. The $\delta Hp/(V-I)$ relation is a non-linear function of $V-I$. For reduction purposes, a linear pseudo-index was defined as given in Table 1.3.2

of *The Hipparcos and Tycho Catalogues* (HIP), Vol. 1 (ESA 1997). This linearization procedure is nearly exact for stars with $V-I$ less than 2, but it may leave reduction residuals of the order of a few percent on individual magnitudes.

Simultaneous observations, both visual estimates by AAVSO observers and photoelectric observations with a CCD camera and classical Geneva photometer, were performed to tie the $Hp$, $V_{CCD}$, $V_G$, and $V_J$ scales for red semiregular and Mira type LPVs. For M, S, and C stars a unique relation exists between $Hp$ and the Johnson $V_J$, as a function of $V-I$. Note that M and C stars have a very distinct behavior in the $(Hp-V)/(B-V)$ or $(Hp-V_T)/(B_T-V_T)$ plane — see Fig. 1.3.7 in *The Hipparcos and Tycho Catalogues* (HIP), Vol. 1 (ESA 1997).

The adopted relation $Hp-V$ versus $V-I$ from the Cousins system is given below:

| $V-I$ : | 2.00 | 2.50 | 3.00 | 3.50 | 4.00 | 4.50 | 5.00 | 5.50 | 6.00 |
|---|---|---|---|---|---|---|---|---|---|
| $Hp-V$ : | 0.08 | 0.02 | −0.09 | −0.28 | −0.53 | −0.81 | −1.10 | −1.38 | −1.66 |

## 3. Light Curves of Carbon Stars

All R, N, and C type stars observed by HIPPARCOS turned out to be variable — either irregular, semi-regular or nearly periodic. Since $\lambda_{\text{eff}}(Hp)$ is larger than $\lambda_{\text{eff}}(V)$, the amplitude in the $Hp$ band is generally smaller than in the $B$ and $V$ bands. This behavior is illustrated in Figure 1 for HIP 59844, the pulsating star BH Cru.

The ratio $Q = A_{Hp}/A_V$ is around 0.7 for early type C and M giants. It shows a slight decrease for the reddest stars (Figure 2).

Large-amplitude periodic variables often show nearly sinusoidal folded light-curves, e.g. HIP 109089 in Figure 3.

A bump near phase 0.7 is present in many light curves as shown for HIP 106583, 99653 or 26753. Semi-regulars and stars pulsating in the first overtone show rather noisy folded light curves due to their varying amplitude and not-so-periodic behavior. For small-amplitude irregulars, the time coverage during the HIPPARCOS mission was sufficient to describe their behavior in terms of peak-to-peak amplitudes and time scales for variations (cf. Eyer & Grenon 1997).

The $Hp$ amplitudes show two regimes for periodic R, N, and C variables. R stars and some C and N stars show a linear relation between the period and the amplitude which may be expressed as $A_{Hp} = 0.0013\times$Period. For the classical C-rich Miras with typical $A_{Hp}$ in the range 1.2 to 2.4 mag, the amplitude shows little if any dependence on the period, ranging between 200 and 480 days. The global behavior is displayed in Figure 4.

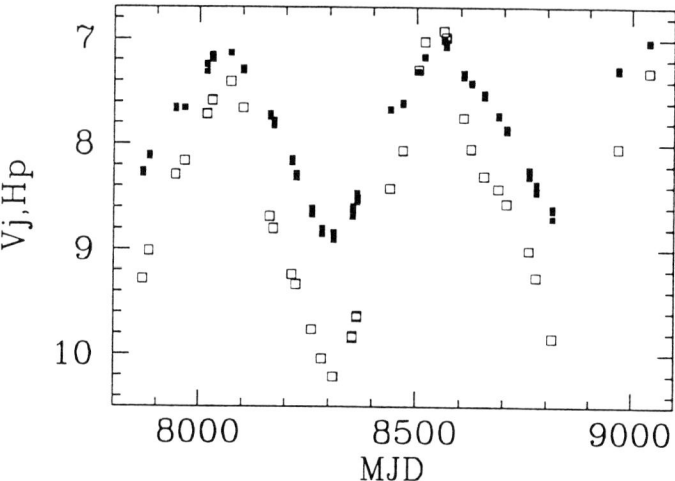

*Figure 1.* The light curve of BH Cru (HIP 59844), of spectral type SC 4,5-8e, in the $Hp$ magnitude (*filled squares*) and in the visual $V$ magnitude (*open squares*) as deduced from AAVSO observations.

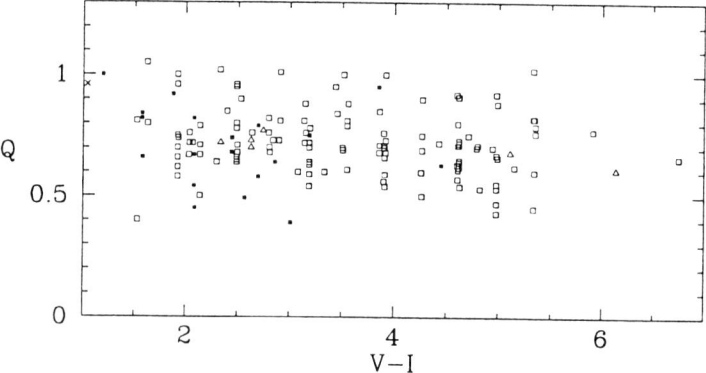

*Figure 2.* Q, the ratio of the $Hp$ amplitude to the $V$ amplitude, as a function of the $V-I$ color for red variables. *Open squares*: M type; *open triangles*: S type; *filled squares*: C type; *crosses*: K and M supergiants.

## 4. HIPPARCOS–AAVSO Light Curves

The visual estimates are obtained by interpolating the brightness of the variable star using a set of reference stars of known magnitude in its field. The difference between the response of the eye and of the $V_J$ band leads to an offset between the visual and the photoelectric $V_J$ magnitude, proportional to the color difference between the comparison stars and the red variable.

The monitoring by AAVSO observers generally produces a dense time

# VARIABILITY OF R, N, AND C STARS    175

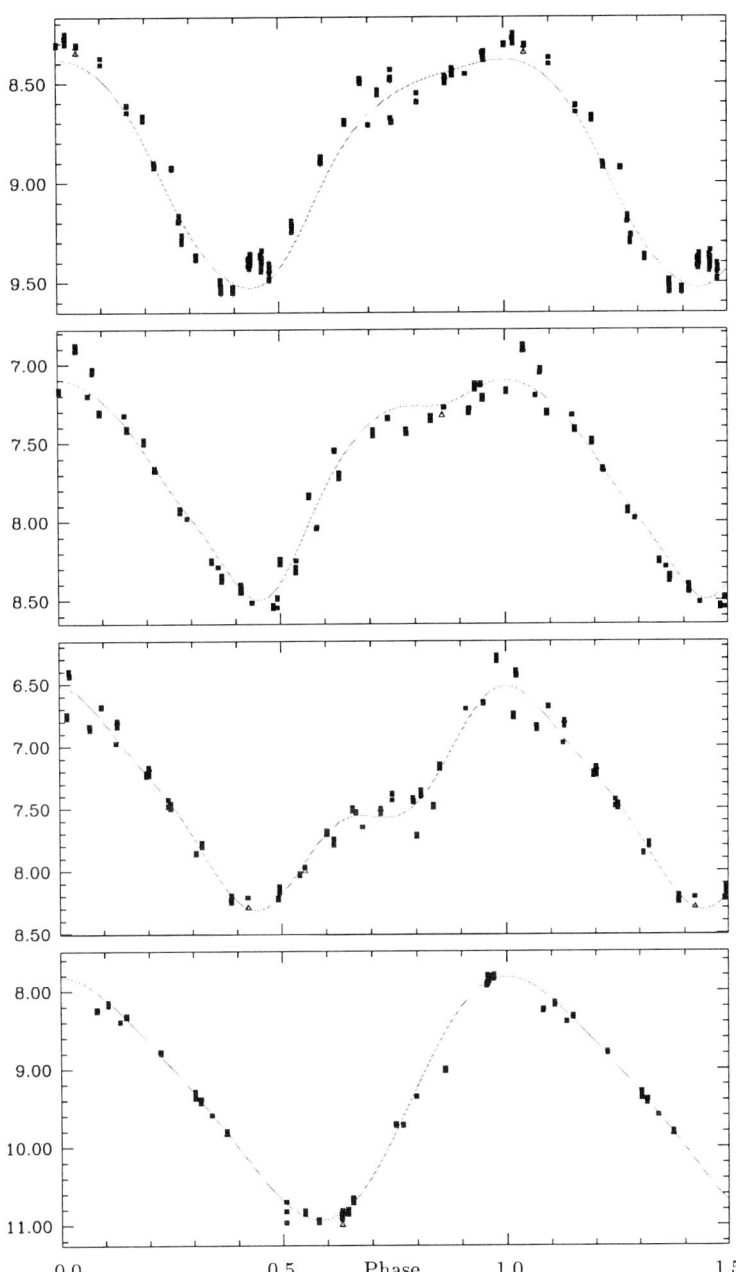

*Figure 3.* Folded light curves of typical C-type periodic variables monitored over 3 to 4 cycles. From top to bottom: HIP 26753, C0e, $P = 326$d; HIP 99653, C5 II, $P = 431$d; HIP 106583, C6 II, $P = 486$d; HIP 109089, C9e, $P = 436$d. Error bars are smaller than the symbol size. The dotted line is the adopted best fit used to derive the epoch and magnitudes at brightness extrema.

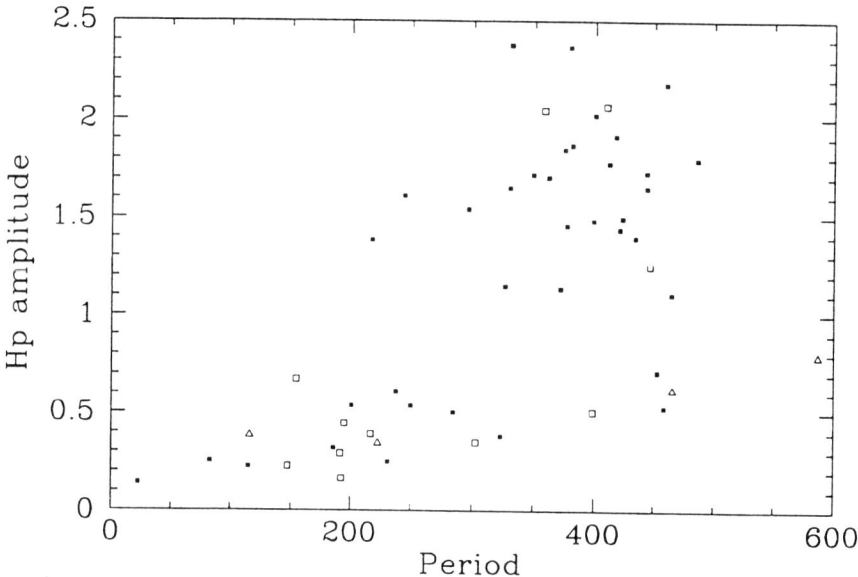

*Figure 4.* The relation between amplitude in $Hp$ and period, for periodic red variables of type C (*filled squares*), N (*open squares*) and R (*open triangles*).

coverage, but because observations of different observers are combined, and because the accuracy of an individual observation is only between 0.1 and 0.3 mag, the light curves are noisy. Thus, average light curves were obtained by fitting curves to the individual observations by Fourier, polynomial or quintic spline methods. These fitted curves were then transformed to $Hp$ magnitudes and HIPPARCOS photometry was then superimposed upon them.

The difference $Hp-V$ is a function of the star's temperature and of the circumstellar and interstellar extinction. Since the color change as a function of the phase is generally unknown, the technique used to reduce AAVSO magnitudes to $Hp$ was to plot $Hp - V_{AAVSO}$ versus $V_{AAVSO}$. For LPVs the relation is often S-shaped, as shown for the C0ev type star HIP 4284 (Figure 5).

The fine structure of $Hp-V_{AAVSO}/V_{AAVSO}$ diagrams depends mainly on the $T_{eff}$ and $\log g$ variations and on the corresponding absorption changes due to TiO, VO or CN, $C_2$, $SiC_2$ molecular bands. Emission lines and dust extinction introduce departures from the mean relation.

Although distinct relations seem to exist for the rising and falling parts of the light curve, a unique third-degree polynomial was used to transform fitted AAVSO magnitudes to $Hp$ magnitudes with an uncertainty of 0.1 to 0.2 mag in most cases. This uncertainty is generally small compared to the peak-to-peak $Hp$ amplitude. Errors in comparison-star magnitudes are

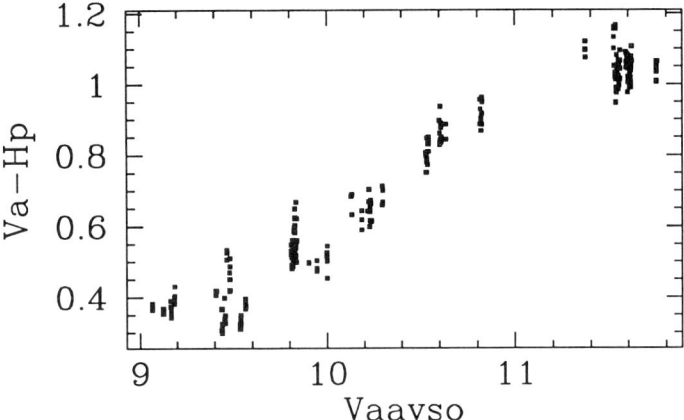

*Figure 5.* Example of the non-linear relation $V_{AAVSO}-Hp$ vs. $V_{AAVSO}$ used to transform visual light curves into $Hp$ light curves.

automatically corrected by this process.

The Atlas of HIPPARCOS–AAVSO light curves, part B (ESA 1997), contains 274 stars in common. Here we show a few representative cases of C-star light curves which would be difficult to interpret with HIPPARCOS data alone. This is especially true for the RCB variable RY Sgr, where the main minima were missed due to the peculiar HIPPARCOS time sampling, and the semiregular variable V Hya, which has periods of 530 days and over 6000 days, and where HIPPARCOS observations were obtained while the star was slowly fading to the minimum of its longer period (see Figure 6).

We sincerely thank variable star observers around the world whose dedicated observations of LPVs provided vital support for this program. We gratefully acknowledge the support of NASA under grant NAGW–1493 which made it possible for the AAVSO to provide data support to the HIPPARCOS mission and the Swiss National Science Foundation for its support of activities at Geneva Observatory.

### References

European Space Agency (ESA) 1997, *The Hipparcos and Tycho Catalogues*, ESA SP-1200, Vols. 1 and 12

Eyer, L. & Grenon, M. 1997, in *Hipparcos – Venice '97*, ESA SP-402, p. 467

Mattei, J. A. 1988, in *Scientific Aspects of the Input Catalogue, Preparation II*, ed. J. Torra and C. Turon, p. 376

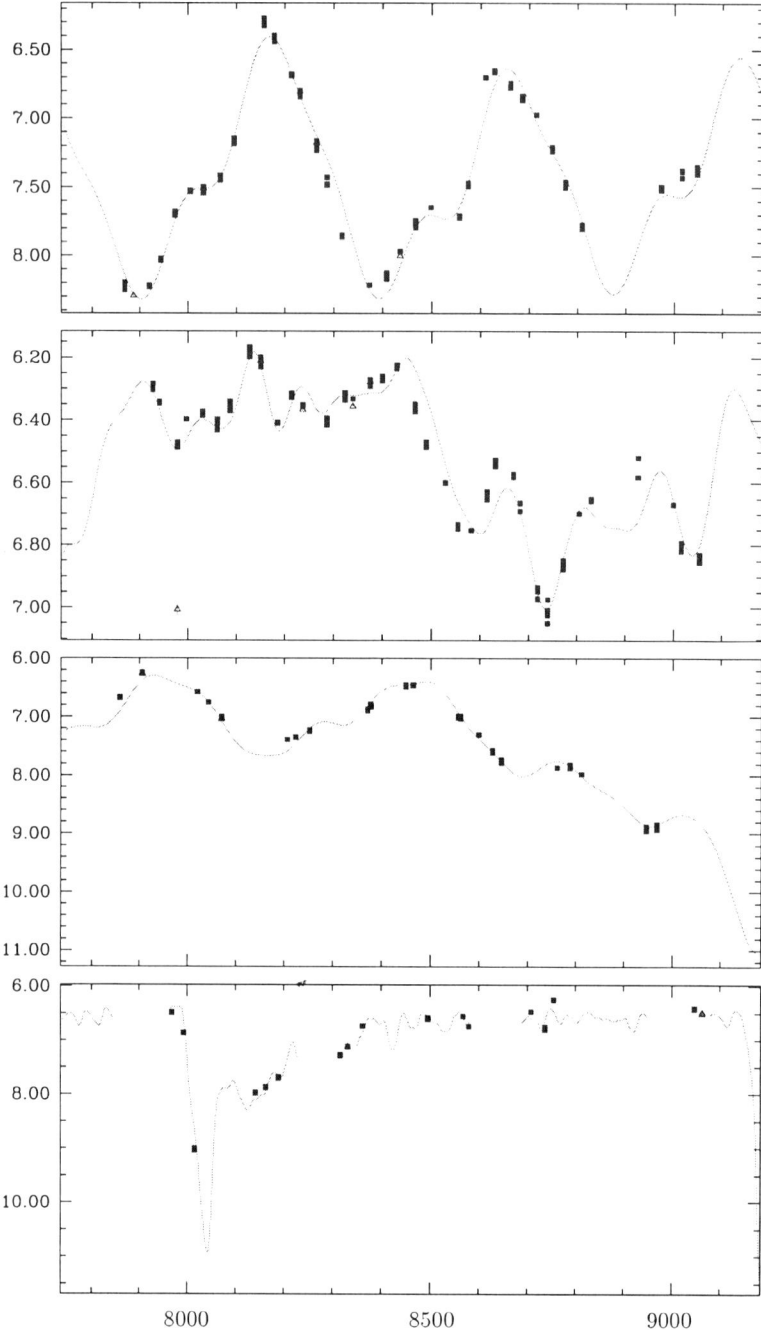

*Figure 6.* Light curves of C-type Mira and semiregular variables. From top to bottom: HIP 106583, C6 II, S Cep; HIP 63152, C7 I, RY Dra; HIP 53085, C9 I, V Hya; and HIP 94730, Cp, RY Sgr, an RCB variable. Error bars are smaller than the symbol size.

# INFRARED LIGHT CURVES OF CARBON-RICH VARIABLES

PATRICIA WHITELOCK
*South African Astronomical Observatory*
*Observatory 7935, South Africa*

**Abstract.** Long-term trends in the infrared ($JHKL$) light curves of various carbon variables are described. Some stars, e.g. the semi-regular variables R Scl and GM CMa, show multiple periodicities; others, particularly the Miras with moderately thick dust-shells, show more erratic long-term changes. The light curves for R For, which have been intensively monitored over 20 years, show a pattern which is reminiscent of that seen for R CrB stars. This pattern is superimposed on regular large-amplitude Mira pulsations. The multi-periodic and erratic behaviour of these stars is compared with the predictions from various models.

## 1. Introduction

Intrinsic carbon stars are the consequence of dredge-up during the helium shell-flash cycle near the top of the Asymptotic Giant Branch (AGB). It is well established that copious mass loss occurs in low and intermediate mass stars towards the end of their AGB evolution. It is also apparent that pulsating stars have particularly high mass-loss rates, strongly suggesting a causal relationship between mass loss and pulsation. It is, however, unclear how either mass loss or pulsation is affected by the changes in stellar structure which accompany the helium shell-flash cycles.

This paper describes some results from a long-term programme in progress at the South African Astronomical Observatory (SAAO) to monitor the infrared magnitudes of carbon stars. The programme is intended to investigate the relationship between pulsation and mass loss over many pulsation cycles and to look for long-term trends. A number of people have been involved in this programme over the years, most notably Michael Feast who initiated the project over 20 years ago. The observations are made through broad-band $JHKL$ filters using the MKII infrared photometer on the 0.75-m telescope at Sutherland in South Africa. It is particularly appropriate and useful to work in the 1 to 4 $\mu$m region when dealing with carbon variables. First, because they are cool stars and the stellar energy distributions usually peak around 1 $\mu$m, we are examining the behaviour of the bolometric output rather than minor fluctuations in temperature. Secondly, most of these stars have significant dust shells, and the light curve in $L$ (3.45 $\mu$m) is principally sampling this dust shell.

One of the principal interests in this work was in studying the properties of the Mira variables, both carbon- and oxygen-rich. Miras are understood to be the last reasonably long-lived phase in the evolution of low and intermediate mass stars. They are at the top of the AGB just prior to the journey across the HR diagram to become white dwarfs. The most informative thing about a Mira is its pulsation period. For oxygen-rich Miras there are various correlations that suggest the pulsation period of the Mira tells us about the population from which it has evolved. In particular the kinematics and scale heights of Miras in the solar neighbourhood tell us that the short-period stars belong to a thick-disk population while the longer period stars are kinematically cooler (Whitelock 1995 and references therein). The Miras obey a period-luminosity relation. Feast et al. (1989) showed that oxygen- and carbon-rich Miras in the LMC obey the same PL relation in $K$ (2.2 $\mu$m). More recently, Groenewegen & Whitelock (1996) established that they also obey the same bolometric period-luminosity relation.

## 2. Infrared Colours

It is instructive to examine the colours of carbon-rich variables in comparison to those of oxygen-rich stars. Figure 1 shows the colours for carbon variables; they have been corrected for reddening using a galactic model of the extinction. These are mean colours; the individual observations often show a good deal of spread. In particular, the colours of stars which undergo obscuration events (see below) are much redder during the faint phases than otherwise. The illustrated stars have not been selected in any systematic way. They include bright local variables which are part of our long-term monitoring program, as discussed below. They also include some carbon-rich stars with infrared excesses that were selected on the basis of their galactic coordinates as potential probes of galactic structure. Some have been very well studied, others much less so. Note that it is not always easy to distinguish between Miras and other types of variable on the basis of limited infrared observations.

Figure 1 shows a complete mixture of Mira and other variables; there is no easy way to separate them on the basis of these colours. This is in marked contrast to similar diagrams for oxygen-rich stars, e.g. Fig. 3 of Whitelock

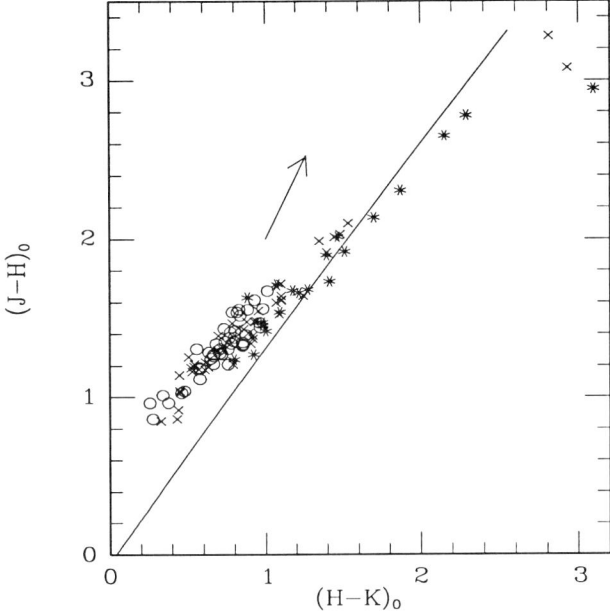

*Figure 1.* The mean infrared colours of carbon variables. The stars discussed below have the following $(H-K)_0$, $(J-H)_0$ colours: R For (1.27, 1.68); TT Cen (0.51, 1.26); IZ Peg (3.10, 2.95); R Scl (0.77, 1.38); GM CMa (0.89, 1.41). *Crosses* are Miras, *open circles* are semi-regular variables, and *asterisks* are Miras with large-amplitude long-period or erratic changes. The line is the locus of blackbodies of various temperatures and the arrow represents a reddening vector of $A_V = 5$ mag.

et al. (1994), where the Miras are clearly separated from the other stars. It is the strong $H_2O$ features in the very extended atmospheres of these O-rich Miras that are the major influence on their colours. The only clear trend among the carbon stars in Fig. 1 is that the stars with the reddest colours, those with the thickest dust shells, are Miras and not small-amplitude variables, and that there is a large fraction of large-amplitude erratic variables among these stars with large excesses. This is of course expected if the erratic and long-term variations are caused by dust obscuration.

## 3. R Fornacis

R Fornacis is a bright, well-studied carbon Mira. Its $J$ light curve is shown in Figure 2. This comprises 153 observations taken over more than 20 years. The early half of this light curve has been discussed by Feast et al. (1984) and by Le Bertre (1988). Both papers conclude that the faint phase in 1983, around JD 2445600, was caused by increased dust obscuration. The combined SAAO and ESO data from JD 2444900 to JD 2447600 were discussed by Winters et al. (1994). They suggest that such dust-obscuration events

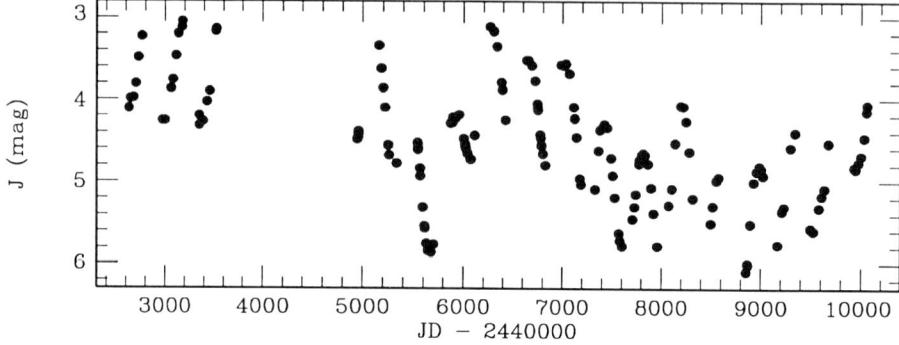

*Figure 2.* The $J$ (1.25 μm) light curve of R For.

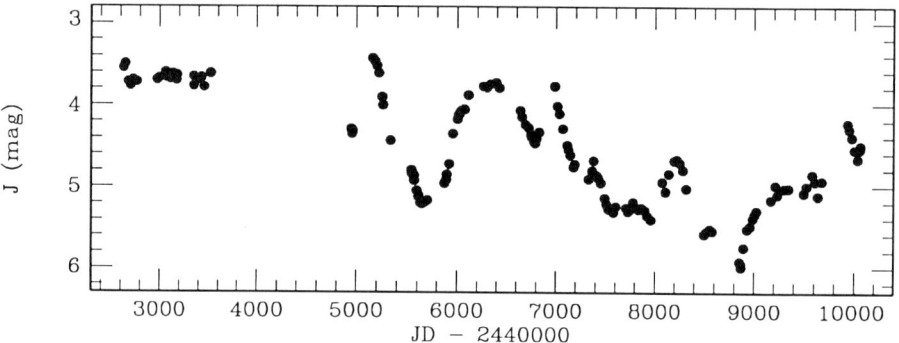

*Figure 3.* The $J$ (1.25 μm) light curve of R For after removing pulsational variations.

could occur periodically.

The GCVS gives the pulsation period of R For as 388.7 days. A Fourier analysis of the $J$ light curve gives $386.6 \pm 1.3$ d for the complete data-set, which is not significantly different from the GCVS value. If the light curve is divided up, then a slightly different period is determined for each of the different time slots: $379.9 \pm 3$ for JD 2442630–3525, $403.1 \pm 6$ for JD 2444947–6828, and $381.4 \pm 3$ for JD 2446986–50062. The early and later data show essentially the same value while the period in between is longer, and the amplitude larger. It is difficult to say how meaningful these differences are and their adoption does not affect the conclusions of this study.

Figure 3 shows the $J$ light curve with the pulsation removed assuming the periods listed above for the different time slots. Although there are a few places where the pulsation appears to have been imperfectly removed (e.g. around JD 2447000), the overall impression is that of a smooth variation of the mean magnitude once the pulsational variations have been removed.

The $H$ and $K$ data are similar to $J$ and are not illustrated here. Figure 4

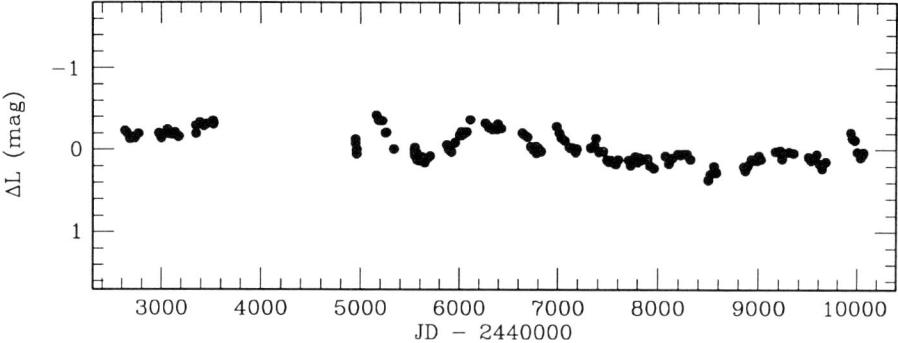

*Figure 4.* The L (3.45 μm) light curve of R For after removing pulsational variations.

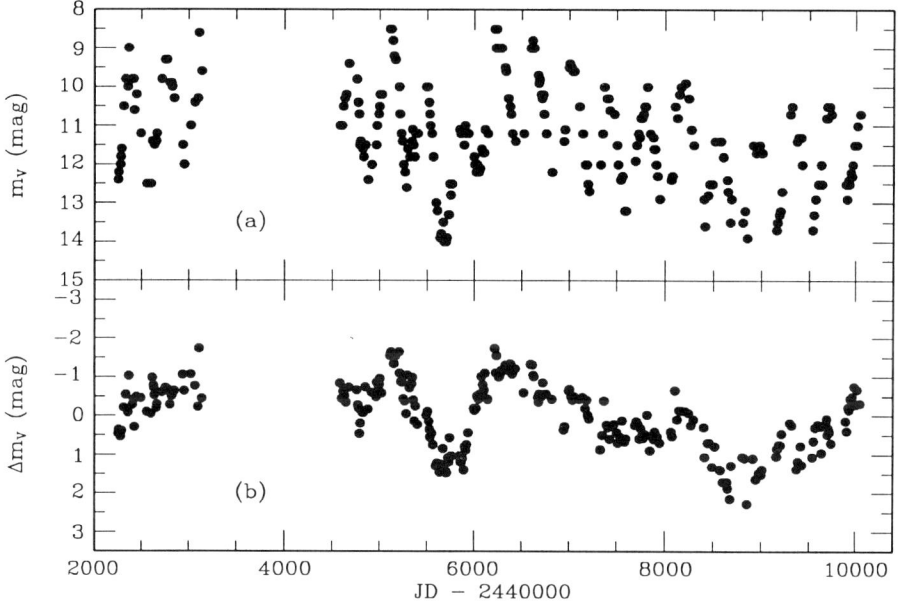

*Figure 5.* (a) Visual light curve for R For from observations by Danie Overbeek. (b) The same after removing pulsational variations.

shows the L data, on the same scale as Fig. 3, after removal of the pulsation in exactly the same way as for the J data. The L curve shows the same effects as the J one, but with a much reduced amplitude.

Figure 5 shows visual magnitude estimates for R For made by Danie Overbeek, an experienced South African variable star observer, over the same time period as the infrared data. His observations are among those reported by the AAVSO. Figure 6 shows these data after removing the

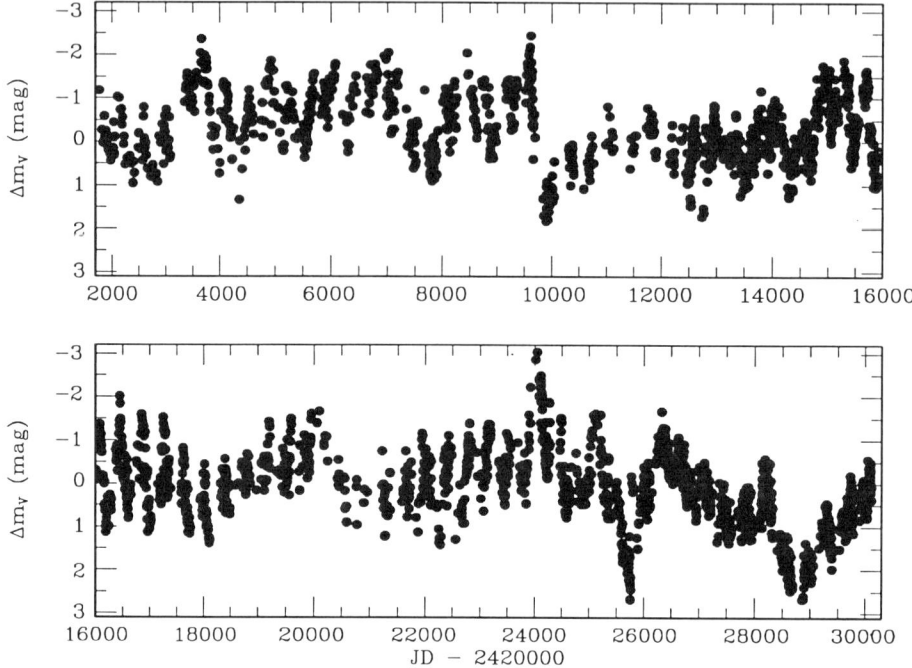

*Figure 6.* 76 years of AAVSO data for R For after removing the pulsational variations.

pulsations with a period of 386.6 d (no attempt was made to use different periods at different times). Although there is a lot of scatter in this diagram, as expected for visual magnitude estimates, it shows the same form as the $J$ light curve.

Feast et al. (1984) discussed the 1983 deep minimum in the visual and infrared light curves. They concluded that the most probable explanation was an increase in the obscuration of the star by a dust cloud containing predominantly large dust grains (diameters of the order of 0.15 $\mu$m). Presumably the more recent broad minimum has a similar explanation.

Winters et al. (1994) have suggested that discrete structures in the circumstellar dust shells may be produced by the formation of new dust which will decisively influence the shapes of the light curves by producing regular deep minima, as is shown in the theoretical light curves for different models illustrated in Figs. 5, 6, and 7 of their paper.

The light curves of R For, after removing the pulsational modulation (Figs. 3, 4, and 5), bear a striking similarity to those of R CrB stars (e.g. Feast 1979). Indeed it seems possible that the explanation is the same, i.e. the star emits puffs of carbon-rich dust in random directions, and a minimum occurs when one of the puffs is emitted in our line of sight to the star. If the R CrB explanation is correct, then we would expect the timing of the events to be random rather than periodic. There is certainly no sign of periodicities in the illustrated data of 20 years ($\sim$ 19 pulsation cycles).

The AAVSO database for R For covers about 76 years and thus provides a much longer time base to examine for periodicities. Figure 6 shows the AAVSO data after the pulsational periodicity has been removed in the same way as it was for the infrared data. There is no sign of periodicity in the occurrences of faint phases even on this longer time base.

Other carbon Miras, in particular R Vol and R Lep, show similar behaviour to R For.

## 4. TT Centauri

We have fewer observations for the Mira TT Cen, but there is enough to suggest a rather different kind of behaviour from that of R For. The GCVS period is 462 days. There is no clear long-term trend at $J$ and a Fourier analysis finds 3 significant periods: 448.6, 222.1, and 149.5 d. The light curve is shown in Figure 7, phased for a period of 448.6 d. The fitted curve assumes this period and also makes use of the first three harmonics. The result is a secondary maximum on the rising branch of the curve. The $L$ data, perhaps surprisingly, do show some sign of fading of the mean light level. Most Miras show $L$ variations that are about half the amplitude of those at $J$. TT Cen is unusual in having comparable amplitudes at $J$ and $L$.

The $J$ light curve of TT Cen bears some resemblance to the theoretical curve produced by Wood (1995; see his Fig. 2) to illustrate the behaviour of the bolometric light curve of a model of an LMC Mira.

RV Cen (not illustrated) shows the same kind of behaviour both at $J$ and visually, but with more of a stand-still than a secondary peak.

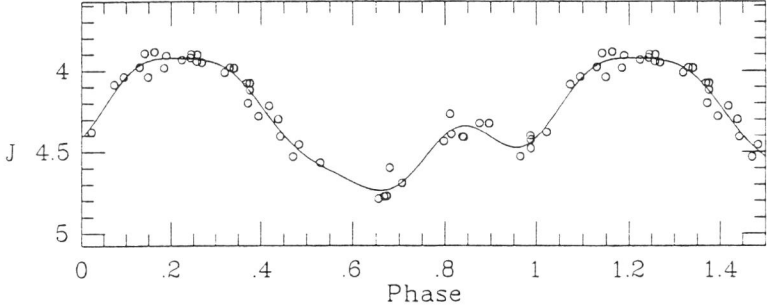

*Figure 7.* The $J$ light curve for TT Cen assuming a period of 448.6 days.

Secondary maxima and stand-stills are common features of the light curves of Miras and many other types of variable star. Such features can be produced by a variety of mechanisms, such as shock-waves or periodic dust formation (e.g. Winters et al. 1994 and references therein) and are not in themselves of great diagnostic value.

## 5. IZ Pegasi

IZ Peg (CRL 3099) is a thick-shelled carbon Mira with very red colours. Whitelock et al. (1994) published a light curve and suggested it had two periods, 488 and 345 d. Recent data do not support the second period, but a reasonable fit can be obtained with a 486-day pulsation and a long-term trend (Figure 8). A longer time base is required to determine the nature of the long-term trend.

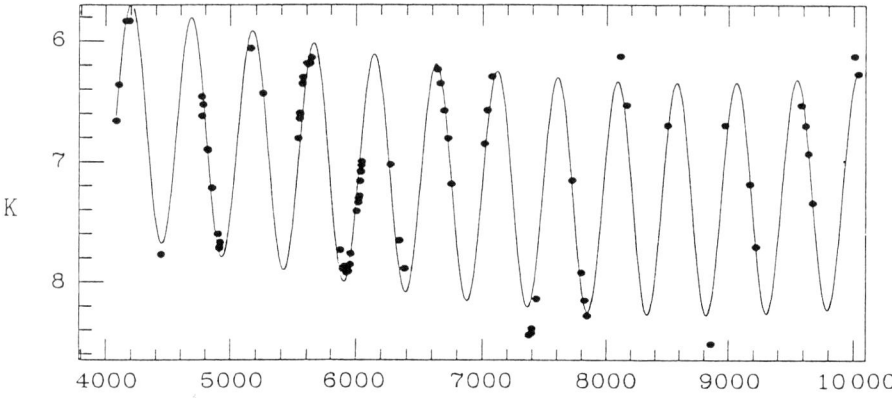

*Figure 8.* The $K$ light curve for IZ Peg; the abscissa is Julian Date minus 2440000. The fitted curve shows the 486-day pulsation and a long-term trend.

## 6. R Sculptoris

R Scl is a well-known bright semi-regular variable. There has been an upsurge of interest in this star since the IRAS survey showed it to have unusual colours and various studies suggested that it was associated with hot and cool dust (e.g. Young et al. 1993). Essentially it has a detached dust shell (confirmed by Olofsson et al. 1996, who mapped the circumstellar CO) and therefore must have had a higher mass-loss rate in the not too distant past than it does at present. If we assume that it was a Mira in its previous high-mass-loss phase, then it seems possible that it may now be in a luminosity minimum following a helium shell flash. Alternatively it has left the Mira instability strip prior to its evolution across the HR diagram to

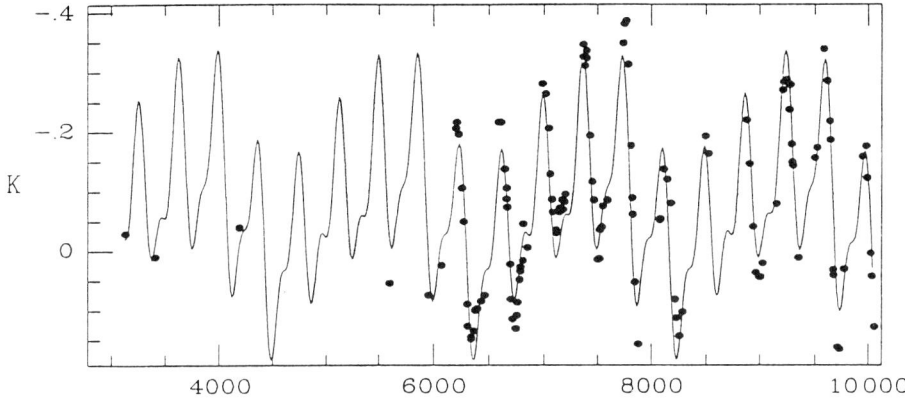

*Figure 9.* The $K$ light curve for R Scl; the abscissa is Julian Date minus 2440000. The fitted curve shows the 374-day pulsations and an additional 1850-day modulation.

become a white dwarf. In any case it is likely to show interesting changes on a relatively rapid time scale. Figure 9 shows a $K$ light curve comprising Le Bertre's (1992) 28 observations and 93 measurements from SAAO (Whitelock et al. 1995, and unpublished).

The curve in Fig. 9 is derived from a 374.3 day pulsation and its first harmonic together with 1850-day variations. There are not enough data here to be sure that the 1850-day changes are really periodic. It is clear, however, that there is something other than the 374-day variations going on. Note that R Scl is a small-amplitude variable, with a range less than 0.4 mag at $K$ which is lower than that of any of the other stars discussed above. This curve has some similarities to the model illustrated in Fig. 7 of Winters et al. (1994).

## 7. GM Canis Majoris

Finally, consider another semiregular, GM CMa, otherwise known as IRC $-20101$. Unlike the stars described above, this is not a well-studied variable and no period has been published for it. I started observing it because its IRAS colours were very similar to those of R Scl. A Fourier analysis shows three significant periods, in order of decreasing amplitude, at 405, 780 and 201 days, and a marginal one at 100 days. The fit shown in Figure 10 is actually for 800 days and the three harmonics at 400, 200 and 100 days. Interestingly in this star the strongest peak is the first harmonic at 400 days which has three times the amplitude of the 800-day variations. This light curve bears some resemblance to those of the RV Tauri stars. Wood (1995) has found that oscillations at twice the primary period sometimes

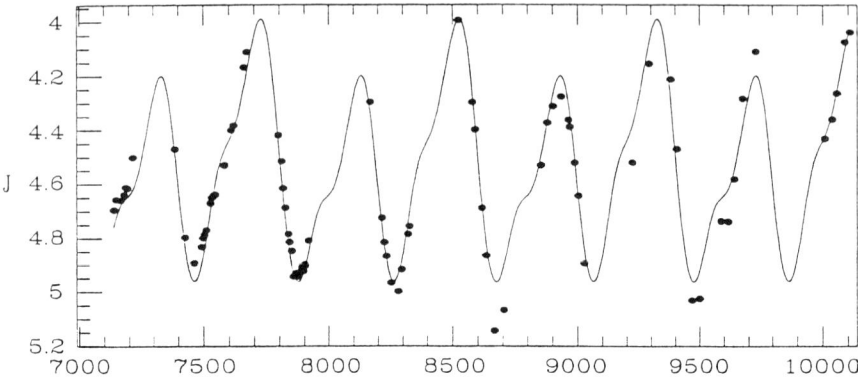

*Figure 10.* The $J$ light curve for GM CMa; the abscissa is Julian Date minus 2440000. The fitted curve has periodicities of 800, 400, 300, 200 and 100 days.

occur in the outer layers of his stellar models. This may explain the 800-day variations found here if 400 days is the stellar pulsation period.

## 8. Conclusions

The infrared light curves of some C-rich variables show striking erratic variations associated with dust obscuration. There is as yet no evidence that these changes are periodic, and they might indicate that R CrB-like phenomena occur in carbon Miras. Other variables show double periods and peculiarities such as stand-stills in their light curves which are predicted by a variety of models. There are as yet insufficient data to be clear what is normal behaviour near the top of the AGB and what requires a more exotic explanation such as binary interaction.

I am very grateful to Danie Overbeek and to my colleagues at SAAO for the use of data prior to publication. I would also like to thank Janet Mattei and the AAVSO for allowing me to use their visual magnitudes for various carbon stars. I am grateful to Michael Feast and John Menzies for helpful discussions.

## References

Feast, M. W. 1979, in IAU Coll. 46: *Changing Trends in Variable Star Research*, ed. F. M. Bateson, J. Smak, and I. H. Urch, Univ. Waikato (Hamilton, New Zealand), p. 246
Feast, M. W., Glass, I. S., Whitelock, P. A. & Catchpole, R. M. 1989, *Mon. Not. Roy. Astron. Soc.*, 241, 375
Feast, M. W., Whitelock, P. A., Catchpole, R. M., Roberts, G. & Overbeek, M. D. 1984, *Mon. Not. Roy. Astron. Soc.*, 211, 331
Groenewegen, M. A. T. & Whitelock, P. A. 1996, *Mon. Not. Roy. Astron. Soc.*, 281, 1347
Le Bertre, T. 1988, *A. & A.*, 190, 79

Le Bertre, T. 1992, *A&A Supp.*, 94, 377
Olofsson, H., Bergman, P., Eriksson, K. & Gustafsson, B. 1996, *A&A*, 311, 587
Whitelock, P., Menzies, J., Feast, M., Marang, F., Carter, B., Roberts, G., Catchpole, R. & Chapman, J. 1994, *MNRAS*, 267, 711
Whitelock, P., Menzies, J., Feast, M., Catchpole, R., Marang, F. & Carter, B. 1995, *MNRAS*, 276, 219
Whitelock, P. A. 1995, in IAU Coll. 155: *Astrophysical Applications of Stellar Pulsation*, ed. R. S. Stobie and P. A. Whitelock, ASP Conf. Ser., 83, 165
Winters, J. M., Fleischer, A. J., Gauger, A. & Sedlmayr, E. 1994, *A&A*, 290, 623
Wood, P. R. 1995, in IAU Coll. 155: *Astrophysical Applications of Stellar Pulsation*, ed. R. S. Stobie and P. A. Whitelock, ASP Conf. Ser., 83, 127
Young, K., Phillips, T. G. & Knapp, G. R. 1993, *ApJ*, 409, 725

## Discussion

**Sedlmayr**: Your material and the theoretical interpretations given are extremely valuable and very much needed in the present state of research. The calculations you mentioned by Winters et al. originally were not intended to model the specific light curves of real astronomical objects (like R For), but were aimed at providing a sufficient theoretical description of the coupled nonlinear processes in the shells of Miras and LPVs, which turned out to result in theoretical light curves exhibiting similar sub-variations. Also, your idea to correlate the irregular variations with a kind of R CrB phenomenon seems very appealing but might raise its own theoretical problems in detail.

**Whitelock**: I think that the paper by Winters et al., and the series of which it is part, represent an important step forward in the understanding of these stars. I specifically mentioned the Winters et al. paper because it does discuss R For and the IR observations. I think that the extensive data we now have for this particular star suggest very strongly that periodic dust formation is not the correct explanation although it may be correct for other stars.

Hard-working Hülya Çalışkan outside the theater at Aspendos. Hülya's behind-the-scenes work on the LOC saved the symposium from possible disaster on several occasions.

# R CrB: PHOTOMETRIC EVIDENCE REGARDING THE NATURE OF ITS PULSATION, AND A PUTATIVE CONNECTION BETWEEN PULSATION AND DEEP DECLINES

J. D. FERNIE
*David Dunlap Observatory, University of Toronto*
*Richmond Hill, ON   Canada*

**Abstract.** It is shown that the historical lightcurve of R CrB contains features that argue against the deep declines being caused by pulsation. As well, recent photoelectric photometry does not show the onset of deep declines occurring at a specific phase of pulsation. The photometry suggests that only one period is present at any time, although this period can drift and/or be replaced by another period. The evidence is strong that the pulsation is radial and obeys relationships found among classical Cepheids.

## 1. Introduction

R CrB is the prototype of a small class of variable supergiants defined as having atmospheres rich in carbon and extremely deficient in hydrogen. Almost all of them show low-amplitude variability of a few tenths of a magnitude in $V$ with unstable periods in the range of ∼35 to 55 days. The class is renowned for the deep declines of up to 8 magnitudes in visible light which occur at apparently random times and which may last for months.

The canonical explanation of this behaviour is that the pulsation generates shock waves that produce mass loss. The material, however, is not emitted as a shell, but rather as a localized puff. This puff, rich in carbon, condenses into dust at some distance from the star, and if the puff was in a direction near our line of sight, the expanding dust cloud will veil the star and cause it to appear much fainter. The slowly expanding cloud eventually dissipates and the star appears restored to its former glory.

The question then arises as to whether puffs are emitted in every pulsation cycle, and if so, at what phase of the cycle. A second question is whether the instability of the period implies that more than one period is present at any given time, and if so, what are the individual periods and how do they relate to the phasing of the puffs. A third question is whether the pulsation is radial or non-radial.

In what follows I review these questions through the use of the historical light curve of R CrB and a database of modern photoelectric photometry of the star. This database is available on the World Wide Web at the URL http://ddo.astro.utoronto.ca/rcrb.html

## 2. The Historical Light Curve

The light curve of R CrB for the 152 years between 1843 and 1995 has been compiled by Mattei et al. (1991, 1996). It contains several features of interest. The first is the consistency with which the star sinks 8 mag to 14th magnitude in a deep decline, but never goes below that level. The latter is probably set by the emission lines that appear in the spectrum at this stage, coupled with a complete veiling of the photosphere by carbon dust. The second feature of interest is the relative frequency of deep and shallow declines. If we define a deep decline as one in which the visual magnitude becomes fainter by 6–8 mag, and a shallow decline as one of 1–2 mag, then the record shows that deep declines outnumber shallow ones by nearly 2:1. (There are, of course, intermediate declines as well.) That this is not due to observational selection is shown by the record of the last 30 years or so, when R CrB has been under almost constant scrutiny by dozens of amateur astronomers whose precision of observation is of order 0.1 mag. A drop of 1–2 mag would not have been missed, yet again there are nearly twice as many deep declines as shallow ones.

This is rather surprising. One would think that deep declines would require puff ejection close to the line-of-sight, and thus be less frequent than the shallow declines which come from puffs ejected at larger angles. Presumably this is an argument for the dust forming rather close to the star and expanding considerably, so that its initial direction is less important. However, the point of interest here is the implication that almost any puffing on the hemisphere of the star facing us will result in a detected decline.

With that in mind we turn to another feature of the historical light curve. Between the years 1924.9 and 1934.9 no declines at all were seen. This interval corresponds to $\sim$100 pulsation cycles, none of which apparently resulted in puffs from the hemisphere facing us. Could they all have happened on the far side of the star? (And rotation, low though it must be, must in the course of 10 years have brought most of the surface of the star into view.) The probability of this happening would be tantamount to the probability of tossing a coin 100 times and having it come down heads every time. Could a way out be found by postulating that the star's pulsation died away during those 10 years, something analogous to the Maunder minimum of sunspots? The answer is no. By coincidence, it was in the middle of this period that Jacchia discovered the low-amplitude pulsations (Jacchia 1933), announcing that the star was pulsating with a visual amplitude of 0.4 mag and period 44 days. Although the difference is hardly significant, this means that, if anything, the star had an even larger amplitude then than it does now (0.25 mag).

Another feature of the historical light curve shows almost the opposite situation. Between 1863.8 and 1873.8 – again an interval of 10 years – R CrB was in a state of almost perpetual veiling. Only once in that interval, and then only very briefly, did it return to its normal maximum brightness. Again it seems improbable that this could happen on the basis of randomly directed puffs.

## 3. Modern Photometry

For the past several decades I have obtained, or had obtained for me, photoelectric photometry of R CrB. These data through 1993 have been published (Fernie & Seager 1994, and references therein); later data are unpublished, but all the data are available at the Web site given above.

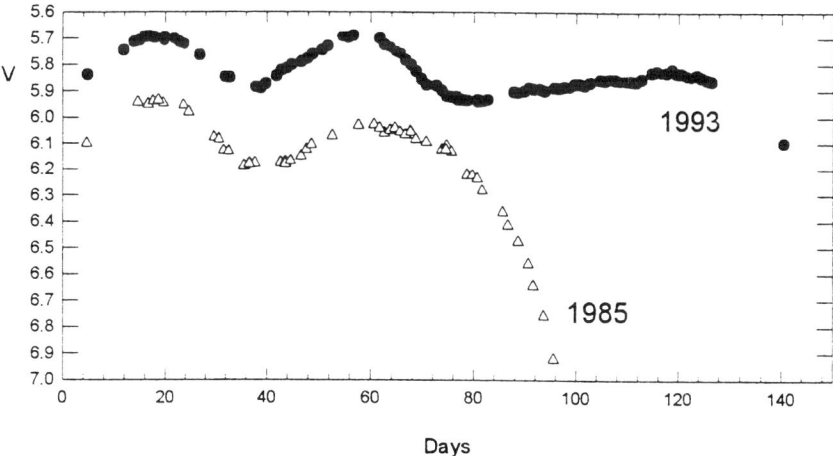

*Figure 1.* The start of two well-observed deep declines. It seems unlikely that they both started at the same phase of the pulsation cycle.

Figure 1 shows two well-documented beginnings of deep declines present in the above photometry. It is difficult to assess precisely when a deep decline begins, but it seems fair to say that the start occurred at different phases of the pulsation in these two cases. In the 1985 case it seems to have started near maximum light, whereas in 1993 the start must have been closer to minimum light. Of course, the delay between the eruption of material through the photosphere and its later condensation into dust is not known; nor is it known whether that delay is the same in all cases, so it is difficult to draw firm conclusions from Fig. 1, but at least it offers no support for the view that puffs are related to the pulsation. Asplund et al. (1997) and Asplund (2000) offer an alternative mechanism for producing the puffs.

### 3.1. NATURE OF THE PULSATION

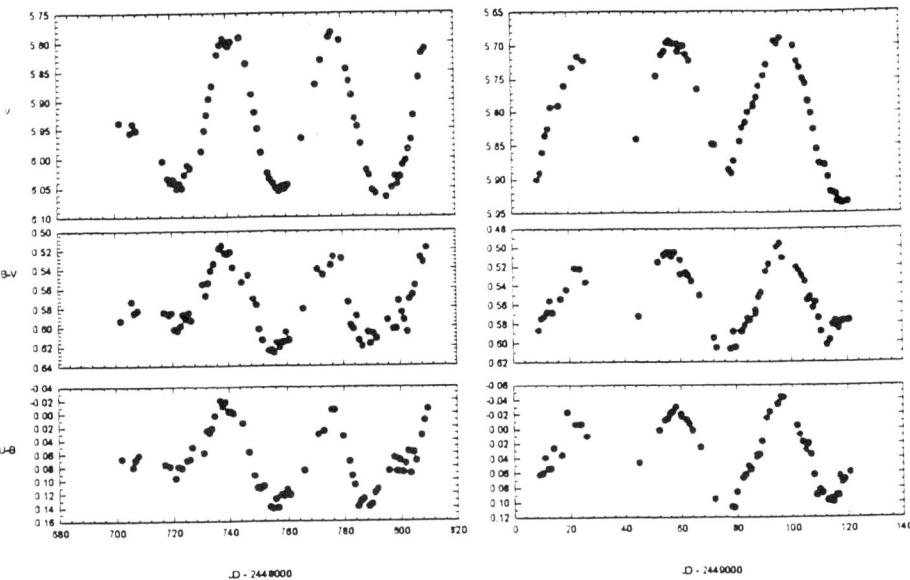

*Figure 2.* Photometry of R CrB in 1992 (*left*) and 1993 (*right*). There is no suggestion of more than one period being present, and times of maximum can be fitted with a simple ephemeris.

Figure 2 shows the $UBV$ photometry of R CrB obtained in 1992 (on the left) and in 1993 (on the right). There is no suggestion in this figure (or in Fig. 1) of there being more than one period present. Not only are the light curves smooth and repeating, but one can fit a simple ephemeris linking the times of maximum across these ten cycles. An even more convincing case was reported in Fernie (1989) where it is shown that a single period of 43.8 days fits all observed times of maxima over 16 cycles in 1985, 1986, and 1987. Moreover, during this interval the star suffered a deep decline and

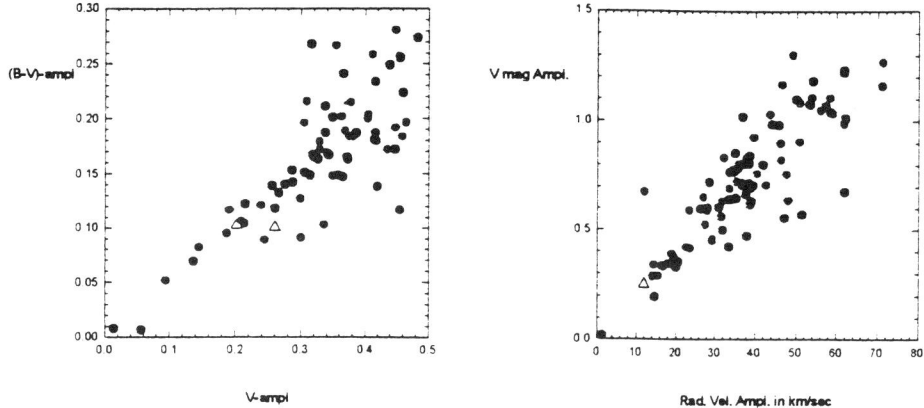

*Figure 3.* The colour/light amplitude relation (*left*) and light/radial velocity amplitude relation (*right*) for classical Cepheids (*circles*) and R CrB (*triangles*). Clearly, R CrB follows the Cepheid relations, implying radial pulsation.

recovery, so that the maintenance of the pulsational phasing strengthens the view that the star itself is not affected by such declines.

It is probably the majority view that the pulsation is radial, but since Stanford et al. (1988) suggested non-radial pulsation it is worth examining the photometric evidence. Figure 2 shows first that the $B-V$ and $U-B$ colour curves are definitely in phase with the $V$ curve, which is a strong indicator of radial pulsation. Secondly, it shows that the colour amplitudes are a substantial fraction of the $V$ amplitude, again strongly favouring radial pulsation. In fact, the left panel of Figure 3 shows the relation of $B-V$ amplitude and $V$ amplitude for classical Cepheids (circles) and for R CrB (triangles) on two occasions, and it is clear that R CrB obeys the Cepheid relation. Fernie & Lawson (1993) have discussed the radial velocities of R CrB, and the right panel of Fig. 3 shows the $V$ amplitude versus the radial velocity amplitude for classical Cepheids (circles) and R CrB (triangle). Again the latter seems to be in accord with Cepheid pulsation. The conclusion must be that R CrB is a radial pulsator.

However, it would be a mistake to believe that R CrB behaves like a Cepheid in all respects. In particular, the period is quite unstable. The well-behaved light curve of 1992 in Fig. 2 has a period near 36 days, which rises to 39 days in 1993, although a linear drift with time easily fits the data. This is significantly different from the constant 43.8 ±0.1 day period of the 16 cycles in 1985–87. The left panel of Figure 4 (the 1994 data) shows how great the period instability can be; the interval between the first two maxima is 52 days, while it is 45 days between the second and third maxima,

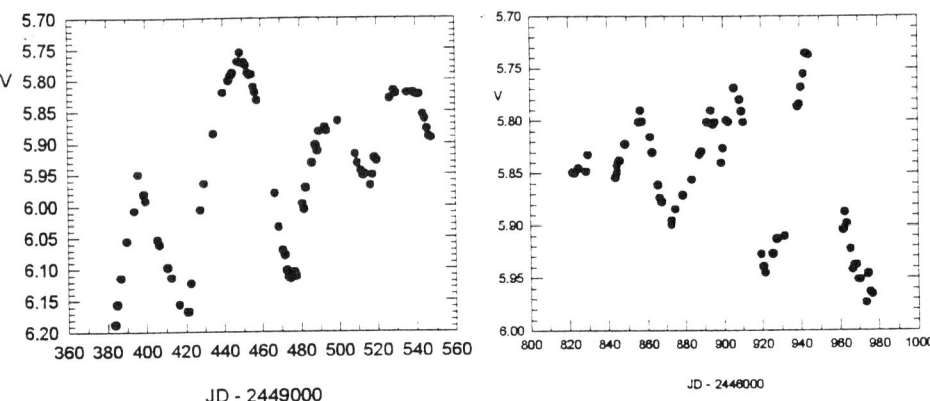

*Figure 4.* Examples of period instability (*left*) and amplitude instability (*right*), showing that some aspects of R CrB's behaviour are decidedly non-Cepheid-like.

and 39 days between the third and fourth peaks. Moreover, not only can there be frequency modulation, but also amplitude modulation. The right panel of Fig. 4 (1987 data) shows a series of cycles with a constant period of 44 days, but with rapidly increasing amplitude. This is probably the signature of strongly non-adiabatic behaviour. It might be thought that partial veiling could be the cause of these various apparent instabilities rather than the star itself, but this is unlikely. The colours continue to follow the $V$ magnitude in just the same way as they do when no modulation is present, and it is improbable that random veiling would produce an apparent period in Fig. 4 (right) that is constant and the same as it had been in the two previous seasons. Finally, given that radial velocities should be unaffected by veiling, one may examine the data of Gorynya et al. (1992). In 1991 they found a well-defined velocity curve with a period of 43.0 days, but, despite the use of the same equipment and procedures, the following season produced only a very ill-defined velocity curve with a period of 34.0 days.

## 4. Conclusions

The historical photometric evidence suggests that a connection between pulsation and mass loss in the form of puffs is unlikely. The fact that deep declines (6–8 mag) in the light outnumber shallow declines (1–2 mag) by nearly a factor of 2 implies that condensation of a puff into dust must happen quite close to the star and expand to veil almost all the photosphere if the puff occurred on the hemisphere facing us, i.e. we should detect almost every puff on that hemisphere. Yet between 1924 and 1934 no declines at all were seen, despite the fact that the star was pulsating at its full amplitude during that time. The probability that all puffs during that decade (~100 pulsation cycles) happened on the far hemisphere is negligibly small. During another decade between 1864 and 1874 the star was veiled almost constantly, again an improbable event if puffs occur randomly.

Recent photoelectric photometry shows that in the two cases observed, the deep decline began once near pulsational maximum light and once near minimum light, suggesting that declines are not associated with any particular phase of pulsation. The photometry also shows that only one period is present at a time, and that this may remain stable for at least sixteen cycles. At other times, both period and amplitude can drift significantly from one cycle to the next. The $B-V$ and $U-B$ curves are closely in phase with the $V$ light curve, and the ratios of colour amplitude to light amplitude and radial velocity amplitude to light amplitude are the same as those for Cepheids. This strongly suggests that the pulsation is radial, as one would expect.

## References

Asplund, M. 2000, in IAU Symp. 177: *The Carbon Star Phenomenon*, ed. R. F. Wing (Kluwer), p. 521
Asplund, M., Gustafsson, B., Kiselman, D. & Eriksson, K. 1997, *A&A*, 318, 521
Fernie, J. D. 1989, *PASP*, 101, 166
Fernie, J. D. & Lawson, W. A. 1993, *MNRAS*, 265, 899
Fernie, J. D. & Seager, S. 1994, *PASP*, 106, 1138
Gorynya, N. A., Rastorguev, A. S. & Samus, N. N. 1992, *Sov. Ast. Lett.*, 18, 142
Jacchia, L. 1933, *Publ. Osserv. Astron. Univ. Bologna*, vol. II, No. 14, p. 241
Mattei, J. A., Waagen, E. O. & Foster, E. G. 1991, *AAVSO Monograph 4*
Mattei, J. A., Waagen, E. O. & Foster, E. G. 1996, *AAVSO Monograph 4, Supp. 1*
Stanford, S. A., Clayton, G. C., Meade, M. R., Nordsieck, K. H., Whitney, B. A., Murison, M. A., Nook, M. A. & Anderson, C. M. 1988, *ApJ*, 325, L9

## Discussion

**Cannon**: You pointed out the remarkable constancy of the depth of the deep minima, with apparently a very stable lower luminosity limit. Is there a simple physical explanation for this?

**Fernie**: The most likely explanation is that it is the spectral emission lines that set the level of the deepest minima, although this begs the question of why they, in turn, are so consistent.

**Asplund**: Even though the modelling by Sedlmayr, Woitke, and collaborators is very promising, the pulsation-induced dust condensation often invoked to explain the declines of RCB stars has some clear problems: not all RCB stars pulsate, and other similar stars pulsate but still do not show any light declines.

**Fernie**: I'm not sure that there are non-pulsating RCB stars; in some cases there is not much data.

**Sedlmayr**: Do you find any noticeable changes in the normal small variations in the quiet state just before a decline event, e.g. amplitudes, frequencies, etc.?

**Fernie**: No, although the data are rather limited.

**Asplund**: The similarity between UU Her and RCB stars only exists for the semi-regular pulsations and not for the light declines and H-deficiency. Therefore I do not expect a relation between these two classes.

**Fernie**: My remark about their similarity referred in fact to the semi-regular pulsations, plus their rarity and galactic distribution. I certainly agree, however, that the pulsation-emission connection is somewhat doubtful.

**Green**: Is there any evidence for increased dust formation above starspots, and could that explain the long gaps without declines as starspot minima and the changes in period as starspot latitude changes?

**Fernie**: Not that I know of. I'm not sure what observational evidence one could expect for this.

**Hrivnak**: You showed a graph of $V$ magnitude vs. time, in which the $V$ amplitude appeared to build up, starting with very low amplitude. At the same time, the low $V$ amplitude corresponds to a large variation in $U - B$. Can you comment on this?

**Fernie**: On reviewing the graph, I think it is more likely observational noise. The APT is only a 25-cm telescope, and the $U$ filter gives the weakest signal.

**Bakker**: If R CrB experiences reddening due to dust ejected in the line of sight, one would also expect dust to be ejected on the other side of the star. In that case blue light is scattered and I would expect an increase in magnitude. Do you observe this?

**Fernie**: No, this is not observed. I don't have a certain explanation for that, but would guess that the dust on the far side is too far away and too thin for backscattering to be detectable.

# THE PULSATIONS AND EVOLUTION OF AGB STARS IN THE LARGE MAGELLANIC CLOUD

Z. G. GONG AND Y. LI

*Yunnan Observatory, Kunming, P. R. China*

**Abstract.** The evolutionary status and pulsational characteristics of AGB stars in the LMC have been studied. It is found that only those stars with a small mixing-length ratio can fit the observed period-luminosity relation. Strong non-adiabatic effects were found in the outer envelope and must be taken into account in future work.

## 1. Introduction

The long-period variable stars (LPVs) are red variables which include the Mira stars, semiregular variable stars, red supergiant variable stars, and periodically pulsating OH/IR sources whose periods are from several tens of days to more than one thousand days (Hughes 1989; Hughes & Wood 1990; Wood, Moore & Hughes 1991; Wood et al. 1992; Whitelock 1995). As all of these stars located in the red and bright part of the HR diagram, their study will be very important for the understanding of late stages of stellar evolution.

The LPVs in theoretical astrophysics are classified into two subgroups. One is for those red supergiant LPVs and some OH/IR sources which are thought to be massive stars in their late helium-burning stages (Li & Gong 1994). The other group includes the intermediate-mass stars on the asymptotic giant branch (AGB), including Miras, semiregular variables, and some other OH/IR sources. In this paper we will focus on the latter subgroup (AGB stars).

As the stars in the Large Magellanic Cloud (LMC) are all at almost the same distance from our observers, and the AGB LPVs in the LMC have been so well studied that a lot of data are available, we will pay most attention to these AGB LPVs.

Theoretical studies of the characteristics of AGB LPVs can be found in some early papers. From these publications we find that, although many aspects of the evolutionary and pulsational status of these stars are known in general, problems and discrepancies still remain, especially as regards the pulsation mode. In the work of Fox & Wood (1982) we can find a full set of pulsational properties of AGB LPVs, but their introduction of a convection – pulsation interaction mechanism, which has not been completely solved and has not been widely used in investigations of other variable stars, in their stellar pulsation calculation is still questionable. And their suggestion of the AGB LPVs being first-overtone pulsators is different from the later conclusion of Ostlie & Cox (1986), who regarded these stars as pulsating in the fundamental mode. The question regarding the work of Ostlie & Cox is that they artificially increased the opacity of the stars by averaging the opacity of upward and downward cells, and such a treatment would seem to need some physical analysis, which was ignored, before its application.

When nonlinear effects are considered in the pulsation calculation of AGB LPVs, the discrepancy in pulsation mode still exists. In Wood (1990) and Bessell et al. (1996), Miras are thought to pulsate in the fundamental mode, while in Barthès & Tuchman (1994) they are first-overtone pulsators.

Bowen (1988), on the other hand, did very helpful work on the nonlinear atmospheric activity of Miras and supported the conclusion that AGB LPVs are fundamental-mode pulsators. However, the need of a pulsational inner boundary condition in his computation implied that the connection between envelope pulsation and this boundary condition must be considered. Some information about the interior is very important in the study of the behavior of the atmosphere, too.

In this paper we will investigate both the evolution and pulsation of intermediate-mass AGB LPVs based on recent physics considerations to get some more ideas about these stars. In §2 we discuss our codes and input physics. In §3 we compare our results with observation, and then make some comments.

## 2. Codes and Input Physics

We evolved our models from the ZAMS to the AGB by using a modified version of the Kippenhahn et al. (1967) stellar structure and evolution code to get static models of AGB LPVs, then calculated their linear non-adiabatic radial pulsation properties under the Li (1992a,b) stellar pulsation code. In this paper we study intermediate-mass AGB LPVs whose initial masses are $3 M_\odot$ and $7 M_\odot$. The mass loss rate was calculated following Nieuwenhuijzen & de Jager (1990), which is valid throughout the HR diagram. The evolutionary tracks of these stars are shown in Figure 1.

## Evolutionary Tracks

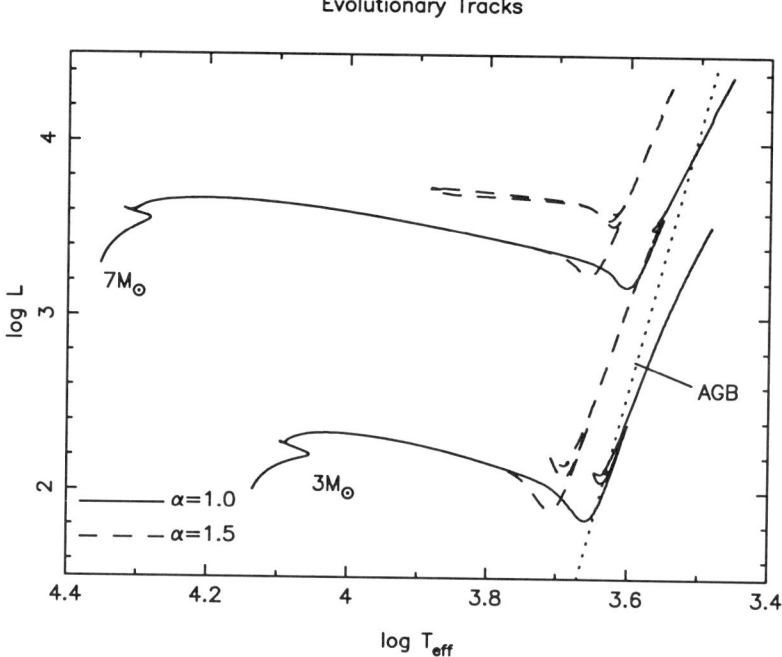

*Figure 1.* Evolutionary tracks of $3\,M_\odot$ and $7\,M_\odot$ stars.

The initial chemical composition was chosen to be $(X, Z) = (0.7, 0.01)$ to fit the requirements of stars in the LMC, whose heavy element abundance is thought to be only one-half that of the solar neighborhood. Recent available opacities are used in our evolution calculation, in which OPAL opacities (Rogers & Iglesias 1992) are used in the high-temperature region ($\log T > 3.85$), while those of Alexander & Ferguson (1994), which include both atomic and molecular opacities, are used in the low-temperature region ($\log T \leq 3.85$). We find that both the data and their derivatives can connect smoothly at this border.

Convective energy transportation was treated according to the mixing length theory of Böhm–Vitense (1958) with the Schwarzschild criterion. The temperature gradient, unlike the treatment of Kippenhahn et al., was calculated by Böhm–Vitense's formalism throughout the whole model, and the effect of overshoot or semiconvection was not included.

In this paper two different pressure scale heights $\alpha = l/H_p$ were tested. One was $\alpha = 1.5$, which is usually used in stellar evolution calculations and is close to the value in the standard solar model (Guzik & Cox 1995). The other one was $\alpha = 1.0$, which is smaller than in most published evolution studies, but it was found that this small value allows the pulsation mode

## P–L Relation

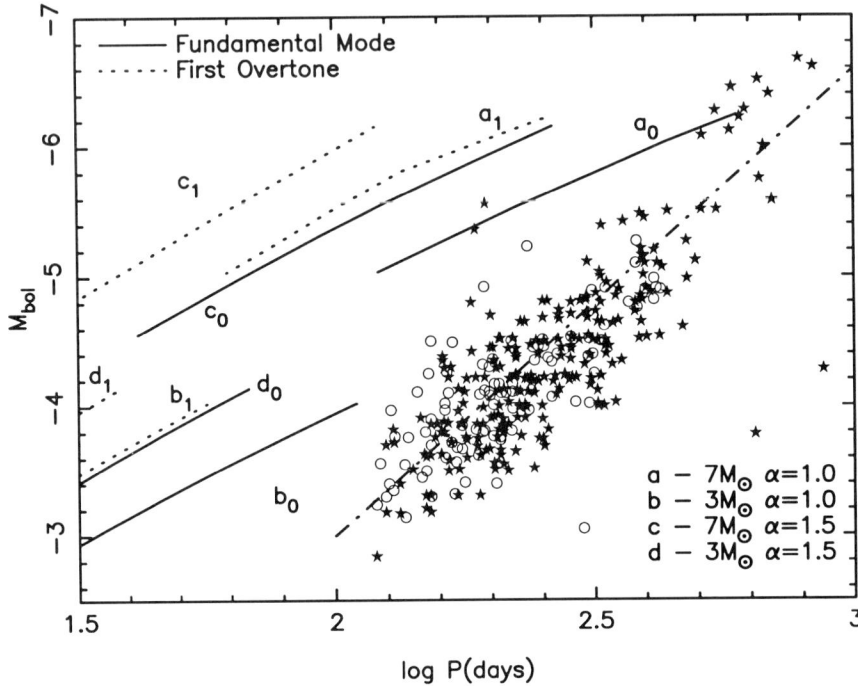

Figure 2. Period–luminosity relation of AGB LPVs in the LMC.

of AGB LPVs to fit the observed data much better.

## 3. Results and Discussion

From our evolution and pulsation calculation it seems that all stars will be pulsationally unstable when they evolve onto the asymptotic giant branch, and the period of such stars will be from several tens of days at the lower AGB to several hundred days at the beginning of thermal pulse phase. The brighter and cooler the stars are, the longer their period will be. This is not difficult to understand because the brighter and cooler stars have a bigger cavity to allow the pulsation wave to travel, and the local sound speed doesn't change very much. Hence, the period will be longer.

When we compare our results with the observational data of AGB LPVs in the LMC (Hughes & Wood 1990) in Figure 2, we find that in the period-luminosity diagram the first-overtone lines, no matter what the mass and $\alpha$ value, all lie far from the observed data, and they show no potential for reaching the region where the observed stars are located. So we suggest that

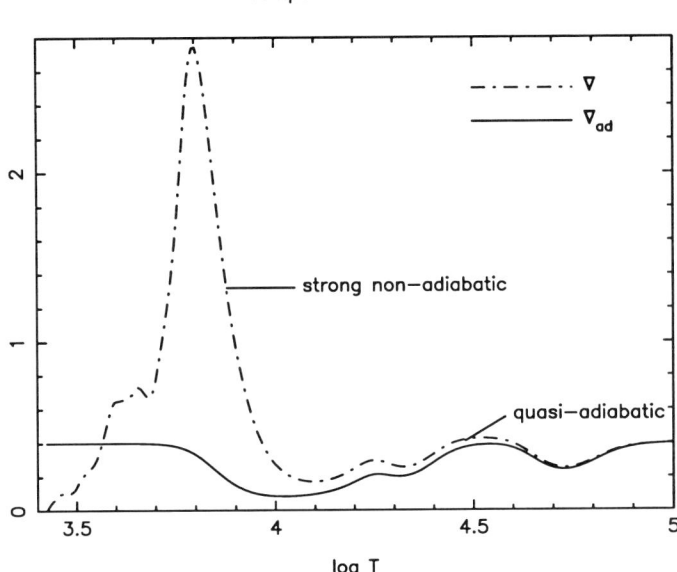

*Figure 3.* Temperature gradient in the outer envelope of AGB LPVs.

the first overtone is not likely to be the first candidate for the pulsation mode of AGB LPVs in the LMC.

For the fundamental mode things are different. Stars with the bigger $\alpha$ value (c & d) lie further from the observed values than those with the smaller $\alpha$ value (a & b). It seems that the fundamental mode of the low $\alpha$ value AGB stars can fit better with the observations. The $b_0$ line does not cross the observed star group because our calculation stopped at the beginning of the thermal pulse phase, and when thermal pulses start the star will evolve to be brighter and cooler and its period will be longer so that this line may cross the observed stars. Our conclusion is that the AGB LPVs may possibly be fundamental mode pulsators.

In our stellar pulsation calculation we find that, in the outer envelopes of intermediate-mass AGB stars ($\log T < 4.0$), the pulsation is strongly non-adiabatic (Figure 3). The quasi-adiabatic stellar pulsation theory that is widely used in solving the non-adiabatic effect in stellar pulsation is not valid in this region. A theory which can include the strong non-adiabatic effect is needed if we want to know more accurately the pulsational characteristics in the outer envelopes of cool stars (Li & Gong 1998).

In conclusion we suggest that AGB LPVs in the LMC are more likely to pulsate in their fundamental mode, and that a small mixing length ratio

$\alpha$ is required in fitting the theoretical results with the observed stars.

The authors thank Dr. Robert Wing for his fruitful help in preparing this paper, and Dr. Nami Mowlavi for helpful discussion. One of the authors (Z.G.) thanks Dr. Robert Wing for the invitation and grants to attend IAU Symposium 177. This work was supported by the Chinese National Natural Science Foundation.

## References

Alexander, D. R. & Ferguson, J. W. 1994, *ApJ*, 437, 879
Barthès, D. & Tuchman, Y. 1994, *A&A*, 289, 429
Bessell, M. S., Scholz, M. & Wood, P. R. 1996, *A&A*, 307, 481
Böhm–Vitense, E. 1958, *Z. Astrophys.*, 46, 108
Bowen, G. H. 1988, *ApJ*, 329, 299
Fox, M. W. & Wood, P. R. 1982, *ApJ*, 259, 198
Guzik, J. A. & Cox, A. N. 1995, *ApJ*, 448, 905
Hughes, S. M. G. 1989, *AJ*, 97, 1634
Hughes, S. M. G. & Wood, P. R. 1990, *AJ*, 99, 784
Kippenhahn, R., Weigert, A. & Hofmeister, E. 1967, *Meth. Comp. Phys.*, 7, 129
Li, Y. 1992a, *A&A*, 257, 133
Li, Y. 1992b, *A&A*, 257, 145
Li, Y. & Gong, Z. G. 1994, *A&A*, 289, 449
Li, Y. & Gong, Z. G. 1998, in IAU Symp. 181: *Sounding the Solar and Stellar Interior*, Poster Volume, ed. F. X. Schmider and J. Provost (Nice: Université de Nice), p. 233
Nieuwenhuijzen, H. & de Jager, C. 1990, *A&A*, 231, 134
Ostlie, D. A. & Cox, A. N. 1986, *ApJ*, 311, 864
Rogers, F. J. & Iglesias, C. A. 1992, *ApJ Supp.*, 79, 507
Whitelock, P. 1995, *Astrophys. & Space Sci.*, 230, 177
Wood, P. R. 1990, in *From Miras to Planetary Nebulae*, ed. M. O. Mennessier and A. Omont (Editions Frontières), p. 67
Wood, P. R., Moore, G. K. G. & Hughes, S. M. G. 1991, in IAU Symp. 148: *The Magellanic Clouds*, ed. R. Haynes and D. Milne (Kluwer), p. 259
Wood, P. R., Whiteoak, J. B., Hughes, S. M. G., Bessell, M. S., Gardner, F. F. & Hyland, A. R. 1992, *ApJ*, 397, 552

# Session V

## Post–AGB Stars

Michael Feast, as remembered by Pierre North.

# POST-AGB VARIABLES AND STELLAR MASS-LOSS

M. W. FEAST
*Astronomy Department*
*University of Cape Town, South Africa*

**Abstract.** A brief review is given of the various types of star which are thought to be in the immediate post-AGB stage of evolution. The paper then concentrates on the properties of the RCB stars and particularly on the mass-loss process in these stars. It is suggested that grain formation takes place over the cool regions of giant convection cells in a super-Eddington outflow and in the form of small clouds or puffs. Attention is drawn to observations which suggest that grain formation in the outer atmospheres of Miras and other cool giants may also take place in puffs rather than in spherical shells. Evidence on the long-term variation of the circumstellar dust emission from RCB stars is summarized.

## 1. Introduction

A wide variety of objects have been considered as belonging to the post-AGB, pre-PN stage, many of them being variables. The present paper gives a very brief survey of some of these classes of objects. It then concentrates on the R Coronae Borealis (RCB) variables and in particular on what we can learn from them about mass-loss and grain-formation processes. The results suggest clues for the understanding of these processes in cool variables at the tip of the AGB.

## 2. Criteria for Post-AGB Stars

Amongst the criteria which have been used to assign stars to the post-AGB phase are the following:

1. Unusual surface chemical abundances;
2. Early or intermediate spectral type together with high luminosity and low mass;

3. Circumstellar material (gas, dust) indicating earlier and/or current mass-loss (e.g. cool dust, detached shells); and
4. Period changes in pulsating variables indicating rapid evolution to higher temperatures.

However there are several uncertainties in applying these criteria:

1. The luminosities of the objects concerned are often quite uncertain;
2. Low masses are frequently inferred from galactic position or kinematics and this can lead to confusion with, for instance, high-mass stars which are in the region of the galactic halo;
3. The evolution of some objects may have been strongly affected by binary interaction leading to some of the characteristics listed above though the stars are not in the post-AGB phase;
4. It is still uncertain whether bipolar symmetry in circumstellar material is always an indication of binary interaction or whether it might occur in single post-AGB stars; and
5. It is often not clear whether observed period variations are of long-term evolutionary significance.

## 3. IRAS Sources

An early criterion for post-AGB status was a cold, probably detached, non-variable dust shell as seen in IRAS data (e.g. van der Veen et al. 1989). However, stars undergoing thermal pulsing at the top of the AGB may only become large-amplitude variables (Miras), with high mass-loss, in the bright phase of a thermal pulse. At other times they might well have only low variability together with a detached shell ejected in the Mira phase (e.g. Hashimoto 1994). Good post-AGB candidates are however found by demanding also that the underlying star be a hot (F–G type) supergiant and that there be unusual surface abundances. Stars selected in this way, for example by Kwok (1993) and Hrivnak (1995), are carbon stars showing evidence for both $C_2$ and $C_3$. There remains some uncertainty for these objects since some at least (e.g. AFGL 2688) have bipolar symmetry. However, recent HST pictures show that this object has ejected multiple shells, possibly connected to thermal pulsing. Little or nothing is known about the possible variability of these objects.

## 4. High-Latitude F–G Type Supergiants

This is a heterogeneous group which overlaps with an IRAS-selected sample. These stars are not generally classified as carbon stars although some have enhanced abundances of C, N, and O (Luck 1993; Luck et al. 1990) which suggest a post-AGB status. Of those with such enhancements, the best studied for variability are 89 Her, which is in a spectroscopic binary system, and HD 161796, which is probably single (Waelkens & Mayor 1993; Waters et al. 1993). These two stars show variations of $\Delta V \approx 0.1$ mag on timescales of 40 to 70 days (e.g. Fernie 1993). Whether these are truly periodic is not certain (they may be similar to the RCB stars discussed below). Radial velocities give no good evidence for radial pulsations in either star (Waelkens & Mayor 1993). Members of a subset of these F–G supergiants have very low overall metallicities. They are interacting binaries and thus not normal post-AGB stars (van Winckel et al. 1995).

## 5. RV Tauri Stars

The RV Tauri stars are also a rather heterogeneous group although they are all probably radial pulsators. One subgroup (Preston type B) consists of carbon stars. RV Tauri stars generally have strong dust excesses in the IRAS data and often in the nearer infrared too. Jura (1986) modelled these excesses in several cases and found that mass-loss had been much greater (perhaps by a factor of 100) several hundred years ago, consistent with these stars being recent entries to the post-AGB phase and also consistent with the presence of silicate-rich dust shells round C-rich RV Tauris (Lloyd Evans 1974). However, different modelling (Raveendran 1989) suggests a more uniform mass-loss continuing to the present.

Stars of the RVb subclass show long-period, $\sim 1000$ day, modulation of their light curves. This is thought to indicate long-period binary motion though the precise mechanism of the modulation is not clear. Several C-rich RV Tauris are in this subclass and have recently been studied extensively by Pollard et al. (1996) using optical photometry. Even RV Tauris which are not in the RVb subclass could still be in binaries. For instance the C-rich star AC Her is classed as RVa (no long-term modulation of the light curve), but radial-velocity work (Waelkens & Waters 1993) indicates that it is in a spectroscopic binary. This suggests that current mass-loss in some, and possibly all, RV Tauris is connected to binary activity and might, for instance, be confined to a disc.

## 6. RCB Variables

The RCB stars (see Feast 1996, Clayton 1996, and references in these papers) are high-luminosity ($M_V \approx -5$), low-mass objects. In our Galaxy they are strongly concentrated to the galactic centre. They have an overabundance of carbon and a hydrogen deficiency which ranges from at least $10^{-8}$ to $10^{-1}$ solar (Lambert & Rao 1994). The RCB class is generally treated as post-AGB although the evolutionary status of these stars (possibly bornagain PN or merged white dwarfs) has not yet been finally settled. RCB stars undergo random drops in optical brightness of 7 mag or more and these have long been attributed to dust obscuration. Near infrared observations show an excess attributable to dust at a mean blackbody temperature of 800–1000 K (Feast & Glass 1973). FG Sge, which brightened visually by $\sim 6$ mag over $\sim 100$ years as its atmosphere expanded and cooled, changing its bolometric correction, has recently undergone obscuration minima and shows carbon bands indicating that it is now an RCB star or at least a related object (e.g. Kipper et al. 1995).

Photometry and radial velocities show the RCB star RY Sgr to be a radial pulsator with a period of $\sim 38$ days (Alexander et al. 1972). With its large light amplitude ($\Delta V \approx 0.5$ mag) and evidence for radial pulsations, this star is atypical of the group. However, its pulsation is particularly useful in establishing a number of basic facts regarding RCB stars. Whilst in RY Sgr and other RCB stars outside obscuration minima most of the radiation at optical and nearer infrared wavelengths (i.e. $1.2\,\mu$m, $J$ band) comes directly from the stellar photosphere, that at $3.5\,\mu$m ($L$ band) and longer wavelengths is mainly from heated circumstellar dust. In RY Sgr the $L$ flux varies in the pulsation period of the star, showing that the pulsating star is responsible for the heating of the dust (e.g. Feast et al. 1977; Feast 1986; Menzies & Feast 1997). During obscuration minima the $L$ observations show that the pulsations go on unchanged, demonstrating that the obscuration has no major effect on the star itself. Also the mean $L$ flux from the shell is unchanged, indicating that the optical obscuration is due to a dust cloud (a puff) of limited size in the line of sight. Obscuration minima of RY Sgr seem normally to begin with a rapid decline, followed immediately by a rapid rise to an intermediate brightness. The minima in these initial dips correspond closely with minima in the pulsation cycle as seen at $L$ and give the optical light curve the appearance of a giant pulsation cycle. This effect can be explained as due to the change, during a pulsation cycle, in the size and surface brightness of the outer regions of the star (the "chromosphere") when the main body of the star is hidden by a dust puff (Feast 1996; Menzies & Feast 1997).

The puffs of dust are driven away from the star at, typically, 200 km s$^{-1}$ by radiation pressure. This is shown by absorption lines from gas entrained with the dust. Since our line of sight cannot be presumed special, we expect dust puffs to be ejected randomly in all directions, and in deep minima we see broad ($\sim 400$ km s$^{-1}$) emission lines which are best interpreted as due to resonance scattering by gas entrained in these puffs (Feast 1996 and earlier papers).

The outer ("chromospheric") regions of RY Sgr are known to be moving outward at $\sim 10$ km s$^{-1}$, and the recent work of Asplund & Gustafsson (1996) suggests that this is a super-Eddington outflow. Mass loss can then be seen as a two-stage process: first, low-velocity gas ejection by radiation pressure; then dust formation in this wind, followed by high-velocity dust ejection (also by radiation pressure).

There are a number of constraints on where in the stellar wind the dust grains are formed:

1. The initial declines of RCB stars tend to be rapid (time-scale of a few days) and the recovery times slow (1 to 3 years). Typical light curves can be reproduced if the dust forms as a small cloud relatively close to the star (say at about two stellar radii above the surface). The rapid decline occurs as the puff, moving radially outwards, expands to cover all the star. The slow recovery takes place as the puff gradually thins optically. It is difficult, though perhaps not entirely impossible, to reproduce the light curves satisfactorily if the dust forms far from the star (say 10 stellar radii or more) (Feast 1986, 1996; Clayton 1996).
2. The blackbody temperature of the dust (800–1000 K) suggests that the bulk of the dust is not very close to the star. One can model the infrared colours of dust formed quasi-continuously in puffs which then move rapidly outward. These colours depend on the temperature at which the bulk of the dust forms. Comparison with the extensive infrared photometry of RCB stars extending over more than 20 years (Feast et al. 1997; Feast 1997) shows that $T_{max} \lesssim 1500$ K. This is consistent with the theoretical upper limit for the formation of carbon grains (Salpeter 1977) and with the observed temperatures of carbon-rich material in other sources (see Feast 1997). Such a temperature is only reached at a distance of 10 stellar radii or more above a 6000 K photosphere.
3. The dust forms in puffs rather than spherical shells.

It appears possible to meet these constraints in the following way. Wdowiak (1975) suggested that dust might form in the cool regions of large photospheric convection cells. Whilst it seems unlikely that these could be cool enough, the existence of a super-Eddington outflow suggests that dust formation at the required condensation temperature could take place close to the star in this wind above such cool regions. Such a model then accounts for (1) the production of dust in puffs, (2) the random distribution of puffs over the star, (3) dust formation close to the star, and (4) dust production at the observed (and expected) temperature.

Since the dust forms above dark areas of the disc, this model explains why in general one does not see the molecules ($C_2$ etc.) which should precede dust formation. However, occasions may occur when the puff covers the entire disc before becoming optically thick. This seems to have been the case for V854 Cen at an obscuration minimum in 1988. One would expect to see strong molecular bands at such times, and these were apparently seen (Kilkenny & Marang 1989; Feast 1997). It might also be expected that in certain conditions molecules would form without subsequent grain formation. When this slowly expanding molecular cloud is projected on the bright areas of the disc, one would see strong molecular absorption near maximum light. Such events were first reported by Espin (1890).

If the convection cell model is basically correct, then we would expect that outside obscuration minima RCB stars will show small-amplitude variations of $\Delta V \approx 0.1$ mag on a time scale of $\sim 1$ month (see Feast 1996). Variations of this kind are found in most RCB stars. It is still not clear how much of this observed variation is due to the convection cell mechanism and how much to underlying pulsation. Clearly pulsation is a main component in RY Sgr. In many other cases, however, the amplitude of variation is small and the evidence for a regular periodicity is rather weak. Feast et al. (1997) give an example in the case of SU Tau where there was a clear variation of the star in the $J$ band, which might well have been taken for a pulsation cycle, but no variation in the dust flux (at $L$), strongly suggesting that this was not a global pulsation.

## 7. Mass Loss at the AGB Tip

Mira variables, both O-rich and C-rich, at the tip of the AGB are losing matter at a copious rate. The process by which they do this is not fully understood although it is of crucial importance in stellar evolution. As with the RCB stars, mass-loss is generally considered as a two-stage process. Gas is raised to sufficient heights above the star to form grains which are then blown away from the star by radiation pressure. The problem has been to understand how the gas is raised to sufficient heights for grain formation. The most promising mechanism for this has seemed to be pulsation, but it is not certain that this can raise gas sufficently (e.g. Wood 1990). However, one might expect that grain formation could, as with RCB stars, take place over the cool regions of the giant convection cells which Schwarzschild (1975) suggested existed in the atmospheres of red giants. Thus the region of grain formation would be much closer to the surface than would otherwise be the case. One then expects this dust to form in puffs. In fact there is considerable direct or circumstantial evidence to support such a model which will lead to surface or circumstellar asymmetries in Mira variables (O- or C-rich). Amongst such evidence is the following:

1. The K I 7699 Å fluorescent emission is highly asymmetrical (Plez & Lambert 1994);
2. The CO distribution around both C- and O-rich Miras is asymmetrical (Stanek et al. 1995);

3. Polarization indicates asymmetry in the scattering dust (e.g. Johnson & Jones 1991, and earlier work);
4. H$_2$O masers show asymmetrical structure (Yates & Cohen 1994);
5. OH masers in U Her, R Cas, and W Hya show that mass loss is "intrinsically chaotic and clumpy" with the clumps retaining their identity as they move outward (Chapman et al. 1994);
6. The lunar occultation of the C star TX Psc (a long-period variable) shows asymmetry due either to a large cold "spot" on the star or to a dust patch close to the star (Richichi et al. 1995); and
7. Optical interferometry of Miras shows that these stars are frequently non-circular in projection, a plausible explanation being large spots on the surface (e.g. Haniff 1995; Tuthill 1995; Lattanzi et al. 1997).

If this mechanism for the production of dust in Miras is correct we might expect to see RCB-type declines in them. Possibly this is what is being seen in such C-type Miras as R For, as suggested by Whitelock (2000). As yet the data are not available to tell whether this is so or whether the declines of these stars are due to the formation of a complete spherical shell. Similar events in O-rich Miras might be expected to be less evident since, in the optical and near infrared, the extinction cross section of silicate grains is about ten times less than that of carbon grains of the same size.

It is interesting to inquire why we apparently see no RCB activity in C-rich RV Tauri variables. There may be two contributory causes for this. First, it is not entirely clear how high the present mass-loss rate of these stars is. Secondly, as mentioned in section 5, many (at least) of the RV Tau stars are in binary systems, and dust production may in those cases be mainly connected to binary interaction and disc formation.

## 8. Long-Term Variations in RCB Dust Formation

The extensive infrared photometry of 12 RCB stars for periods extending in several cases over more than 20 years (Feast et al. 1997) has allowed the study of long-term variations in the dust production in RCB stars. In no case is this constant. Several types of variation have been distinguished:

1. Large-amplitude changes in the infrared flux from the dust ($\Delta L \approx 2$ mag) on time scales of from one to several thousand days (e.g. UW Cen, WX CrA, RY Sgr, GU Sgr, RS Tel);
2. Smaller amplitude changes ($\Delta L \approx 0.5$ mag) on a time scale of hundreds of days (e.g. S Aps, V854 Cen, V CrA, RZ Nor);
3. Secular variations ($\Delta L \approx 0.5$ mag) over $\sim 10{,}000$ days (e.g V CrA);
4. Periodic variations; R CrB itself seems to stand alone as showing some evidence of a real periodicity (1260 days) rather than simply a general time scale for variations.

The physical mechanisms behind these various time scales are not understood although ~1000 days is the average time between obscuration events in several RCB stars and may therefore be taken as the average time for the renewal of the entire circumstellar environment (Feast et al. 1997).

## References

Alexander, J. B., Andrews, P. J., Catchpole, R. M., Feast, M. W., Lloyd Evans, T., Menzies, J. W., Wisse, P. N. J. & Wisse, M. 1972, *MNRAS*, 158, 305
Asplund, M. & Gustafsson, B. 1996, in *Hydrogen-Deficient Stars*, ed. C. S. Jeffery and U. Heber, ASP Conf. Ser., 96, 39
Chapman, J. M., Sivagnanam, P., Cohen, R. J. & Le Squeren, A. M. 1994, *MNRAS*, 268, 475
Clayton, G. C. 1996, *PASP*, 108, 225
Espin, T. E. 1890, *MNRAS*, 51, 12
Feast, M. W. 1986, in IAU Coll. 87: *Hydrogen-Deficient Stars and Related Objects*, ed. K. Hunger, D. Schönberner and N. K. Rao (Reidel), p. 151
Feast, M. W. 1996, in *Hydrogen-Deficient Stars*, ed. C. S. Jeffery and U. Heber, ASP Conf. Ser., 96, 3
Feast, M. W. 1997, *MNRAS*, 285, 339
Feast, M. W., Carter, B. S., Roberts, G., Marang, F. & Catchpole, R. M. 1997, *MNRAS*, 285, 317
Feast, M. W., Catchpole, R. M., Lloyd Evans, T., Robertson, B. S. C., Dean, J. F. & Bywater, R. A. 1977, *MNRAS*, 178, 415
Feast, M. W. & Glass, I. S. 1973, *MNRAS*, 161, 293
Fernie, J. D. 1993, in *Luminous High-Latitude Stars*, ed. D. D. Sasselov, ASP Conf. Ser., 45, 253
Haniff, C. 1995, in IAU Coll. 155: *Astrophysical Applications of Stellar Pulsation*, ed. R. S. Stobie and P. A. Whitelock, ASP Conf. Ser., 83, 270
Hashimoto, O. 1994, *A&A Supp.*, 107, 445
Hrivnak, B. J. 1995, *ApJ*, 438, 341
Johnson, J. J. & Jones, T. J. 1991 *AJ*, 101, 1735
Jura, M. 1986, *ApJ*, 309, 732
Kilkenny, D. & Marang, F. 1989, *MNRAS*, 238, 1P
Kipper, T., Kipper, M. & Klochkova, V. G. 1995, *A&A Supp.*, 297, L33
Kwok, S. 1993, *Ann. Rev. Astr. Astrophys.*, 31, 63
Lambert, D. L. & Rao, N. K. 1994, *J. Astrophys. Astron.*, 15, 47
Lattanzi, M. G., Munari, U., Whitelock, P. A. & Feast, M. W. 1997, *ApJ*, 485, 328
Lloyd Evans, T. 1974, *MNRAS*, 167, 17P
Luck, R. E. 1993, in *Luminous High-Latitude Stars*, ed. D. D. Sasselov, ASP Conf. Ser., 45, 87
Luck, R. E., Bond, H. E. & Lambert, D. L. 1990, *ApJ*, 357, 188
Menzies, J. W. & Feast, M. W. 1997, *MNRAS*, 285, 358
Plez, B. & Lambert, D. L. 1994, *ApJ*, 425, L101
Pollard, K. R., Cottrell, P. L., Kilmartin, P. M. & Gilmore, A. C. 1996, *MNRAS*, 279, 949
Raveendran, A. V. 1989, *MNRAS*, 238, 945
Richichi, A., Chandrasekhar, T., Lisi, F., Howell, R. R., Meyer, C., Rabbia, Y., Ragland, S. & Ashok, N. M. 1995, *A&A*, 301, 439
Salpeter, E. E. 1977, *Ann. Rev. Astr. Astrophys.*, 15, 267
Schwarzschild, M. 1975, *ApJ* 195, 137
Stanek, K. Z., Knapp, G. R., Young, K. & Phillips, T. G. 1995, *ApJ Supp.*, 100, 169
Tuthill, P. G. 1995, Ph. D. thesis, Cambridge University

van der Veen, W. E., Habing, H. J. & Geballe, T. R. 1989, in IAU Symp. 131: *Planetary Nebulae*, ed. S. Torres-Peimbert (Kluwer), p. 445
van Winckel, H., Waelkens, C. & Waters, L. B. F. M. 1995, *A&A*, 293, L25
Waelkens, C. & Mayor, M. 1993, in *Luminous High-Latitude Stars*, ed. D. D. Sasselov, ASP Conf. Ser., 45, 318
Waelkens, C. & Waters, L. B. F. M. 1993, in *Luminous High-Latitude Stars*, ed. D. D. Sasselov, ASP Conf. Ser., 45, 219
Waters, L. B. F. M., Waelkens, C., Mayor, M. & Trams, N. R. 1993, *A&A*, 269, 242
Wdowiak, T. J. 1975, *ApJ*, 198, L139
Whitelock, P. A. 2000, in IAU Symp. 177: *The Carbon Star Phenomenon*, ed. R. F. Wing (Kluwer), p. 179
Wood, P. R. 1990, in *Confrontation Between Stellar Pulsation and Evolution*, ed. C. Cacciari and G. Clementini, ASP Conf. Ser., 11, 355
Yates, J. A. & Cohen, R. J. 1994, *MNRAS*, 270, 958

## Discussion

**Cherchneff**: You've mentioned large convective cells in which dust could condense. Do you have a feeling for the temperatures of these cells compared to the effective temperature of the star?

**Feast**: The suggestion is that the dust condenses in the quasi-steady Eddington outflow (which is observed in the $\sim 10$ km s$^{-1}$ "chromospheric" lines) above cool regions of convection cells, i.e. at $\sim 2$ stellar radii or so from the star.

**Gustafsson**: Your proposal that the puffs are related to and triggered by convection cells is interesting. If the downdrafts really cool the gas above them and cause dust formation, one might possibly see tendencies for low-excitation spectral lines in the red to be redshifted (or to have red asymmetrical wings) shortly before the decline, and this redshift would then soon vanish as the downdraft gets occulted. Conversely, if the rising hot granules push the gas across the Eddington limit, the corresponding phenomena should appear in the blue wings of the high-excitation lines. Spectral monitoring of these stars at high resolution could be rewarding.

**Asplund**: For RCB stars convection occurs in the region of the He ionization, which takes place in rather deep atmospheric layers ($\tau > 10$). Therefore I'm not convinced that large temperature differences due to convection cells still exist on the stellar surface, which has been proposed to cause the dust condensation events.

**Feast**: Evidently realistic atmospheric models are required. At present I think that the suggestion of large convection cells is quite reasonable. As I said, this predicts low-amplitude variations on time scales of about a month near maximum light, and this may well explain the observed small-scale variations.

**Asplund**: You also mentioned that before the dust condensation event one expects a strengthening of molecular features. Of course few spectra have been taken immediately before a decline, but in fact this has been observed: the $C_2$ bands became stronger shortly before a decline in R CrB, as reported by Rao and collaborators at IAU Colloquium 106.

**Feast**: There is good evidence of essentially neutral-color declines at several minima of RCB stars showing that there was no very strong additional molecular absorption. This does not preclude occasional $C_2$ production and the "Espin effect."

**Magalhães**: I would like to comment that when one observes in the IR, one is basically measuring the dust mass, i.e. favoring larger grains. These grains will typically scatter poorly in the optical; here, smaller grains play a role. This would probably have to be taken into account while interpreting IR and optical measurements and inferring where the grains are with respect to the star, as we may be looking at different grains.

# LINES OF CIRCUMSTELLAR $C_2$, CN, AND $CH^+$ IN THE OPTICAL SPECTRA OF POST-AGB STARS

ERIC J. BAKKER AND DAVID L. LAMBERT
*Department of Astronomy and McDonald Observatory*
*University of Texas, Austin TX, U.S.A.*

AND

EWINE F. VAN DISHOECK
*Leiden Observatory, University of Leiden, The Netherlands*

**Abstract.** Recent optical spectra of post–AGB stars show the presence of $C_2$, CN, and $CH^+$ originating in the circumstellar shell. We present here new, higher resolution spectra which provide constraints on the physical parameters and information on the line profiles. An empirical curve of growth for the $C_2$ Phillips and CN Red system lines in the spectrum of HD 56126 yields $b = 0.50^{+0.59}_{-0.23}$ km s$^{-1}$. $CH^+$ (0,0) emission lines in the spectrum of the Red Rectangle have been resolved with a FWHM $\approx 8.5 \pm 0.8$ km s$^{-1}$. The circumstellar CN lines of IRAS 08005–2356 are resolved into two separate components with a velocity separation of $\Delta v = 5.7 \pm 2.0$ km s$^{-1}$. The line profiles of CN of HD 235858 have not been resolved.

## 1. Introduction

Post–AGB stars are in a transition stage between the Asymptotic Giant Branch (AGB) and the planetary nebulae (PN) stage. During the early stage of post–AGB evolution the star is obscured by material expelled during the AGB phase (the AGB ejecta). As this ejecta slowly moves away from the central star, the optical depth decreases and the star can be detected in the optical region. When the star reaches high enough temperatures, the AGB ejecta is ionized and is observable as a planetary nebula.

We have studied optically bright post–AGB stars (spectral type A to G supergiants) which show circumstellar molecular line absorption ($C_2$ and CN, or $CH^+$) or emission ($CH^+$) in their optical spectra. The radial velocities and low excitation temperatures of the molecules (Bakker et al. 1996,

1997) identify them as circumstellar rather than photospheric or interstellar (see also Hrivnak 1995). The excitation of $C_2$ is generally described by rather high temperatures, $T_{ex} = 43 - 399$ K, whereas the CN excitation is found to be much lower, $18 - 50$ K. The reason for this difference is discussed in section 2.2. The observed abundances and excitation of the molecules can lead to a determination of the mass-loss rate.

We have used the $C_2$ ($A^1\Pi_u - X^1\Sigma_g^+$) and CN ($A^2\Pi - X^2\Sigma^+$) data of HD 56126 to study optical depth effects by means of the curve of growth, and we present the first results of a survey to observe these optical molecular bands at high spectral resolution ($R \geq 120\,000$). Our primary goal is to resolve the line profiles and to determine the Doppler parameter $b$ and the chemical (e.g. abundances) and physical (e.g. expansion velocities and temperatures) conditions of the circumstellar shell.

## 2. Curve of Growth Analysis for HD 56126

### 2.1. THE EMPIRICAL CURVE OF GROWTH

A curve of growth (CoG) has been empirically determined for the $^{12}C^{12}C$ Phillips ($A^1\Pi_u - X^1\Sigma_g^+$) ($v' = 1, 2, 3$, $v'' = 0$, $J'' \leq 24$) and the $^{12}C^{14}N$ Red system ($A^2\Pi - X^2\Sigma^+$) bands ($v' = 1, 2, 3$, $v'' = 0$, $N'' \leq 3$) of HD 56126 (Figure 1). Our motive for this investigation is to decide whether the assumption of optically thin lines for the weaker bands (e.g. 3,0) is valid. The equivalent widths are taken from the work of Bakker et al. (1996, 1997). The range of line oscillator strengths is $6.67 \times 10^{-5}$ to $1.44 \times 10^{-3}$ and $7.19 \times 10^{-5}$ to $9.79 \times 10^{-4}$ for $C_2$ and CN, respectively.

Because three $C_2$ Phillips bands have been observed, the CoG has for each lower energy level $J''$ up to nine transitions (a P, Q, and R branch for each band). This redundancy allows the determination of the CoG. All transitions from a given $J''$ level are fitted to the CoG by changing the column density of that level $N(J'')$. For the CN Red system the redundancy is higher due to spin-doublet splitting of the lower and upper electronic state and $\Lambda$-type doubling of the upper electronic state. There are twelve transitions from each $N''$ level: six main and six satellite branches. Three bands were used, which gives in total 36 allowed transitions per $N''$ level ($F_1$ and $F_2$).

After all observed $J''$ (or $N''$) levels have been fitted to the CoG (Fig. 1, left panels) an optical-depth-corrected absolute rotational diagram (Fig. 1, middle panels) gives the absolute population for each $J''$ (or $N''$) level. Under the assumption of a Boltzmann distribution, a linear fit to the diagram gives the average rotational temperature $T_{rot}$. Finally the rotational temperature from two successive energy levels can be determined as a function

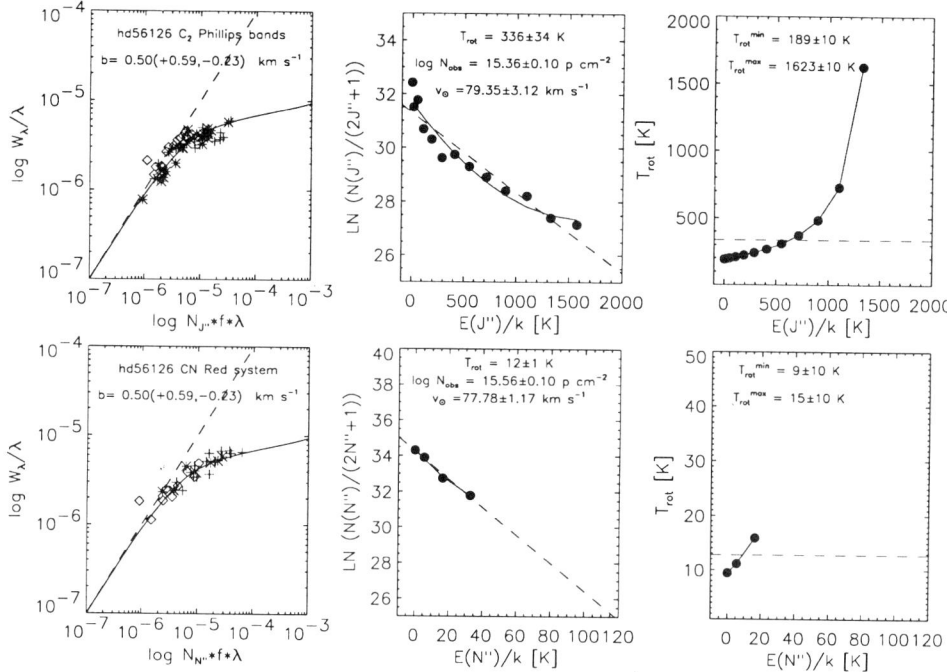

*Figure 1.* Curve of growth analysis for three $C_2$ Phillips and three CN Red system bands in the optical spectrum of HD 56126 (diamond: $v' = 3$, asterisk: $v' = 2$, plus: $v' = 1$, all for $v'' = 0$). The upper three panels concern $C_2$ and the lower three panels concern CN. *Left panels:* CoG with the best theoretical CoG over-plotted. *Middle panels:* optical-depth-corrected rotational diagram with a fit to the data assuming a Boltzmann distribution (dashed line) and a smooth non-Boltzmann distribution (solid line). *Right panels:* the rotational temperature determined from two successive energy levels (the dashed line is the average $T_{\rm rot}$).

of the lower energy level (Fig. 1, right panels). The parameters derived from this analysis are given in Table 1.

2.2. RESULTS AND INTERPRETATION

The empirical CoG (Fig. 1) shows clearly that the observations cover the optically-thin and saturated parts of the CoG. The theoretical CoG has a Doppler parameter $b = 0.50^{+0.59}_{-0.23}$ km s$^{-1}$ and $\tau \approx 1$ is reached at an equivalent width of $EW(\tau \approx 1) = 18$ mÅ. Since the strongest lines in the (3,0) transitions of $C_2$ and CN have equivalent width of 38 and 34 mÅ, respectively, the optically thin approximation is only valid for the weaker lines in these bands. We note that all the lines of the $C_2$ $(v',v'') = (3,0)$ band seem to be offset with respect to the other lines. This might indicate that the band oscillator strength used is somewhat too low.

TABLE 1. Results from the curve of growth analysis of $C_2$ and CN in the optical spectrum of HD 56126

| | $C_2$ $A^1\Pi_u - X^1\Sigma_g^+$ | CN $A^2\Pi - X^2\Sigma^+$ | |
|---|---|---|---|
| $b$ | $0.50(+0.59, -0.23)$ | $0.50(+0.59, -0.23)$ | km s$^{-1}$ |
| $T_{\rm rot}$ average | $336 \pm 34$ | $12 \pm 1$ | K |
| $T_{\rm rot}$ minimum | $189 \pm 10$ | $9 \pm 10$ | K |
| $T_{\rm rot}$ maximum | $1623 \pm 10$ | $15 \pm 10$ | K |
| log $N_{\rm obs}$ | $15.38 \pm 0.10$ | $15.56 \pm 0.10$ | p cm$^{-2}$ |
| $v_{\rm helio}$ | $79.35 \pm 3.12$ | $77.78 \pm 1.17$ | km s$^{-1}$ |

The right panel of Figure 1 clearly shows that the rotational temperature for $C_2$ is not constant. The molecule is therefore not in local thermodynamic equilibrium (LTE) and the population distribution over the rotational energy levels is non-Boltzmann. Van Dishoeck & Black (1982) have shown that interstellar $C_2$ is radiatively pumped. $C_2$ is a homonuclear molecule and does not have allowed pure rotational or vibrational transitions and can therefore not cool radiatively: $T_{\rm rot} \geq T_{\rm kin}$. For low $J''$ levels the rotational temperature reaches the kinetic temperature, while for very high $J''$ levels the rotational temperature is expected to reach the color temperature of the local radiation field. CN on the other hand can effectively cool: $T_{\rm rot} \leq T_{\rm kin}$. Based on the population ratio between the $C_2$ Phillips band $J'' = 0$ and $J'' = 2$ levels we find: $T_{\rm kin} = 189 \pm 10$ K. Combining this with the derived $b$ yields $v_{\rm microturb} = 0.34 \pm 0.6$ km s$^{-1}$. The measured $b = 0.50^{+0.59}_{-0.23}$ km s$^{-1}$ gives a line profile with FWHM $= 2\sqrt{\ln 2} \times b = 0.83$ km s$^{-1}$. In order to resolve these lines a spectral resolution of $R \geq 360\,000$ is needed. In the presence of macroturbulence the lines are broader and can be resolved at a lower spectral resolution. The microturbulence is very likely due to a velocity gradient in the line of sight.

## 3. Line Profiles

### 3.1. CH$^+$ EMISSION LINES OF THE RED RECTANGLE

The CH$^+$ $A^1\Pi - X^1\Sigma^+$ (0,0) emission band in the optical spectrum of the Red Rectangle (HD 44179) has been observed at a resolution of $R \approx 120\,000$ using the 2.7-m telescope of McDonald Observatory (Figure 2). These lines originate from levels which are 34 000 K above the ground level. We have resolved the line profile of the strongest CH$^+$ emission lines and find a FWHM $\approx 8.5 \pm 0.8$ km s$^{-1}$ (Table 2). The intensities of the emission lines fall below the detection limit for FWFM $\approx 20.0 \pm 1.0$ km s$^{-1}$. The line

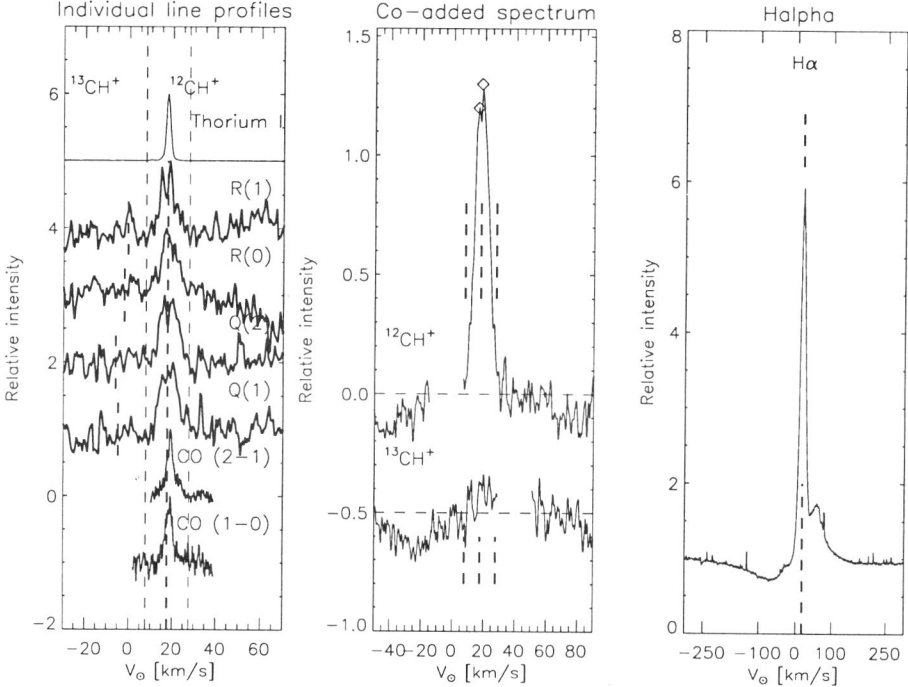

*Figure 2. Left:* normalized line profiles of the strongest CH$^+$ A$^1\Pi$–X$^1\Sigma^+$ (0,0) emission lines and CO radio emission lines (after Jura et al. 1995) of the Red Rectangle. The thick dashed lines at 17.7 km s$^{-1}$ gives the average velocity, with on both sides (at a velocity offset of 10.0 km s$^{-1}$) a line where the intensity of the emission line has fallen below the detection limit. The dashed line at approximately $-0.7$ km s$^{-1}$ ($J''$ dependent) marks the location where $^{13}$CH$^+$ is predicted. *Middle:* A non-detection of $^{13}$CH$^+$ after the signals of several lines have been co-added on a velocity axis. It seems that the $^{12}$CH$^+$ lines are a disk-like feature. *Right:* the H$\alpha$ profile with the central spike at the system velocity.

profiles of the R(1), Q(1), and Q(2) lines suggest a central absorption or the presence of two emission components at 16.0 and 18.1 km s$^{-1}$. The location of $^{13}$CH$^+$ is at approximately $-0.7$ km s$^{-1}$ ($J''$ dependent). To investigate the presence of weak isotopic lines we have added the signals of all lines into one single profile (middle panel of Fig. 2). We have a non-detection of $^{13}$CH$^+$ which yields an isotopic ratio of $^{12}$C/$^{13}$C $\geq 22$. This is consistent with the progenitor being a carbon star (Smith & Lambert 1990).

The line profile of CO is interpreted as due to a broad and narrow component (Jura et al. 1995, 1997). The broad component has a width comparable to that of the CH$^+$ lines which might suggest that they are formed in the same region (the circumbinary disk). The H$\alpha$ and CO profiles shows a strong central emission at the system velocity. Jura et al. argue that the spike is formed in an extended region of ionized gas.

TABLE 2. Results for the Red Rectangle, HD 235858, and IRAS 08005−2356

|  | Red Rectangle CH$^+$ (0,0) | HD 235858 CN Red (2,0) | IRAS 08005−2356 CN Red (2,0) |  |
| --- | --- | --- | --- | --- |
| Date | 1995 Dec. 13 | 1995 Sep. 17 | 1995 Dec. 12 |  |
| HJD | 2450064.7428 | 2449977.6607 | 2450065.8708 |  |
| $v_{*,\mathrm{helio}}$ | $27.7 \pm 1.4$ | $-34.1 \pm 0.1$ | $64.5 \pm 1.0$ | km s$^{-1}$ |
| $v_{\mathrm{mol,helio}}$ | $17.7 \pm 0.1$ | $-48.2 \pm 0.1$ | $19.4 \pm 0.2$ (A) | km s$^{-1}$ |
|  |  |  | $25.2 \pm 1.0$ (B) |  |
| $v_{\mathrm{FWHM}}$ | $8.5 \pm 0.8$ | $2.2 \pm 0.1$ | $3.1 \pm 0.9$ (A) | km s$^{-1}$ |
|  |  |  | $3.4 \pm 2.0$ (B) |  |
| $v_{\mathrm{FWFM}}$ | $20.0 \pm 1.0$ |  |  | km s$^{-1}$ |
| $^{12}$C/$^{13}$C | $\geq 22$ | $\geq 11$ | $\geq 11$ |  |

### 3.2. CN ABSORPTION IN HD 235858 AND IRAS 08005−2356

HD 235858 has been observed at a resolution of $R \approx 120\,000$. Figure 3 shows a part of the spectrum which contains three different categories of molecular features. The broad absorption lines are due to photospheric CN ($T_{\mathrm{eff}} = 5500$ K), the narrow absorption lines are circumstellar CN ($T_{\mathrm{eff}} = 20$ K and $\log N = 15.30$ p cm$^{-2}$), and the three strongest features are telluric H$_2$O. There is one photospheric atomic line (N I) present in this spectrum. The circumstellar CN lines are not resolved. Since Začs et al. (1995) did not notice photospheric CN absorption in their spectrum, it seems that this pulsating star has only photospheric molecular absorption when the stellar effective temperature is sufficient low (when the star is largest).

In an earlier report on the detection of circumstellar CN in IRAS 08005−2356 we reported the presence of only one broad component (Bakker et al. 1997). The new spectra (Fig. 3, $R \approx 120\,000$) have resolved the broad component into two separate resolved components with a velocity separation of $\Delta v = 5.7 \pm 2.0$ km s$^{-1}$. The two absorption components could be due to wind material moving at two different velocities (multiple shells), or to two photodissociation fronts at different expansion velocities — possibly one due to the stellar and one to the interstellar radiation field.

## 4. Summary

We have investigated optical depth effects for the circumstellar C$_2$ and CN lines in the optical spectrum of HD 56126. From a curve of growth analysis we find a Doppler parameter $b = 0.50^{+0.59}_{-0.23}$ km s$^{-1}$. CN (2,0)

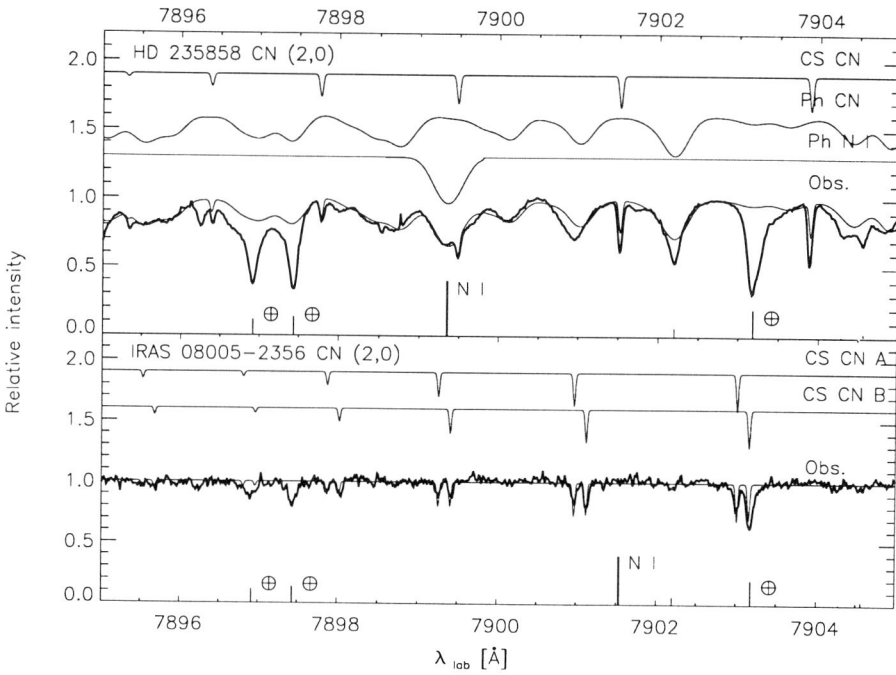

*Figure 3.* Upper: a part of the optical spectrum of HD 235858 which shows the CN Red system (2,0) band lines. Lower: for IRAS 08005–2356, a synthetic spectrum using photospheric and circumstellar CN lines, and N I (7898.985 Å) is over-plotted (thin line). The abscissa is in the rest frame of the molecule.

lines of IRAS 08005–2356 and HD 235858 and CH$^+$ of the Red Rectangle have been observed at a resolution $R \geq 120\,000$. CH$^+$ has been resolved. The CN lines of IRAS 08005–2356 are resolved into two separate resolved components, and the CN of HD 235858 has not been resolved.

## References

Bakker, E. J., Waters, L.B.F.M., Lamers, H.J.G.L.M., Trams, N. R. & Van der Wolf, F.L.A. 1996, *A&A*, 310, 893
Bakker, E. J., van Dishoeck, E. F., Waters, L.B.F.M. & Schoenmaker, T. 1997, *A&A*, 323, 469
Hrivnak, B. J. 1995, *ApJ*, 438, 341
Jura, M., Balm, S. P. & Kahane, C. 1995, *ApJ*, 453, 721
Jura, M., Turner, J. & Balm, S. P. 1997, *ApJ*, 474, 741
Smith, V. V. & Lambert, D. L. 1990, *ApJ Supp.*, 72, 387
van Dishoeck, E. F. & Black, J. H 1982, *ApJ*, 258, 533
Začs, L., Klochkova, V. G. & Panchuk, V. E. 1995, *MNRAS*, 275, 764

## Discussion

**Frogel:** Have you tried to observe the very strong $C_2$ band in the $H$ window?

**Bakker:** No.

# CHEMICAL COMPOSITION AND EVOLUTION OF POST–AGB STARS

M. PARTHASARATHY
*Indian Institute of Astrophysics*
*Bangalore, India*

**Abstract.** Analysis of the chemical compositions of post–AGB stars reveals the following abundance patterns: ($i$) Post–AGB stars which are extremely underabundant in Fe and other refractory elements, but which have nearly normal abundances of C, N, O, S, and Zn. The depleted refractory elements are locked up in circumstellar dust grains. Formation of dust close to the star, and dust–gas separation and dust-driven mass loss driving out mostly the dust may explain the abundances of these stars. ($ii$) High-latitude hot post–AGB stars which show an underabundance of carbon, indicating that they left the AGB before the third dredge-up occurred. ($iii$) Post–AGB stars with overabundances of carbon and $s$-process elements, indicating that they have gone through the third dredge-up and carbon-star phase on the AGB. The overabundance of Li, Al, C and $s$-process elements in some post–AGB stars indicate that they have gone through the dredge-up and Hot Bottom Burning nucleosynthesis at the base of the convective envelope. The observed characteristics of post–AGB stars indicate an evolutionary sequence in the transition region from the tip of the AGB into the young planetary nebula stage.

## 1. Introduction

The low- and intermediate-mass stars that initially have masses between 1 and 5 $M_\odot$ (perhaps up to 8 $M_\odot$) in the late stages of their evolution go through the AGB and planetary nebula stage and become white dwarfs with masses of $\sim 0.6 M_\odot$. These stars shed their massive outer envelopes during the mass loss on the AGB leaving a nuclear-processed white-dwarf-like core with a thin outer envelope. We do not know exactly where and when the strong AGB mass loss ceases. It is normally assumed that this happens when the envelope mass is in the range of 0.001 $M_\odot$, corresponding to an effective temperature around 5000 K (Schönberner 1989). In practice the mass loss will cease when the mass of the hydrogen-rich envelope becomes so small that the pulsational amplitude decreases or the pulsation stops altogether. Stars close to the end of their AGB evolution are completely obscured by optically thick, relatively cool dust shells as a consequence of high mass-loss rates. The non-variable OH/IR stars most likely belong to the very early phases of post–AGB evolution (Bedijn 1987). With the advent of the IRAS a completely new class of stars was detected. These stars are found to be in rapid transition from the tip of the AGB into the planetary nebula region (Parthasarathy & Pottasch 1986). These stars often have cold detached circumstellar dust shells with far-infrared colors, flux distributions, dust temperatures and dust masses similar to the dust shells of planetary nebulae. They have supergiant-like spectra in the optical region extending from cool K,G,F supergiant types to hot A,B supergiant types (Parthasarathy & Pottasch 1986; Parthasarathy 1993a,b). Parthasarathy & Pottasch (1986) proposed that the dust shells around these stars are the result of severe mass loss during their AGB stage of evolution. It is likely that these objects are in a hitherto unseen post–AGB phase of the stellar evolution. The characteristics of molecular envelopes around these stars further confirms their post–AGB status (Likkel et al. 1987; Omont et al. 1993). Theoretical calculations through the AGB with inclusion of mass loss and also through the following evolutionary stages down to the white-dwarf sequence have become available (Schönberner 1989, 1993). These calculations predict evolutionary lifetimes of several thousand years in the transition region between the tip of the AGB and the planetary nebula region. Thus a direct comparison between theory and observation now appears possible in the very early phase of this post–AGB evolution.

Recent studies in the optical, infrared, millimeter and radio regions indicate that the following types of stars are in the post–AGB stage of evolution: ($i$) high galactic latitude supergiants, ($ii$) IRAS sources with far–IR colors similar to planetary nebulae and indentified with B, A, F, G, or K supergiants, ($iii$) RV Tauri and UU Her stars and Type II Cepheids like ST Pup, and ($iv$) UV–bright stars in globular clusters.

## 2. Observed Characteristics

Most of the post–AGB stars are IRAS sources with warm and/or cold circumstellar dust shells. Their spatial distribution shows little concentration to the galactic plane, indicative of an older stellar population. Many are high galactic latitude, high-velocity, metal-poor and low-gravity stars. Their spectral types range from K I supergiants to B I supergiants. Observations of CO, OH, and HCN lines in the millimeter and radio regions have revealed molecular envelopes with expansion velocities and mass-loss rates similar to those of evolved stars. Also these observations show that some have oxygen-rich and some have carbon-rich envelopes (Omont et al. 1993). Several of these objects have been found to show unusual emission features at 3 $\mu$m and 21 $\mu$m which are probably due to PAH molecules. Most of them show H$\alpha$ emission or a peculiar H$\alpha$ profile. The hotter ones also show [N II], [S II] and several Balmer lines in emission indicating the presence of a very low excitation nebula.

From an analysis of the IRAS point-source catalogue a large number of post–AGB stars have been detected (Parthasarathy & Pottasch 1986, 1989; Lamers et al. 1986; Pottasch & Parthasarathy 1988; Hrivnak et al. 1989; Kwok 1993, and references therein). Detailed studies of these objects in wavelength regions from ultraviolet to radio are providing clues to understand the role of thermal pulses, nucleosynthesis, mixing, mass loss, and the transition from oxygen-rich to carbon-rich stars. The chemical composition analysis of several post–AGB stars in recent years reveals the following main abundance patterns: post–AGB stars which are extremely underabundant in Fe, high latitude hot post–AGB stars which are underabundant in carbon, and post–AGB stars with overabundances of carbon and $s$-process elements. An analysis and discussion of the chemical compositions of several post–AGB stars is presented in this paper.

## 3. Extremely Iron-Deficient Post–AGB stars

Recent studies have established the existence of a class of post–AGB candidates with extremely low iron-group abundances. HR 4049 ([Fe/H] = –4.8), HD 52961 ([Fe/H] = –4.8), HD 44179 ([Fe/H] = –3.1), and BD +39°4926 ([Fe/H] = –1.6) belong to this group. With [Fe/H] = –4.8, HR 4049 and HD 52961 are among the most iron-poor stars known so far in our Galaxy (Waelkens et al. 1991). The photospheric abundance patttern of these stars is similar to the gas-phase abundances of the interstellar medium. The refractory elements Fe, Mg, Si, Al, Ti, Ca etc. are depleted and the abundances of volatile elements C, N, O, S, and Zn are nearly normal. The photospheric abundances of these stars correlates with the grain condensation temperatures (Bond 1991; Van Winckel et al. 1992; Parthasarathy et al. 1992). Van Winckel et al. (1992) found a nearly normal abundance of Zn in the extremely Fe-poor post–AGB star HD 52961. S and Zn have very low condensation temperatures and therefore do not easily condense into dust grains. The detection of nearly normal amounts of Zn ([Zn/Fe] = +3.1) in HD 52961 is convincing evidence for the fractionation hypothesis. The presence of circumstellar dust around most of these stars lends strong support to the idea that the depleted refractory elements are locked up in circumstellar dust grains. Bond (1991) suggested that there is a selective removal of the metals from the photosphere through grain formation and mass loss.

Parthasarathy et al. (1992) suggested that during the AGB and/or post–AGB stage the outer atmospheres of these stars had expanded and cooled to the limit of the condensation temperature of refractory elements. Formation of cores of dust grains close to the stars and the resulting dust-driven mass loss, driving out mostly the dust, may be able to explain the photospheric abundances of the extremely iron-poor post–AGB stars.

Hoyle & Wickramasinghe (1962) were the first to suggest that dust grains tend to be formed in the atmospheres of cool red giants and supergiants at temperatures less than 2700 K. The dust grains thus formed have a significant effect on the photospheric opacity causing the photospheric density to decrease very markedly as the temperature falls towards 2000 K. It is this fall of density that allows the grains to be repelled outward by radiation pressure and to leave the star altogether.

Recently Van Winckel et al. (1995) found conclusive evidence that all the known extremely iron-deficient post–AGB A – F supergiants (HR 4049, HD 44179, HD 52961, HD 46703 and BD +39°4926) are single-lined spectroscopic binaries with periods of the order of one or two years. Post–AGB stars in binary systems with a narrow range of orbital periods and circumbinary disks resulting from mass loss and mass transfer processes may result in the separation of gas and dust, casuing the photosphere of the post–AGB supergiant companion to be depleted in refractory elements. In fact, Parthasarathy & Pottasch (1986) suggested that some of the post–AGB stars may be low-mass, long-period binaries with a circumbinary disk.

Recently Gonzalez & Wallerstein (1996) found the Type II Cepheid ST Pup ([Fe/H] = −1.47) to be a binary ($P = 410$ d) with a chemical composition similar to that of very iron-poor post–AGB stars. A few carbon-rich RV Tauri stars — IW Car (Giridhar et al. 1994), DY Ori, EP Lyr, AR Pup, and R Sge (Gonzalez et al. 1997) — show chemical compositions simliar to those of very iron-poor post–AGB A – F supergiants. It is likely that the RV Tau stars depleted of refractory elements are all binary stars with periods of the order of one to two years similar to those of high-latitude metal-depleted post–AGB stars. These results show that some of the RV Tau stars are not intricsically metal-poor. The metals are depleted due to fractionation and all these stars are IRAS sources with warm dust shells. The presence of circumstellar dust shells and the similarity in chemical compostion of RV Tau stars to the extremely iron-deficient post–AGB A and F supergiants indicates that RV Tau stars and Type II Cepheids like ST Pup are most likely in the post–AGB stage of evolution.

The processes that produce photospheres with depleted refractory elements in λ Boo stars and in post–AGB A – F supergiants may be the same. Formation of dust close to the star (during the pre-main-sequence phase in the case of λ Boo stars, and during the AGB and/or post–AGB mass-loss phase in the case of post–AGB stars) and subsequent gas and dust separation and dust-driven mass loss (driving out mostly the dust) may have taken place resulting in the depletion of refractory elements in the photospheres of these stars (Parthasarathy 1994). The presence of a companion and/or disk may help in the separation of gas and dust.

The depletion of Fe and other refractory elements seems to range from extremely Fe–depleted stars like HR 4049 and HD 52961 ([Fe/H] = −4.8) to stars with mild depletions. Recently Reddy (1996) found HD 70379 to be an F6 I post–AGB supergiant with a cold detached dust shell and far-infrared colors similar to those of planetary nebulae and high-latitude F supergiants like HD 161796. An analysis of the spectrum of HD 70379 shows that [Fe/H] = −0.4 and the abundances of S and Zn are almost solar (Reddy 1996).

Recently Hambly et al. (1996) have found a high latitude B–type star ($T_{\rm eff}$ = 25,000 K) to be a binary with an abundance pattern similar to that of stars with depleted refractory elements. They find normal He, marginally enhanced C, N, and O, and a deficiency of 0.4 to 0.6 dex in Mg, Si, Fe, etc. The abundance of sulphur is normal.

In a few high-latitude low-gravity B stars, Ca seems to be very underabundant compared to Mg and Fe. In ROA 24, a post–AGB star in $\omega$ Cen, Gonzalez & Wallerstein (1992) found an extreme deficiency of Al. The condensation chemistry is quite complicated. The depletions of various refractory elements may depend on many factors and on the physical conditions in the grain-forming regions of these stars.

## 4. Carbon-Rich Post–AGB Stars

Post–AGB stars which have gone through the carbon-star stage on the AGB are expected to show the products of the triple-alpha process, CN and ON cycling, and the s-process on the surface as a result of thermally-pulsing AGB evolution and the third dredge-up phenomenon. Recent chemical composition analysis of several post–AGB stars reveals that they are overabundant in carbon and s-process elements, indicating that the triple-alpha products have been brought to the surface. The chemical compositions of the following post–AGB stars show that they are carbon-rich and overabundant in s-process elements:

- HD 56126   (F5 I)    (Parthasarathy et al. 1992; Klochkova 1995)
- HD 179821  (G5 I)    (Parthasarathy et al. 1998)
- HD 187885  (F3 I)    (Van Winckel 1996; Parthasarathy et al. 1998)
- SAO 34504            (Začs et al. 1995)

- IRAS 05341+0852  (Reddy et al. 1997)
- IRAS 18095+2704  (Klochkova 1995; Reddy 1996)

The iron abundance in most of these stars is around [Fe/H] = –1.0. The low iron abundance in these stars is intrinsic and not due to fractionation. Some of these stars are at high latitudes and have high radial velocities. All of them are IRAS sources with far-infrared colors similar to planetary nebulae. The CO and HCN data and infrared spectra indicate that the dust shells around these stars are also carbon-rich. These results indicate that they are low-mass metal-poor stars which have gone through the carbon-star stage on the AGB.

All of these stars show overabundances of $s$-process elemets. SAO 34504 and IRAS 05341+0852 show overabundances of lithium also. In the above list of stars HD 56126, SAO 34504 and IRAS 05341+0852 show the 21 $\mu$m emission feature. The large ratio of HCN/CO millimeter emission, the presence of $C_2$, $C_3$, and CN molecular absorption features in their optical spectra, and the overabundance of $s$-procees elements indicate that these stars have evolved from the carbon-star phase on the AGB. Recently Webster (1995) suggested that the 21 $\mu$m emission observed in these carbon-rich post–AGB stars is due to fulleranes $C_{60}H_m$ ($m = 1$ to 60).

IRAS 18095+2704 (F3 Ib) is a high latitude post–AGB F supergiant. The iron and carbon abundances are found to be [Fe/H] = –1.0 and [C/Fe] = +0.5, respectively (Klochkova 1995; Reddy 1996). However, the $s$-process elements are more underabundant. A similar abundance pattern is also observed in the high-latitude post–AGB star HD 133656 (Van Winckel et al. 1996): [Fe/H] = –1.0 and [C/Fe] = +0.3. However, in another high latitude A supergiant, HD 105262, Reddy, Parthasarathy & Sivarani (1996) find C/O = 1.2 and [Fe/H]= –2.2. It had previously been classified as a Field Horizontal Branch (FHB) star. The hydrogen line profiles suggest a lower surface gravity and higher luminosity. Its very high galactic latitude (+72°) and large proper motion (0.″057 yr$^{-1}$), as well as its overabundance of carbon, makes this star an important one for further study.

## 4.1. IRAS 05341+0852 (F6 I)

Recently Reddy et al. (1997) derived the photospheric abundances of the F–type post–AGB supergiant IRAS 05341+0852. This object shows the 3.3 and 21 $\mu$m emission features which are attributed to carbon-rich molecules such as PAHs and fullerenes, indicating that the circumstellar dust is carbon rich. From an analysis of high-resolution spectra we find that the star is carbon-rich (C/O = 2.2) and metal-poor ([Fe/H] = –1.0). Lithium is overabundant (log Li = 2.5) by a factor of about 100 relative to that observed in normal giants and supergiants. Carbon, nitrogen, oxygen, aluminum and silicon are found to be overabundant. Most importantly we find large overabundance of $s$-process elements: [Y/Fe] = 1.8, [Ba/Fe] = 2.58, [La/Fe] = 2.86, [Ce/Fe] = 2.95, [Pr/Fe] = 2.27, [Nd/Fe] = 1.97 and [Sm/Fe] = 0.86. The result that [S/Fe] = 0.07 indicates that the low Fe abundance is intrinsic and is not due to fractionation in this case. The overabundance of Li, CNO, and $s$-process elements and the large C/O ratio, high galactic latitude, detached cold dust shell, and supergiant-type spectrum indicate that IRAS 05341+0852 evolved from the carbon-rich AGB phase and is now in the post–AGB phase of evolution.

So far IRAS 05341+0852 is the only post–AGB star showing overabundances of Li, C, Al and $s$-process elements which are all in general agreement with the predictions of the third dredge-up and Hot Bottom Burning (HBB) AGB evolutionary models (Lattanzio 1993; Lattanzio et al. 2000). However, these theoretical models suggest that Li and Al are produced in significant amounts during HBB in massive AGB stars. The low iron abundance and high galactic latitude suggest that IRAS 05341+0852 is a low-mass star. The overabundance of Li, Al, C, and $s$-process elements in IRAS 05341+0852 suggests that HBB and third dredge-up occurs in low-mass stars also.

4.2. HD 179821 (= AFGL 2343) (G5 Ia)

HD 179821 is an IRAS source with a cold detached dust shell and far–IR colors similar to planetary nebulae. Pottasch & Parthasarathy (1988) concluded that HD 179821 is a post–AGB star. Recently Hawkins et al. (1995) have found that HD 179821 is surrounded by a dusty nebula 4–5″ in diameter at 10.5 and 12.5 $\mu$m. Hawkins et al. (1995) conclude that HD 179821 is an extremely massive star at a distance of about 6 kpc. They estimate the dust-shell mass to be 8 $M_\odot$. Kastner & Weintraub (1995) also conclude that HD 179821 is a massive post–red-supergiant with a dust-shell mass of 5 $M_\odot$. They infer that it is evolving towards the LBV or W–R stage.

Recently we have determined the chemical composition of HD 179821. We find it to be metal-poor ([Fe/H] = −1.0) and overabundant in carbon and $s$-process elements. Its high radial velocity (+100 km s$^{-1}$), underabundance of metals, overabundance of carbon and $s$-process elements, cold detached dust shell, and CO, OH molecular envelope indicate that HD 179821 is a low-mass high-velocity star in the post–AGB stage of evolution.

## 5. Hot Post–AGB Stars

Not all high galactic latitude OB stars have compositions similar to those of Pop. I stars. Several metal-poor high galactic latitude B stars have now been found and identified as post–AGB stars. Conlon et al. (1991), McCausland et al. (1992), Kendal et al. (1994), and Hambly et al. (1996) have determined the chemical compositions of several high-latitude B stars. These high latitude hot post–AGB stars do not have the compositions of Population II dwarfs and red giants. The high latitude hot post–AGB stars are found to be metal-poor and also significantly underabundant in carbon. The chemical composition of these stars indicates that they left the AGB before the third dredge-up occurred. They left the AGB before or at the beginning of the thermal pulsing stage.

Hambly et al. find that the high-latitude B star CPD −61°455 is a hotter analogue of very metal-poor post–AGB stars like HR 4049. CPD −61°455 shows normal He, marginally enhanced CNO, a metal deficiency of 0.4 to 0.6 dex and normal abundance of sulphur. Several hot post–AGB stars have been found by Parthasarathy & Pottasch (1989), Parthasarathy (1994), and Oudmaijer (1996). Chemical composition analysis of all these stars is clearly important.

## 6. Post–AGB Stars in Globular Clusters

The chemical composition of Barnard 29 in M13 ($V = 13$, $T_{\text{eff}} = 20,000\,\text{K}$, $\log g = 0.3$, [Fe/H] = −1.46, $\log L = 3.25\,L_\odot$) has been determined by Conlon et al. (1994). Barnard 29 shows a severe carbon deficiency of more than 2 dex which has also been observed in a number of high-latitude low-gravity B–type stars. This implies that these stars left the AGB before undergoing the third dredge-up. The derived CNO abundances are compatible with the products of hydrogen burning having been brought to the surface during the first and second dredge-ups. Carbon is depleted and nitrogen is enhanced. Oxygen, magnesium and silicon are not enhanced, indicating no evidence of $\alpha$-capture processing. Low-mass stars (0.8 to 1 $M_\odot$) may lose their envelopes and evolve blueward before thermal pulsing begins.

The chemical composition of star No. 1412 in M4 ($V = 10.1$, $T_{\text{eff}} = 4125\,\text{K}$, $\log g = 0.5$, [Fe/H] = −1.45) was determined by Brown et al. (1990) and Whitmer et al. (1995). It is grossly deficient in carbon and overabundant in nitrogen compared to other stars in M4. It lies about 1 mag above the AGB in M4. It is a late-type analogue of Barnard 29. These results indicate that globular clusters contain low-mass post–AGB stars which left the AGB before undergoing the third dredge-up.

The chemical composition of ROA 24 (= HD 116745), an F0 Ibp star in $\omega$ Cen ($V = 10.82$, $T_{\text{eff}} = 6950\,\text{K}$, $\log g = 1.15$, [Fe/H] = −1.77, $M_V$ = −3.66), was determined by Gonzalez & Wallerstein (1992). In ROA 24 the C, N, O, Na and $s$-process elements are significantly enhanced relative to the other giants in $\omega$ Cen. The large C abundance implies that triple-alpha products have been brought to the surface. The abundance pattern indicates that material that has experienced the triple-alpha process, s-process, and possibly Ne–Na, CN and ON cycles has reached the surface.

Thus chemical composition analysis of post–AGB stars in globular clusters is important. Often these are called UV–bright stars (Zinn 1974). There are at least half a dozen luminous non-variable F supergiants known in globular clusters (Harris et al. 1983). Recent surveys have revealed the presence of several hot post–AGB stars in several different globular clusters. The deatiled study of these stars will enable us to better understand the post–AGB evolution of low-mass stars.

## 7. Evolution

Zuckerman & Aller (1986) find that 62 % of planetary nebulae are carbon-rich. They find that a clear majority (62 %) of planeatry nebulae with reasonably reliably determined C/O ratios have C/O > 1. This result suggests that more than half of all intermediate and low mass main-sequence stars go through the carbon-rich phase during their AGB and post–AGB evolutionary stage. The percentage of carbon-rich planetary nebulae is in agreement with the relative numbers of carbon-rich and oxygen-rich red giant stars with large mass-loss rates. The planetary nebulae with [WC] central stars are carbon-rich. However, IC 4997 and SwSt 1 show oxygen-rich nebulae and carbon-rich central stars. A similar phenomenon is observed in the post–AGB star Roberts 22. More recently Zijlstra et al. (1991) have found the young planetary nebula IRAS 07027–7934 with a [WC11] central star to show a strong 1612 MHz OH maser as well as weak CO emission. PAH features suggest that the ionized region is carbon-rich, and the outer region where the OH maser is situated is oxygen-rich. This star appears to have transformed from an OH/IR star to a carbon star within the last few hundred years. Parthasarathy (1993a) suggested that the [WC11] central stars of planetary nebulae are the hotter analogues of carbon-rich post–AGB supergiants. The carbon-rich post–AGB supergiants during their evolution to higher temperatures may turn into [WC11] central stars. The carbon-rich post–AGB supergiants with 21 $\mu$m emission and the planetary nebulae with [WC11] nuclei show similar PAH and UIR emission features between 3 and 12 $\mu$m. The characteristics of the circumstellar dust around both these types of objects are similar. It is likely that the carbon-rich post–AGB supergiants may evolve into planetary nebulae with [WC11] nuclei.

### 7.1. SAO 244567 (= Hen 1357)

SAO 244567 (= Hen 1357) is an IRAS source with far–IR colors similar to planetary nebulae. The optical spectrum of this star obtained by Henize around 1950 shows only the H$\alpha$ line in emission. The optical spectrum obtained by Kilkenny in 1971 shows that it was a B1 supergiant at that time. Optical spectra obtained since 1990 show strong forbidden emission lines corresponding to a low-excitation and young planetary nebula. It has turned into a planetary nebula within the last 20 years (Parthasarathy et al. 1993, 1995).

HST planetary camera imaging in H$\beta$ and [O III] 5007 Å revealed a 2″ nebula around the central star (Bobrowsky 1994). The IUE ultraviolet spectra obtained during the last seven years show that the central star is rapidly evolving. It is found that the central star of this young planetary has faded by a factor of 2.83 within the last seven years. The terminal velocity of the stellar wind has decreased from $-3500$ km s$^{-1}$ in 1988 to almost zero in 1994.

We derive the parameters of the nebula and the central star to be the following: radius of the nebula 0.02 pc, expansion age 2700 years, luminosity = 3000 $L_\odot$, core mass = 0.55 $M_\odot$. The B-type supergiant spectrum in 1971 suggests the effective temperature of the star was around 20,000 K at that time. However, the 1995 IUE high-resolution spectrum of this star and the nebular emission lines indicate that the effective temperature of the central star is now around 50,000 K. The time scale of evolution appears to be very rapid. For such a fast evolution a core mass of 0.8 $M_\odot$ or even higher is required. However, the observed luminosity of the central star does not suggest a high core mass. The estimated distance to SAO 244567 may be uncertain. Further observations of this young planetary nebula may shed new light on the evolution of post–AGB stars.

The B1 supergiant-like spectrum of SAO 244567 in 1971 shows that post–AGB stars, before they turn into planetary nebulae, have extended atmospheres and may mimic the spectra of supergiants. It also confirms the evolutionary sequence of post–AGB supergiants from cooler to hotter and into young planetary nebulae.

I am thankful to Prof. Robert F. Wing for his kind encouragement and support which enabled me to participate in this conference. I also thank the IAU for partial travel support.

## References

Bedijn, P. J. 1987, *A&A*, 186, 136
Bobrowsky, M. 1994, *ApJ*, 426, L47
Bond, H. E. 1991, in IAU Symp. 145: *Evolution of Stars: The Photospheric Abundance Connection*, ed. G. Michaud and A. Tutukov (Kluwer), p. 341
Brown, J. A., Wallerstein, G. & Oke, J. B. 1990, *AJ*, 100, 1561
Conlon, E. S., Dufton, P. L., Keenan, F. P. & McCausland, R. J. H. 1991, *MNRAS*, 248, 820
Conlon, E. S., Dufton, P. L. & Keenan, F. P. 1994, *A&A*, 290, 897
Giridhar, S., Rao, N. K. & Lambert, D. L. 1994, *ApJ*, 437, 476

Gonzalez, G., Lambert, D. L. & Giridhar, S. 1997, *ApJ*, 479, 427
Gonzalez, G. & Wallerstein, G. 1992, *MNRAS*, 254, 343
Gonzalez, G. & Wallerstein, G. 1996, *MNRAS*, 280, 515
Hambly, N. C., Dufton, P. L., Keenan, F. P. & Lumsden, S. L. 1996, *MNRAS*, 278, 811
Harris, H. C., Nemec, J. M. & Hesser, J. E. 1983, *PASP*, 95, 256
Hawkins, G. W., Skinner, C. J., Meixner, M. M., Jernigan, J. G., Arens, J. F., Keto, E. & Graham, J. R. 1995, *ApJ*, 452, 314
Hoyle, F. & Wickramasinghe, N. C. 1962, *MNRAS*, 124, 417
Hrivnak, B. J., Kwok, S. & Volk, K. M. 1989, *ApJ*, 346, 265
Kastner, J. H. & Weintraub, D. A. 1995, *ApJ*, 452, 833
Kendall, T. R., Brown, P. J. F., Conlon, E. S., Dufton, P. L. & Keenan, F. P. 1994, *A&A*, 291, 851
Klochkova, V. G. 1995, *MNRAS*, 272, 710
Kwok, S. 1993, *Ann. Rev. Astron. Astrophys.*, 31, 63
Lamers, H.J.G.L.M., Waters, L.B.F.M., Garmany, C. D., Perez, M. R. & Waelkens, C. 1986, *A&A*, 154, L20
Lattanzio, J. C. 1993, in IAU Symp. 155: *Planetary Nebulae*, ed. R. Weinberger and A. Acker (Kluwer), p. 235
Lattanzio, J. C., Frost, C. A., Cannon, R. C. & Wood, P. R. 2000, in IAU Symp. 177: *The Carbon Star Phenomenon*, ed. R. F. Wing (Kluwer), p. 449
Likkel, L., Omont, A., Morris, M. & Forveille, T. 1987, *A&A*, 173, L11
McCausland, R. J. H., Conlon, E. S., Dufton, P. L. & Keenan, F. P. 1992, *ApJ*, 394, 298
Omont, A., Loup, C., Forveille, T., te Lintel Hekkert, P., Habing, H. & Sivagnanam, P. 1993, *A&A*, 267, 515
Oudmaijer, R. D. 1996, *A&A*, 306, 823
Parthasarathy, M. 1993a, in *Luminous High-Latitude Stars*, ed. D. D. Sasselov, ASP Conf. Ser., 45, 173
Parthasarathy, M. 1993b, *ApJ*, 414, L109
Parthasarathy, M. 1994, in *The MK Process at 50 Years: A Powerful Tool for Astrophysical Insight*, ed. C. J. Corbally, R. O. Gray and R. F. Garrison, ASP Conf. Ser., 60, 261
Parthasarathy, M., Garcia-Lario, P., de Martino, D., Pottasch, S. R., Kilkenny, D., Martinez, P., Sahu, K. C., Reddy, B. E. & Sewell, B. T. 1995, *A&A*, 300, L25
Parthasarathy, M., Garcia Lario, P. & Pottasch, S. R. 1992, *A&A*, 264, 159
Parthasarathy, M., Garcia-Lario, P., Pottasch, S. R., Manchado, A., Clavel, J., de Martino, D., Van de Steene, G. C. M. & Sahu, K. C. 1993, *A&A*, 267, L19
Parthasarathy, M. & Pottasch, S. R. 1986, *A&A*, 154, L16
Parthasarathy, M. & Pottasch, S. R. 1989, *A&A*, 225, 521
Parthasarathy, M., Reddy, B. E. & Garcia-Lario, P. 1998, in IAU Symp. 187: *Cosmic Chemical Evolution*, in press
Pottasch, S. R. & Parthasarathy, M. 1988, *A&A*, 192, 182
Reddy, B. E. 1996, Ph.D. thesis, Bangalore Univ., Bangalore, India
Reddy, B. E., Parthasarathy, M., Gonzalez, G. & Bakker, E. J. 1997, *A&A*, 328, 331
Reddy, B. E., Parthasarathy, M. & Sivarani, T. 1996, *A&A*, 313, 191
Schönberner, D. 1989, in IAU Coll. 106: *Evolution of Peculiar Red Giant Stars*, ed. H. R. Johnson and B. Zuckerman (Cambridge), p. 348
Schönberner, D. 1993, in IAU Symp. 155: *Planetary Nebulae*, ed. R. Weinberger and A. Acker (Kluwer), p. 415
Van Winckel, H., Mathis, J. S. & Waelkens, C. 1992, *Nature*, 356, 500
Van Winckel, H., Waelkens, C. & Waters, L.B.F.M. 1995, *A&A*, 293, L25
Van Winckel, H., Waelkens, C. & Waters, L.B.F.M. 1996, *A&A*, 306, L37
Waelkens, C., Van Winckel, H., Bogaert, E. & Trams, N. R. 1991, *A&A*, 251, 495
Webster, A. 1995, *MNRAS*, 277, 1555
Whitmer, J. C., Beck-Winchatz, B., Brown, J. A. & Wallerstein, G. 1995, *PASP*, 107, 127
Začs, L., Klochkova, V. G. & Panchuk, V. E. 1995, *MNRAS*, 275, 764

Zijlstra, A. A., Gaylard, M. J., te Lintel Hekkert, P., Menzies, J., Nyman, L.-Å. & Schwarz, H. E. 1991, *A&A*, 243, L9
Zinn, R. 1974, *ApJ*, 193, 593
Zuckerman, B. & Aller, L. H. 1986, *ApJ*, 301, 772

# OPTICAL SPECTROSCOPIC MONITORING OF THE CARBON–RICH POST–AGB STAR HD 56126:

*Pulsation and Shock Waves*

AGNÈS LÈBRE AND NICOLAS MAURON
*GRAAL, Université de Montpellier*
*34095 Montpellier cedex 5, France*

DENIS GILLET
*Observatoire de Haute Provence*
*04870 Saint Michel l'Observatoire, France*

AND

DOMINIQUE BARTHÈS
*GRAAL, Université de Montpellier*
*34095 Montpellier cedex 5, France*

**Abstract.** A spectroscopic monitoring of the post–AGB star HD 56126 was performed as regularly as possible over a 14–month interval in order to study atmospheric motions that could be associated with shock wave propagation through the stellar atmosphere. Some spectral features are strongly variable on a timescale of a few days. Radial velocity variations are also good evidence for complex atmospheric dynamics, in agreement with the recently found photometric variability. The data point to a pulsating nature for HD 56126, with a main period of 27.3 days.

## 1. Introduction

In order to investigate internal structure and atmospheric motions in post–AGB stars, we performed regular spectroscopic observations of HD 56126 = SAO 96709 = IRAS 07134+1005. Parthasarathy et al. (1992) and Klochkova (1995) have analysed its atmospheric chemical composition. Both derive a moderate metal deficiency ([Fe/H] = −1.0) and a large excess of $s$-process and CNO elements. The object displays a strong excess in the IRAS bands and was later found to have peculiar mid–infrared features including the 3.3 $\mu$m band often attributed to PAH (Kwok et al. 1990), emissions in the 6–8 $\mu$m region (Buss et al. 1990), and very strong emission bands near 21 and 30 µm (Kwok et al. 1989; Omont et al. 1995). These bands are seen in objects with carbon-rich material. Zuckerman et al. (1986) and Bujarrabal et al. (1992), from molecular observations of the circumstellar envelope of HD 56126 (CO, HCN), also infer such a C-rich nature. All these findings on photospheric abundances, envelope dust and molecules support the case for a genuine post–AGB star.

Until very recently, this F5 I star was not considered to be variable. Bogaert (1994) discovered the first evidence for the photometric variability of HD 56126, presenting a light curve with a very small and irregular amplitude ($\Delta V = 0.06 - 0.15$) and a period of about 50 days, although it is very difficult to estimate the period confidently from these data.

A few optical spectroscopic studies have already been devoted to hydrogen lines in HD 56126. Oudmaijer and Bakker (1994) obtained high-signal-to-noise observations taken at a two-month interval and showed that long-term (but not very fast) variability of the H$\alpha$ line is present, in agreement with an expected period in the range 30–96 days.

## 2. The Spectroscopic Monitoring

We present the first high-resolution optical spectroscopic monitoring of HD 56126, performed over a 14 month period (January 1991 to April 1992) with as regular intervals as possible. This work is based on observations carried out at the 1.52-m telescope of the Observatoire de Haute Provence (CNRS), France and a more detailed report is given in Lèbre et al. (1996). The AURELIE spectrometer was used at coudé focus with a 1800 line/mm grating blazed at 5000 Å. Twenty-one spectra at the H$\alpha$ line and seventeen of the Na D doublet were secured, with central wavelengths of 6582 and 5885 Å and resolving power $R = \lambda/\Delta\lambda$ of 41 000 and 36 000, respectively (i.e. about 8 km s$^{-1}$ in velocity resolution). All our spectra are presented after reduction to the stellar rest frame (SRF), for which we adopted the value of 86.1 km s$^{-1}$ on the heliocentric scale (centroid velocity of molecular CO and HCN emissions: $V_{LSR} = 72.0$ km s$^{-1}$), obtained by Bujarrabal et al. (1992).

## 3. The Spectral Line Profile Variations

In Figure 1, we present the results of our H$\alpha$ monitoring over 14 months. The H$\alpha$ line is strongly variable (on timescales of the order of days) and presents several types of profiles: P–Cygni profile, reverse P–Cygni profile, shell–type profile, or asymmetric absorption. Within only a few days, the H$\alpha$ line can change from one type of profile to another. Our time resolution (about 5–15 days) seems a bit too coarse to follow the evolution of these variations with good accuracy, especially the appearance/development/dis-

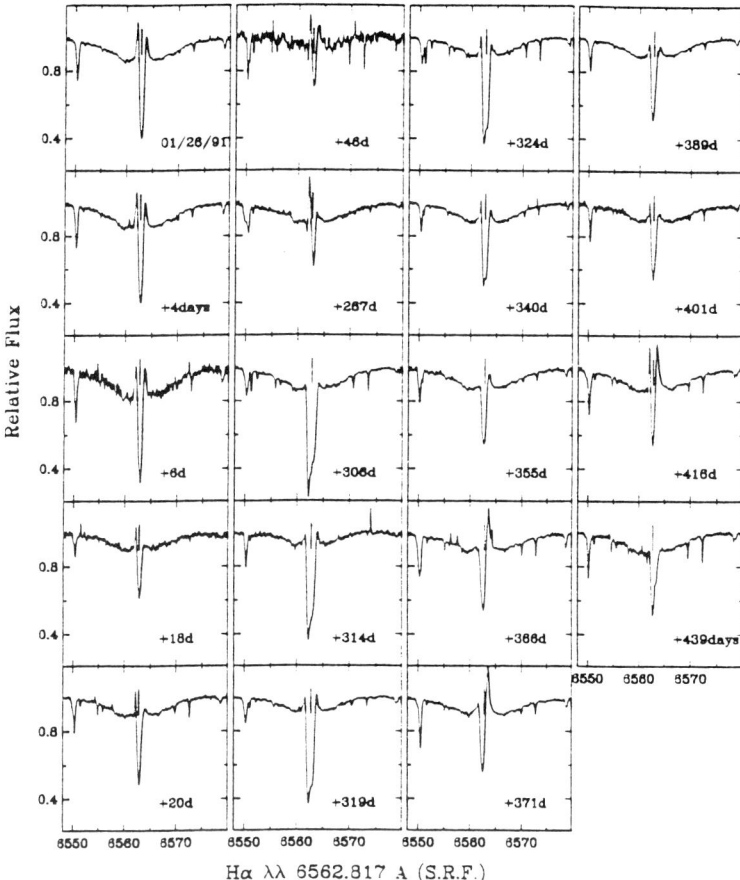

*Figure 1.* Spectroscopic observations of the Hα line region. The laboratory wavelength is indicated by the short straight line. The date of the observation is indicated through the progression in days from the first spectrum.

appearance of emission in the blue and/or red wings of the hydrogen line. It is consequently difficult to derive a reliable period for that Hα variability.

One of the most striking observational features is the permanent presence of the Hα emission component. Because the emission width is narrow, it is probably not a shell type emission, which would be formed in a very extended volume far away from the photosphere. Instead, it is better explained by a strong shock wave propagating in the atmosphere. In this case, the emission would be formed within the de-excitation region of the shock wake, i.e. in a very narrow shell above the photosphere. Because at some phases the blueshifted emission component, or the redshifted one, or both, are clearly above the continuum, we expect that we have a single broad shock emission mutilated by an almost central absorption. Depending on

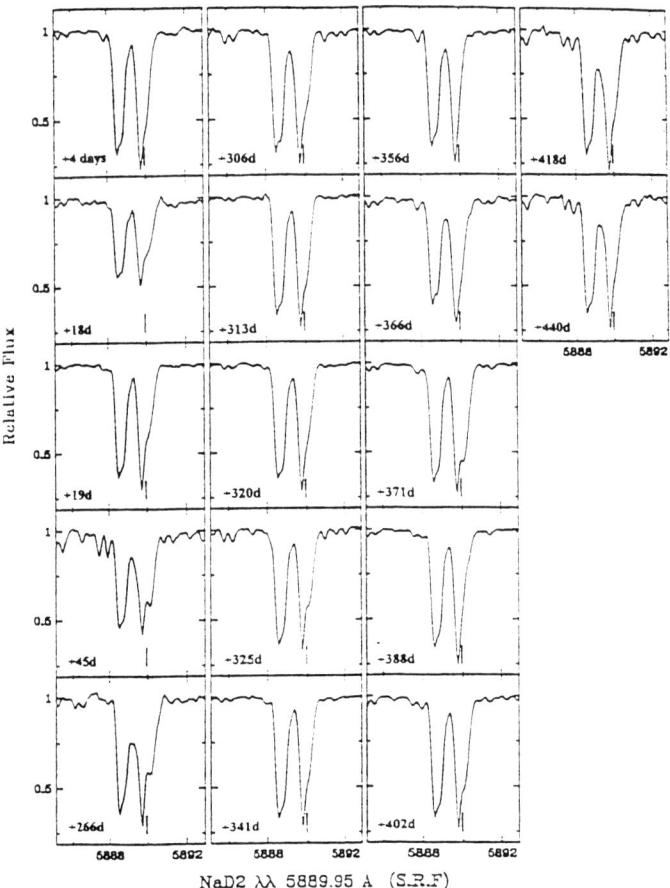

*Figure 2.* Spectroscopic observations of the Na D2 line. The short straight line indicates the line wavelength (5889.95 Å) in the stellar rest frame.

the wavelength position of absorptions with respect to the shock emission, the observed double-peak emission can appear with a stronger blueshifted component or *vice versa*. This kind of double-peaked profile is well observed in RV Tauri and W Virginis stars as reported by Lèbre & Gillet (1991 and 1992, respectively). Fokin (1991) has calculated similar double-peak emissions for the pulsating star W Vir. He shows that absorption components are caused by a self-absorption mostly due to scattering in the pre-shock layers and confirms that the emission is formed behind the shock front. Since HD 56126 is a post–AGB star, we can also expect that it is an irregular pulsating star, probably showing large variations from one cycle to another.

In Figure 2 we show the profile variations of the Na D2 line at 5889.95 Å (the same profile structure and variation being observed at D1). The most blueshifted component (around 5888.65 Å) is centered at $V_{LSR} = 6$ km s$^{-1}$ ± 2 km s$^{-1}$ with a full width of about 50 km s$^{-1}$. It is probably of interstellar origin because no significant profile variation is seen and an interstellar CO radio emission at $V_{LSR} = 10$ km s$^{-1}$ has been detected along the line of sight (Zuckerman et al. 1986).

The other component located near the rest velocity has at least a partially stellar origin. Obviously, this blended line shows a profile that varies in time, but on the other hand its bluest absorption peak (around 5889.77 Å) remains very stable in shape and radial velocity ($V_{SRF} = -10$ km s$^{-1}$, or a $V_{LSR}$ of 62 km s$^{-1}$) and is thought to be of circumstellar origin.

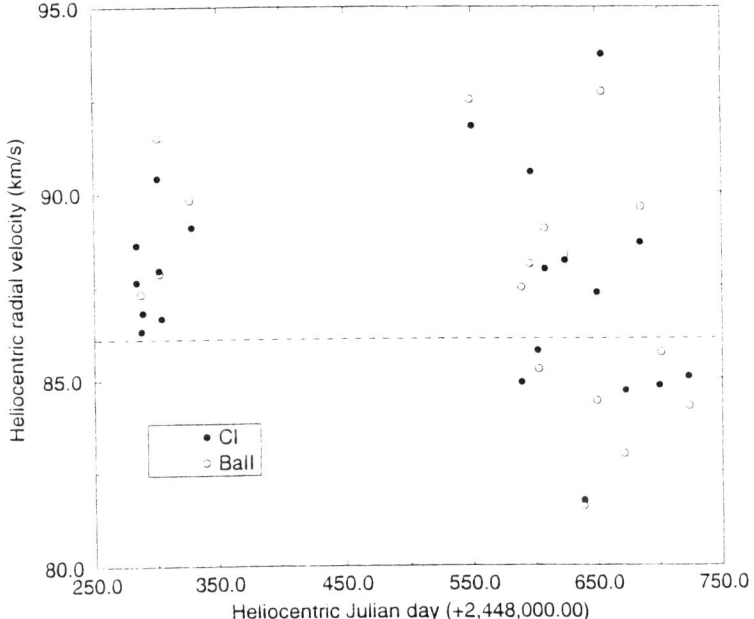

*Figure 3.* Heliocentric radial velocity curves for the C I and Ba II lines. The centroid radial velocity of molecular circumstellar profiles, $V_{LSR} = 72$ km s$^{-1}$ ($V_{\text{helio}} = 86$ km s$^{-1}$), is indicated by the horizontal dashed line.

Two other strong lines, Ba II λ 5853.688 Å and C I λ 6587.622 Å, are useful. They are not blended with any other spectral feature and seem always very symmetric. We first noted a strong agreement in the Doppler shifts of the Ba II, C I, and stellar sodium lines, all being strongly blueshifted or redshifted at the same dates. More quantitatively, we achieved Gaussian fits on both the Ba II and C I lines, and we measured a total of 38 independant radial velocities which are plotted versus time in Figure 3. The Ba II and C I measurements made on the same night generally agree within 1.5 km s$^{-1}$, suggesting synchronous dynamics for the atmospheric layers where

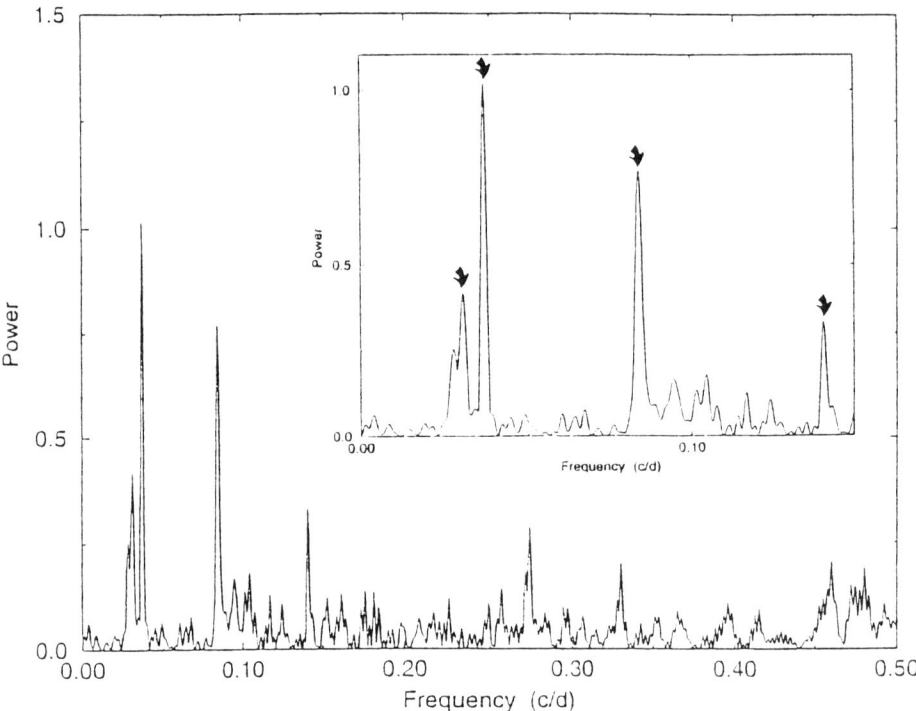

*Figure 4.* CLEAN power spectrum from the heliocentric radial velocity measurements on the C I and Ba II lines. The relevant frequencies (see text) are indicated with arrows.

these lines are formed. The peak-to-peak amplitude of the radial velocity variations is $\sim 15$ km s$^{-1}$.

## 4. Period Searching

From the radial velocity measurements on the Ba II and C I lines we have searched for characteristic periods through Fourier analysis. The CLEAN power spectrum (Roberts et al. 1987) is displayed in Figure 4. With a high degree of confidence, four peaks are detected, corresponding to a primary period of 27.3 days and three secondary periods of 32.9, 11.9, and 7.1 days. The frequency accuracy is about $4 \times 10^{-4}$ c/d for each peak, that is $\pm 0.3$ days on the 27.3 days period. The dominant period ($P = 27.3$ days) is half as large as the one we can roughly derive from the photometric data ($\sim 50$ days). This indeed is reminiscent of the atmospheric dynamics of the RV Tauri star R Sct, with two shock waves propagating through the stellar atmosphere over a light cycle of 142 days.

## 5. Conclusions

In order to look for variations in the Hα and Na D line profiles that could be related to the propagation of a shock wave through the stellar atmosphere, we performed a spectroscopic monitoring on the well-known carbon–rich infrared post–AGB candidate HD 56126. On a timescale of days we observed important changes in the Hα profile as well as synchroneous radial-velocity variations in the Na D, Ba II and C I photospheric lines. Searching for periods with the CLEAN procedure yields a main period of $27.3 \pm 0.3$ days and three other probable ones around 32.9, 11.9, and 7.1 days. From the recently-found photometric variability of this star ($V$ amplitude of 0.15 mag), and from our results, we infer that HD 56126 has atmospheric dynamics which remind us of those observed for RV Tauri stars. Linear and non-linear modelling of HD 56126 (Jeannin et al. 1996) also reconciles strong shock-wave propagation in the stellar atmosphere with the small light amplitude, suggesting that HD 56126 is a first-overtone pulsator with a period of 30 days.

Clearly, more photometric and spectroscopic observations are needed to investigate the evolutionary phase of HD 56126 and related objects. For that purpose, we have undertaken with the 1.93-m telescope of OHP a second spectroscopic monitoring covering a wider spectral region (from 3800 to 6800 Å) with better resolution ($R = 42\,000$) and better timescale sampling (one observation per week from October 1995 to April 1996). We hope to report new results soon, with improvement on the period determination.

## References

Bogaert, E. 1994, Ph.D. thesis, Univ. Louvain
Bujarrabal, V., Alcolea, J. & Planesas, P. 1992, A&A, 257, 701
Buss, R. H. Jr., Cohen, M., Tielens, A.G.G.M., Werner, M. W., Bregman, J. D., Witteborn, F. C., Rank, D. & Sandford, S. A. 1990, ApJ, 365, L23
Fokin, A. B. 1991, MNRAS, 250, 258
Jeannin, L., Fokin, A. B., Gillet, D. & Baraffe, I. 1996, A&A, 314, L1
Klochkova, V. G. 1995, MNRAS, 272, 710
Kwok, S., Hrivnak, B. J. & Geballe, T. R. 1990, ApJ, 360, L23
Kwok, S., Volk, K. M. & Hrivnak, B. J. 1989, ApJ, 345, L51
Lèbre, A. & Gillet, D. 1991, A&A, 246, 490
Lèbre, A. & Gillet, D. 1992, A&A, 255, 221
Lèbre, A., Mauron, N., Gillet, D. & Barthès, D. 1996, A&A, 310, 923
Omont, A., Moseley, S. H., Cox, P., Glaccum, W., Casey, S., Forveille, T., Chan, K.-W., Szczerba, R., Loewenstein, R. F., Harvey, P. H. & Kwok, S. 1995, ApJ, 454, 819
Oudmaijer, R. D. & Bakker, E. J. 1994, MNRAS, 271, 615
Parthasarathy, M., Garcia Lario, P. & Pottasch, S. R. 1992, A&A, 264, 159
Roberts D. H., Lehár J. & Dreher J. W. 1987, AJ, 93, 968
Zuckerman, B., Dyck, H. M. & Claussen, M. J. 1986, ApJ, 304, 401

## Discussion

**Luttermoser**: Is the Hα line *optically* thin or *effectively* thin? This will be very important when trying to deduce velocity fields from this line. The technique of deducing macroscopic gas velocities by noting velocity "widths" and velocity "shifts" from fitted Gaussian profiles over a "two-peaked" emission line, as has been done for past Mira observations, is only valid if the line is *optically* thin. *Effectively* thin lines typically display "two-peaked" emission features in their cores (e.g. Ca II in late-type stars), even if the atmosphere is static and plane parallel (e.g. the solar Ca II lines).

**Lèbre**: Following the work of Dr. Fokin, we expect that the hydrogen lines are produced in atmospheric regions of high opacity. Consequently the observed Hα profile is very probably composed of a single broad emission with an almost central absorption caused by re-absorption of the cool hydrogen located in the high atmospheric layers. The resulting profile is a two-peaked visible emission. The optically thin emitting possibility is certainly not acceptable, but only a quantitative modeling of hydrogen profiles can provide the definitive answer. This needs a full nonlinear non-adiabatic pulsating model of post–AGB stars.

**Cherchneff**: With shock strengths of $47\,\mathrm{km\,s^{-1}}$, you should expect the formation of a precursor in the pre-shock gas. Do you have evidence for such a phenomenon?

**Lèbre**: Indeed a strong shock is a radiative shock, with a precursor in the pre-shock gas. But we do not have any observational features or typical variations that could be associated with it as direct evidence.

**Van Winckel**: How do you generate a shock when the photometry does not give evidence for pulsation?

**Lèbre**: Photometric observations of HD 56126 have a very small amplitude. This can be reconciled with the presence of strong shocks in the atmosphere when taking into account a radiative shock and at the limit an isothermal shock. Consequently we must expect that the stellar luminosity $L_*$ always dominates over the shock luminosity (by a factor of 10), independently of the weak amplitude variation of $L_*$. Nevertheless, some luminosity contribution from the shock can be envisaged in the blue spectral range at some phases because the "effective" temperature of the recombination region of the shock wake is around $15{,}000\,\mathrm{K}$.

# COLLIMATED OUTFLOW FROM STARS: THE PLANETARY NEBULA ABELL 78

PARIS PİŞMİŞ
*Instituto de Astronomía*
*UNAM, México*

The planetary nebula Abell 78 is slightly oval in an H$\alpha$ image taken with a focal reducer attached to the 2.1-m telescope of San Pedro Mártir Observatory. Jacoby (1979) noted that the morphology of this PN in the [O III] $\lambda$5007 line is quite different from that shown on the POSS red plates.

The velocity points in the H$\alpha$ line can be clearly divided into positive and negative sections around a line passing through the central star in the direction of the minor axis. This velocity structure has suggested a model for the formation of the nebular images as follows: matter is ejected from a spot at a latitude of 8° from a rotating massive progenitor. The two roundish images that one observes as an oval image in the H$\alpha$ figure are inclined to the line of sight causing the slight difference in the velocity of the planetary.

Later ejections coming from deeper layers of the star are mostly composed of heavy elements.

## Epilogue

Most of the structure in velocity-field and morphology of ejected matter from stars can be explained by assuming that it is not from spherical outflows, but ejections from specific areas of the surface of the star.

## Reference

Jacoby, G. H. 1979, *PASP*, 91, 754

# Session VI

## BINARITY

Archaeological site of the ancient city of Perge, less than 20 km from Antalya. Columns date from the Roman period.

Conference participants listening attentively to their guide at the thermal baths of Perge. In the foreground are Robert McClure, Steve and Irene Little, and Sophie Van Eck.

# THE ROLE OF BINARIES IN THE CARBON STAR PHENOMENON

ROBERT D. McCLURE
*Dominion Astrophysical Observatory*
*Herzberg Institute of Astrophysics*
*National Research Council of Canada*
*Victoria, BC   Canada*

**Abstract.** This presentation reviews the role that binaries play in the production of Barium, CH, and S stars. New radial velocity observations confirming the binary nature of subgiant CH stars are also discussed. Evidence is presented that the early R–type carbon stars exhibit **no** binaries. It is suggested that they were once **all** binaries, but having small separations, a coalescing companion has caused them to mix near the helium core flash.

## 1. Introduction

In a review of the roles of the various carbon stars, Scalo (1976) pointed out that the classical carbon stars lie above the luminosity of the onset of helium shell flashing on the asymptotic giant branch (AGB), as shown by Crabtree et al. (1976). But he noted that most Barium stars are too faint to be identified with double-shell models. Along with CH stars, their population II equivalents, they did not fit into the AGB scenario for production of carbon stars, even though they exhibit many of the same abundance characteristics. It seemed that they must be associated with the helium core flash, but standard stellar models did not predict mixing of the core flash material.

## 2. The Barium, CH, and S stars

### 2.1. EVIDENCE FOR BINARIES

My own involvement with this subject began in 1979 with a program to measure radial velocities of Barium stars using the new radial velocity spectrometer in Victoria, and within a year it was obvious that Barium stars are mostly binaries (McClure, Fletcher & Nemec 1980). Over the course of the next few years, it was found that the CH stars as well as the Barium stars are likely all binaries (McClure 1983, 1984; McClure & Woodsworth 1990). It was suggested that the Barium and CH stars have received material contaminated with carbon and s-process elements from a companion star undergoing helium shell flashes on the AGB. The companion has since become a non-visible white dwarf.

The CORAVEL group proved beyond doubt that the binary nature of the Barium stars could not be ignored. Jorissen & Mayor (1988) found that in a sample of 27 southern Barium stars, 89% exhibited variable radial velocities above the $3\sigma$ level. But in addition they showed that eight out of a sample of nine non-variable S stars are also binaries.

The discovery of the binary nature of a subset of the S stars is one of the true success stories in the investigation of the carbon star phenomenon. The possibility of an evolutionary link between Barium and S stars has been suggested, going back as far as Burbidge & Burbidge (1957), on the basis of the spectral similarity between the coolest Barium stars and the hottest S stars. However, the S stars were thought to be undergoing helium shell-flashing in a sequence going from M to S to C stars.

Interest in duplicity among the S stars stemmed from surveys for the presence or absence of the radioactive s-process element Tc. Beginning with a spectral survey by Little–Marenin & Little (1979), and discussed by Scalo & Miller (1981), it was shown that some 30% of S stars exhibit no Tc lines. This quickly led several groups to suggest that the Tc-poor S stars may be binary analogs to the Barium stars (Iben & Renzini 1983; Johnson & Ake 1984; Perry 1985; Smith & Lambert 1986, 1988; Little et al. 1987). Tc, which has a half-life of $2 \times 10^5$ years, a good fraction of the star's lifetime on the giant branch, should be present if the S stars are thermally pulsating AGB stars in the M–S–C evolutionary sequence.

Thus it appears that the S stars can be divided into two groups: the "intrinsic" S stars are true AGB stars undergoing thermal pulses, whereas the "extrinsic" S stars are just binaries that have undergone a mass exchange process like the Barium stars. Brown et al. (1990), Johnson et al. (1993), and Jorissen et al. (1993) presented further evidence which fully supports this dual nature of the S stars.

2.2. PRECURSORS TO THE BARIUM STARS

The mass-transfer hypothesis for the origin of Barium and CH stars was not without difficulties, as pointed out in papers by Luck & Bond (1982), Dominy & Lambert (1983), and Luck & Bond (1991). Although evidence had been found from IUE spectra for white-dwarf companions to one or two Barium stars (Böhm–Vitense 1980; Dominy & Lambert 1983; Böhm–Vitense et al. 1984), in fact, these programs had met with very limited success. Dominy & Lambert (1983) and Bond (1984) found no additional evidence for white dwarfs among numerous Barium and subgiant CH stars. Since the time-scale for white dwarfs to cool below the IUE detection limit exceeds the evolutionary time-scale for red giants by a considerable factor, then mass transfer must have occurred while the present visible star was on the main sequence. The paucity of dwarf Barium stars, except for G77–61 (Dahn et al. 1977) and perhaps some of the sgCH stars in a limited spectral range near G0, did not favour this. So just where were the precursors to the Barium stars?

Slowly, the evidence has been accumulating that indeed, we do find these stars if we search hard enough. Besides the sgCH stars near spectral type G0, more difficult to detect F–type counterparts were observed by Tomkin et al. (1989), North & Duquennoy (1991), and North, Berthet & Lanz (1994). Despite the previous apparent lack of late-type dwarf analogues to the Barium and CH stars, it now appears that they too exist in abundance (Green & Margon 1994; de Kool & Green 1995; Green 2000).

2.3. MASS EXCHANGE MECHANISM

Webbink (1986) and McClure & Woodsworth (1990) pointed out that the eccentricities of Barium star orbits are lower on average than for normal giants. This orbital dissipation suggested that mass exchange may have taken place onto barium stars through Roche lobe overflow. After examining orbits for a good sample, however, Boffin & Jorissen (1988) suggested that wind accretion rather than Roche lobe overflow must account for the mass exchange.

The mass exchange mechanism has been discussed further by Jorissen & Boffin (1992), who point out several problems with the Roche lobe overflow scenarios. Začs (1994) finds a correlation between orbital period and $s$-process abundance anomalies for barium star binaries. Boffin & Začs (1994) show that this correlation agrees with predictions of wind accretion models. Recently, Han et al. (1995) have concluded from a theoretical investigation that mass transfer was likely accomplished in about 70% of Barium stars through a stellar wind, but that RLOF and common envelope evolution are important for the remaining significant fraction of cases. They successfully explain the distribution of orbital periods, mass functions, and numbers of observed Barium and CH stars.

## 3. Binaries among the sgCH Stars

For the rest this presentation, I will report on further radial velocity observations of my own for sgCH stars and early R stars. Smith & Demarque (1980) first suggested a binary mass-transfer origin for the sgCH stars on the basis that they were difficult to explain by mixing of helium core flash material. Very preliminary velocity results have been published for these stars in several conference proceedings (e.g. McClure 1985, 1989). The conclusion was reached that a large fraction of sgCH stars are binaries, but this was based on very limited data, and in the intervening years, I have collected enough velocities to set this result on a much firmer basis. In eight out of ten stars there is definitely evidence for duplicity, and perhaps evidence for velocity variations in one further star. It is reasonable to conclude that all sgCH stars are binaries, and that Luck & Bond's suggestion that they are the precursors to the Barium and CH stars is correct.

## 4. The R Stars – Carbon Stars of a Different Kind!

The R stars, I believe, represent the outstanding problem today in the subject of the origin of carbon stars. It is interesting, however, that the study of these stars has been almost completely ignored in the last decade, and as far as I can tell, there are no other papers at this conference that deal with them. This is despite the observation that the spectrum exhibited on all the advertisements for the conference appears to be that of an R star! [The spectrum shown on the conference poster and T-shirts is indeed that of the R star HD 156074, classified C–R2 by Barnbaum et al. (1996) – Ed.] Dominy (1984) summarizes the characteristics of R stars, and Lloyd Evans' (1986) paper is very informative. Many are not bright enough to be on the AGB, and they do not exhibit significantly enhanced $s$-process elements, signatures of carbon-related stars that have received their peculiar abundances on the AGB or from mass transfer from an AGB star. Dominy suggests that the carbon likely has to come from the helium core flash, but it is not obvious how this carbon can be conveyed out to the surface of the star.

As for the case of the sgCH stars, I have drawn preliminary conclusions in previous conferences concerning the frequency of binaries among the R stars. However, in this case, these conclusions may not have been accurate! I have suggested that the incidence of binaries among the R stars is low, but that it is probably normal for giant stars. With velocity data covering up to 16 years, I have now taken a more careful look at this question in preparation for the present conference.

When I first picked my sample years ago, I was not concerned about including stars which might turn out not to be true R stars. I intended to study mostly R0–R4 stars, since these are more normal, intrinsically fainter, and less likely to have unstable atmospheres; however, I included a few later R5–R8 stars. Some stars in the sample have turned out to be N stars according to more modern classifications and better data (e.g. the spectral atlas of Barnbaum, Stone & Keenan 1996). Still others have been classified as CH or CH–like stars (Yamashita 1975). Of course all CH stars were classified as R under the old HD classification.

In my original sample of 38 stars, five are N stars, and seven are R5–R8 stars. All these show velocity variations considerably larger than the observational errors, but no evidence for binary motion. This velocity "jitter" is not unusual for very high luminosity giants.

The problem of misclassification of R stars versus CH stars is a very difficult one, the differences sometimes being subtle at low dispersion. Those with strong CH bands that have high velocity can be assumed to be CH stars. However, Yamashita (1975) has suggested that there are some low velocity CH stars also, which he refers to as "CH–like". It is very important that these low velocity CH stars be removed from the sample, since CH stars appear to be binaries (McClure 1984), whereas the frequency of binaries for the R stars (which we are trying to determine here) is low.

HD 16115 has been classified as a CH–like star by Yamashita (1975) and by Keenan (1993). I have kept it in the R star sample because Dominy's (1984) analysis shows that the $s$-process element abundances are near normal, which is a primary distinction separating the R stars from the CH stars. Four other CH–like stars, however, have been excluded from my final sample. HD 85066, although classified by Yamashita (1972) as R, has a very strong CH index and more recently has been classified as a CH star by Hartwick & Cowley (1985). This star is a definite binary with a period of 2902 days. There is evidence for duplicity in BD +2°3336, for which I have determined a good orbit with a 446-day period. BD +29°95 exhibits very erratic velocities; it may be a binary, but this is uncertain. Although HD 197604 has been classified by Yamashita (1975) as a CH–like star, there is no evidence for radial velocity variations.

Finally, Figure 1 shows velocities for the 22 remaining R0–R4 stars in the sample. There is no evidence whatever for duplicity in any of these observations. This seems very surprising, given that significant numbers of normal G and K giants are binaries. For example, Gunn & Griffin (1979) comment that in their velocity data about 30% of field giants exhibit multiplicity. The largest program of velocity monitoring among giants is that by Mermilliod & Mayor (1992), who find 187 binaries among 905 giant stars in open clusters, or a frequency of ∼21%. I have monitored a much smaller sample of 39 randomly picked field K giants and found six binaries. Our finding of zero binaries in a sample of 22 R stars can be compared with this

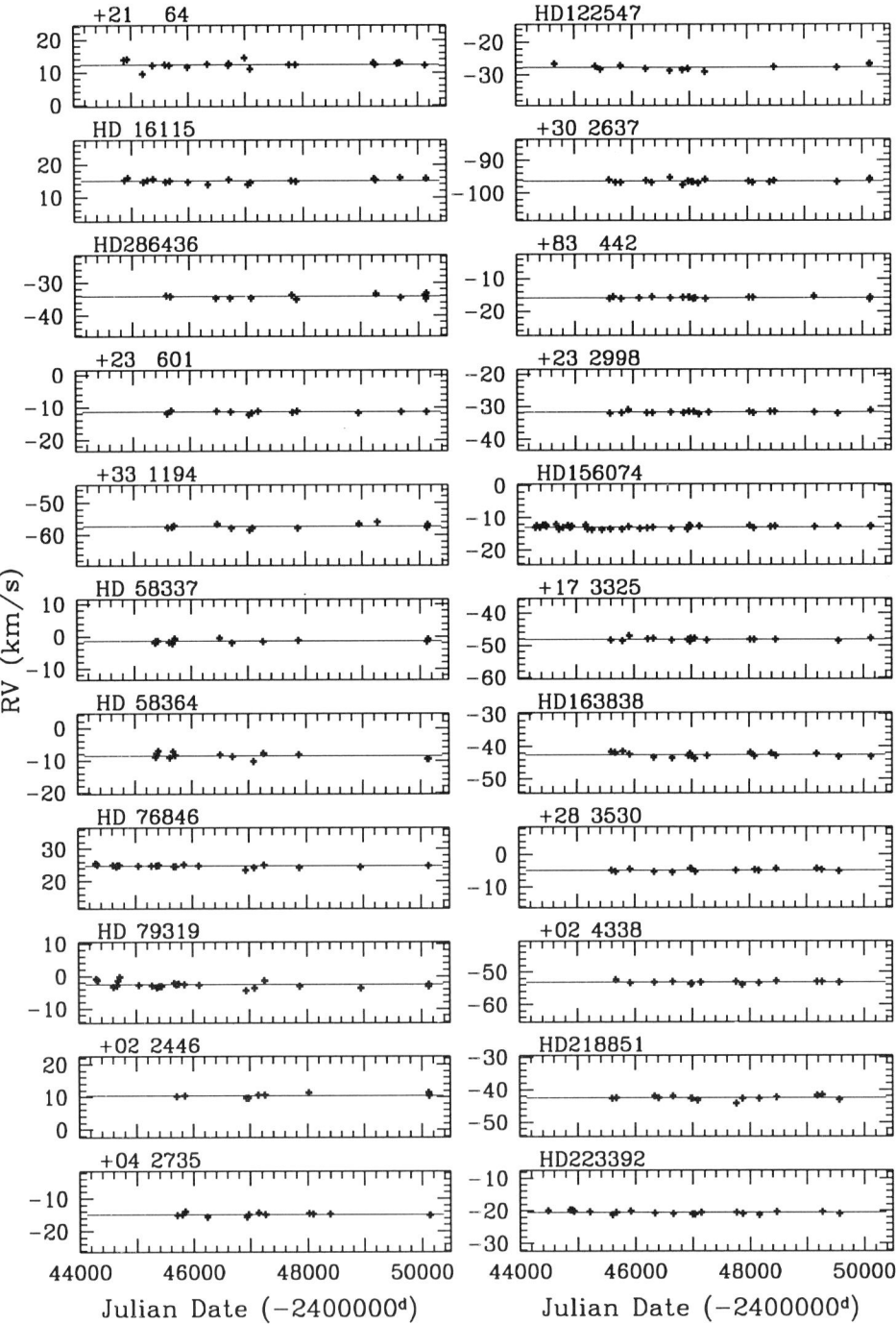

Figure 1. Velocities versus Julian Date for R0–R4 stars.

large combined sample of normal giant stars. A $\chi^2$ test in a $2 \times 2$ contingency table indicates that the possibility that the R stars have a normal binary frequency can be rejected at the 98% significance level!

Now, it is much more difficult to prove that a class of stars has *no* binaries, than it is to prove that a class of stars such as the Barium stars are *all* binaries. In the latter case, if a star or two are found to have constant velocity, this can be reconciled by a high orbital inclination and/or large separation, for example. In contrast, normal giants have a rather low frequency themselves ($\sim$15–25%). Therefore, it only takes two or three binaries to turn up amongst R stars to remove the significance of the result. However, suppose we take the statistics at face value. How can we reconcile the result that the R stars appear to lack binaries? A possibility is that the R stars were once all close binaries, with separations too small to survive the expansion as the star evolved up the giant branch. We have talked a lot about binary mass-transfer from an AGB star to form a Barium or a CH star. What happens, however, to binaries that have separations that are too small? Various authors have discussed common-envelope evolution. It could be that a companion spiraling into the center of an evolving giant star, either during the helium core flash or afterwards, could cause this material to be mixed to the surface.

The assumption that the carbon that is polluting the atmospheres of the R stars came from the helium core flash (e.g. Dominy 1984) is reasonable. The R stars do not show the enhanced *s*-process material that is a signature of all other types of carbon stars originating on the AGB. The R stars, it appears, were created in some other way. In addition, there do not seem to be main-sequence precursors to R stars. The Barium and CH stars appear to be just evolved sgCH stars. Green & Margon (1994) have reported enhanced abundances for *s*-process elements in all six dwarf carbon stars that they observed. The R stars appear to be ones that are polluted with carbon at a more evolved state, therefore, and it is very reasonable that this might be at the helium core flash. It appears difficult to comprehend, however, just how this carbon can be mixed out to the surface under conventional stellar evolution.

I wish to thank Chad Hogan for help in preparing for this presentation, and Chris Tout for very helpful discussions.

## References

Barnbaum, C., Stone, R. P. S. & Keenan, P. C. 1996, *ApJ Supp.*, 105, 419
Boffin, H. M. J. & Jorissen, A. 1988, *A&A*, 205, 155
Boffin, H. M. J. & Začs, L. 1994, *A&A*, 291, 811
Böhm–Vitense, E. 1980, *ApJ*, 239, L79
Böhm–Vitense, E., Nemec, J. & Proffitt, C. 1984, *ApJ*, 278, 726
Bond, H. E. 1974, *ApJ*, 194, 95

Bond, H. E. 1984, in *Future of Ultraviolet Astronomy Based on Six Years of IUE Research*, ed. J. M. Mead, R. D. Chapman and Y. Kondo (NASA CP–2349), p. 289
Brown, J. A., Smith, V. V., Lambert, D. L., Dutchover, E. Jr., Hinkle, K. H. & Johnson, H. R. 1990, *AJ*, 99, 1930
Burbidge, E. M. & Burbidge, G. R. 1957, *ApJ*, 126, 357
Crabtree, D. R., Richer, H. B. & Westerlund, B. E. 1976, *ApJ*, 203, L81
Dahn, C. C., Liebert, J., Kron, R. G., Spinrad, H. & Hintzen, P. M. 1977, *ApJ*, 216, 757
de Kool, M. & Green, P. J. 1995, *ApJ*, 449, 236
Dominy, J. F. 1984, *ApJ Supp.*, 55, 27
Dominy, J. F. & Lambert, D. L. 1983, *ApJ*, 270, 180
Green, P. J. 2000, in IAU Symp. 177: *The Carbon Star Phenomenon*, ed. R. F. Wing (Kluwer), p. 27
Green, P. J. & Margon, B. 1994, *ApJ*, 423, 723
Gunn, J. E. & Griffin, R. F. 1979, *AJ*, 84, 752
Han, Z., Eggleton, P. P., Podsiadlowski, P. & Tout, C. A. 1995, *MNRAS*, 277, 1443
Hartwick, F. D. A. & Cowley, A. P. 1985, *AJ*, 90, 2244
Iben, I. Jr. & Renzini, A. 1983, *Ann. Rev. Astron. Astrophys.*, 21, 271
Johnson, H. R. & Ake, T. B. 1984, in *Cool Stars, Stellar Systems and the Sun*, ed. S. L. Baliunas and L. Hartmann (Springer–Verlag), Lecture Notes in Physics No. 193, p. 362
Johnson, H. R., Ake, T. B. & Ameen, M. M. 1993, *ApJ*, 402, 667
Jorissen, A. & Boffin, H. M. J. 1992, in *Binaries as Tracers of Star Formation*, ed. A. Duquennoy and M. Mayor, (Cambridge Univ. Press), p. 110
Jorissen, A., Frayer, D. T., Johnson, H. R., Mayor, M. & Smith, V. V. 1993, *A&A*, 271, 463
Jorissen, A. & Mayor, M. 1988, *A&A*, 198, 187
Keenan, P. C. 1993, *PASP*, 105, 905
Little, S. J., Little–Marenin, I. R. & Bauer, W. H. 1987, *AJ*, 94, 981
Little–Marenin, I. R. & Little, S. J. 1979, *AJ*, 84, 1374
Lloyd Evans, T. 1986, *MNRAS*, 220, 723
Luck, R. E. & Bond, H. E. 1982, *ApJ*, 259, 792
Luck, R. E. & Bond, H. E. 1991, *ApJ Supp.*, 77, 515
McClure, R. D. 1983, *ApJ*, 268, 264
McClure, R. D. 1984, *ApJ*, 280, L31
McClure, R. D. 1985, *Cool Stars with Excesses of Heavy Elements*, ed. M. Jaschek and P. C. Keenan (Reidel), p. 315
McClure, R. D. 1989, in IAU Coll. 106: *Evolution of Peculiar Red Giant Stars*, ed. H. R. Johnson and B. Zuckerman (Cambridge Univ. Press), p. 196
McClure, R. D., Fletcher, J. M. & Nemec, J. M. 1980, *ApJ Supp.*, 238, L35
McClure, R. D. & Woodsworth, A. W. 1990, *ApJ*, 352, 709
Mermilliod, J.-C. & Mayor, M. 1992, in *Binaries as Tracers of Star Formation*, ed. A. Duquennoy and M. Mayor (Cambridge Univ. Press), p. 183
North, P., Berthet, S. & Lanz, T. 1994, *A&A*, 281, 775
North, P. & Duquennoy, A. 1991, *A&A*, 244, 335
Perry, B. F. Jr. 1985, in *Cool Stars with Excesses of Heavy Elements*, ed. M. Jaschek and P. C. Keenan (Reidel), p. 333
Scalo, J. M. 1976, *ApJ*, 206, 474
Scalo, J. M. & Miller, G. E. 1981, *ApJ*, 246, 251
Smith, J. A. & Demarque, P. 1980, *A&A*, 92, 163
Smith, V. V. & Lambert, D. L. 1986, *ApJ*, 311, 843
Smith, V. V. & Lambert, D. L. 1988, *ApJ*, 333, 219
Tomkin, J., Lambert, D. L., Edvardsson, B., Gustafsson, B. & Nissen, P. E. 1989, *A&A*, 219, L15
Webbink, R. F. 1986, in *Critical Observations versus Physical Models for Close Binary Systems*, ed. K.-C. Leung and D. S. Zhai (Gordon and Breach), p. 403

Yamashita, Y. 1972, *Ann. Tokyo Astr. Obs.*, 13, 169
Yamashita, Y. 1975, *PASJ*, 27, 325
Začs, L. 1994, *A&A*, 283, 937

## Discussion

**Feast**: Is it possible that the R stars are very wide binaries which have undergone only a small amount of mass transfer?

**McClure**: We can observe binaries with periods of 15 years or so. I don't think you will get much mass transfer with so widely separated stars.

**Frogel**: From Ba star orbits you can probably set limits on the size of orbits needed to get mass transfer. You can also make estimates for sizes of R stars. Then you can see if these two numbers are consistent with the hypothesis that R stars engulf their stellar companions.

**McClure**: I have only done all this analysis of my velocities very recently while preparing for this meeting, so I haven't looked into details such as you mention.

**Lloyd Evans**: Perhaps it is time to consider alternatives to nuclear astrophysics, such as chemical fractionation in circumstellar material (such as exists around the J-silicate stars), to explain the properties of the R stars.

258

Zhigang Gong, who single-handedly represented the PRC at the symposium, ponders the ruins of Perge.

This stork calls Perge home.

# BARIUM STARS AND Tc–POOR S STARS: BINARY MASQUERADERS WITHIN THE CARBON–STAR FAMILY

A. JORISSEN AND S. VAN ECK
*Institut d'Astronomie et d'Astrophysique,*
*Université Libre de Bruxelles, Belgium*

**Abstract.** Our current understanding of the origin of barium and S stars is briefly reviewed, based on new orbital elements and binary frequencies.

## 1. The relation of barium and S stars to carbon stars

Since the last conference devoted to chemically-peculiar red giant stars (IAU Coll. 106, *Evolution of Peculiar Red Giant Stars*, ed. Johnson & Zuckerman, 1989), much progress has been made in understanding how barium and S stars relate to the other kinds of peculiar red giants. The discovery of the binary nature of barium stars (McClure et al. 1980; McClure 1983) suggested from the beginning that mass transfer was likely to play a key role in the formation of the barium syndrome. As far as S stars are concerned, it has become clear that Tc-rich and Tc-poor S stars form two separate families with similar chemical peculiarities albeit of very different origins (Iben & Renzini 1983; Little et al. 1987; Jorissen & Mayor 1988; Smith & Lambert 1988; Brown et al. 1990; Johnson 1992; Jorissen & Mayor 1992; Groenewegen 1993; Johnson et al. 1993; Jorissen et al. 1993; Ake, this conference). Tc-rich (or 'intrinsic') S stars are genuine thermally-pulsing AGB stars where the $s$-process operates in relation with the thermal pulses, and where the third dredge-up brings the freshly synthesized $s$-elements (including Tc) to the surface (e.g. Iben & Renzini 1983; Sackmann & Boothroyd 1991). By contrast, Tc-poor (or 'extrinsic') S stars are believed to be the cool descendants of barium stars.

Figure 1 summarizes our current understanding of the relationship between the different families of peculiar red giant stars. This general picture raises several questions that will be addressed briefly in this paper:

– Is binarity a necessary condition to produce a barium star?

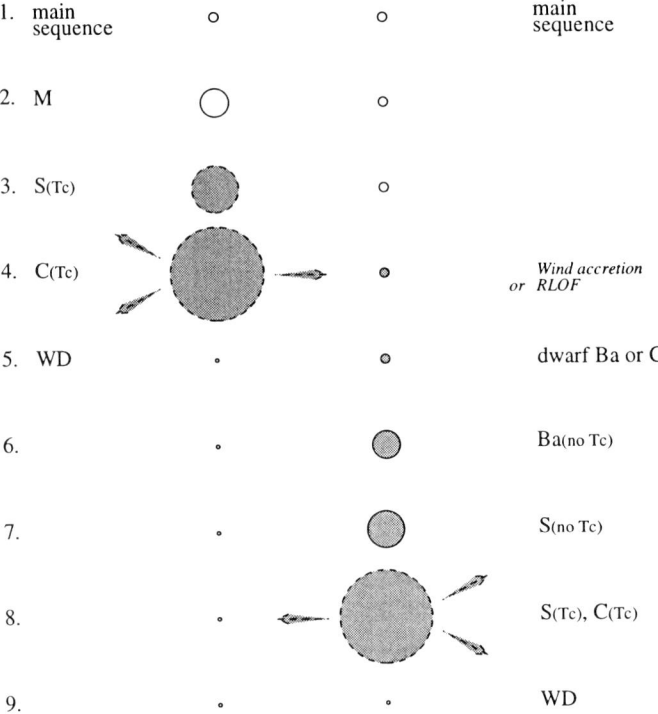

Figure 1. Relationship between several families of peculiar red giant stars. Grey symbols represent heavy-element-rich stars, and dashed boundaries indicate Tc-rich stars. The left column depicts the normal (i.e. not requiring binarity) M–S–C evolution on the AGB, whereas the right column represents the evolution of a companion star. Note in particular the possibility that this companion itself evolves into a Tc-rich S star on the AGB, after having first shown up as a Tc-poor S star.

- What is the mass transfer mode [wind accretion or Roche lobe overflow (RLOF)] responsible for their formation?
- Do barium stars form as dwarfs or as giants?
- Do barium stars evolve into Tc-poor S stars?
- What is the relative frequency of Tc-rich and Tc-poor S stars?
- Are the abundances in the mass-loser star (i.e. the AGB progenitor of the present white dwarf companion) compatible with those presently observed in the barium or extrinsic S star?

We refer to Jorissen & Boffin (1992), Han et al. (1995) and Busso et al. (1995) for a detailed discussion of the last item.

## 2. Is binarity a necessary condition to form a barium star?

To answer that question, barium stars with *strong* anomalies (including *all* Ba4 and Ba5 stars on the scale devised by Warner 1965 from the list of Lü et al. 1983) have been monitored either with the CORAVEL spectrovelocimeter (Baranne et al. 1979) or by McClure & Woodsworth (1990). HD 19014 and HD 65854 are the only stars in that sample that do not show any sign of binary motion. For a fictitious population of binaries observed with the same time sampling and the same internal errors as the real sample of barium stars, and having eccentricity and mass-function distributions matching the observed ones, a Monte-Carlo simulation yields a binary detection rate of between 96% (35.4/37) and 98% (36.2/37), depending on whether the observed period distribution is extrapolated or not towards periods as long as $2 \times 10^4$ d [see Jorissen et al. 1998 for more details]. *Binarity is thus a necessary condition to produce strong barium stars.*

In a comparison sample of 40 *mild* barium stars (i.e. with Ba1 and Ba2 indices) randomly selected from the list of Lü et al. (1983) and monitored in a similar way as the strong barium stars, 34 (85%) are definitely spectroscopic binaries, 3 (8%) are probably binaries, and 3 (HD 50843, HD 95345, HD 119185) show no sign of radial velocity variations at the level of 0.3 km s$^{-1}$ r.m.s. after more than 10 years of monitoring. Detailed spectroscopic abundance analyses performed on HD 95345 (Sneden et al. 1981) and HD 119185 (Začs et al. 1997) confirm the existence of mild heavy-element overabundances ($[s/Fe] = 0.2$ to 0.3 dex) for these stars with constant radial velocity. This frequency of constant stars is again consistent with the binary detection rate predicted for that sample by a Monte-Carlo simulation, provided that the period distribution of mild barium stars extends up to $2 \times 10^4$ d. In these conditions, there is no need to invoke any formation mechanism other than mass transfer in a binary system to produce mild barium stars. On the contrary, an alternative formation scenario (such as galactic fluctuations of the $s/Fe$ ratio; Williams 1975, Sneden et al. 1981, Edvardsson et al. 1993) may be required to account for a population of non-binary stars found among *dwarf* mild barium stars (North et al., this conference).

Is binarity a *sufficient* condition to produce a barium star? Probably not, since binary systems consisting of a *normal* red giant and a WD companion with Ba-like orbital parameters do exist (Jorissen & Boffin 1992). DR Dra (= HD 160538) is probably the best example, with $P = 904$ d, $e = 0.07$ (compare with Fig. 3) and a hot WD companion detected by Fekel et al. (1993). Berdyugina (1994) finds a metallicity close to solar and normal Zr and La abundances in the giant. Začs et al. (1997) basically confirm that result.

Metallicity may be the other key parameter, besides binarity, controlling the formation of barium stars. The $s$-process efficiency, expressed in terms of the neutron irradiation, seems to be larger in low-metallicity stars (Kovács 1985; Busso et al. 1995). Clayton (1988) provides a theoretical foundation

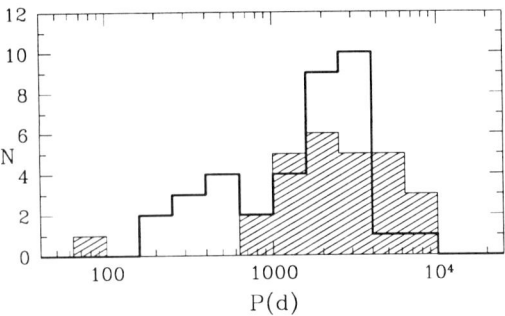

*Figure 2.* The distribution of orbital periods for mild barium stars (shaded histogram) and strong barium stars (thick line) (from McClure & Woodsworth 1990 and Jorissen et al. 1998). The distribution is complete up to about 4000 d.

for that empirical finding, provided that $^{13}C(\alpha, n)^{16}O$ is the neutron source for the s-process. Barium stars would therefore be easier to produce in a low-metallicity population.

## 3. Inferring the mass transfer mode from the orbital elements: Wind accretion and/or Roche lobe overflow?

Synthetic binary evolution models (Han et al. 1995; de Kool & Green 1995) suggest that the bimodal period distribution exhibited by strong barium stars (Fig. 2) reflects the operation of two distinct mass-transfer modes, RLOF in the short-period mode (peaking around 500 d) and wind accretion in the long-period mode (around 3000 d).

This general picture actually faces three major difficulties: first, the threshold period (about 1000 d) between the RLOF and wind-accretion modes is much too short to accomodate the large radii reached by AGB stars. Second, the period – eccentricity diagram (Fig. 3) reveals that not all orbits in the short-period (i.e. post-RLOF) mode are circular, although tidal effects are expected to circularize the orbit in the phase of large radius just preceding RLOF (e.g. Zahn 1977). A similar problem exists for the orbits of dwarf barium stars (see North et al., this conference). Third, RLOF from AGB stars with a deep convective envelope is dynamically unstable ('unstable case C RLOF'; e.g. Tout & Hall 1991), with the ensuing common envelope stage generally accompanied by dramatic orbital shrinkage leading to the formation of a cataclysmic binary with a period much shorter than that of barium stars (e.g. Meyer & Meyer-Hofmeister 1979). To solve these problems, Han et al. (1995), Livio (1996) and Jorissen et al. (1998) propose avenues to explore. One of these involves Eggleton's CRAP (Companion-Reinforced Attrition Process; Eggleton 1986), speculating that larger mass-loss rates for AGB stars in binary systems may reverse the mass ratio of the system prior to RLOF, thus stabilizing the mass transfer process (Tout

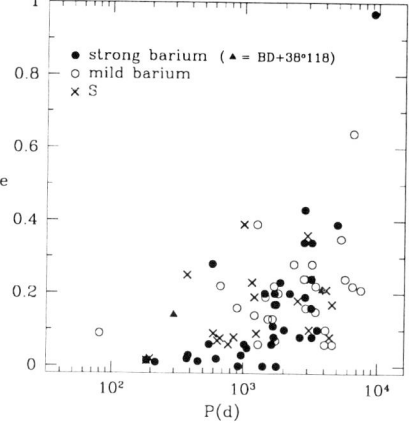 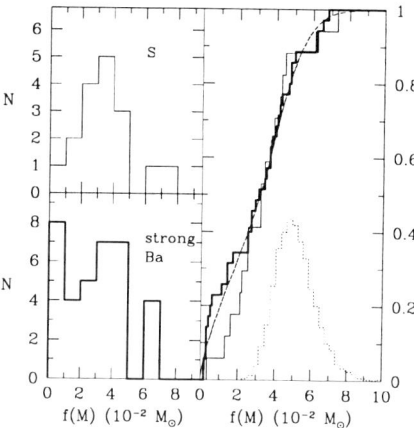

*Figure 3.* The $(e, \log P)$ diagram for barium and S stars (from Jorissen et al. 1998). BD+38°118 is a triple hierarchical system, with the close inner binary and the orbit of the third star around the center of mass of the inner binary represented by filled triangles.

*Figure 4.* The mass-function distributions of strong barium stars (thick line) and S stars (thin line) from Jorissen et al. (1998), excluding the peculiar S systems HDE 332077 and HD 191589 (see text). The dashed line in the right panel is the fit obtained with the $Q$ distribution shown (dotted line).

& Eggleton 1988; Han et al. 1995).

## 4. Do barium stars form as dwarfs or as giants?

In Fig. 1, it is assumed that the mass transfer responsible for the barium syndrome occurred when the barium star was still on the main sequence. Because the stellar lifetime is longer on the main sequence than in the giant phase, that possibility indeed appears more probable than the formation of the barium star directly as a giant star. However, as pointed out by Iben & Tutukov (1985), the mismatch between the thermal time scale of the dwarf's envelope and that of the mass-losing AGB star may prevent the formation of dwarf barium stars. A main-sequence star would indeed be driven out of thermal equilibrium in case of rapid mass accretion from its giant companion, and would swell to giant dimensions (e.g. Kippenhahn & Meyer-Hofmeister 1977), leading to a common envelope stage with possibly dramatic consequences on the fate of the binary system (see e.g. Meyer & Meyer-Hofmeister 1979, and Sect. 3). Dwarf barium stars long remained elusive, until Luck & Bond (1982, 1991) and North et al. (1994) recognized that some of the CH subgiants previously identified by Bond (1974), as well as some of the F dwarfs previously classified by Bidelman (1985) as having 'strong Sr $\lambda$ 4077', had the proper abundance anomalies, gravities and galactic frequencies to be identified with the long-sought Ba dwarfs. A large fraction of binaries (about 90%) has been found among the stars with strong anomalies, as expected (McClure 1985; North & Duquennoy 1992; North et al., this conference). The very existence of binary dwarf Ba stars, in spite of Iben & Tutukov's argument, is another indication that, if RLOF does indeed occur in these systems, it does not have the catastrophic consequences generally associated with unstable case C RLOF. The question of whether these dwarf barium stars will eventually evolve into giant barium stars is addressed by North et al. in these Proceedings.

The formation of a barium star directly as a giant, though probably less frequent, is by no means excluded. The barium star HD 165141 may be such a case. HD 165141 is unique in sharing properties of barium and RS CVn systems (Fekel et al. 1993; Jorissen et al. 1996). Its rapid rotation ($V \sin i = 14$ km s$^{-1}$) and X-ray flux (probably from a hot corona) are typical of RS CVn systems. However, the spin-up of that star (and the concomittant RS CVn properties) cannot be attributed to tidal effects synchronizing the stellar rotation with the orbit, as is the case for RS CVn systems, since the orbital period (about 5200 d) is much too long. That puzzle may be solved if the wind accretion episode responsible for the barium syndrome spun the star up, as suggested by detailed hydrodynamical simulations (Theuns & Jorissen 1993; Theuns et al. 1996). Since magnetic braking is generally faster than the stellar lifetime on the giant branch, wind accretion and concomittant spin-up must have occurred when HD 165141 was already a giant star. Strong support for that hypothesis comes from the fact that HD 165141 has a hot WD companion (Fekel et al. 1993) whose cooling time scale is shorter than the lifetime of HD 165141 on the red giant branch. Finally, note that Jeffries & Stevens (1996) have reported more cases of WIRRing (Wind-Induced Rapidly Rotating) stars among binaries involving a hot WD.

## 5. Do barium stars evolve into Tc-poor S stars?

Figure 3 shows that strong barium stars and Tc-poor S stars occupy the same region of the $(e, \log P)$ diagram. The distributions of the mass function $f(M)$ presented in Fig. 4 [where $f(M) = M_2^3 \sin^3 i/(M_1 + M_2)^2 \equiv Q \sin^3 i$, $M_1$ and $M_2$ being the masses of the giant and of the WD, respectively] for the two families are compatible with the hypothesis that they are extracted from the same parent population. Following the usual analysis (Webbink 1986; McClure & Woodsworth 1990) of the mass function distribution in terms of a peaked distribution of mass ratios $Q$ convolved with randomly inclined orbits, an average ratio $Q = 0.045 \ M_\odot$ is found for the two classes, translating into a giant mass of 1.6 $M_\odot$ when adopting $M_2 = 0.6\ M_\odot$ for the WD companion. These two results thus provide strong support for the hypothesis that strong barium and Tc-poor S stars represent successive stages in the evolutionary path sketched in Fig. 1.

Note, however, that the above comparison of the mass functions does not include two Tc-poor S stars (HD 191589 and HDE 332077) with main sequence companions detected with the *International Ultraviolet Explorer* satellite (Ake & Johnson 1992; Ake et al. 1992). The evolutionary status of these stars is currently unknown.

## 6. The relative frequency of intrinsic/extrinsic S stars

The evaluation of the relative frequency of intrinsic/extrinsic S stars faces two difficulties: (*i*) one needs an efficient criterion for distinguishing extrinsic from intrinsic S stars, and (*ii*) the frequency evaluation must be corrected for selection bias, since extrinsic and intrinsic S stars follow different galactic distributions (Jorissen et al. 1993). As far as (*i*) is concerned, the defining criterion of intrinsic/extrinsic S stars based on the presence/absence of Tc, respectively, may be difficult to apply to a complete sample of S stars such as Henize's (see below), since it involves many faint stars for which high-resolution spectroscopy is difficult to secure. Binarity may be an alternative, since the binary paradigm for S stars states that all Tc-poor S stars should be binaries (Brown et al. 1990; Johnson 1992). However, some binaries must be expected among Tc-rich S stars as well, as in any class of stars. Binary intrinsic S stars with main sequence companions (case 3 in Fig. 1) include the close visual binary $\pi^1$ Gru (Feast 1953) and stars with composite spectra like T Sgr, W Aql, WY Cas (Herbig 1966; Culver & Ianna 1975), and possibly S Lyr (Merrill 1956). The situation is further confused by extrinsic S stars reaching the AGB phase and eventually becoming Tc-rich (case 8 in Fig. 1). $o^1$ Ori, a Tc-rich binary S star with a WD companion (Ake & Johnson 1988), may be such a case.

The CORAVEL $Sb$ parameter, measuring the average line width (see Jorissen & Mayor 1988 for a more detailed definition), offers an interesting and efficient alternative to identify extrinsic/intrinsic S stars. In cool red giants where macroturbulence is the main line-broadening factor, the $Sb$ parameter may be expected to be a sensitive function of the luminosity, as is macroturbulence (e.g. Gray 1988). But at the same time, bright giants exhibit large velocity jitters probably caused by envelope pulsations (e.g. Mayor et al. 1984). A correlation between $Sb$ and the radial velocity jitter must thus be expected, as observed in Fig. 5 for barium, intrinsic and extrinsic S stars (Jorissen & Mayor 1992; Jorissen et al. 1998).

All Tc-poor S stars are binary stars, as expected, but moreover, they are restricted to $Sb < 5$ km s$^{-1}$. That criterion has been used to identify

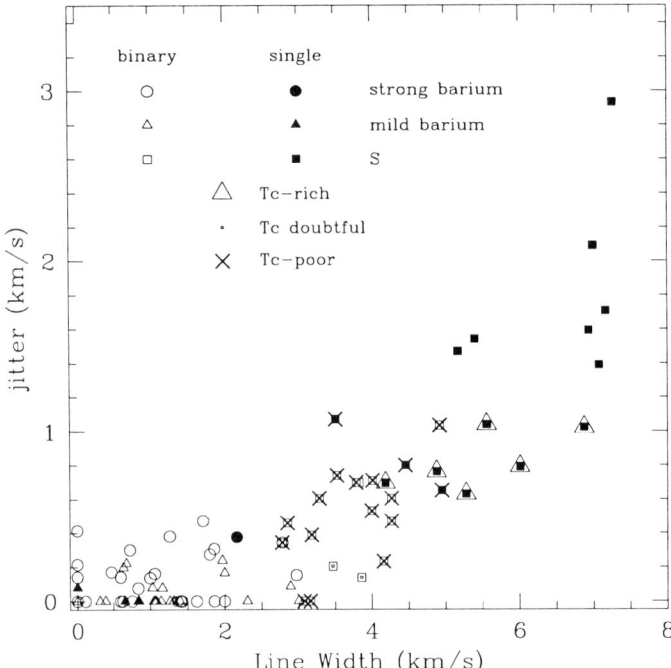

*Figure 5.* The jitter $(\sigma^2 - \bar{\epsilon}_1^2)^{1/2}$ (where $\bar{\epsilon}_1$ is the average error on one measurement, and $\sigma$ is the standard deviation of the radial velocity for single stars, and of the $O-C$ residuals around the computed orbit for binary stars) vs. the CORAVEL line-width parameter $Sb$ (see text).

intrinsic S stars among the Henize sample (Henize 1960). That sample covers the sky south of declination $-25°$ uniformly to red magnitude 10.5, and 205 S stars were found from their ZrO $\lambda 6345$ band on red-yellow spectra with a dispersion of 450 Å mm$^{-1}$ at H$\alpha$. The galactic distribution of the Henize sample is presented in Fig. 6. Intrinsic S stars are clearly more concentrated towards the galactic plane than extrinsic S stars. Correcting for the uneven sampling of galactic latitudes, the frequency of intrinsic S stars (based on the $Sb > 5$ km s$^{-1}$ criterion) then amounts to at least $62 \pm 5\%$ (in a magnitude-limited sample).

It is our pleasure to thank M. Mayor and the CORAVEL team at the Observatoire de Genève for making possible the long-term radial-velocity monitoring discussed here. A.J. is Research Associate, *Fonds National de la Recherche Scientifique* (Belgium); S.V.E. is *Boursier F.R.I.A.* (Belgium). We thank the *Fonds National de la Recherche Scientifique* (Belgium), the *Communauté Française de Belgique* and the symposium Organizing Committee for financial support.

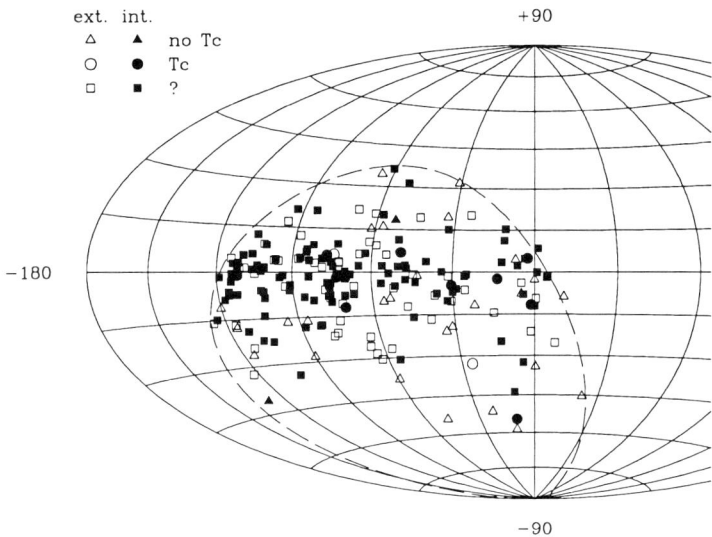

*Figure 6.* Galactic distribution of S stars from the Henize sample.

## References

Ake, T. B. & Johnson, H. R. 1988, *ApJ*, 327, 214
Ake, T. B. & Johnson, H. R. 1992, in *Cool Stars, Stellar Systems and the Sun*, ed. M. S. Giampapa and J. A. Bookbinder, ASP Conf. Ser., 26, 579
Ake, T., Jorissen, A., Johnson, H., Mayor, M. & Bopp, B. 1992, *Bull. Am. Astron. Soc.*, 24, 1280
Baranne, A., Mayor, M. & Poncet, J. L. 1979, *Vistas in Astronomy*, 23, 279
Berdyugina, S. V. 1994, *Pi'sma Astron. J.*, 20, 910
Bidelman, W. P. 1985, *AJ*, 90, 341
Bond, H. E. 1974, *ApJ*, 194, 95
Brown, J. A., Smith, V. V., Lambert, D. L., Dutchover, E. Jr., Hinkle, K. H. & Johnson, H. R. 1990, *AJ*, 99, 1930
Busso, M., Lambert, D. L., Beglio, L., Gallino, R., Raiteri, C. M. & Smith, V. V. 1995, *ApJ*, 446, 775
Clayton, D. D. 1988, *MNRAS*, 234, 1
Culver, R. B. & Ianna, P. A. 1975, *ApJ*, 195, L37
de Kool, M. & Green, P. J. 1995, *ApJ*, 449, 236
Edvardsson, B., Andersen, J., Gustafsson, B., Lambert, D. L., Nissen, P. E. & Tomkin, J. 1993, *A&A*, 275, 101
Eggleton, P. P. 1986, in *The Evolution of Galactic X-ray Binaries*, ed. J. Trümper, W. H. G. Lewin and W. Brinkmann (Reidel), p. 87
Feast, M. W. 1953, *MNRAS*, 113, 510
Fekel, F. C., Henry, G. W., Busby, M. R. & Eitter, J. J. 1993, *AJ*, 106, 2370
Gray, R. F. 1988, *Lectures on Spectral-line Analysis: F, G and K stars*, (Arva, Ontario: The Publisher)
Groenewegen, M. A. T. 1993, *A&A*, 271, 180
Han, Z., Eggleton, P. P., Podsiadlowski, P. & Tout, C. A. 1995, *MNRAS*, 277, 1443

Henize, K. G. 1960, *AJ*, 65, 491
Herbig, G. H. 1966, *AJ*, 71, 779
Iben, I. Jr. & Renzini, A. 1983, *Ann. Rev. Astron. Astrophys.*, 21, 271
Iben, I. Jr. & Tutukov, A. V. 1985, *ApJ Supp.*, 58, 661
Jeffries, R. D. & Stevens, I. R. 1996, *MNRAS*, 279, 180
Johnson, H. R. 1992, in IAU Symp. 151: *Evolutionary Processes in Interacting Binary Stars*, ed. Y. Kondo, R. F. Sistero, and R. S. Polidan (Kluwer), p. 157
Johnson, H. R., Ake, T. B. & Ameen, M. M. 1993, *ApJ*, 402, 667
Jorissen, A. & Boffin, H. M. J. 1992, in *Binaries as Tracers of Stellar Formation*, ed. A. Duquennoy and M. Mayor (Cambridge), p. 110
Jorissen, A., Frayer, D. T., Johnson, H. R., Mayor, M. & Smith, V. V. 1993, *A&A*, 271, 463
Jorissen, A. & Mayor, M. 1988, *A&A*, 198, 187
Jorissen, A. & Mayor, M. 1992, *A&A*, 260, 115
Jorissen, A., Schmitt, J. H. M. M., Carquillat, J. M., Ginestet, N. & Bickert, K. F. 1996, *A&A*, 306, 467
Jorissen, A., Van Eck, S., Mayor, M. & Udry, S. 1998, *A&A*, 332, 877
Kippenhahn, R. & Meyer-Hofmeister, E. 1977, *A&A*, 54, 539
Kovács, N. 1985, *A&A*, 150, 232
Little, S. J., Little-Marenin, I. R. & Bauer, W. H. 1987, *AJ*, 94, 981
Livio, M. 1996, in *Evolutionary Processes in Binary Stars*, ed. R. A. M. J. Wijers, M. B. Davies, and C. A. Tout (Kluwer), p. 141
Lü, P. K., Dawson, D. W., Upgren, A. R. & Weis, E. W. 1983, *ApJ Supp.*, 52, 169
Luck, R. E. & Bond, H. E. 1982, *ApJ*, 259, 792
Luck, R. E. & Bond, H. E. 1991, *ApJ Supp.*, 77, 515
Mayor, M., Imbert, M., Andersen, J., Ardeberg, A., Benz, W., Lindgren, H., Martin, N., Maurice, E., Nordström, B. & Prévot, L. 1984, *A&A*, 134, 118
McClure, R. D. 1983, *ApJ*, 268, 264
McClure, R. D. 1985, in *Cool Stars with Excesses of Heavy Elements*, ed. M. Jaschek and P. C. Keenan (Reidel), p. 159
McClure, R. D., Fletcher, J. M. & Nemec, J. M. 1980, *ApJ*, 238, L35
McClure, R. D. & Woodsworth, A. W. 1990, *ApJ*, 352, 709
Merrill, P. W. 1956, *PASP*, 68, 162
Meyer, F. & Meyer-Hofmeister, E. 1979, *A&A*, 78, 167
North, P., Berthet, S. & Lanz, T. 1994, *A&A*, 281, 775
North, P. & Duquennoy, A. 1992, in *Binaries as Tracers of Stellar Formation*, ed. A. Duquennoy and M. Mayor (Cambridge), p. 202
Sackmann, I.-J. & Boothroyd, A. I. 1991, in IAU Symp. 145: *Evolution of Stars: The Photospheric Abundance Connection*, ed. G. Michaud and A. Tutukov (Kluwer), p. 275
Smith, V. V. & Lambert, D. L. 1988, *ApJ*, 333, 219
Sneden, C., Lambert, D. L. & Pilachowski, C. A. 1981, *ApJ*, 247, 1052
Theuns, T., Boffin, H. M. J. & Jorissen, A. 1996, *MNRAS*, 280, 1264
Theuns, T. & Jorissen, A. 1993, *MNRAS*, 265, 946
Tout, C. A. & Eggleton, P. P. 1988, *MNRAS*, 231, 823
Tout, C. A. & Hall, D. S. 1991, *MNRAS*, 253, 9
Warner, B. 1965, *MNRAS*, 129, 263
Webbink, R. F. 1986, in *Critical Observations vs. Physical Models for Close Binary Systems*, ed. K. C. Leung and D. S. Zhai (Gordon and Breach), p. 403
Williams, P. M. 1975, *MNRAS*, 170, 343
Začs, L., Musaev, F. A., Bikmaev, I. F. & Alksnis, O. 1997, *A&A Supp.*, 122, 31
Zahn, J.-P. 1977, *A&A*, 57, 383

# BINARITY AMONG BARIUM DWARFS AND CH SUBGIANTS: WILL THEY BECOME BARIUM GIANTS?

PIERRE NORTH
*Institut d'Astronomie de l'Université de Lausanne*
*CH-1290 Chavannes-des-Bois, Switzerland*

ALAIN JORISSEN
*Institut d'Astronomie et d'Astrophysique*
*Université Libre de Bruxelles*
*B-1050 Bruxelles, Belgium*

AND

MICHEL MAYOR
*Observatoire de Genève*
*CH-1290 Sauverny, Switzerland*

**Abstract.** We report the results of monitoring the radial velocities of 31 confirmed Ba dwarfs and CH subgiants, showing that about 27 of them have a variable radial velocity. Therefore, mass transfer in a binary most probably explains their chemical anomalies, as for the Ba giants. The orbital parameters of the 14 systems having sufficient phase coverage are very similar to those of the Ba giants. Whether the CH subgiants and Ba dwarfs may become Ba giants is discussed briefly. Most of the marginal Ba dwarfs listed by Edvardsson et al. (1993) are single, so that their chemical peculiarity, if real, cannot be due to binary evolution.

## 1. Introduction

The first barium stars discovered were all giants. The CH giants seem to be the metal-poor analogs of the Ba giants. Later on, Bond (1974) discovered what he called the CH subgiants, which have chemical anomalies very similar to those of the Ba and CH giants but are less evolved. Luck & Bond (1982) recognized that some CH subgiants at least belong to the main sequence, according to the high surface gravities obtained from high-resolution spectroscopic analyses. This became even clearer when Tomkin et al. (1989) discovered the F5 V star HR 107 to be a mild Ba star. Since then, still other main sequence Ba stars have been identified, so that the designation "Ba dwarf" seems justified and will be used here as an equivalent to "CH subgiant".

The questions this work tries to answer are quite simple:

– Are the Ba dwarfs the result of a mass-transfer event, like their giant counterparts?

– If so, is there an evolutionary link between them and the Ba giants? In other words, will the Ba dwarfs evolve into Ba giants?

A radial-velocity survey of 51 candidate and confirmed Ba dwarfs has been done in order to answer these questions.

## 2. The Sample

We have gathered our stars from a variety of sources, so that some of them are confirmed Ba dwarfs while others are not. Many have published abundances derived from high-resolution observations; those which did not were observed by us in the $\lambda 5100$ Å region to check their overabundances of $s$-process elements. However, a detailed abundance analysis of the latter is not yet available. We have divided our sample into two parts, according to the strength of their peculiarity:

1. 31 "strong" Ba dwarfs, confirmed by high-resolution spectroscopy.
   These had been selected from the following sources:
   - 9 stars from Bidelman (1981, 1983, 1985) and confirmed as Ba dwarfs by North et al. (1994).
   - 6 Ba stars with a spectral type earlier than G2 from the extensive list of Lü et al. (1983), which contains essentially Ba giants; the early spectral type was considered as betraying their dwarf rather than giant nature. Some of them have been confirmed by North et al. (1994) and by Luck & Bond (1991).
   - 5 F stars having a remark of the kind "Strong Sr 4077" in the Michigan catalogue (Houk & Cowley 1975), and confirmed by us.
   - 1 F star, HR 107, from Tomkin et al. (1989).
   - 10 CH subgiants from Luck & Bond (1991).
2. 20 mild or marginal Ba dwarfs, including
   - 10 candidates from Lü (1991), which were classified on the basis of low-resolution spectra (43 Å mm$^{-1}$) as having a very slight Ba anomaly: Ba $\leq 1$ on the usual scale running from 1 to 5 (Warner 1965). Our attempts to confirm them with high-resolution spectroscopy have failed, indicating that these stars cannot be considered real Ba dwarfs. However, we consider them here a useful

comparison sample, since radial velocity measurements have been done for them as well.

- 9 mild Ba dwarfs from Edvardsson et al. (1993). These objects have typically $[s/\text{Fe}] \approx 0.2$ and are therefore quite marginal.
- 1 candidate from Houk & Cowley (1975), HD 9529, in which yttrium seems slightly overabundant, judging from the ratio of the lines Y II $\lambda 5087.43$/Fe I $\lambda 5090.78$.

The spectral types range from about F3 to K5 for the whole sample, but most objects have types in the interval F5 – G2. The CH subgiants already monitored by McClure (1985) have not been included in our sample, except for HD 182274 which had been selected on the basis of its classification "F str $\lambda 4077$" by Bidelman (1985).

## 3. Observations

Most radial velocities have been obtained using CORAVEL scanners (Baranne et al. 1979) attached, in the northern hemisphere, to the 1-m Swiss telescope at Observatoire de Haute-Provence, France, and in the southern hemisphere to the 1.54-m Danish telescope at ESO, La Silla, Chile. The measurements of the stars selected from Bidelman's papers began as early as 1987, even before their Ba nature was suspected, while the other stars, most of which can only be observed from the South, have been measured not earlier than 1992.

High-resolution spectra have been taken with the CES spectrograph fed by the 1.4-m CAT telescope at ESO in 1989, 1990, 1992 and/or 1993; the Short Camera was used, together with the RCA CCD #9 (15 $\mu$m pixels), the resolving power being 60 000 in most cases. Although the primary goal was to obtain equivalent widths and abundances, good-quality radial velocities could be obtained as well, as a by-product. They have an accuracy comparable to that of CORAVEL, even if no radial-velocity standard were measured, thanks to the high resolution and excellent stability of the instrument.

## 4. Results

The rate of binaries we obtain is given below for the different categories of stars described above. In each case, we have made an estimate of the detection rate by generating artificial binaries at random, and "observing" them at the same epochs as the real ones, and with the same random error. Orbital elements typical of the giant Ba stars were assumed, as well as a mass of the secondary of 0.6 $M_\odot$.

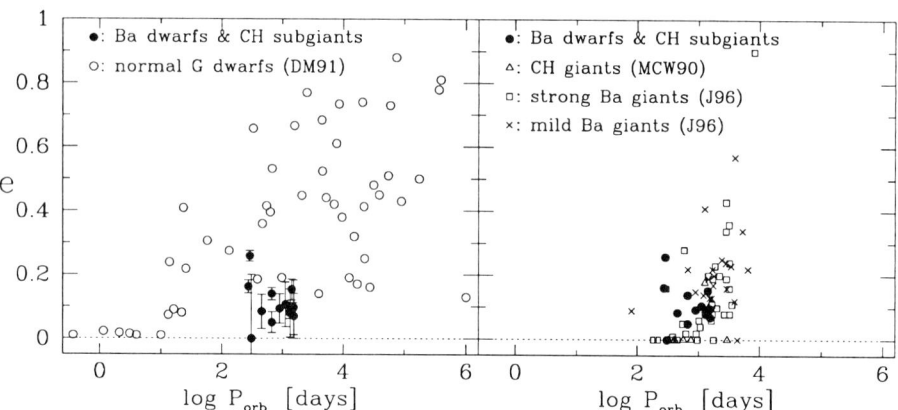

*Figure 1.* Left: e − log P diagram for the 14 orbits of Ba dwarfs (black dots) compared with the orbits of normal G dwarfs (open dots). Right: Comparison between Ba dwarfs and Ba giants, at the same scale; notice the bias against long periods for the dwarfs.

- Among the 10 candidates of Lü (1991), only 3 are binaries; one of these is an SB2 system, and the other two are SB1 binaries, one of which is only marginally detected.
- Among the 10 mild Ba dwarfs of Edvardsson et al. (1993) and HD 9529, we find only one SB1 system. Taking together these stars and Lü's, one obtains a rate of binaries of only 20%, while the detection rate is better than 98%. Therefore, the rate of binaries here is typical of normal rather than Ba stars.
- Among the 31 "strong" Ba dwarfs, we have found 28 binaries, one of which, HD 26455, is an SB2 system. Unfortunately, we do not have enough data to tell whether there is a third companion (the expected white dwarf) in this system or not. Excluding this star from the sample leads to a rate of 90%, while the detection probability is 99%.

Among the binaries, a very interesting system is HD 48565, which has the shortest period known among all Ba stars (giants and dwarfs): only 73 days, which is too short to allow the former primary to reach the typical radius of an AGB star. Recent measurements have shown a clear trend of the O–C residuals, betraying the presence of a third component which is very probably the white dwarf. Since the short period is probably not linked with the WD but rather with some faint main-sequence star, we have not plotted this system on Figure 1.

Among the 28 binaries found in the sample of the 31 "strong" Ba dwarfs, we could determine only 14 orbits, all of which have long periods and small eccentricities, as shown in Figure 1. The binaries among normal G dwarfs studied by Duquennoy & Mayor (1991) are shown for comparison. The

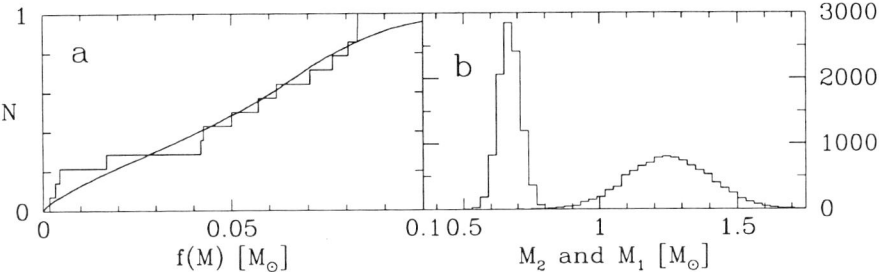

Figure 2. (a) Observed and simulated cumulative distribution of the mass functions. (b) Gaussian distributions of the masses of the secondary (left) and of the primary (right) used in the simulation. The distribution centered on 1.25 $M_\odot$ reproduces approximately the observed one.

contrast is striking: some dissipation mechanism must have acted upon the orbits of the Ba dwarfs to reduce their excentricity. On the other hand, the orbital elements of the Ba dwarfs are very similar to those of the Ba giants, although the former suffer from a bias against long periods.

A very interesting possibility offered by the Ba dwarfs is that of a direct, though statistical, estimate of the average mass of their companion. Contrary to the case of the Ba giants, the mass of the primary can be determined using a photometric estimate of the effective temperature and spectroscopic estimates of the surface gravity and metallicity. An interpolation in the evolutionary tracks of Schaller et al. (1992), Schaerer et al. (1993) and Charbonnel et al. (1993) gives the mass. Then we compare the observed cumulative distribution of the mass functions with a simulated distribution, where we assume random orientations of the orbits and approximate the observed distribution of the primary's masses by a gaussian. The best fit, shown in Figure 2, occurs for an average secondary mass of

$$M_2 = 0.67 \pm 0.09 M_\odot \qquad (1)$$

where the uncertainty corresponds to the 95% confidence level. This is the first direct estimate of the average mass of the companion in a sample of Ba stars. Our result is entirely compatible with the masses of white dwarfs resulting from 2–3 $M_\odot$ progenitors (Reid 1996).

## 5. Conclusion

We have shown to a high degree of probability that Ba dwarfs, like Ba giants, result from a mass-transfer event in a binary. Will they become Ba giants after they have left the main sequence? From the point of view of the abundances, the answer is yes, according to Smith et al. (1993). From the "orbital" point of view, the answer is yes for the most massive dwarfs of our sample, but no for the others. Indeed, the mass functions of the Ba dwarfs are much larger than those of the Ba giants, and are closer to those of the CH giants. Therefore the Ba giants are statistically more massive than the Ba dwarfs, consistent with the estimate of 1.65 $M_\odot$ (instead of $\sim$1.25 $M_\odot$ for the dwarfs) by Jorissen et al. (1998). But this difference might be due to a detection bias against massive Ba dwarfs: the Am phenomenon is frequently present for spectral types A5–F0 and includes overabundances of $s$-process elements, but due to radiative diffusion (Michaud 1991). Perhaps some massive Ba dwarfs are concealed among Am–Fm stars.

## References

Baranne, A., Mayor, M. & Poncet, J. L. 1979, *Vistas in Astron.*, 23, 279
Bidelman, W. P. 1981, *AJ*, 86, 553
Bidelman, W. P. 1983, *AJ*, 88, 1182
Bidelman, W. P. 1985, *AJ*, 90, 341
Bond, H. E. 1974, *ApJ*, 194, 95
Charbonnel, C., Meynet, G., Maeder, A., Schaller, G. & Schaerer, D. 1993, *A&A Supp.*, 101, 415
Duquennoy, A. & Mayor, M. 1991, *A&A*, 248, 485
Edvardsson, B., Andersen, J., Gustafsson, B., Lambert, D. L., Nissen, P. E. & Tomkin, J. 1993, *A&A*, 275, 101
Houk, N. & Cowley, A. P. 1975, *University of Michigan Catalogue of Two-Dimensional Spectral Types for the HD Stars*, Vol. 1, University of Michigan, Ann Arbor
Jorissen, A., Van Eck, S., Mayor, M. & Udry, S. 1998, *A&A*, 332, 877
Lü, P. K. 1991, *AJ*, 101, 2229
Lü, P. K., Dawson, D. W., Upgren, A. R. & Weis, E. W. 1983, *ApJ Supp.*, 52, 169
Luck, R. E. & Bond, H. E. 1982, *ApJ*, 259, 792
Luck, R. E. & Bond, H. E. 1991, *ApJ Supp.*, 77, 515
McClure, R. D. 1985, in *Cool Stars with Excesses of Heavy Elements*, ed. M. Jaschek and P. C. Keenan (Reidel), p. 159
Michaud, G. 1991, in IAU Symp. 145: *Evolution of Stars: The Photospheric Abundance Connection*, ed. G. Michaud and A. Tutukov (Kluwer), p. 111
North, P., Berthet, S. & Lanz, T. 1994, *A&A*, 281, 775
Reid, I. N. 1996, *AJ*, 111, 2000
Schaller, G., Schaerer, D., Meynet, G. & Maeder, A. 1992, *A&A Supp.*, 96, 269
Schaerer, D., Meynet, G., Maeder, A. & Schaller, G. 1993, *A&A Supp.*, 98, 523
Smith, V. V., Coleman, H. & Lambert, D. L. 1993, *ApJ*, 417, 287
Tomkin, J., Lambert, D. L., Edvardsson, B., Gustafsson, B. & Nissen, P. E. 1989, *A&A*, 219, L15
Warner, B. 1965, *MNRAS*, 129, 263

## Discussion

**Feast**: What constraints do the relative space densities place on the question of whether or not Ba dwarfs evolve into variable giants?

**North**: The frequency of the Ba dwarfs belonging to binaries is about 1–2% of all mid–F to early G dwarfs, which is roughly the same as the frequency of Ba giants among the G–K giants. Therefore it is reasonable to assume that Ba dwarfs evolve into Ba giants.

**Little-Marenin**: What is the origin of the non-binary mild Ba dwarfs?

**North**: I am not sure; possibly primordial.

**Jorissen**: In a sample of 28 mild Ba giants (Ba 1 stars) monitored for about 10 years with CORAVEL, at least 3 (HD 50843, HD 95345 and HD 119185) do not seem to be binaries. Sneden et al. (1981, *ApJ*, 247, 1052) analyzed HD 95345, showing that it is definitely a barium star, though with moderate overabundances. Galactic fluctuations in the primordial Ba/Fe ratio may perhaps account for this small fraction ($\sim 10\,\%$) of non-binary mild Ba stars.

**Gustafsson**: One should, of course, be careful when comparing main-sequence Ba stars with Ba-rich giants, in particular when the previous ones are F-type dwarfs having comparatively shallow convection zones. It may be easy to pollute one of those, but hard to keep this pollution visible when the convection zone deepens and includes more mass. I presume the precursors of the Ba giants must have been lower-mass dwarfs with deep convection zones which, by accretion, have moved up along the main sequence. Alternatively, other mixing processes were active at the mass transfer.

**North**: The abundance anomalies do not necessarily remain confined within the outer convective zone. On the contrary, the $\mu$ inversion implies a mixing in the radiative zone, and on a rather short timescale, as shown by Proffitt & Michaud (1989, *ApJ*, 345, 998).

# THE CHEMICAL COMPOSITION AND ORBITAL PARAMETERS OF BARIUM STARS

LAIMONS ZAČS
*Radioastrophysical Observatory*
*Latvian Academy of Sciences*
*LV-1527 Riga, Latvia*

**Abstract.** The observational relation between $s$-process abundance anomalies and orbital periods for barium stars is discussed and compared with mass-transfer simulations. Recent detailed abundance analyses of a large sample of single-lined long-period binaries provide evidence that all giants with white dwarf companions are likely to have abundance anomalies.

## 1. Introduction

In 1987 when as an undergraduate student I started my study of barium stars, it seemed to me that all was known about these peculiar objects, because abundance analyses confirmed the overabundance of carbon and $s$-process elements in their atmospheres (see Lambert 1985, and references therein), and their binarity (McClure et al. 1980; McClure 1983) and the white dwarf (WD) nature of the companion (Böhm-Vitense 1980; Böhm-Vitense et al. 1984) indicated that a transfer of heavy-element-rich matter from an AGB companion (now a WD) to the pre-barium star is a good explanation for the chemical peculiarities of the latter. Only some time later I saw that there are several critical questions that need additional observational tests.

## 2. Is There a Difference between Variable and Non-Variable Radial-Velocity Examples among Barium Stars?

The previous differentiation between classical (strong) and mild barium stars suggested by McClure et al. (1980) and Sneden et al. (1981) is not clear physically. Furthermore, McClure and Woodsworth (1990) found several barium stars that have shown no radial velocity (RV) variations. It is clearly of interest to know if these two groups of barium stars also have

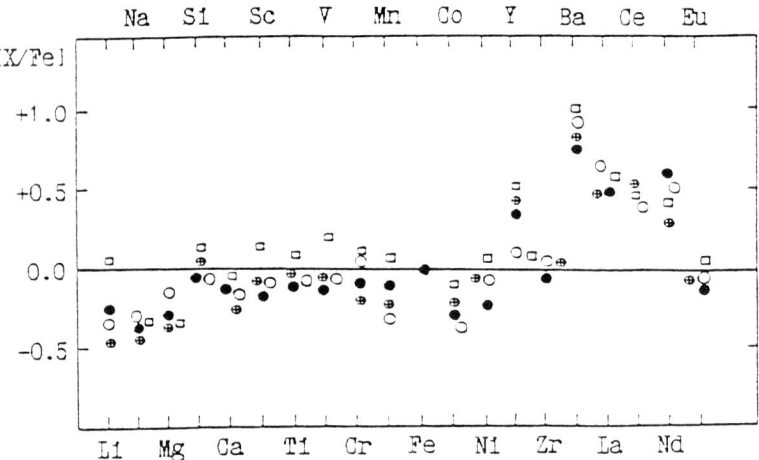

*Figure 1.* Differential abundance comparisons for two barium stars with variable radial velocity (crossed circles: HD 205011; squares: HD 131670) and two non-variable examples (filled circles: HD 65854; open circles: HD 104979).

different physical properties (Začs 1994). A comparison of the abundance patterns for four radial velocity variable and nonvariable examples with similar peculiarity levels is shown in Figure 1. There is no significant difference in the abundances of these stars. The absolute magnitudes and atmospheric parameters are similar as well. Thus the quantitative analyses show that both groups of barium stars apparently belong to a single family of peculiar giants. This means that the mechanism responsible for the abundance peculiarities is probably common to both.

Two RV-nonvariable mild barium stars from Jorissen's list (Jorissen 1994) have been analyzed recently (Začs et al. 1997) using high S/N CCD spectra covering the wide spectral region 5000–7200 Å. A significant enhancement of $s$-process elements was found for these two stars: HD 130255 (G4 IV, Ba 1.0, [Ba II/Fe II] = +0.79 dex) and HD 119185 (K0 III, Ba 1.0, [Ba II/Fe II] = +0.35 dex). These observations confirm the conclusion that RV-nonvariable barium stars have either very long orbital periods or highly inclined orbital planes. Certainly, it is possible that some normal red giants have been classified by mistake as mild barium stars.

## 3. The Relation between Level of Chemical Anomalies and Orbital Period

If a transfer of heavy-element-rich matter from an AGB companion to the pre-barium star is the explanation for the chemical peculiarities of barium stars, there should be a correlation between the level of chemical anomalies

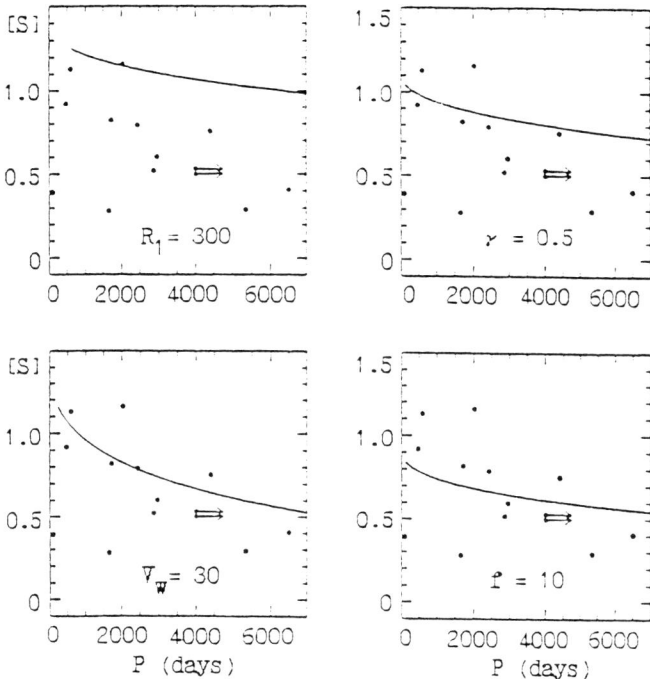

*Figure 2.* Comparison between the wind accretion model prediction (solid line) and the observational data (points) for different sets of parameters (Boffin & Začs 1994).

and the orbital period (Boffin & Jorissen 1988; Han et al. 1995). This prediction has been tested on the basis of mean abundance data of high homogeneity for a large sample of barium stars with known orbital elements (Začs 1994). To minimize observational errors, the overabundances of $s$-process elements in barium stars are characterized by an index: $[s] = ([Y/Fe] + [Ba/Fe] + [La/Fe] + [Ce/Fe] + [Nd/Fe])/5$. The relation between the orbital period and the observed overabundance of $s$-process elements is confirmed, although it is clear that a one-to-one relation does not exist: for a given period, various overabundances may occur.

## 4. Mass Transfer Mechanism

The first quantitative confrontation of observational data for a homogeneous sample of barium stars with theoretical predictions for the wind accretion scenario (Boffin & Jorissen 1988) confirmed that the global trend of the relation (abundance peculiarity–orbital period) can be obtained with the wind accretion model (Boffin & Začs 1994). Better agreement with the observations is obtained either by increasing the wind velocity or by de-

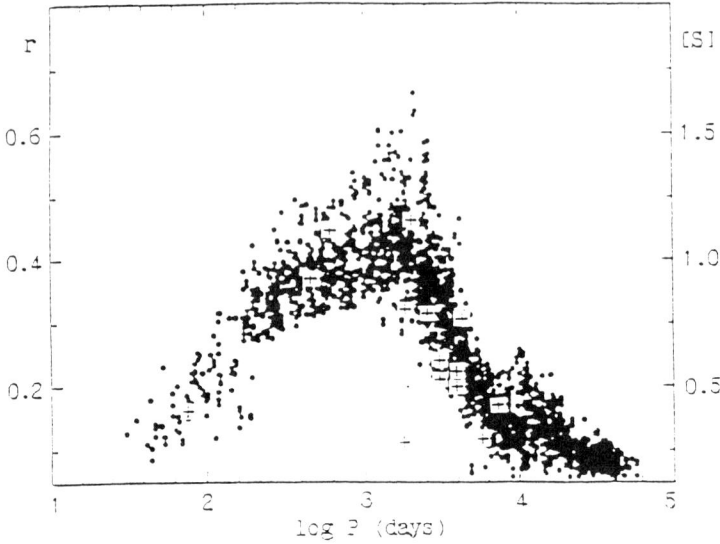

*Figure 3.* Comparison between the combined model predictions (points) and the observational data (plus signs). $P$ is the orbital period, and $r$ is the pollution factor (Han et al. 1995).

creasing the initial overabundance of $s$-process elements in the atmosphere of the AGB star (Figure 2). Recent smoothed particle hydrodynamics simulations (Theuns, Boffin & Jorissen 1996) have confirmed the efficiency of wind accretion to account for the chemical peculiarities for orbital periods up to about 90 years; however, this scenario faces difficulties for systems with short orbital periods. The confrontation of observations (Začs 1994) with simulations for the combined mass transfer scenario which includes wind accretion, wind exposure, stable Roche lobe overflow, and common-envelope ejection (Han et al. 1995) is shown in Figure 3. There is excellent agreement for both long and short period systems.

The confrontation of observations with mass-transfer simulations provides strong evidence that long-period ($P > 3$ years) barium stars are mainly the result of wind accretion (and probably wind exposure) from an AGB companion. Roche lobe overflow and common-envelope ejection are probably the most important evolutionary channels for short period barium stars, although additional observations are required for more secure conclusions.

## 5. Barium-Star-Like Long-Period Single-Lined Binaries

Many spectroscopic long-period binaries with a red giant primary and an unseen secondary (probably a WD in some cases) whose orbits have similar characteristics to those of barium stars have in all probability solar abundances of the heavy elements, including four normal giants ($\xi^1$ Cet, DR Dra, HD 21120, HD 81817) with directly observed WD companions (Jorissen & Boffin 1992; Boffin et al. 1993; Fekel et al. 1993). The existence of such giants led the authors to conclude that the existence of a WD companion in a barium-star-like system is not sufficient to produce a barium star. However, small abundance anomalies are difficult to detect, especially on the basis of a spectrum in a narrow spectral region. For example, Pilachowski (1977) and Začs (1994) suggested that $\xi^1$ Cet (HD 13611) has a mild s-process enhancement. An important step toward a better understanding of the phenomenon of barium stars is a detailed abundance analysis of spectroscopic binaries with barium-star-like orbital elements (Začs et al. 1997), especially normal giants with directly observed WD companions.

## 5.1. ARE THERE NORMAL GIANTS WITH WHITE DWARF COMPANIONS?

Unfortunately, the answer to this question depends for the time being on the accuracy required in the definition of "normal". Certainly, there are at least some "normal giants" with WD companions having atmospheric anomalies less than 0.4 dex (see Jorissen & Boffin 1992). However, only two such giants have been analyzed in detail using high resolution spectra in a wide spectral region. These analyses provide strong evidence that the atmospheres of $\xi^1$ Cet and DR Dra (Začs et al. 1997) are *slightly peculiar*! The question concerning the peculiarity of other candidates for normal giants remains unanswered, because s-process anomalies at the level of 0.3 dex can be confirmed only on the basis of spectra over a wide spectral region, ensuring high accuracy in the atmospheric parameters and abundance pattern obtained. A recent abundance analysis of long-period single-lined binaries with orbital elements similar to barium stars (Začs et al. 1997) indicated that all giants in the (eccentricity ($e$), log $P$) plane (Figure 4) in the area occupied by typical barium stars have slight (0.2–0.3 dex) abundance anomalies. Thus it seems likely that all giants with WD companions in wide systems have s-process peculiarities. The level of peculiarity probably depends on the efficiency of the s-process and mass transfer in each specific binary system.

## 5.2. CAN THE PRESENCE OF A COMPANION INFLUENCE THE INTERNAL STRUCTURE OF THE BARIUM STAR?

It has sometimes been suggested that the phenomenon of barium stars may be connected to an unseen companion indirectly – not via mass transfer to a pre-barium star. In principle, a companion may have affected the pre-

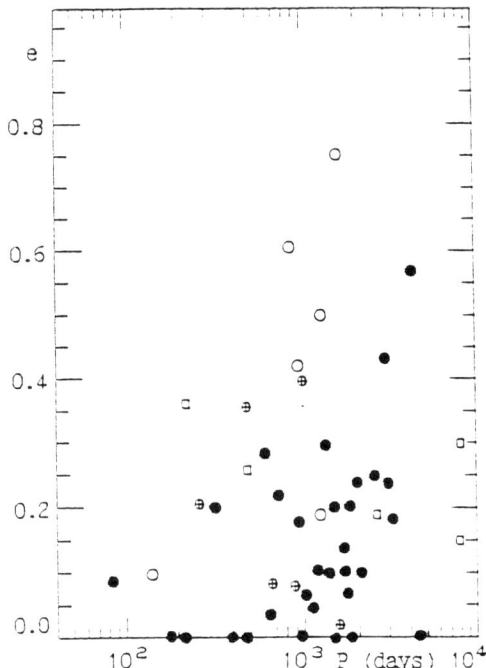

*Figure 4.* The (e, log P) diagram for single-lined spectroscopic binaries with barium-star-like orbital elements. Filled circles: barium stars from Jorissen & Boffin (1992); open circles: giants showing normal barium abundance; crossed circles: giants showing mild barium enhancement; squares: dwarfs.

barium star's interior in some way which causes an extra degree of mixing (McClure 1984). This assumption has been tested for long-period binaries (Začs et al. 1997) with characteristics similar to barium stars but containing a main-sequence companion. The analysis indicates that a main-sequence companion does not have a significant influence on the internal structure (chemical composition) of the primary star.

5.3. WHAT ARE THE SUFFICIENT CONDITIONS FOR THE FORMATION OF ABUNDANCE PECULIARITIES?

In recent years there has been a tendency to consider that the presence of a WD companion in long-period systems is not sufficient for the barium star phenomenon, and additional parameters have been sought to explain the absence of abundance anomalies. Metallicity in particular has been considered as one such extra parameter. However, we need to explain why there exist a large number of barium giants such as HD 121447 ([Fe/H] = +0.05, [Ba/Fe] = +0.57), HD 100503 ([Fe/H] = +0.05, [Ba/Fe] = +0.56), and CPD−64°4333 ([Fe/H] = +0.05, [Ba/Fe] = +0.62) (see Smith 1984).

In my opinion, if one or more barium stars were found to have a solar metallicity, this would mean that it is not necessary for the pre-barium system to be metal-deficient to produce the peculiarity. The observational fact that a relation exists between [Ba/Fe] and [Fe/H] (Kovacs 1985) — i.e. more extreme barium stars are more iron-deficient (older) than the stars with small overabundances of barium — does not mean that a low metallicity is one of the parameters which determines whether a normal giant becomes peculiar. We should take into account that there is a relation between abundance peculiarity and orbital period for barium stars. For example, the two stars HR 774 and HD 178717 have a difference in orbital periods (848 days) corresponding to a difference of $\sim 0.3$ dex in $[s]$ (see Fig. 3), which is in agreement with the ([Ba/Fe], Fe/H]) relation (see Fig. 10 in Kovacs 1985). It is clearly of interest to investigate the influence of binarity (strong barium stars have larger radial velocity amplitudes than mild barium stars) on the space velocity dispersion (strong barium stars have a larger velocity dispersion, see Lu 1991). In any case, abundance studies (Začs et al. 1997) give strong evidence that the existence of a white dwarf companion in binary systems with barium-star-like characteristics is not sufficient to produce a *strong* barium star, but it is perhaps premature to conclude that the existence of a WD in a barium-star-like system is not sufficient to create abundance peculiarities.

## 6. Future Developments

First, it will be very important to check whether the giants with directly confirmed WD companions (for example, HD 21120 and HD 81817) really do not have any s-process anomalies. Since WD companions have been confirmed only for some barium stars, investigations of "normal" long period giants where the companion is suspected to be a WD ($e < 0.1$, $P > 1000$ days; see Jorissen & Boffin 1992) would be very useful. For further improvement, high S/N spectra over wide spectral regions are necessary to test in detail the element-by-element composition.

Second, it is important to look on the barium stars as a phenomenon caused by polluted mass transfer in binary systems, independent of the present evolutionary status of the primary. In this connection we should mention the recent detailed abundance analysis of AG Dra (Cunha et al. 2000) and the Monte Carlo simulations by Han et al. (1995) which try to explain a large variety of peculiar stars including cataclysmic variables and symbiotics on the basis of the mass-transfer mechanism. Detailed abundance analyses of peculiar stars such as HR 1105 (S3.5/2, $P = 596.21$ days, $e = 0.09$, $f(m) = 0.0375$) and $\tau$ UMa (Am, $P = 1062.4$ days, $e = 0.48$, $f(m) = 0.0044$) will enrich our knowledge of the barium star phenomenon.

If it has sometimes been suggested that the phenomenon of barium stars may possibly be the result of a combined process of mass transfer and stellar evolution, so too the phenomenon of long period metallic-line stars may be the result of mass transfer and diffusion in the atmospheres of A stars. On the other hand, examination of the correlation between abundance anomalies and orbital periods for barium-related stars (CH stars, subgiant CH stars) may allow us to find the general principle of the barium star phenomenon.

Third, an examination of s-process parameters and the neutron source is needed for the families of barium-related stars. Although many advances have been made in recent years, further progress on a large scale may depend on the confrontation of the averaged $s$-process abundance patterns for homogeneous samples of peculiar red giants with $s$-process simulations.

I am grateful to my teachers and collaborators V. E. Panchuk, V. G. Klochkova, I. F. Bikmaev, H. M. J. Boffin, and F. A. Musaev. My work is partially supported by the ESO C&EE Committee (through grants B-01-012 and E-06-002). I thank the IAU and the symposium SOC (chaired by R. Wing) for travel support.

## References

Boffin, H. M. J., Cerf, N. & Paulus, G. 1993, A&A, 271, 125
Boffin, H. M. J. & Jorissen, A. 1988, A&A, 205, 155
Boffin, H. M. J. & Začs, L. 1994, A&A, 291, 811
Böhm-Vitense, E. 1980, ApJ, 239, L79
Böhm-Vitense, E., Nemec, J. & Proffitt, C. 1984, ApJ, 278, 726
Cunha, K., Smith, V. V. & Jorissen, A. 2000, in IAU Symposium 177: *The Carbon Star Phenomenon*, ed. R. F. Wing (Kluwer), p. 103
Fekel, F. C., Henry, G. W., Busby, M. R. & Eitter, J. J. 1993, AJ, 106, 2370
Han, Z., Eggleton, P. P., Podsiadlowski, P. & Tout, C. A. 1995, MNRAS, 277, 1443
Jorissen, A. 1994, private communication
Jorissen, A. & Boffin, H. M. J. 1992, in *Binaries as Tracers of Stellar Formation*, ed. A. Duquennoy and M. Mayor (Cambridge Univ. Press), p. 110
Kovacs, N. 1985, A&A, 150, 232
Lambert, D. 1985, in *Cool stars with Excesses of Heavy Elements*, ed. M. Jaschek and P. C. Keenan (Reidel), p. 191
Lu, P. K. 1991, AJ, 101, 2229
McClure, R. D. 1983, ApJ, 268, 264
McClure, R. D. 1984, PASP, 96, 117
McClure, R. D., Fletcher, J. M. & Nemec, J. M. 1980, ApJ, 238, L35
McClure, R. D. & Woodsworth, A. W. 1990, ApJ, 352, 709
Pilachowski, C. A. 1977, A&A, 54, 465
Smith, V. V. 1984, A&A, 132, 326
Sneden, C., Lambert, D. L. & Pilachowski, C. A. 1981, ApJ, 247, 1052
Theuns, T., Boffin, H. M. J. & Jorissen, A. 1996, MNRAS, 280, 1264
Začs, L. 1994, A&A, 283, 937
Začs, L., Musaev, F. A., Bikmaev, I. F. & Alksnis, O. 1997, A&A Supp., 122, 31

# BINARY "POST–AGB" STARS

HANS VAN WINCKEL AND CHRISTOFFEL WAELKENS
*Instituut voor Sterrenkunde, Katholieke Universiteit Leuven*
*Heverlee, Belgium*

AND

LAURENS B. F. M. WATERS
*Astronomisch Instituut "Anton Pannekoek,"*
*Univ. van Amsterdam, Amsterdam, The Netherlands*
*and*
*SRON Laboratory for Space Research*
*Groningen, The Netherlands*

**Abstract.** In this contribution we report on our radial-velocity monitoring of optically bright, high-latitude supergiants that appear to be in a post–AGB evolutionary stage. Binarity is a widespread phenomenon among our sample stars. More precisely: *all* objects with a near-IR excess in their energy distribution turn out to be binaries while the fraction of binaries in our program stars with only a far-IR excess is very small. The orbital periods, the often non-zero eccentricities, and the sometimes large mass functions set strong constraints on the previous evolution in which mass transfer must have been an important ingredient. We have accumulated observational evidence that the presence of a circum-binary dusty disk has an important dynamical and sometimes even chemical influence on the binary and its evolution. Some objects with a high mass function still defy an explanation.

## 1. Introduction

Systematic searches for post–AGB stars of low initial mass have concentrated mainly on optically bright objects with an IR excess due to circumstellar dust (Hrivnak et al. 1989; Pottasch & Parthasarathy 1988; Oudmaijer et al. 1992). This IR excess, the high or intermediate galactic latitude, the on average low metal content, and in some cases the high space motion are observational indications for the old and evolved population of these objects. Detailed studies of the spectral energy distribution (SED) have

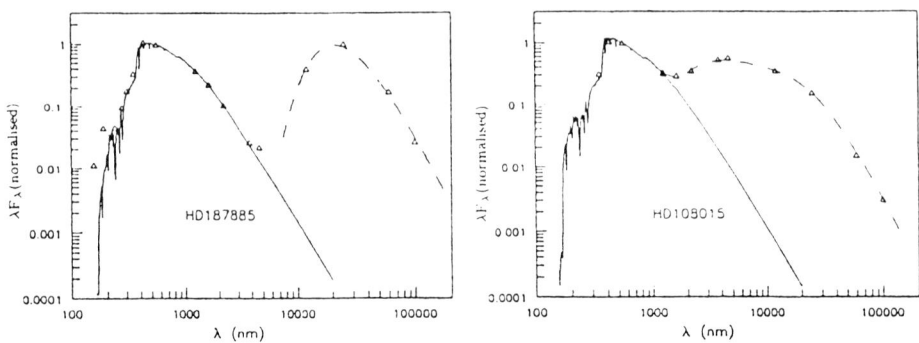

*Figure 1.* The SEDs of representatives of the two classes of optically bright post–AGB stars: HD 187885 shows the presence of only cold dust, while HD 108015 has hot dust as well.

shown that the objects can be divided into two groups depending on the shape of the IR excess (e.g. Trams et al. 1991; van der Veen et al. 1994; Bogaert 1994). Sources with an IR excess starting already at the near-IR have both hot and cool dust in their circumstellar dust shells, while others with only a far-IR excess show the presence of only cool dust (Figure 1).

In this contribution we report on our ongoing radial velocity monitoring and strengthen the earlier suggestions (Waters et al. 1993b) that *all* objects with a near-IR excess are binaries. We will focus on observational evidence that the presence not only of a companion but also of a circum-binary dust-disk has important influence on the stellar evolution.

## 2. Orbital Parameters

We have been accumulating radial velocity measurements using the CO-RAVEL spectrometer at ESO and OHP complemented with observations obtained with the CES at ESO, La Silla. A detailed report of this ongoing program will be published elsewhere; here we limit ourselves to the most important results. In Table 1 we list the objects for which we have found conclusive evidence of binarity. The orbital periods, eccentricities and mass-functions are given for the objects for which we have enough data-points to analyze the radial-velocity curve in detail. In the last column we indicate whether the circumstellar environment is carbon or oxygen rich. For none of the stars with only a far-IR excess in our sample (6 stars in total) did we find evidence for a binary companion in our radial velocity data!

Most of the periods we found are of the order of one to a few years. Noticeable exceptions are SAO 173329 with a period of only 116 days (!) and the RV Tauri stars HD 131356, AC Her and U Mon which have longer

TABLE 1. Overview of the orbital elements

| Name | Period (d) | e | $F(M)$ ($M_\odot$) | CS chemistry |
|---|---|---|---|---|
| *Chemically peuliar* | | | | |
| HR 4049[1] | 429 | 0.31 | 0.143 | C-rich |
| HD 44179[1] | 318 | 0.39 | 0.049 | C-rich |
| HD 52961[1] | a few years | > 0 | ? | C-rich |
| BD +39°4926[2] | 775 | ? | ? | - |
| HD 213985 | 259 | 0 | 0.97 | C-rich |
| HD 46703[3] | 610 | > 0 | ? | ? |
| SAO 173329 | 115.9 | 0 | 0.026 | ? |
| 89 Her[4] | 288.4 | 0.19 | 0.00083 | O-rich |
| HD 108015 | 938 | ? | 0.0029 | ? |
| HD 95767 | ± 2050 | ? | ± 0.3 | ? |
| HD 131356 | 1150 | ? | ± 0.52 | ? |
| AC Her | ±1230 | ? | ? | ? |
| U Mon[5] | 2597 | 0.43 | ? | ? |

1) Van Winckel, et al. 1995, *A&A*, 293, L25
2) Kodaira, K., et al. 1970, *ApJ*, 159, 485
3) Hrivnak & Lu 2000 (these proceedings)
4) Waters, et al. 1993, *A&A*, 269, 242
5) Pollard, K. R. 1995, ASP Conf. Ser., 83, 409

periods. For the RV Tauri stars this is probably a selection effect as the shorter period orbits are very hard to detect due to the large pulsational velocity amplitudes.

The objects are thought to be post–AGB objects: their kinematics point to low masses, the CS dust (for some objects C-rich!) and high luminosity suggest a terminal phase of evolution. Current evolutionary tracks suggest then a post–AGB stage of evolution. However, for the shorter period binaries (periods smaller than a few years), the orbit is too small to accommodate an AGB star. On the AGB, an object with the same luminosity and orbital parameters would experience severe Roche-lobe overflow leading not only to circularization of the orbit, but also to spiral-in. This would result in a circular short period binary, contrary to what is observed. Note that the same problem is encountered in the short period Ba stars and Tc-poor S stars (see Jorissen & Van Eck 2000); these stars are also post–AGB objects since the actual white dwarf has been an AGB star in the past.

Most remarkable are, however, the non-zero eccentricities in certainly 3 of the shorter period objects (HR 4049, HD 44179 and 89 Her). As can be seen on the $e$–log$P$ diagram shown by Jorissen & Van Eck (2000), evolved binaries with orbital periods in the range of our program stars have circular orbits due to dissipative processes during red giant and asymptotic giant evolution. We can conclude that especially these shorter period binaries do not follow standard stellar evolution.

## 3. Circumbinary Disk

For a thorough review of the observational evidence for the presence of a circumbinary disk around these binaries we refer to Waters et al. (1993a,b) and Waelkens et al. (1991, 1996). We stress that the best evidence comes from the three objects (HR 4049, HD 44179 and HD 213985) where the photometric brightness and color variations we observe are in phase with the orbital motion. This behavior can best be explained by variable extinction by a thick circumbinary dust-disk that is viewed at a certain inclination: when the supergiant is nearest to us the obscuration by the dust ring is maximal so that the light is minimal and reddest (Waelkens et al. 1991).

With the dust trapped in a Keplerian thick disk around the system, it is prevented from cooling down rapidly. The dust in the inner part of the disk will remain in thermal equilibrium with the radiation field of the binary and remains hot. This gives a natural explanation for the correlation between the presence of hot dust and the binary nature of the objects. How such disks are formed in the wide period range given in Table 1 is still an open question.

## 4. Binary-Disk Interactions

### 4.1. CHEMICAL

In the top part of Table 1, the chemically peculiar objects are listed. In these extremely metal-deficient objects (with iron abundances as low as [Fe/H] = $-4.8$) the photospheric chemical patterns are acquired by accretion of pure circumstellar gas, separated from the dust (see Waters et al. 1992, and references therein). Recently other evolved objects have been found that display the same, albeit less extreme, chemical patterns (see Giridhar 2000). Some of them are again known to be binaries.

There is growing evidence that the presence of a disk is essential for the fractionation process to be effective. According to Waters et al. (1992), a stationary disk offers the best environment for two essential conditions to be fulfilled: a density low enough for the drag of the dust grains on the gas to be minimal, and a slow accretion rate.

## 4.2. DYNAMICAL

It is interesting to note that Artymowicz et al. (1991) concluded that the binary eccentricity distribution on the main sequence is determined in the pre-main-sequence phase of evolution by tidal interaction between the binary and the circum-binary dust disk, which is present as a relic from the star formation process. They conclude that for stars with a mass ratio smaller than five, and a small but non-zero eccentricity, a resonant interaction between the binary and the circum-binary disk can cause very rapid eccentricity growth. Very similar conditions are now present in the short-period stars of Table 1: these objects are also surrounded by a thick dusty circumbinary disk. It may be then that resonant behavior between the circum-system disk and the binary has caused rapid eccentricity growth leading to high eccentricities. Clearly this process should be explored in more detail.

## 4.3. TIDALLY ENHANCED MASS-LOSS

Also for HD 213985 the period is much too short to accommodate an AGB star. For this object with a C-rich circumstellar environment, we have the best evidence that severe mass transfer occurred since the (probably) unevolved secondary has a mass which is larger that the main-sequence mass of the primary (the mass function is high). It is, however, not clear how the system avoided spiral-in on the AGB. It could be that severe mass transfer took place, but only after the system reversed its mass-ratio. In that case, Roche-lobe overflow (RLO) is stable. Tout & Eggleton (1988) and Han et al. (1995) describe a scenario where the mass-ratio of a binary system can be reversed before RLO occurs due to a tidally enhanced stellar wind. These authors invoke this scenario in order to explain the existence and the period distribution of several groups of binaries. Observational evidence that such an enhanced mass-loss mechanism exists is, however, difficult to find. SAO 173329 may be in this respect a most interesting object: it shows a strong P-Cygni H$\alpha$ line-profile with an expansion velocity of some 300 km s$^{-1}$ which is highly remarkable for a F-type object. At the same time, the object turned out to be a binary star with the shortest period of our sample, only 116 days. The short orbital period and the high mass loss rate may be related: SAO 173329 may the best observational evidence that the companion can induce a high mass-loss rate by a tidally enhanced wind.

## 5. Conclusions

The optically bright post–AGB stars with a near IR-excess have turned out to be binaries. The observed periods, eccentricities and mass-functions

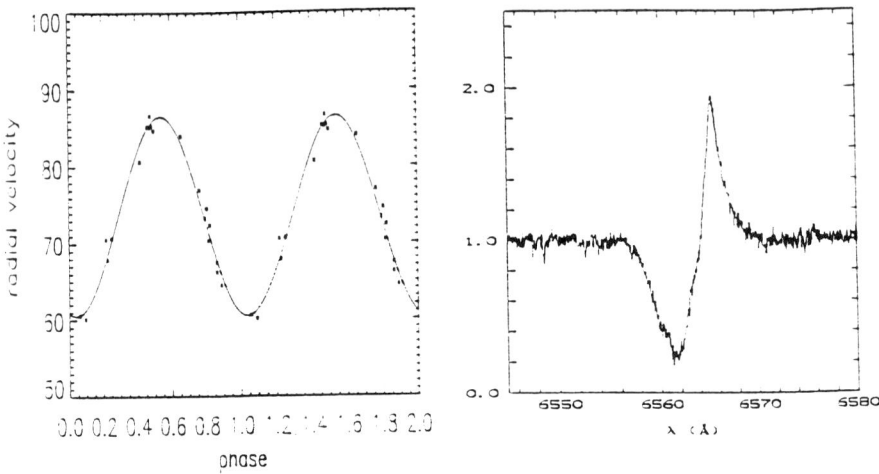

*Figure 2. Left*: The radial velocity curve of SAO 173329 folded on the 115.9-day period. *Right*: The Hα line profile.

imply that they do not follow standard evolutionary scenarios and non-standard phenomena connected to the specific binary nature have to be invoked in order to understand these systems.

## References

Artymowicz, P., Clarke, C. J., Lubow, S. H. & Pringle, J. E. 1991, *ApJ*, 370, L35
Bogaert, E. 1994, Ph.D. dissertation, K. U. Leuven University
Giridhar, S. 2000, in IAU Symp. 177: *The Carbon Star Phenomenon*, ed. R. F. Wing (Kluwer), p. 117
Han, Z., Eggleton, P. P., Podsiadlowski, P. & Tout, C. A. 1995, *MNRAS*, 277, 1443
Hrivnak, B. J., Kwok, S. & Volk, K. M. 1989, *ApJ*, 346, 265
Hrivnak, B. J. & Lu, W. 2000, in IAU Symp. 177: *The Carbon Star Phenomenon*, ed. R. F. Wing (Kluwer), p. 293
Jorissen, A. & Van Eck, S. 2000, in IAU Symp. 177: *The Carbon Star Phenomenon*, ed. R. F. Wing (Kluwer), p. 259
Oudmaijer, R. D., van der Veen, W.E.C.J., Waters, L.B.F.M., Trams, N. R., Waelkens, C. & Engelsman, E. 1992, *A&A Supp.*, 96, 625
Pottasch, S. R. & Parthasarathy, M. 1988, *A&A*, 192, 182
Tout, C. A. & Eggleton, P. P. 1988, *ApJ*, 334, 357
Trams, N. R., Waters, L.B.F.M., Lamers, H.J.G.L.M., Waelkens, C., Geballe, T. R. & Thé, P. S. 1991, *A&A Supp.*, 87, 361
van der Veen, W.E.C.J. Waters, L.B.F.M., Trams, N. R. & Matthews, H. E. 1994, *A&A*, 285, 551
Waelkens, C., Lamers, H.J.G.L.M., Waters, L.B.F.M., Rufener, F., Trams, N. R., Le Bertre, T., Ferlet, R. & Vidal-Madjar, A. 1991, *A&A*, 242, 433
Waelkens, C., Van Winckel, H., Waters, L.B.F.M. & Bakker, E. J. 1996, *A&A*, 314, L17
Waters, L.B.F.M., Trams, N. R. & Waelkens, C. 1992, *A&A*, 262, L37
Waters, L.B.F.M., Waelkens, C., Mayor, M. & Trams, N. R. 1993a, *A&A*, 269, 242
Waters, L.B.F.M., Waelkens, C. & Trams, N. R. 1993b, in *Mass Loss on the AGB and Beyond*, ed. H. E. Schwarz (European Southern Observatory), p. 298

## Discussion

**Giridhar**: You mentioned mass transfer in the case of HD 213985. Would you like to elaborate on it?

**Van Winckel**: The mass function is high, and if we adopt a typical mass for a post–AGB star of $0.6\,M_\odot$, the companion must be around $2\,M_\odot$ or even higher, depending on the inclination. The mass of the secondary is then higher than the main-sequence mass of the actual high-latitude supergiant primary.

**Giridhar**: Is the binarity of U Mon confirmed on the basis of radial-velocity variation?

**Van Winckel**: Yes. In the paper of Pollard, the radial velocity measurements are shown with a long-timescale trend which they interpret as due to orbital motion.

**Molster**: Is there enough mass in the circumbinary disk to increase the eccentricity of the orbit?

**Van Winckel**: Yes. The amount of mass is not too important in this scenario of Artymowicz. The location of the dust shell is, however, crucial.

**Bakker**: The star BD +39°4926 is in your sample of post–AGB binaries but does not exhibit an infrared excess. Could you comment on this?

**Van Winckel**: This star is more distant than the others. For example, an IR excess similar to what IRAS measured for HR 4049 would be below the detection limit for BD +39°4926. We have a proposal in for ISO to try to detect the star and see whether there is still circumstellar material.

**Linsky**: Do you have an estimate for the mass-loss rate of SAO 173329 from the profile of the H$\alpha$ line? Will this mass-loss rate alter the evolution of the binary system?

**Van Winckel**: It is about $10^{-7}\,M_\odot$/year. It is the only star in our sample of F supergiants with a P-Cygni profile and such a high mass-loss rate. Increased mass loss in the post–AGB phase will speed up the evolution drastically. We believe this star is very good evidence that tidally enhanced mass loss indeed exists and can be very efficient. It might be crucial to flip over the mass ratio of the binary system, before Roche-lobe overflow.

# LIGHT AND VELOCITY VARIABILITY OF POST–AGB STARS

BRUCE J. HRIVNAK AND WENXIAN LU

*Dept. of Physics and Astronomy*
*Valparaiso University, Valparaiso, IN 46383, U.S.A.*

**Abstract.** We have monitored velocity variability in nine proto-planetary nebulae (PPN) over a 5 year interval and have monitored light variability in 40 PPN over a 2 year interval. We find all nine of the objects to vary in velocity and almost all of the 40 to vary in light. Three of the objects display a clear periodicity in their velocity variations, and in all three the light varies with the same period as the velocity. Periodic light variations are found in six other objects. We interpret these as due to pulsation in the PPN.

## 1. Introduction

Our knowledge of post–AGB stars has expanded greatly as a result of IRAS and follow-up ground-based observations. The present study is part of a larger program to discover and study proto-planetary nebulae (PPN), objects in transition from the asymptotic giant branch (AGB) to the planetary nebula (PN) phase (see Hrivnak, Kwok & Volk 1989; Kwok 1993).

Our candidate objects were selected primarily from IRAS objects which peak in the 25 $\mu$m band (due to re-emission from circumstellar dust). The associations of the optical counterparts were in most cases discovered or confirmed by ground-based observations at 10 $\mu$m; in a few cases the identification is based simply on the very close positional association of the IRAS source and the optical star.

The spectral energy distributions of the objects show a characteristic double peak, with about equal amounts of energy emitted in the visible and near-infrared (from the reddened photosphere) and the mid-infrared (re-emission from circumstellar dust). Examples are shown by Hrivnak et al. (1989).

## 2. Variability: Why Study?

Variability can be a source of additional information about the nature of these objects. It can be due to a binary companion or can be intrinsic to the star, such as pulsation.

*Binarity*: Many PN are bipolar. One of the mechanisms which can lead to this is a binary companion, which can cause the AGB star to lose mass preferentially in the orbital plane, thereby producing a torus which retards later mass loss from a fast wind. This is supported by the discovery of several binary PN nuclei, almost all of short period ($<3\,\mathrm{d}$; Bond 1995). Longer-period binaries are harder to discover in PNe, since the "reflection effect" is no longer significant in causing light variations and since small velocities variations are difficult to measure in the intrinsically broad spectral lines of these hot stars. However, by searching PPN of spectral types F and G, one can make use of their many and sharper lines to search for longer-period binaries. From this one can hope to learn more about the mechanisms shaping the nebulae and the properties of binary PN nuclei which may not pass through a common envelope phase.

*Pulsation*: It is known that some (perhaps most) post–AGB objects vary in light and velocity due to pulsations, such as RV Tauri stars (spectral types G–K; $P = 50-150\,\mathrm{d}$) and UU Her stars (type F). The study of these pulsations can help us to learn more about the physical properties of the star.

No previous study of variability in a sample of PPN has been published. Since we had access to both a good list of candidates and to telescopes well-suited to make these studies, we have undertaken this project.

## 3. Radial Velocity Study

Radial velocity observations were made over five seasons from 1991 to 1995 (especially 1991–92) at the Dominion Astrophysical Observatory (DAO) in Victoria, Canada. The 1.2-m Coudé telescope was used, equipped with the Radial Velocity Spectrometer and an F–star mask. The typical precision was 0.65 km s$^{-1}$. S. Morris and A. Woodsworth of the DAO collaborated in the observations. Nine PPN of spectral types F–G and magnitudes $V = 7-10$ were observed about 35 times each.

All nine are found to vary, with an average peak-to-peak variation of 10 km s$^{-1}$. Examples are shown in Figure 1. For three of them a consistent period was found: IRAS 18095+2704 (F3 Ib; $P = 109\,\mathrm{d}$), 22223+4327 (G0 Ia; 89 d), and 22272+5435 (G5 Ia; 127 d). These are shown in Figure 2. In each case, while a clear periodicity is seen in the velocity curves, the amplitude appears to vary. Only a few observations deviate greatly from the cyclical patterns.

# LIGHT AND VELOCITY VARIABILITY OF POST–AGB STARS

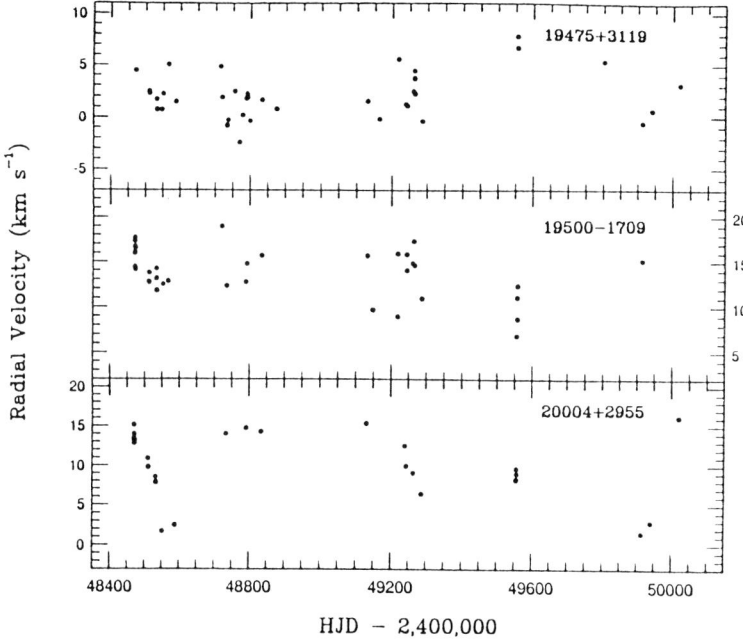

*Figure 1.* Examples of the radial velocity variation in several PPN. No periods are found for these variations.

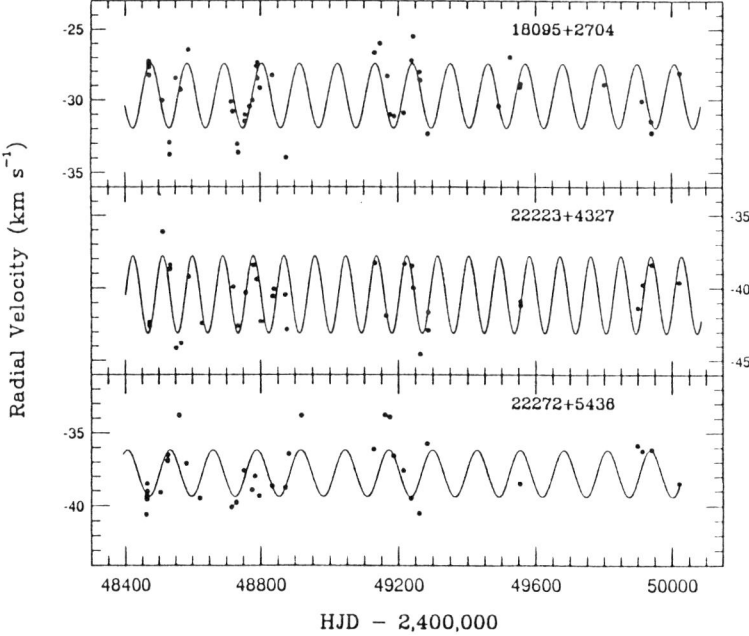

*Figure 2.* The radial velocity curves for three PPN which show periodic variability: IRAS 18095+2704 (109 d), 22223+4327 (89 d), and 22272+5435 (127 d).

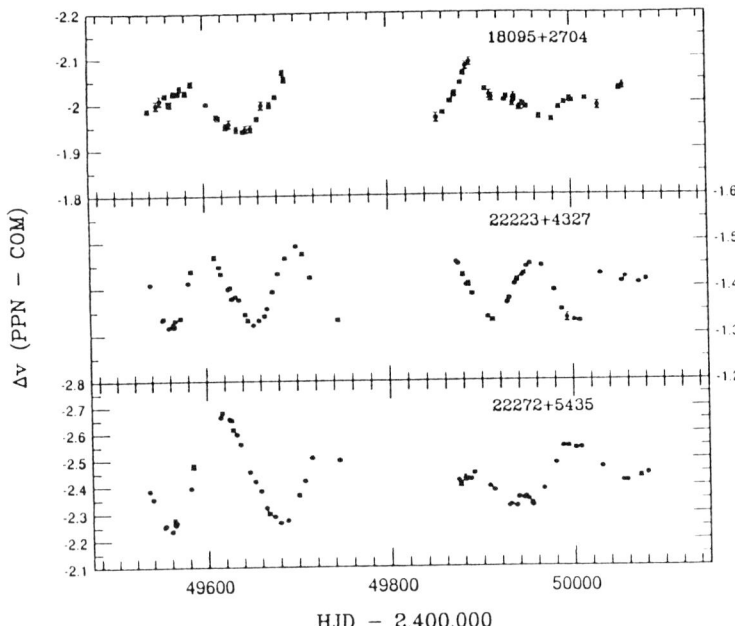

Figure 3. Differential light curves for the three PPN with periodic velocity variability.

## 4. Light–Curve Study

The light-curve study has been carried out over the last two seasons, 1994–1996, using the 0.4-m telescope at the Valparaiso University Observatory. A CCD (Photometrics Star I) detector is used with a standard $V$ (and occasionally $R$) filter. The precision is $< 0.01$ mag, except for the faintest stars. The program includes 40 PPN, with a range in magnitude of $V = 7 - 14$ and in spectral type of O to M. Included among these are the nine for which we have studied the radial velocity. Four Valparaiso University undergraduate students have assisted in the observing during the summer months. We plan to continue this light-curve study, and additional observations are being made this summer.

The three PPN with periodic radial velocity curves all have light curves showing light variations with the same period as the velocity variations. However, the variability does not have a simple periodic form, but varies in amplitude for each of the three. These are shown in Figure 3. The variability ranges from 0.15 mag in IRAS 18095+2704 to 0.50 mag in IRAS 22272+5435.

While all three exhibit similar periods in the velocity and light variations, the relative phasing of their velocity and light variability differs among the three: IRAS 22272+5435 is brightest when it is at its average size and expanding and faintest when at its average size and contracting;

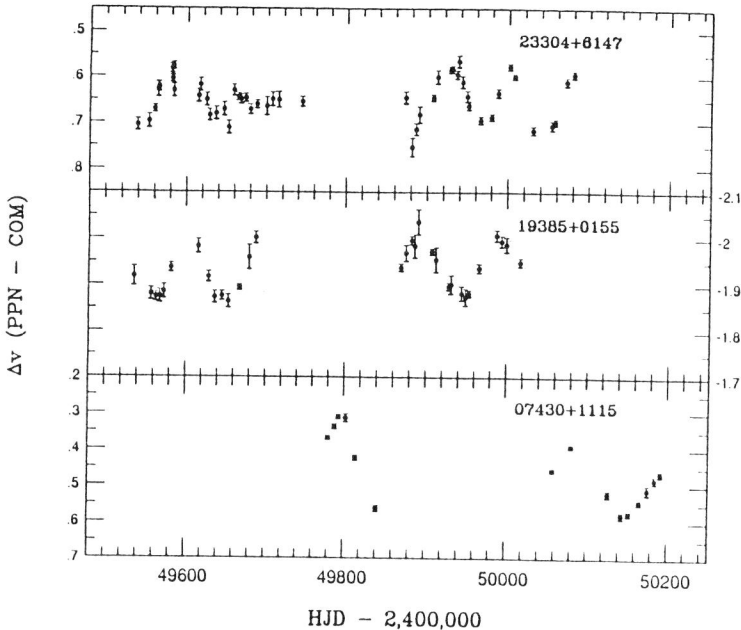

*Figure 4.* Examples of other PPN candidates with periodic light variability: IRAS 23304+6147 (84 d), 19385+0155 (96 d), and 07430+1115 (146 d).

IRAS 22223+4327 is brightest when smallest, faintest when largest; and IRAS 18095+2704 is brightest when smaller than average size and contracting, and faintest when larger than average size and expanding.

We interpret this variability as due to pulsation in the stars, rather than a binary nature. There are several reasons for this interpretation: (a) the amplitudes of the variation change, (b) the periods are short for a binary companion to orbit a giant star, and (c) the periods of the light and velocity are the same, while in a binary system in which the light variation was due to tidal distortion the light would have two maxima in each orbit.

Most of the others among the 40 sources show light variations, although in only a few others is a periodicity found. The light variation in $V$ is typically 0.15 to 0.35 mag. The following is a preliminary summary of the light variations detected:

- Periodic variability was found in nine PPN (including the three discussed above), with periods ranging from 25 to 146 days, and with varying amplitudes. These objects have spectral types F–G. Examples are shown in Figure 4.
- Short-time-scale variability ($< 10$ d) was found in many of the objects, especially those of early spectral types, O–B.
- Several have unusual light curves, such as a periodic variation that appears in one season and not the next, or an unusual shape.

## 5. Conclusions

As a result of this study of the velocity and light variations in PPN, we have found the following:

1. (Almost) all PPN vary in light and velocity.
2. This variability is due to pulsation; no binaries were discovered.
3. The variability does not have a simple periodic form; rather, it varies in amplitude or period, or there may exist multiple periods.
4. In the three cases with periodic velocity variability, the light varies with the same period as the velocity. However, the phase relationship between the light and velocity curves is different in each case.
5. There appears to be a general trend of period with spectral type, with later spectral types (G) possessing the longer periods (80–150 d), middle spectral types (A–F) possessing the middle periods (25–100 d), and early spectral types (O–B) showing short-time-scale variability.

Support for this research has been provided by the National Science Foundation (AST–9018032 and AST–9315107) and the NASA JOVE Program (NAG8–232), and is gratefully acknowledged.

## References

Bond, H. E. 1995, in *Asymmetrical Planetary Nebulae*, ed. A. Harpaz and N. Soker (Jerusalem: Israel Physics Society), p. 61

Hrivnak, B. J., Kwok, S. & Volk, K. M. 1989, *ApJ*, 346, 265

Kwok, S. 1993, *Ann. Rev. Astron. Astrophys.*, 31, 63

# THE CASE FOR S STAR BINARIES

THOMAS B. AKE
*Computer Sciences Corporation/Goddard Space Flight Center*
*Greenbelt, MD 20771, U.S.A.*

**Abstract.** Several lines of evidence point to a scenario in which Tc-poor S stars are the cooler analogs to the Ba II stars, i.e. they are binary systems where the peculiar atmospheric composition of the primary star is due to mass accreted from a secondary star long ago. Cases have been found where such S stars have WD or main-sequence companions, but an increasing number are found to be in interactive, symbiotic-like systems. Evidence of wind shocks, gas-streaming, and/or accretion disks in these systems attest to the current proximity of the components and provide striking evidence that accretion can be an important mechanism in their chemical evolution. Interactive effects are not as prominent in Ba II stars, presumably because of the lack of a strong wind from the primary star.

*No text received*

## Discussion

**Frantsman:** What is your opinion about the evolutionary phase of Tc-poor S stars? You mentioned that these stars and Ba II stars have similar properties (spectral characteristics, binarity). However it is known that their mean luminosities and effective temperatures are quite different.

**Ake:** The Tc-poor S stars are the cooler analogs of the Ba II stars. As the Ba II stars evolve up the giant branch, they become these S stars. The development of a wind, then, causes the S star binaries to show signs of interaction between the components, which is not normally seen in the Ba II star binaries.

**Frantsman**: Why can't these stars be in the early-AGB phase, especially taking into account that for stars of a given mass this stage lasts considerably longer than the TP-AGB phase?

**Ake**: The S stars show enhancements of $s$-processing, except for Tc. Thus the surface abundance peculiarities indicate that the material was modified during the TP-AGB phase, but long ago.

**Green**: Just a technical question: What advantage is there to the GHRS over the FOS for the detection of white dwarf companions to the extrinsic S stars?

**Ake**: The GHRS doesn't have the scattered light problem of the FOS.

# Session VII

## Mass–Loss, Winds, and Formation of Dust Grains

Four participants. Visual impressions by Pierre North.

# INFERRING MASS LOSS RATES FOR COOL LUMINOUS STARS FROM HIGH-RESOLUTION GHRS SPECTRA

J. L. LINSKY, G. M. HARPER AND J. VALENTI
*JILA, University of Colorado and NIST*
*Boulder CO, 80309-0440 U.S.A.*

AND

P. D. BENNETT AND A. BROWN
*CASA, University of Colorado*
*Boulder CO, 80309-0389 U.S.A.*

**Abstract.** We discuss GHRS spectra of single and binary late-type stars and describe in detail the spectra of $\alpha$ TrA and of $\zeta$ Aurigae obtained at ten orbital phases. The wind properties of $\alpha$ TrA are derived using a complete redistribution radiative transfer code, and we describe the properties of a new code, PRISMA, that we are building to fit line profiles using partial redistribution in a spherically-symmetric geometry. The $\zeta$ Aur spectra show that the mass loss process is variable on the timescale of several months, the wind density structure does not repeat from orbit to orbit, and the wind ionization structure is complex.

## 1. Introduction

The rates at which stars lose matter to the interstellar medium play an important role in stellar and Galactic evolution. Mass loss can change stellar evolution when the mass loss time scale is comparable to the core evolution time scale or when the cumulative loss of mass is significant compared to the initial stellar mass. Both conditions can be important for O–type and Wolf-Rayet stars and for late-type stars near the tip of the asymptotic giant branch, the region of the HR diagram where the carbon star phenomenon begins. Here mass loss peels off the outer hydrogen-rich layers, revealing the carbon-rich inner layers where dredge-up processes have altered the initial chemical composition.

Understanding stellar mass loss is even more important for modeling the chemical evolution of the Galaxy, as the metal enrichment of the interstellar medium, out of which the next generation of stars will emerge, depends on both steady and explosive mass loss processes.

Despite the critical importance of stellar mass loss, empirical measurements of mass loss rates for individual stars are very uncertain and, except for the O–type stars, we lack an accepted theory of mass loss with predictive power. Mass loss rates have been inferred from free-free continuum radio and far-infrared emission, infrared dust emission, emission in CO and other molecular lines, and blue-shifted circumstellar absorption features in ultraviolet and optical resonance lines. For a review of these techniques, see Drake (1986) and Dupree (1986).

Here we summarize what the Hubble Space Telescope spectra are telling us about mass loss rates from late-type stars, and we describe a new code for inferring mass loss rates from the analysis of such high-resolution spectra. Our intention is to compute accurate mass loss rates for representative late-type stars with which one can derive a new mass loss rate prescription based on fundamental stellar parameters ($\dot{M} = f(L, M, R)$; e.g. Nieuwenhuijzen & de Jager 1990). Unfortunately, the emerging picture is that the phenomenology of mass loss is very complex and no simple mass loss formula may be realistic.

## 2. Representative GHRS Spectra

Since 1978, IUE has obtained spectra of the Mg II h and k lines in a large number of late-type stars of all luminosity classes with a spectral resolution $R = \lambda/\Delta\lambda \approx 10,000$. The IUE archive now contains high-resolution spectra of more than 400 G–M giants and supergiants. Robinson & Carpenter (1995) show representative examples of such spectra. IUE spectra of Fe II and O I lines are also useful for studying stellar winds.

Since 1990, the GHRS has obtained echelle spectra with the higher resolution ($R \approx 90,000$) needed to distinguish interstellar from wind absorption features and with the higher signal/noise needed to study the nearly black wind absorption features far to the blue of line center. Examples of GHRS spectra of such late-type stars as $\alpha$ Ori, $\gamma$ Cru, $\gamma$ Dra, and $\alpha$ Tau may be found in Robinson & Carpenter (1995) and Carpenter et al. (1995). The N–type carbon star UU Aur was also studied with the GHRS echelle (Johnson et al. 1995). The high spectral resolution of the GHRS data is especially useful in identifying circumstellar absorption by Mn I and Fe I that distort the Mg II profiles of cool supergiants like $\alpha$ Ori. Carpenter et al. (1995) find evidence for wind acceleration in $\gamma$ Cru from the increasing blueshift of absorption lines formed with increasing height.

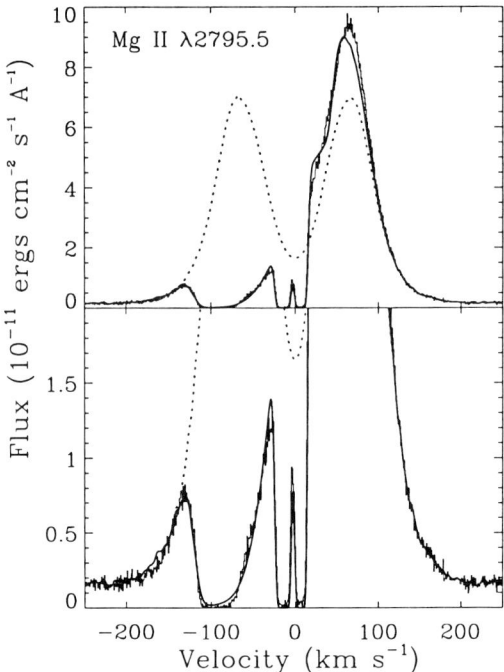

*Figure 1.* Upper panel: The observed α TrA echelle spectrum of the Mg II k line (noisy line), the assumed intensity profile at the base of the chromosphere (dashed line), and the best model fit (thick solid line). Lower panel: An enlargement of the lower portion of the line profile. From Harper et al. (1995).

## 3. Analysis of the GHRS Spectrum of α TrA

We cite here the analysis of an excellent GHRS echelle spectrum of the Mg II resonance lines that demonstrates what can be learned empirically about the wind properties of late-type stars. Harper et al. (1995) obtained very high quality spectra of the hybrid-chromosphere star α TrA (K4 II) (see Figure 1) that show the blue-shifted absorption and red-shifted emission (the so-called P Cygni line profile) characteristic of scattering in an optically thick, geometrically extended wind. The profile shows total extinction by the wind near $-100$ km s$^{-1}$ and absorption by the interstellar medium (two components near 0 km s$^{-1}$).

Harper et al. (1995) assumed the specific intensity at the base of the spherically-symmetric chromosphere (before the acceleration begins) and complete redistribution (CRD) of the scattered photons. They then solved the transfer equation for a two-level atom using an accelerated lambda iteration scheme and the velocity law $V(R) = V_\infty(1 - R_\star/R)^\beta$, where $V_\infty$ is the wind terminal velocity and $\beta$ is an unknown parameter characterizing the scale length over which the wind acceleration occurs. Despite the physical and computational limitations of the analysis methodology, they obtained an excellent fit to the line profile with $\dot{M} \geq 1.8 \times 10^{-10}\ M_\odot\ \mathrm{yr}^{-1}$, $V_\infty = 100$ km s$^{-1}$, $V_{\mathrm{turb}} = 24$ km s$^{-1}$, and $\beta \sim 1$. These results indicate a much lower mass loss rate than some previous estimates (e.g. Hartmann et al. 1981) and a large nonthermal pressure gradient near the base of the wind where it is needed, although future analyses of P Cygni–like profiles will require computer codes based on more physically realistic assumptions. In particular, the spectrum at the base of the chromosphere must be determined self-consistently.

## 4. A New Radiative Transfer Code: PRISMA

The interesting wind parameters that emerged from our rather simplistic analysis of the $\alpha$ TrA Mg II resonance lines stimulated us to develop a new radiative transfer code to infer mass loss rates from GHRS and IUE spectra. Our review of existing codes identified a number of desirable characteristics that a code to study the winds of late-type stars should include:

(a) *A physically accurate treatment of scattering.* For optically thick resonance lines like those of Mg II, most interactions between line photons and ions result in scatterings that are coherent in the atomic rest frame rather than collisional de-excitation or elastic scattering by collisions. Partial redistribution (PRD) codes include the physics of coherent scattering in the atomic frame, Doppler redistribution of the emitted photon due to the atom's motion, elastic scattering when collisions perturb the upper state of the transition, and photon destruction during the rare events when collisions lead to de-excitation before the atom can re-emit a line photon. PRD codes are especially important for analyzing the spectra of late-type stars since the wind velocities are usually only a few times the Doppler width. The Sobolev approximation, which is useful when there are large velocity gradients in the wind, is not usually valid for late-type stars.

(b) *A realistic geometry.* The presence of P Cygni–type features (blue-shifted absorption and red-shifted emission) in the resonance line profiles of luminous late-type stars provides unmistakable evidence that the winds are geometrically extended compared to the stellar photosphere. Thus useful codes must be able to solve the transfer equation in spherical geometry. As we shall see, wind geometries for binaries are almost certainly more complex than axisymmetric.

(c) *A multi-level atom.* Codes should solve the statistical equilibrium equations for multi-level atoms to properly include non-LTE ionization, recombination, and transitions between bound levels. The ionization equilibrium of important species should not be assumed constant, but computed in a self-consistent manner.

(d) *A self-consistent atmospheric model.* Rather than specifying a temperature/density structure, a code should derive semi-empirical thermodynamic parameters of the wind that best fit the line profiles.

(e) *Time-variability.* Since the time scales for ionization and recombination can be comparable to that of advection, codes should eventually include time-varying properties in the wind.

No existing radiative transfer code includes all of these desirable properties. For example, Drake & Linsky (1983) developed a PRD code that solves the transfer equation in the co-moving frame of the wind in a spherically symmetric extended atmosphere, but they used a pre-specified thermal structure, ionization, and velocity law for a two-level Mg II ion. They were able to show schematically the changes in a line profile when the wind velocity and atmospheric extension are varied. Using this code, Drake (1985) showed that the wind of Arcturus (K2 III) observed in the Mg II line is very extended and estimated a mass loss rate of $2 \times 10^{-10}$ $M_\odot$ yr$^{-1}$. Hartmann & Avrett (1984) developed a code using an approximate escape probability formalism that led to an estimated mass loss rate of $1.4 \times 10^{-6}$ $M_\odot$ yr$^{-1}$ for $\alpha$ Ori (M2 Iab), but they could not fit the shape of the Mg II lines well. Luttermoser et al. (1994) analyzed IUE spectra of the M6 III star g Her using the PANDORA code with the partial coherent scattering approximation to PRD, but they considered the atmosphere to be static and plane-parallel with an expanding circumstellar envelope.

We have therefore developed a new radiative transfer code PRISMA (Partial Redistribution In Spherical Moving Atmospheres) that rests on the heritage of the MULTI code written by Carlsson (1986). MULTI handles departures from LTE in multi-level atoms well, but it is written for a plane-parallel geometry and does not include PRD. MULTI has already been modified to include PRD (Uitenbroek 1989) and spherical geometry (Harper 1994) separately, but the new code will include both. We are now testing a preliminary version of PRISMA that solves the transfer equation in the observer's frame using the PRD technique developed by Uitenbroek (1989) and a global Scharmer operator (Scharmer & Carlsson 1985). This version of the code will be revised later to the co-moving frame with a different PRD technique (e.g. Hubeny & Lites 1995).

## 5. The Ugly Truth about Mass Loss from Real Stars: $\zeta$ Aur

The true complexity of stellar mass loss begins to emerge when one dissects a wind using an empirical probe. We are now analyzing GHRS spectra of the $\zeta$ Aur eclipsing binary system (K4 Ib + B5 V) observed at ten orbital

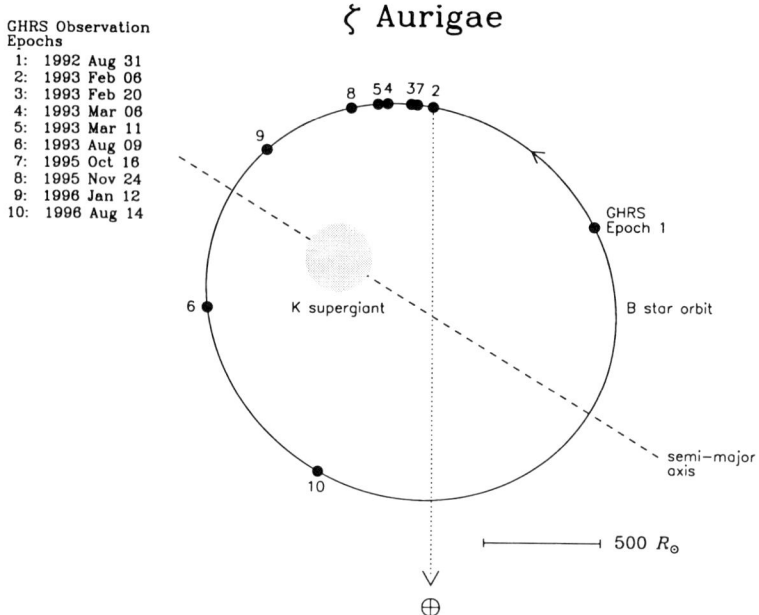

*Figure 2.* The orbit of ζ Aurigae drawn to scale. The positions of the secondary at the GHRS observation epochs are indicated, and the semi-major axis and the direction of the line of sight from the Earth are also shown. The size of the K supergiant primary is shown to scale, but the actual size of the B star secondary is much smaller than indicated in this diagram.

phases. We initiated this intense observing campaign to exploit the slow passage of the small B star behind the K supergiant, which provides an excellent UV–bright searchlight with which to probe the physical properties of the K star wind. Figure 2 shows the circumstances of these observations that extend over two orbits ($P_{\rm orb} = 972^{\rm d}$).

Analysis of the UV spectrum of the B star, optical spectra of the K star, Mark III optical long-baseline interferometry, and eclipse photometry led Bennett et al. (1996) to derive very precise values for the stellar and orbital parameters of the system. Comparison of these parameters with stellar evolution models indicates an age of $(80 \pm 15) \times 10^6$ yr if the initial abundances are solar. Thus the K star ($M = 5.8 \pm 0.2\,M_\odot$, $R = 148 \pm 3\,R_\odot$) probably lies near the tip of the red giant branch.

The GHRS spectra include lines of Fe II, Ni II, S II, Si II, and Al II. Typically one sees absorption from ions in the K star wind superimposed on the B star continuum and broad emission from B star photons scattered in the wind (Baade et al. 1996). Figure 3 shows the spectrum at epoch 1. Analysis of each spectrum leads to the column density for each ion as a function of velocity and then to the inferred hydrogen column density $N_{\rm H}(v)$ through

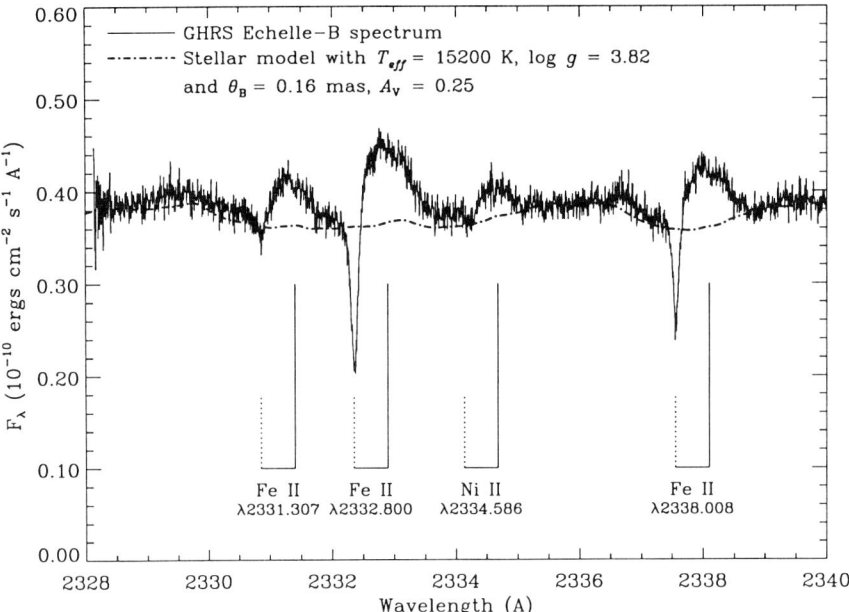

*Figure 3.* The spectrum of ζ Aurigae B observed with the GHRS Ech-B at epoch 1. A synthetic spectrum of the photosphere of the B star has been superimposed to illustrate the extent of the emission in the wind lines of Fe II and Ni II. This spectrum was computed using the TLUSTY and SYNSPEC codes of Hubeny (1988). The synthetic spectrum has been shifted by the radial velocity of the B star at epoch 1 (+30.6 km s$^{-1}$). The lines of Fe II and Ni II are shown at the systemic velocity of the ζ Aur system (+12.2 km s$^{-1}$, Griffin 1995) as a solid line, and at the terminal wind velocity of –70 km s$^{-1}$ relative to systemic as a dotted line.

the K star wind for each line of sight. The shapes of these columns differ from one phase to another, leading to the following conclusions:

(a) *The ionization structure in the wind is complex.* Some lines of sight show little or no Fe II at certain velocities. A schematic model in which the second ionization of Fe occurs in two regions — an ionized spherical volume around the B star, and an ionized spherical shell around the K star — can qualitatively account for the observed column density structure. At epoch 1, the Fe II ionization shell about the K star is very large, extending out to 45 K–star radii. At closer lines of sight, the outer boundary of the Fe II ionization shell moves in to about 8–10 K–star radii. Sulfur and silicon remain mostly singly ionized throughout the wind.

(b) *The properties of the wind are time-dependent.* This is shown by the different column densities at epochs 3 and 7, which are at the same orbital phase but are one orbit apart. At epoch 3 the column density peaks near 0 km s$^{-1}$ (near the radial velocities of both stars) with a smaller peak near $-70$ km s$^{-1}$, whereas at epoch 7 the column density peaks near $-65$ km s$^{-1}$ with no absorption near 0 km s$^{-1}$. The $N_H(v)$ column density at epoch 1, which samples a line of sight where the wind is near terminal velocity, is fit by a mass loss rate of $8.0 \times 10^{-9}$ $M_\odot$ yr$^{-1}$ and a terminal velocity of 70 km s$^{-1}$. However, the column densities observed at epochs 2 and 3, which sample the wind much closer to the stellar surface, are consistent with a mass loss rate reduced by factors of 3-5 from that of epoch 1. We conclude that while the shape of the observed column density structure $N_H(v)$ is consistent with a spherical wind with $\dot{M} = 8.0 \times 10^{-9}$ $M_\odot$ yr$^{-1}$ and $v_\infty = 70$ km s$^{-1}$, variations in the mass loss rate by up to a factor of 5 occur on a timescale of months. The column density structure does *not* repeat from orbit to orbit, suggesting that intrinsic mass loss variability is an important effect. Observations during the next orbit are needed to further quantify the extent and timescale of the wind variability.

Thus the commonly used assumptions of time-independent, homogeneous ionization winds are inconsistent with the $\zeta$ Aur data. Future studies of stellar winds should be guided by this ugly truth.

This work is supported by grants AR–6383, GO–3626 and GO–6069 from the Space Telescope Science Institute, and grants S–56460–D and NAG5–23007 from NASA.

## References

Baade, R., Kirsch, T., Reimers, D., Toussaint, F., Bennett, P. D., Brown, A. & Harper, G. M. 1996, *ApJ*, 466, 979
Bennett, P. D., Harper, G. M., Brown, A. & Hummel, C. A. 1996, *ApJ*, 471, 454
Carlsson, M. 1986, Uppsala Astron. Obs. Report No. 33
Carpenter, K. G., Robinson, R. D. & Judge, P. G. 1995, *ApJ*, 444, 424
Drake, S. A. 1985, in *Progress in Stellar Spectral Line Formation Theory*, ed. J. E. Beckman and L. Crivellari (Reidel), p. 351
Drake, S. A. 1986, in *Cool Stars, Stellar Systems, and the Sun*, ed. M. Zeilik and D. M. Gibson (Springer-Verlag), p. 369
Drake, S. A. & Linsky, J. L. 1983, *ApJ*, 273, 299
Dupree, A. K. 1986, *Ann. Rev. Astron. Astrophys.*, 24, 377
Griffin, R. F. 1995, private communication
Harper, G. M. 1994, *MNRAS*, 268, 894
Harper, G. M., Wood, B. E., Linsky, J. L., Bennett, P. D., Ayres, T. R. & Brown, A. 1995, *ApJ*, 452, 407
Hartmann, L. & Avrett, E. H. 1984, *ApJ*, 284, 238
Hartmann, L., Dupree, A. K. & Raymond, J. C. 1981, *ApJ*, 246, 193
Hubeny, I. 1988, *Comp. Phys. Comm.*, 52, 103
Hubeny, I. & Lites, B. W. 1995, *ApJ*, 455, 376
Johnson, H. R. et al. 1995, *ApJ*, 443, 281
Luttermoser, D. G., Johnson, H. R. & Eaton, J. A. 1994, *ApJ*, 422, 351
Nieuwenhuijzen, H. & de Jager, C. 1990, *A&A*, 231, 134
Robinson, R. D. & Carpenter, K. G. 1995, *ApJ*, 442, 328

Scharmer, G. B. & Carlsson, M. 1985, *J. Comp. Phys.*, 59, 56
Uitenbroek, H. 1989, *A&A*, 213, 360

## Discussion

**Whitelock**: As you have said, "pulsation is important," and we find, from IR observations, that low-mass, large-amplitude pulsators lose mass at between $10^{-7}$ and $10^{-4}$ $M_\odot \, \text{yr}^{-1}$. These are the mass-loss rates that modify stellar evolution, enrich the interstellar medium, and give us C stars. Is there a possibility of your applying the very impressive techniques you have described to these stars?

**Linsky**: Our objective is to use our new radiative transfer code to derive mass-loss rates for all types of stars for which there will be high quality data, including AGB stars with high mass-loss rates. Unfortunately these high-mass-loss-rate stars will be difficult to analyze as a result of complex circumstellar absorption by other lines superimposed on the resonance lines that we are studying, and the mass loss will likely be asymmetric and time-dependent. It is therefore prudent to start with the simpler cases, such as giants with small mass-loss rates like Arcturus. Once we are confident that the code is debugged and provides reliable results, we will proceed on to the more difficult and interesting cases.

**Little–Marenin**: Given the complexity of the $\zeta$ Aur system, do you think we will be able to make some general statement about winds and mass-loss rates of stars in general?

**Linsky**: We have no alternative but to try to generalize the results that come from our study of $\zeta$ Aur and other binary systems. The ability to observe a K supergiant wind in absorption against a bright point source that moves behind the K star is an extraordinary opportunity that must be exploited. At the same time one must not forget the added complexities introduced by a binary companion — ionization of the K star's wind by the B star, interaction of the two winds that can lead to a shock front, changed gravitational field, and perhaps other effects.

**Luttermoser**: Do you want to speculate on what mechanism drives the winds of these stars?

**Linsky**: What acceleration mechanisms drive winds in different types of stars remains an open question despite many years of theoretical and observational studies. I think that mechanical energy input from pulsations and acoustic or magnetoacoustic waves probably plays a major role for most late-type giants and supergiants, but other mechanisms (e.g. thermally-driven winds, dust, and perhaps radiation pressure on molecules and atoms) probably also contribute to the acceleration somewhere in the flow. It is important to include multiple acceleration processes.

**Feast**: Has anyone attempted to observe wind characteristics spectroscopically from the white dwarf companion to Mira Ceti?

**Linsky**: Not to my knowledge, but this would be an interesting observing program.

# DUST FORMATION IN STELLAR PHOTOSPHERES

*The Case of Carbon Stars from Dwarfs to AGB Stars*

TAKASHI TSUJI
*Institute of Astronomy, The University of Tokyo*
*Mitaka, Tokyo, 181 Japan*

**Abstract.** We examine whether dust forms in the photospheres of carbon-rich stars by referring to the case of red and brown dwarfs for which some observational clues on dust formation are now known. Dust may form in the photospheres of dwarf carbon stars and produce significant effects on both their structure and spectra. In carbon-rich asymptotic giant branch stars, dust probably forms in the photosphere, if not in the circumstellar envelope, and radiation pressure on dust is sufficient to expel the matter directly from the photosphere. This fact may play some role in mass-loss from cool luminous stars in general, including non-pulsating stars for which no successful mechanism of mass-loss was known.

## 1. Introduction

The idea that dust may form in stellar photospheres is not necessarily new, but it has not been known how dust forms in photospheres. For this reason, few attempts have yet been made to include dust in modeling the stellar photosphere. It is only recently that some observational evidence of dust in the photosphere has been recognized in very low-mass stars (VLMSs) and in brown dwarfs. These observations revealed that dust formed in the photospheres of VLMSs may survive as long as the lifetime of these long-lived stars, while the dust formed in brown dwarfs may follow a different fate (Tsuji et al. 1996a,b). These observations provide important insight on how dust forms and evolves in the stellar photosphere (Sect. 2).

An interesting case of dust formation in photospheres may be that of the dwarf carbon (dC) stars, the discovery of which is undoubtedly having an important impact in our understanding of the carbon star phenomenon, especially if the space density of dC stars is much larger than that of carbon-rich asymptotic giant branch (AGB) stars (Green, this volume). The dC star may have been formed by binary mass-transfer from an AGB primary, and it has preserved a direct probe into the evolution of intermediate mass stars to the AGB stage in the metal-poor era. The large space density of dC stars implies that their progenitors should have contributed significantly to the Galactic chemical evolution, especially of such key elements as C, N, and heavy elements whose origins are by no means well understood yet (Gustafsson, this volume). To clarify the nature of the dC star, the method of model atmospheres should serve as a useful guide, and dust formation will play an important role especially in view of the rather high condensation temperatures of the dust species expected in the carbon-rich case (Sect. 3). In fact, dusty models should be needed for dC stars with $T_{\text{eff}} < 3000\,\text{K}$ even for the low metallicities typical of halo objects (Sect. 4). Although no observation of this is known yet, dust will have appreciable effects on the spectra of cool dC stars. On the other hand, the limited observations now available on relatively warm dC stars can reasonably be understood by our dust-free models (Sect. 5).

We also found that the necessary condition for dust formation is well met in the photospheres of carbon-rich AGB stars with $T_{\text{eff}} < 3000\,\text{K}$. This possibility that dust may form already in the photospheres of AGB stars rather than in the circumstellar envelope has important implications. For example, radiation pressure on dust may provide a long-sought driving force for mass-loss directly from the photospheres of AGB stars. Also, dust formed in photospheres will produce noticeable observable effects (Sect. 6).

## 2. An Empirical Approach to Dust Formation in Stellar Photospheres

The possible presence of dust in stellar photospheres has been shown first for late M dwarfs, where predicted spectral energy distributions (SEDs) based on models including dust agree rather well with observed ones (Tsuji et al. 1996a). A more clear demonstration of the presence of dust was found in a brown dwarf candidate GD 165B discovered by Becklin & Zuckerman (1988). In this object, the SED shows evidence of severe extinction by dust, and the molecular bands such as those of $H_2O$ appear to be too weak to be explained by any model without dust. These observations could be explained first by our dusty models in which dust formed in accordance with the thermodynamical condition of condensation is uniformly mixed with gas throughout (dust-gas homogeneous mixture model). On the other hand, no clear evidence of dust could be seen in the cooler brown dwarf Gl 229B discovered by Nakajima et al. (1995), and its spectrum is dominated by volatile molecules such as methane (Oppenheimer et al. 1995). The observed data for Gl 229B could never be understood by our dusty model but instead can be well explained by our dust-free model (Tsuji et al. 1996b).

These contrasting observations can be regarded as unique experiments on dust formation in astronomical "laboratories" under different conditions, and they can be interpreted as follows: When the temperature is relatively high as in GD 165B ($T_{\text{eff}} \approx 1800\,\text{K}$), the dust is still smaller than the critical radius $r_{\text{cr}}$ below which the dust once formed cannot grow and dissolves back to molecules. In this case, small dust grains and gaseous molecules are in detailed balance and hence a certain amount of dust allowed by the thermodynamical equilibrium is always present in the photosphere. This provides a rationale to our simple dust-gas homogeneous mixture model and also a reason why dust can survive as long as the lifetime of VLMSs. On the other hand, if the temperature is relatively low as in Gl 229B ($T_{\text{eff}} \approx 1000\,\text{K}$), the dust can be larger than the critical radius $r_{\text{cr}}$ and the dust once formed begins to grow. Then, the detailed balance between dust and gas can no longer be maintained because of the onset of this irreversible process. For proper treatment of such a case, not only the non-equilibrium kinetic theory of dust growth such as has been developed for cool stellar winds (Sedlmayr & Winters, this volume) but also the methods of meteorology may be needed. Anyhow, dust grows and finally segregates from the dust-gas homogeneous mixture to form a dust aggregation or cloud. Then, volatile molecules can appear through the gaps between dust clouds, as has been observed in Gl 229B, and the dust will no longer be so effective in blocking the emergent radiation as in the dust-gas homogeneous mixture model which applied well to GD 165B.

Now, one important by-product of the recent efforts to clarify the nature of brown dwarfs is a finding that there should be a regime where dust may be formed in LTE and remain in detailed balance with the gas so long as the thermodynamical condition of condensation is fulfilled at the relatively high temperatures of the stellar photosphere. We assume in the following that the photosphere of carbon-rich stars from dwarfs to AGB stars can also be in this dust-gas detailed balance regime, although we will not speak of carbon-rich brown dwarfs for the moment.

## 3. Physical Properties of Carbon-Rich Mixture at High Density

We begin with a simple computation of chemical equilibrium in which condensation is taken into account. An example of a carbon-rich mixture at $\log P_g = 6.0$ is shown in Figure 1. A noticeable feature in the chemical equilibrium at high density, in contrast to the more familiar case of low density relevant to AGB carbon stars, is the predominance of $CH_4$ instead of CO. This has important consequences: *First*, CO can no longer be the major species of carbon. Then, as oxygen in CO is released, a large amount of $H_2O$ must be formed. This unexpected but important result that $H_2O$

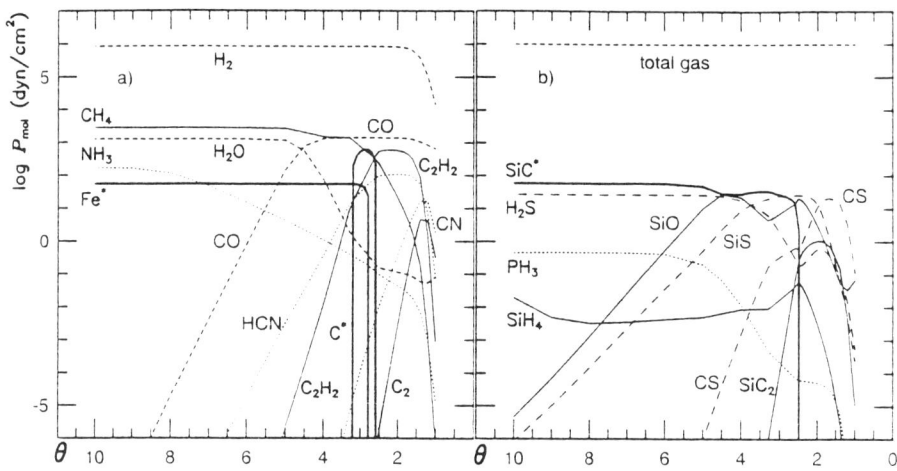

*Figure 1.* Chemical equilibrium of a carbon-rich mixture (C is increased so that C/O = 2 in the solar composition mixture) at log $P_g$ = 6.0. Partial pressures of selected molecules (asterisks indicate fictitious pressures of monomers composed of condensed species) are plotted against $\theta$ = 5040/T. (a) Molecules composed of H, C, N, and O. Note that $H_2O$ dominates at T < 1000 K even for C > O. (b) Molecules including S, Si, and P.

dominates even in a carbon-rich mixture at low temperature and high pressure seems to have been overlooked until the present. Thus, a high density, cool atmosphere is dominated by $H_2O$ as well as by $CH_4$ even if C > O, and CO formation no longer plays such a special role as to characterize oxygen-rich and carbon-rich atmospheres. *Second*, although graphite is formed at relatively high temperature (about 2500 K), it soon disappears under high pressure (at about 1500 K under log $P_g$ = 6.0), while graphite can survive to lower temperatures under low pressure. This somewhat unexpected result, however, is not a new one but has been noticed before (Tsuji 1964).

Now, in extending our opacity code for carbon-rich mixtures to high densities, we identify three major problems: *First*, we must consider the infrared bands of almost all the molecules that appear in Fig. 1, namely, CO, $H_2O$, $C_2H_2$, $CH_4$, HCN, $NH_3$, $SiH_4$, SiO, SiS, CS, $H_2S$, and $PH_3$ together with the electronic bands of CH, $C_2$, CN, CaH, MgH, and FeH. We treat all these molecular opacities by the band model method throughout (Tsuji 1994). For the reason already mentioned, we should include $H_2O$ opacity even for C > O at high density. *Second*, abundant $H_2$ is now an important source of opacity at high density because of the collision-induced absorption (CIA), whose absorption coefficients have been updated by Borysow (1994). The importance of $H_2$ CIA relative to other sources of opacity is essentially the same as in the oxygen-rich case (Tsuji & Ohnaka 1995). *Third*, we must represent the effects of dust opacities by amorphous carbon (C), silicon

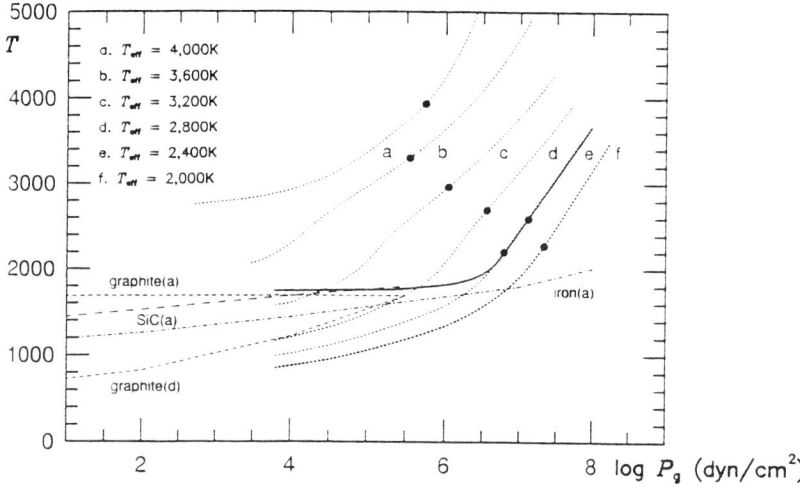

*Figure 2.* Dust-free model atmospheres of dC stars with $Z/Z_\odot = 0.1$ and C/O = 2 are shown by dotted lines for $T_{\rm eff}$ between 2000 and 4000 K (log $g$ = 5.0 and $v_{\rm micro}$ = 1.0 km s$^{-1}$ throughout). The appearance-lines for graphite, silicon carbide, and iron, as well as the disappearance-line for graphite, are indicated. An example of a dusty model is shown by the solid line for $T_{\rm eff}$ = 2400 K. Dusty models of other $T_{\rm eff}$ show essentially the same structure. Filled circles indicate the onset of convection.

carbide (SiC), and iron (Fe). As we are assuming that the size of the dust species is still smaller than the critical radius $r_{\rm cr}$ (Sect. 2), the extinction coefficient can be evaluated by a series expansion of the Mie formula (van de Hulst 1957).

## 4. Photospheres of Dwarf Carbon Stars with and without Dust

We first considered gaseous opacities alone and constructed a small grid of model atmospheres for dC stars which covers $T_{\rm eff}$ between 2000 and 4000 K with a step of 400 K (log $g$ = 5.0 and $v_{\rm micro}$ = 1.0 km s$^{-1}$ throughout). We assumed metallicities of $Z/Z_\odot$ = 1, 0.1, and 0.01 in which the relative abundances are solar except for C/O = 2 (C increased). The resulting models in radiative-convective equilibrium, for the case of $Z/Z_\odot = 0.1$ as an example, are shown by the dotted lines in Figure 2. The appearance temperatures for graphite, silicon carbide, and iron, as well as the disappearance temperatures for graphite, are plotted against log $P_g$ in Fig. 2. As discussed in Sect. 3, graphite condenses only in a limited regime of the phase diagram. Figure 2 reveals that only the models for $T_{\rm eff}$ = 4000 and 3600 K can be dust-free, while all the models with $T_{\rm eff} < 3200$ K are not self-consistent in that dust formation is neglected, contrary to the thermodynamical result. The results are not much different for other metallicities, although dust tends to appear in models of higher $T_{\rm eff}$ at higher metallicity.

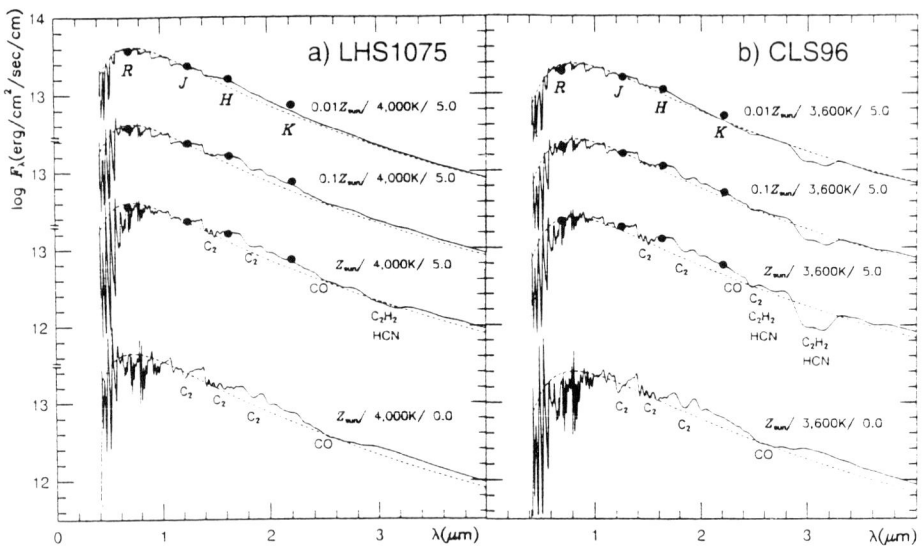

*Figure 3.* (a) Predicted SEDs based on our models for a dC star with $T_{\text{eff}} = 4000$ K are shown by solid lines (dashed lines are blackbody curves for 4000 K) for $Z/Z_\odot = 0.01$, 0.1, and 1 (C/O = 2 throughout), and compared with the observed SED (filled circles) of the dC star LHS 1075 based on $R$ (Green et al. 1991) and $JHK$ (Bothun et al. 1991) photometry. The predicted SED for a carbon-rich AGB star of the same $T_{\text{eff}}$ is also shown at the bottom, and a significant luminosity effect can be seen ($Z/T_{\text{eff}}/\log g$ is indicated for each SED). (b) The same for the case of $T_{\text{eff}} = 3600$ K and the dC star CLS 96.

Now, dust should form in the photospheres of dC stars with $T_{\text{eff}} <$ 3200 K. Modeling a dusty photosphere, however, was more difficult in the carbon-rich case than in the oxygen-rich case. In fact, the heating by highly absorbing forms of dust such as graphite is so large that the dust appears only at the very surface of the photosphere, in marked contrast to the oxygen-rich case where dust could form deeper in the photosphere (Tsuji et al. 1996a). The resulting dusty model for a dC star of $T_{\text{eff}} = 2400$ K in radiative-convective equilibrium is shown by the solid line in Fig. 2. A small amount of dust condensing at the very surface gives a drastic heating of nearly 1000 K.

## 5. Some Observable Properties of Dwarf Carbon Stars

The predicted SEDs based on our models for $T_{\text{eff}} = 4000$ and 3600 K for the cases of $Z/Z_\odot = 0.01$, 0.1, and 1 are shown in Figure 3 by the solid lines. Compared with the SED based on a model of an AGB carbon star of the same $T_{\text{eff}}$ shown at the bottom of each panel, molecular bands such as those of CO (2.3 $\mu$m) are much weaker in the model for dC stars. This prediction agrees nicely with recent observations of the infrared spectra of dC stars

DUST FORMATION IN PHOTOSPHERES OF C-RICH STARS    319

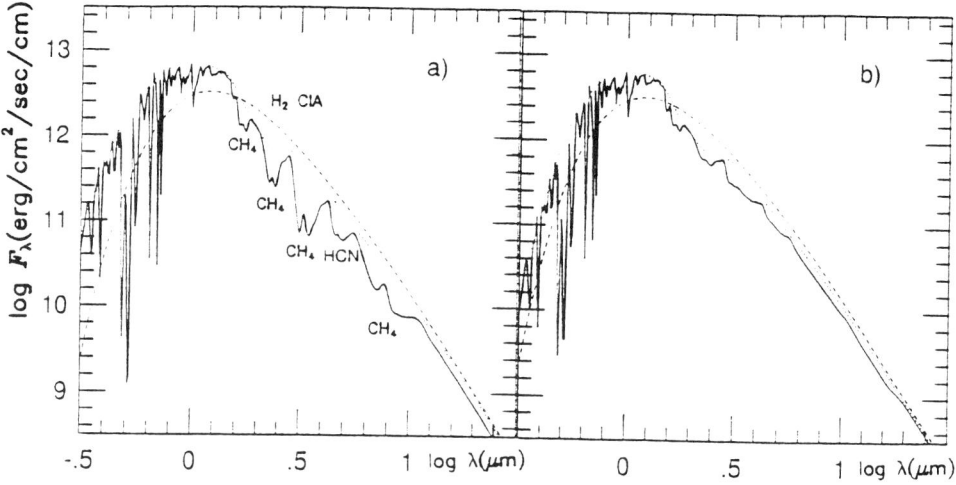

*Figure 4.* Predicted SEDs for (*a*) dust-free and (*b*) dusty models of a dC star with $T_{\text{eff}} = 2400\,\text{K}$ ($Z/Z_\odot = 0.1$, C/O = 2, $\log g = 5.0$, and $v_{\text{micro}} = 1.0\,\text{km s}^{-1}$) are shown by the solid lines. The dotted and dashed lines are the emergent fluxes for continuous opacities alone (including $H_2$ CIA; dust opacities are also included in the dusty case) and blackbody curves, respectively.

(Joyce, this volume). On the other hand, absorption bands due to HCN and $C_2H_2$ (3.1 $\mu$m) appear already at $T_{\text{eff}} = 4000\,\text{K}$ ($Z = Z_\odot$) and are much stronger at $T_{\text{eff}} = 3600\,\text{K}$ (even with low $Z$) in dC star models, while they are invisible in the AGB models of $T_{\text{eff}} = 4000$ and $3600\,\text{K}$ ($Z = Z_\odot$). Thus the infrared spectra show significant luminosity effects in carbon-rich stars. The observed SEDs of the dC stars LHS 1075 and CLS 96 (filled circles) are compared with the predicted ones in Figs. 3*a* and 3*b*, respectively. It appears that $T_{\text{eff}}$ may be close to 4000 K for LHS 1075 and close to 3600 K for CLS 96. In both dC stars, the case of $Z = 0.1 Z_\odot$ provides the best overall fit. It may still be premature to suggest that the metallicity can be estimated this way, especially because of the possibility that the relative abundances of the elements may be highly non-solar in dC stars.

As an example of cooler dC stars, the SED predicted by the dust-free model for $Z = 0.1 Z_\odot$ and $T_{\text{eff}} = 2400\,\text{K}$ is shown in Figure 4*a* by the solid line. Now, the infrared region ($\lambda > 1.5 \mu$m) shows a severe depression by strong molecular bands, mostly of $CH_4$ (note that HCN and $C_2H_2$ are already rather weak). Further, quasi-continuous $H_2$ CIA plays an important role and this effect is more important in more metal-poor cases. As a result of the large opacities at $\lambda > 1.5 \mu$m, the SED shows a large excess shortward of 1.5 $\mu$m and peaks at about 1 $\mu$m. Because of the increasing importance of $CH_4$ and $H_2$ CIA at the lower luminosity, the spectra again show significant luminosity effects (compare Figures 4 and 6 with the difference in

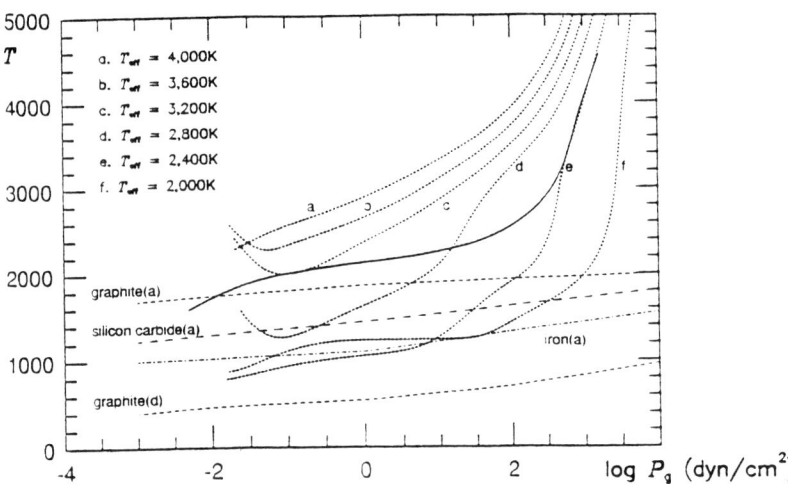

*Figure 5.* Dust-free model atmospheres of carbon-rich AGB stars with $Z/Z_\odot = 1.0$ and $C/O = 2$ are shown by dotted lines for $T_{\text{eff}}$ between 2000 and 4000 K (log $g = 0.0$ and $v_{\text{micro}} = 3.0$ km s$^{-1}$ throughout). An example of a dusty model in radiative equilibrium is shown by the solid line for $T_{\text{eff}} = 2400$ K. Dusty models of other $T_{\text{eff}}$ show essentially the same structure. As to other details, see the legend to Fig. 2.

metallicity in mind), but the effects are somewhat different from those in the hotter case noted above. If dust formation is introduced in this model, the SED shows a drastic change as shown by the solid line in Fig. 4b. In particular, molecular bands are now much weaker, reflecting the higher photospheric temperature. However, the SED still shows a large excess in the 1 $\mu$m region. At present, no observation of such a cool dC star is known, but we hope that our prediction can be of some use as a guide for future searches for such objects.

## 6. Dust Formation in Photospheres of Carbon-Rich AGB Stars

The belief that high density is more favorable for dust formation is not necessarily true, at least for graphite (Sect. 3). In fact, the thermodynamical condition for condensation is well fulfilled for cool AGB stars (solar abundance except for $C/O = 2$, log $g = 0.0$, and $v_{\text{micro}} = 3.0$ km s$^{-1}$) with $T_{\text{eff}} < 3000$ K, as shown in Figure 5, and dust-free models can no longer be self-consistent for them. It is possible to recover the thermodynamical equilibrium by including dust formation in our modeling, and an example of a dusty model in radiative equilibrium is shown by the solid line in Fig. 5 for $T_{\text{eff}} = 2400$ K. Again, a small amount of dust formed at the very surface produces an enormous effect on the thermal structure of the photosphere.

What is more interesting, the effective gravity turns out to be negative ($g_{\text{eff}} = g - g_{\text{rad}} < 0$) due to radiation pressure on the dust in the dusty model

DUST FORMATION IN PHOTOSPHERES OF C-RICH STARS    321

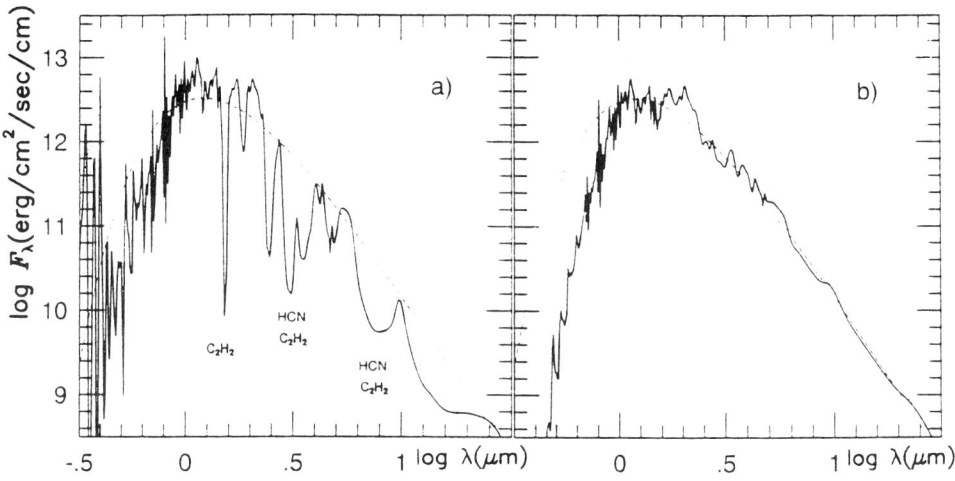

*Figure 6.* Predicted SEDs for (*a*) dust-free and (*b*) dusty models of a carbon-rich AGB star with $T_{\text{eff}} = 2400$ K ($Z/Z_\odot = 1.0$, C/O = 2, $\log g = 0.0$, and $v_{\text{micro}} = 3.0$ km s$^{-1}$) are shown by the solid lines. As to other details, see the legend to Fig. 4.

photosphere of AGB stars. This fact implies that a stable photosphere can no longer be possible and that a mass-loss outflow initiates directly from the photosphere. It is now generally believed that radiation pressure on dust is the major driving force for mass-loss from cool evolved stars. However, dust is generally assumed to be formed in the circumstellar envelope where the temperature may be below the condensation temperature but the density may be too low for dust to form. This difficulty has been solved for the case of Mira variables by levitating the mater to the condensation point by pulsation (Fleischer et al., this volume). So far, however, no mechanism of mass-loss for general cases other than the pulsating Miras is known, since dust can no longer be formed in the circumstellar envelope.

We now find that the surface temperatures of carbon stars with $T_{\text{eff}} < 3000$ K can be very low ($\sim 1000$ K, see Fig. 5) due to their highly non-gray molecular opacities, and dust can easily be formed in the photosphere. Actually, the formation of dust results in a strong heating of the photosphere, but a small amount of dust can always be formed in radiative equilibrium, since otherwise the temperature drops again drastically. Then the outward acceleration due to radiation pressure on the dust is sufficient to overcome the stellar surface gravity in AGB stars, and an outflow originating directly from the photosphere can provide a simple yet physically clear mechanism for stellar mass-loss from cool evolved stars, for general cases other than the pulsating Miras. Previously, radiation pressure on molecules had been considered as an outward driving force, but it appeared not to be sufficient for stellar mass-loss either in O-rich (Tsuji 1978) or in C-rich (Jørgensen & Johnson 1992) cases. Now, dust can do easily what molecules could not.

In Figure 6a, the SED predicted by a dust-free model for $Z = Z_\odot$, C/O = 2, and $T_{\text{eff}} = 2400$ K is shown by the solid line. A noticeable feature here is the enormously deep absorption due to HCN and $C_2H_2$. However, no observation of such strong absorption is known in real carbon stars. If dust formation is introduced in this model, the SED shows a drastic change as shown by the solid line in Fig. 6b. The molecular bands are now much weaker, reflecting the higher photospheric temperatures and extinction by the dust. Although we have not yet incorporated the dynamics of mass outflow, a small amount of dust formed in the photosphere will give more or less similar observable effects. Such an SED can be more compatible with the observations so far known, and this can be taken as observational evidence that dust forms already in the photosphere. We hope that this problem can be clarified by future observations, especially by ISO.

## 7. Concluding Remarks

We have investigated the characteristic features of dust formation in the photospheres of carbon-rich stars, where thermodynamics indicates that dust condenses in a wide range of physical conditions. However, once the constraint of radiative equilibrium is introduced, the temperature of the photosphere rises enormously and only a small amount of dust appears at the very surface of the photosphere. This is fatal for the absorbing dust species expected in the carbon-rich case, in contrast to the non-absorbing silicates in the oxygen-rich case. Nevertheless this small amount of dust, if formed in the photosphere, will have significant effects on both the photospheric structure and the spectra, not only of dwarf carbon stars but also of AGB stars. In AGB stars, however, hydrodynamics must be introduced already in the photosphere because of the negative effective gravity due to radiation pressure on the dust. This in turn opens the interesting possibility that a radiatively-driven wind can be realized in cool evolved stars other than the Mira variables, since dust can be formed in the photosphere even if not formed in the circumstellar envelope. Whether this can explain the large mass-loss observed in cool luminous stars in general remains to be examined further.

## References

Becklin, E. E. & Zuckerman, B. 1988, *Nature*, 336, 656
Borysow, A. 1994, in *Molecules in the Stellar Environment*, ed. U. G. Jørgensen (Springer), p. 209
Bothun, G., Elias, J. H., MacAlpine, G., Matthews, K., Mould, J. R., Neugebauer, G. & Reid, I. N. 1991, *AJ*, 101, 2220

Green, P. J., Morgan, B. & MacConnell, D. J. 1991, *ApJ*, 380, L31
Jørgensen, U. G. & Johnson, H. R. 1992, *A&A*, 265, 168
Nakajima, T., Oppenheimer, B. R., Kulkarni, S. R., Golimowski, D. A., Matthews, K. & Durrance, S. T. 1995, *Nature*, 378, 464
Oppenheimer, B. R., Kulkarni, S. R., Matthews, K. & Nakajima, T. 1995, *Science*, 270, 1478
Tsuji, T. 1964, *Ann. Tokyo Astron. Obs.*, 2nd Ser., 9, 1
Tsuji, T. 1978, *A&A*, 62, 29
Tsuji, T. 1994, in *Molecules in the Stellar Environment*, ed. U. G. Jørgensen (Springer), p. 79
Tsuji, T. & Ohnaka, K. 1995, *ASP Conf. Series*, 78, 69
Tsuji, T., Ohnaka, K. & Aoki, W. 1996a, *A&A*, 305, L1
Tsuji, T., Ohnaka, K. Aoki, W. & Nakajima, T. 1996b, *A&A*, 308, L29
van de Hulst, H. C. 1957, *Light Scattering by Small Particles* (J. Wiley & Sons)

## Discussion

**Dorfi**: In the case of dust formation in AGB atmospheres, a massive stellar outflow will develop which modifies the density and temperature structure of the outer atmosphere. How can you calculate spectra of these models assuming an atmosphere in hydrostatic equilibrium?

**Tsuji**: As I mentioned, the effective gravity becomes negative ($g_{\text{eff}} < 0$) if dust forms in an AGB photosphere, and the atmosphere is no longer in hydrostatic equilibrium. In our very preliminary test model of a dusty AGB photosphere, we switched off the radiation pressure term and artificially recovered hydrostatic equilibrium. The reason why I showed this very preliminary spectral energy distribution for a dusty AGB photosphere was just to illustrate qualitatively the possible enormous effects of a small amount of dust forming in the photosphere (not in the circumstellar envelope), especially because this may relax one of the major difficulties in our interpretation of the observed spectral energy distributions of very cool carbon stars. A more self-consistent treatment is certainly needed to see the full consequences of dust formation in the photospheres of AGB stars, both on their spectral energy distributions and on mass outflows.

[*Note added in proof:* After the meeting, we examined the dynamical effect by keeping the radiation pressure term in the momentum equation. Because of the dust formed in the photosphere, the sonic point is found within the photosphere, and it turns out that the wind velocity is several hundred km s$^{-1}$ at the stellar surface but the mass-loss rate is only of the order of $10^{-10} M_\odot$ per year. From this result, we think it more difficult for the massive wind to originate directly from the photosphere even if dust forms in the photosphere. On the other hand, a less massive but high-speed wind may be possible. Such a wind may not necessarily be directly observable but may have some effects on both the atmospheric structure and mass-loss of AGB carbon stars.]

**Lloyd Evans**: A possibly relevant observation is that, in the dwarf carbon star LHS 1075, though the *JHK* colors and spectrum at 5000 Å indicate it is a warm carbon star like the CH stars in $\omega$ Cen, at 4000 Å it has the weak CN bands and Ca II H & K lines of the coolest giant carbon stars (Swings-Struve effect). But this might result instead from a low abundance of elements such as Ca and N.

**Tsuji**: It is very interesting to hear some details about the spectra of dwarf carbon stars. As for the CN bands and Ca II H & K lines, these may show a significant luminosity effect as can be inferred from the oxygen-rich case, but I have not examined this yet. I also agree with you that the metallicity plays some role, and the possibility that the relative abundances are highly non-solar makes the problem more complicated. As to the violet spectra of very cool carbon stars, the so-called UV depression in these stars is not yet fully understood, and I suppose that extinction by the dust that may form in the photospheres of very cool carbon stars may play some role.

# DUST-DRIVEN WINDS OF ROTATING CARBON STARS

ERNST A. DORFI AND SUSANNE HÖFNER
*Institut für Astronomie der Universität Wien*
*Vienna, Austria*

**Abstract.** A new mechanism is proposed to explain an asymmetric mass loss of carbon-rich AGB stars where slow stellar rotation modifies a wind due to the non-linear behavior of the dust formation process. This effect leads to a preferential mass loss with higher velocities in the equatorial plane and also provides a simple explanation for the widely-observed asymmetries in the shapes of planetary nebulae.

## 1. Introduction

Stellar rotation is a ubiquitous phenomenon at all stages of stellar evolution, but older single stars are expected to exhibit only small rotational velocities at the photosphere due to magnetic braking occurring during their evolution and due to an expansion of the outer layers (e.g. Slettebak 1970). However, a small number of fast-rotating AGB objects like V Hya (Barnbaum et al. 1995) seem to exist where the outer envelope is interacting with a close companion. In the case of a planetary system around an expanding AGB star it is plausible that these outer stellar layers also can accumulate some of the angular momentum of the orbiting planets. Hence, it may well be that a larger fraction of AGB stars than expected exhibit rotation of a few $km\,s^{-1}$ in their photospheric layers without showing evidence for binarity. Since most of these stars are variable and maintain a modulated stellar wind of about 15 $km\,s^{-1}$, it will be difficult to detect photospheric rotational velocites of a few $km\,s^{-1}$.

Taking into account some average observational properties, we construct dust-driven winds of slowly rotating carbon-rich AGB stars where only tiny deviations from spherical symmetry are present in the extended atmospheres, allowing a quasi-spherical approximation of the stellar structure (e.g. Kippenhahn et al. 1970). However, these small asymmetries between the equator and pole are sufficient to introduce significant changes in the dust condensation process through non-linearities and in the subsequently

TABLE 1. Parameters of Dust-Driven Winds of Rotating AGB Stars

| $L$ [$L_\odot$] | $\varepsilon_C/\varepsilon_O$ [C/O] | $\omega$ [s$^{-1}$] | $v \sin i$ [km s$^{-1}$] | $\dot{M}$ [$10^{-6} M_\odot$ yr$^{-1}$] | $\dot{M}_e/\dot{M}_p$ | $u_{\infty,e}/u_{\infty,p}$ |
|---|---|---|---|---|---|---|
| $10^4$ | 2.3 | $2 \times 10^{-8}$ | 6.9 | 1.38 | 25.8 | 3.4 |
| $10^4$ | 2.3 | $1.5 \times 10^{-8}$ | 5.1 | 0.74 | 7.1 | 2.0 |
| $10^4$ | 2.3 | $10^{-8}$ | 3.4 | 0.44 | 2.6 | 1.4 |
| $10^4$ | 2.3 | $5 \times 10^{-9}$ | 1.7 | 0.31 | 1.3 | 1.1 |
| $1.2 \times 10^4$ | 2.0 | $1.5 \times 10^{-8}$ | 5.6 | 1.24 | 9.7 | 2.4 |
| $1.2 \times 10^4$ | 2.0 | $10^{-8}$ | 3.8 | 0.65 | 3.2 | 1.6 |
| $1.2 \times 10^4$ | 2.0 | $10^{-9}$ | 0.4 | 0.38 | 1.0 | 1.0 |

developing dust-driven outflow. The inner boundary of our dust-driven winds can be obtained from simple models of slowly rotating stars (e.g. Tassoul 1978), and the inclusion of small rotation rates already modifies the effective gravity and the scale height of the atmosphere, as well as the radiative flux. To get a quantitative picture of the dust-driven winds of a rotating AGB star we have solved the time-dependent system of grey radiation hydrodynamics together with the equations governing the dust formation in carbon-rich atmospheres (cf. Höfner et al. 1995 and Dorfi & Höfner 1996 for further references and details).

## 2. Outflow Properties and Mass Loss of Rotating AGB Stars

The process of dust formation depends strongly on the density and temperature stratification in an extended atmosphere where even small changes — e.g. in the density at a given temperature — can result in a totally different behavior of the stellar wind. Any small stellar surface fluctuations transported outwards are amplified in the dust-forming region, and therefore it is very likely that clumps are formed in the wind.

Some properties of the rotating dust-driven winds are summarized in Table 1 where the last three columns state the total mass loss rate $\dot{M}$, the ratio of equatorial to polar mass loss $\dot{M}_e/\dot{M}_p$, and the ratio of equatorial to polar final velocity $u_{\infty,e}/u_{\infty,p}$, respectively. It is evident that an increase in rotational velocity results in larger total mass-loss rates. This mass loss becomes more concentrated towards the equatorial plane and the angular difference in the outflow velocities also increases. The quantitative value clearly depends on the stellar parameters determining the effective gravity in the atmospheric layers as well as on the amount of carbon available to condense into grains. The two blocks of models (cf. Table 1) illustrate that the mass loss of the more extended model ($R = 540 R_\odot$) with a higher

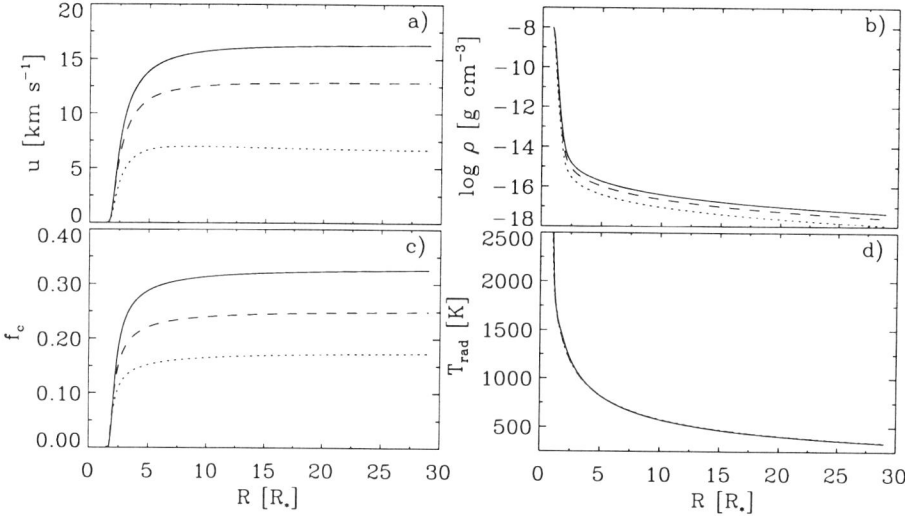

*Figure 1.* The radial structure of a rotating stationary dust-driven wind ($L = 1.2 \times 10^4$ $L_\odot$, $M = 1\,M_\odot$, $T_{\text{eff}} = 2600\,\text{K}$, $\omega = 1.5 \times 10^{-8}\,\text{s}^{-1}$) plotted for three different angles: equatorial wind (solid line), 45° (dashed line), and polar wind (dotted line); (*a*) the gas velocity in units of $\text{km}\,\text{s}^{-1}$ between 1 and $30\,R_*$; (*b*) the gas density; (*c*) the degree of condensation of carbon dust grains; (*d*) the radiation temperature.

luminosity of $L = 1.2 \times 10^4 L_\odot$ ($M = 1\,M_\odot$ and $T_{\text{eff}} = 2600\,\text{K}$) is more sensitive to the rotational velocity than the lower luminosity model $L = 10^4\,L_\odot$ corresponding to $R = 493\,R_\odot$. The stationary solutions obtained without rotation yield $\dot{M} = 2.9 \times 10^{-7} M_\odot\,\text{yr}^{-1}$ with $u_\infty = 12\,\text{km}\,\text{s}^{-1}$ and $\dot{M} = 3.8 \times 10^{-7}\,M_\odot\,\text{yr}^{-1}$ with $u_\infty = 9.5\,\text{km}\,\text{s}^{-1}$. In the stationary solutions discussed here the carbon-to-oxygen ratio $\varepsilon_C/\varepsilon_O$ has to be large enough to drive a purely dust-driven wind, whereas the time-dependent solutions can be obtained with much smaller values of $\varepsilon_C/\varepsilon_O$ (e.g. Höfner & Dorfi 1997). In the latter case the stellar pulsations deposit kinetic energy in the atmospheric layers, lifting more condensable material into the dust formation zone.

A typical spatial structure of a dust-driven wind generated by a rotating carbon-rich AGB star is plotted for different angles in Figure 1. The mass-loss rate, velocity, and degree of condensation $f_c$ increase towards the equatorial plane (solid lines). Since the radiation fields depicted by the radiation temperature $T_{\text{rad}}$ in Fig. 1*d* are almost identical, the angular wind variation is mainly caused by the difference in the gas density (cf. Fig. 1*b*) which can strongly amplify the dust condensation, controlling the outflow.

The mass loss as a function of the polar angle $\theta$ is plotted in Figure 2 where the lines correspond to a fit by $\dot{M}(\theta) = \dot{M}_p(1 + \varepsilon \sin^n \theta)$. Kahn and West (1985) have assumed such a dependence of the mass-loss rate

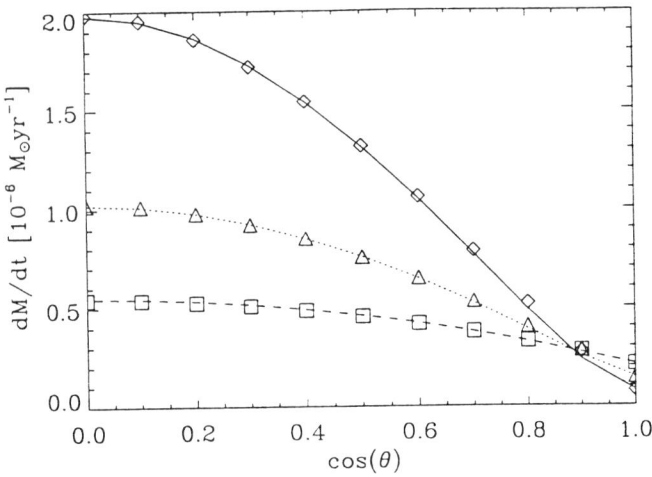

*Figure 2.* The mass-loss rates for different rotational velocities $\omega$ of an AGB star with $M = 1\,M_\odot$, $L = 10^4 L_\odot$ and $T_{\text{eff}} = 2600\,\text{K}$ (diamonds: $\omega = 2 \times 10^{-8}\,\text{s}^{-1}$, triangles: $\omega = 1.5 \times 10^{-8}\,\text{s}^{-1}$, squares: $\omega = 10^{-8}\,\text{s}^{-1}$) as a function of the polar angle $\theta$. The equatorial plane is denoted by $\cos\theta = 0$. The different lines are analytical fits according to Kahn & West (1985).

on the angle $\theta$ to explain the shaping of planetary nebulae in their two-wind model. The subsequently developing spherical wind of the white dwarf blows into the angle-dependent, slowly outflowing material of the former red giant. Our calculations yield values of $n \simeq 2 \ldots 3$ where $n = 2$ is valid for the slowest angular velocities, i.e. $\omega = 10^{-9}\,\text{s}^{-1}$. Most of the variation of the wind as a function of the rotation period is contained in the second parameter and given by $\varepsilon = \dot{M}_\text{e}/\dot{M}_\text{p} - 1$. So far we have not obtained a simple relation to estimate the polar mass loss rate $\dot{M}_\text{p}$, the exponent $n$ and the ratio $\dot{M}_\text{e}/\dot{M}_\text{p}$ as a function of the fundamental stellar parameters. The overall mass loss of a star evolving along the AGB will also depend on the angular momentum distribution inside the stellar mass zones which gradually become the stellar surface when the star loses more and more of its mass. Without having detailed stellar models of rotating AGB stars, it is therefore impossible to calculate the exact shaping of planetary nebulae based on such a two-wind model.

## 3. Discussion

The stellar evolution on the AGB (e.g. Iben & Renzini 1983; Lattanzio 1986) complicates this simplfied picture of a dust-driven wind in several ways. First, as the star produces its heavier elements by thermonuclear reactions in the interior, the chemical composition of the outer layers changes due to several dredge-up events, and accordingly the condensation process of dust particles yields different grains which are then accelerated by a luminosity which increases during the AGB phase. Second, in the case of faster rotations than, for example, taking the solar rotation rate and expanding the star to AGB radii while conserving angular momentum, we run into the problem that the mass-loss rate depends strongly on the polar angle. There exists a lower limit for dust-driven stationary winds (Gail & Sedlmayr 1987), and in the fastest rotating models of Table 1 the polar mass-loss rate of about $\dot{M}_p \simeq 5 \times 10^{-8} M_\odot \, \mathrm{yr}^{-1}$ is already close to this limit. In such cases the nature of the polar wind must be different and cannot be supported by the radiation pressure on newly formed dust grains. Third, the stationary solutions presented are possible only in a limited parameter range due to the so-called dust-induced $\kappa$–mechanism leading to non-stationary dust-driven outflows (Fleischer et al. 1995; Höfner et al. 1995). This instability is more likely to start in the denser equatorial regions, whereas the polar wind is still maintained by a stationary wind. Fourth, the overall flow structure can contain a number of clumps since the time scale for dust condensation is very sensitive to density fluctuations and varies for different polar angles. Furthermore, it is not clear how the angular momentum is redistributed in a rotating star having a preferential mass loss (and therefore clumps) in the equatorial region. If the rotation rate decreases as more and more atmospheric layers are blown away, the overall mass loss becomes more spherical and further surface fluctuations may allow the formation of clumps also at higher latitudes. Fifth, in calculating our dust-driven models we have assumed the most conservative case of keeping the angular momentum constant within the expanding flow. If weak magnetic fields permeate the stellar atmosphere and the dust formation zone, causing a slower decrease of the centrifugal acceleration than $a_\omega \propto r^{-3}$ by angular momentum transfer, the effect of preferential mass loss towards the equatorial plane is strongly enhanced. However, to quantify this statement the exact topology of the magnetic field has to be known.

## 4. Conclusions

On the basis of a dust-driven wind generated by a slowly rotating carbon-rich red giant (e.g. $T_{\mathrm{eff}} = 2600 \, \mathrm{K}$, $M = 1 M_\odot$, $L = 10^4 L_\odot$) with a photospheric rotational velocity of about $2 \, \mathrm{km \, s^{-1}}$, we obtain a total mass loss increased by at least a factor of 1.5 relative to the mass loss of the non-rotating star. The mass loss in the equatorial plane is about 2 times larger than in the polar direction and is accompanied by an increase of the terminal velocity in going from the pole towards the equator. Note that these changes are due to the non-linear modifications of the dust formation process introduced by small deviations from spherical symmetry. Such an asymmetric mass loss during the AGB phase could explain the bipolar appearance of many planetary nebulae according to interacting wind models (e.g. Kwok 1982; Kahn & West 1985). A density contrast of a factor of 2 can easily be generated by a very small stellar rotation and is already sufficient to produce a bipolar structure. Since this proposed new mechanism is by no means restricted to time-independent solutions, we are currently investigating time-dependent models of dust-driven winds of pulsating and slowly rotating AGB stars.

This work is supported by the Österreichischer Fonds zur Förderung der wissenschaftlichen Forschung (FWF) under project number S7305–AST.

## References

Barnbaum, C., Morris, M. & Kahane, C. 1995, *ApJ*, 450, 862
Dorfi, E. A. & Höfner, S. 1996, *A&A*, 313, 605
Fleischer, A. J., Gauger, A. & Sedlmayr, E. 1995, *A&A*, 297, 543
Gail, H.-P. & Sedlmayr, E. 1987, *A&A*, 177, 186
Höfner, S. & Dorfi, E.A. 1997, *A&A*, 319, 648
Höfner, S., Feuchtinger, M. U. & Dorfi, E.A. 1995, *A&A*, 297, 815
Iben, I. Jr. & Renzini, A. 1983, *Ann. Rev. Astron. Astrophys.*, 21, 271
Kahn, F.D. and West, K. A. 1985, *MNRAS*, 212, 837
Kippenhahn, R., Meyer-Hofmeister, E. & Thomas H. C. 1970, *A&A*, 5, 155
Kwok, S. 1982, *ApJ*, 258, 280
Lattanzio, J. C. 1986, *ApJ*, 311, 708
Slettebak, A. (ed.) 1970, IAU Coll. on *Stellar Rotation* (D. Reidel)
Tassoul, J.-L. 1978, *Theory of Rotating Stars* (Princeton Univ. Press)

## Discussion

**Cherchneff**: How do you describe the dust formation process in your model? Is it the same description at the pole and at the equator?

**Dorfi**: We take the dust formation and growth formalism developed by Gail & Sedlmayr and Gauger et al. (1988) to calculate the dust-driven winds. We used the same description for the polar and equatorial regions.

**Jorissen**: Would your model be able to account for the fast outflow ($\sim 200$ km s$^{-1}$) observed in the rotating ($v \sin i \approx 15$ km s$^{-1}$) carbon star V Hya?

**Dorfi**: In the case of V Hya, I expect from our model a very asymmetric wind driven by condensation of C grains. However it will be very difficult to drive the polar wind since the mass-loss rate goes down in the polar regions. Hence I think that a mechanism other than radiation pressure on C grains is responsible for this high-velocity outflow.

**Plez**: The fast wind of V Hya was first detected by T. Lloyd Evans through observation of Fe II emission lines. Long-slit spectroscopy of the K I 7699 Å line by D. L. Lambert and myself shows the bipolar structure of the fast component at about 150 km s$^{-1}$ (Plez & Lambert 1994, *ApJ*, 425, L101).

# NUCLEATING DUST IN CARBON–RICH AGB STARS

I. CHERCHNEFF
*Physics Department, UMIST*
*Manchester W60 1QD, United Kingdom*

**Abstract.** Chemical models of the inner circumstellar envelope of a typical carbon-rich AGB star are presented. The effect of pulsation-driven shocks on the gas close to the stellar photosphere is considered. The chemistry of dust condensation nuclei formation is described and applied to the gas layers close to the star. We derive formation yields for polycyclic aromatic hydrocarbon (PAH) species and their dimers and discuss their role in the condensation of dust.

## 1. Introduction

AGB stars are characterized by very extended circumstellar envelopes due to the onset of a strong stellar wind during the thermally-pulsating AGB phase. These envelopes are very rich in molecular species, as evidenced by IR, millimetre, and sub-millimetre observations. They are also fascinating astrophysical environments because the physical parameters of the gas flow change very quickly over the full extent of the wind. Therefore the chemical processes responsible for the formation of molecules are highly dependent on the distance to the star. Close to the photosphere, the gas temperature and density are high and molecular formation is governed by termolecular and bimolecular reactions. These regions are where dust grains condense. Further out in the wind we expect the gas and the solid phases to interact, and molecular formation triggered by surface chemistry may occur. Also, some molecular species may be depleted from the gas onto dust grains. At distances of many stellar radii (i.e. in the outer envelope) the physical conditions are similar to those encountered in molecular clouds, and the resulting chemical processes are governed mainly by neutral-neutral, ion-molecule, and photodissociation/ionization reactions.

The inner and outer envelopes are closely linked and it is therefore crucial to examine the various processes occurring in the inner region in order to assess the molecular input to the outer envelope and to confront the theoretical molecular abundances with those derived from IR observations. We present models of the carbon chemistry responsible for dust nucleation in the inner envelope of a typical carbon-rich AGB star.

## 2. A Description of the Inner Envelope

Many AGB stars are Long-Period Variables with typical pulsation periods of about a year. The pulsations result in the formation of shocks just outside the photosphere and these periodic shocks dramatically alter the layers of gas close to the star (Bowen 1988; Cherchneff & Tielens 1994). Indeed, energy is communicated to the gas by the shock and the gas heats up. The gas then cools via various microscopic processes (molecular dissociation, ionization, recombination, etc...) and via work, i.e. expansion (Willson & Bowen 1986). As a result, the gas layers close to the photosphere are accelerated upwards, decelerate under the influence of the stellar gravitational field, and eventually fall back toward their initial position. This cycle will repeat itself with the next pulsation and the gas layers will experience oscillatory motions with a period equal to the pulsation period of the star.

In order to investigate the formation of molecules in the inner envelope of a typical C-rich AGB star, we use the standard stellar model of Cherchneff et al. (1993) and study the chemistry occurring in the shocks and in the excursions experienced by the gas. In both cases, the chemistry can be decoupled from the hydrodynamics of the flow because the chemical species considered do not have, as a first approximation, a strong effect on the cooling of the gas and on the dynamics of the wind (Cherchneff et al. 1991; Allain et al. 1997).

### 2.1. THE SHOCKS

The pulsation-driven shocks in AGB stars are low-velocity (typically 20 km s$^{-1}$) molecular shocks. The cooling in the post-shock region is due mainly to the collisional dissociation of molecular hydrogen by atomic hydrogen (Fox & Wood 1985). Radiative processes are not efficient in cooling the post-shock gas because the ionization and excitation in the post-shock region are due to collisions with H atoms and the resultant emission intensities are very low. Therefore we describe the shock profiles assuming the Rankine-Hugoniot jump conditions in the shock front and we consider only H$_2$ dissociation as the main cooling process of the post-shock gas. We investigate several shock velocities ranging from 20 to 13 km s$^{-1}$. These values correspond to a 20 km s$^{-1}$ photospheric shock whose strength is damped as the shock travels through the envelope (Cherchneff & Tielens 1996).

## 2.2. THE EXCURSIONS

The gas excursions are modelled following a semi-analytical description derived by Bertschinger & Chevalier (1985). We consider strictly periodic motions, i.e. the parcel of gas accelerated by the shock is moving upwards but falls back to its initial position. Such an assumption is adequate for the regions close to the photosphere where gravity has a strong effect. As the dust formation zone in C-rich AGB stars is believed to correspond to a narrow region at a few stellar radii from the star (Danchi & Bester 1995), the approximation of periodic motions can apply.

## 3. The Chemistry of PAHs and PAH Dimers

C-rich AGB stars emit a strong infrared excess due to the presence of dust grains in their extended winds. This excess is well fitted with a mixture of amorphous carbon (AC) and silicon carbide (SiC) particles (Groenewegen 1995) and we investigate the formation processes of AC grains in the inner, shocked regions of AGB winds. The gas in the inner envelope of C-rich AGB stars is molecular and is rich in $H_2$, H, He, CO, and $C_2H_2$. A large fraction of the available carbon is locked in CO which survives the hot environment of the inner regions due to its high binding energy. The remaining carbon is in the form of acetylene ($C_2H_2$), and this species will react with atomic hydrogen to start an active hydrocarbon chemistry. Such a chemistry is found on Earth in flames rich in acetylene and in many combustion processes involving $C_2H_2$, and we use a chemical scheme based on combustion chemistry to describe the formation of hydrocarbon molecules. As for acetylenic flames, we expect polycyclic aromatic hydrocarbons (PAHs) to form as important intermediates in the dust condensation process (Frenklach & Feigelson 1989; Cherchneff et al. 1993). Therefore we describe the nucleation of AC grains by forming large PAHs up to coronene ($C_{24}H_{12}$), and we study the first step of dust condensation by forming PAH dimers, i.e. non-planar molecules.

The nucleation steps to large PAHs have been described previously (Cherchneff et al. 1993) and consist of various chemical reactions involving hydrocarbons leading to the closure of the first aromatic ring to form phenyl ($C_6H_5$). The subsequent growth of aromatic rings to form PAHs can be described by a set of three reactions: addition of $C_2H_2$, H abstraction to form a radical, and a second addition of acetylene followed by ring closure. These three reactions are important because they determine a gas temperature window of 900 − 1100 K in which PAH growth and the formation of large dust precursors is possible (Frenklach & Wang 1991). Therefore, PAH molecules will be able to form in regions where this temperature window is found. To help meet this requirement, we apply the inverse greenhouse effect on small PAHs (Cherchneff et al. 1991).

We consider the formation of PAH dimers up to $C_{24}H_{11}C_{24}H_{11}$. The reaction rates for dimer formation from their PAH parents are not available and we base our analysis on the reaction of benzene with phenyl

$$C_6H_6 + C_6H_5 \to C_6H_5C_6H_5 + H$$

for which a rate has been measured by Fahr et al. (1988). We have $k_1 = 5 \times 10^{-13} \exp(-15.7/RT)$ (in cm$^3$ mol$^{-1}$ s; kJ mol$^{-1}$ where R is the perfect gas constant and T the gas temperature). The rates for the formation of PAH dimers larger than biphenyl should be greater than $k_1$ because the van der Waals force between molecules increases with their size. This force can then act as a temporary glue which could hold the adduct together while the chemical bond forms. Miller et al. (1984) have estimated the van der Waals forces for several PAHs and found that the ratio between the force and the number of carbon atoms of the PAH reactants is roughly constant. We use this result to estimate the van der Waals forces for the PAHs involved in this calculation and scale the reaction rates with respect to the force values.

It is important to mention that no destruction processes have been considered for dimers in th present calculations, and the formation yields presented in § 4 are then upper limits.

## 4. PAH Dimer Formation Yields

The formation yield of PAHs and PAH dimers is defined as the total number of carbon atoms in PAHs and PAH dimers divided by the total number of carbon atoms initially in the form of hydrocarbons. The yield was calculated for the cycles of shock+excursion described in § 2. Each shock strength corresponds to a certain position in the inner envelope and the correspondence is given in Table 1 along with relevant pre-shock gas parameters.

TABLE 1. Shock Strengths and Pre-shock Gas Parameters

| Radius ($R_\star$) | Shock Strength (km s$^{-1}$) | Mach Number (K) | Gas Temperature (K) | Gas Concentration (cm$^{-3}$) |
|---|---|---|---|---|
| 1.3 | 20.0 | 6.2 | 1965 | $1.70 \times 10^{13}$ |
| 1.9 | 16.5 | 5.7 | 1565 | $9.84 \times 10^{11}$ |
| 2.5 | 14.4 | 5.4 | 1327 | $1.55 \times 10^{11}$ |
| 3.1 | 13.0 | 5.2 | 1167 | $4.18 \times 10^{10}$ |

TABLE 2. PAH and PAH Dimer Formation Yields

| Radius ($R_\star$) | Shock Strength (km s$^1$) | PAH Yield | PAH Dimer Yield |
|---|---|---|---|
| 1.3 | 20.0 | 0 | 0 |
| 1.9 | 16.5 | $3\times10^{-5}$ | $10^{-8}$ |
| 2.5 | 14.4 | $7\times10^{-5}$ | $2\times10^{-8}$ |
| 3.1 | 13.0 | $7\times10^{-5}$ | $2\times10^{-8}$ |

Preliminary results are summarized in Table 2. In the 20 km s$^{-1}$ shock and in the excursion following, no PAH/PAH dimers can form because the 900 – 1100 K temperature window necessary to the growth of PAHs cannot be reached in the flow. PAH formation starts further out at radii of $\sim 2$ $R_\star$, and the yields range between $10^{-5}$ and $10^{-4}$ at r $\geq$ 2 $R_\star$. The yields for the dimers follow the same trend as for PAHs but are three orders of magnitude smaller.

We have not considered the destruction of PAH dimers, although these compounds will undoubtedly be destroyed in the shocks. However, the results show that what happens in the shocks does not have an impact on what happens in the excursions following. The gas has no time to reach equilibrium in the post-shock region but does so at the beginning of each excursion because the time scales are much longer. Therefore, we expect the dimers to be destroyed in the shocks but to reform in the excursions and to be gradually expelled in the flow.

Dust contents of C-rich AGB envelopes have been derived by Knapp (1985) using long-wavelength infrared excesses and assumed models for the dust. Knapp obtained an average dust-to-gas ratio of $5.2 \times 10^{-13}$. Then assuming a population of dust grains with an average radius of $a = 500$ Å, the ratio of dust grain number density to acetylene number density is $\sim 5 \times 10^{-9}$. This value is in very good agreement with the PAH dimer formation yields found in our calculations and supports the fact that PAH dimers can be considered as condensation nuclei in the dust condensation process.

# References

Allain, T., Sedlmayr, E. & Leach, S. 1997, *A&A*, 323, 163
Bertschinger, E. & Chevalier, R. A. 1985, *ApJ*, 299, 167
Bowen, G. H. 1988, *ApJ*, 329, 299
Cherchneff, I., Barker, J. R. & Tielens, A.G.G.M. 1991, *ApJ*, 377, 541
Cherchneff, I., Barker, J. R. & Tielens, A.G.G.M. 1993, *ApJ*, 413, 445

Cherchneff, I. & Tielens, A.G.G.M. 1994, in *Circumstellar Media in the Late Stages of Stellar Evolution*, ed. R. E. S. Clegg, I. R. Stevens and W. P. S. Meikle (Cambridge Univ. Press), p. 232
Cherchneff, I. & Tielens, A.G.G.M. 1996, in preparation
Danchi, W. C. & Bester, M. 1995, *Astrophys. Space Sci.*, 224, 339
Fahr, A., Mallard, W. G. & Stein, S. E. 1988, in *21st Symposium (International) on Combustion*, The Combustion Institute, Pittsburgh, p. 825
Fox, M. W. & Wood, P. R. 1985, *ApJ*, 297, 455
Frenklach, M. & Feigelson, E. D. 1989, *ApJ*, 341, 372
Frenklach, M. & Wang, H. 1991, in *23rd Symposium (International) on Combustion*, The Combustion Institute, Pittsburgh, p. 1559
Groenewegen, M. A. T. 1995, *A&A*, 293, 463
Knapp, G. R. 1985, *ApJ*, 293, 273
Miller, J. H., Mallard, W. G. & Smyth, K. C. 1984, *J. Phys. Chem.*, 88, 4963
Willson, L. A. & Bowen, G. H. 1986, in *Cool Stars, Stellar Systems, and the Sun*, ed. M. Zeilik and D. M. Gibson, Lecture Notes in Physics, vol. 254 (Springer-Verlag), p. 385

## Discussion

**Ake**: Hydrogen plays an important role in the nucleation process you describe. What happens in hydrogen-deficient carbon stars? For example, R CrB stars are known to generate dust.

**Cherchneff**: For H-deficient environments such as the winds of R CrB stars or Wolf-Rayet stars, dust grains form and the chemistry involves a pure carbon phase. We believe that small carbon chains are formed first ($C_2$, $C_3$ ... $C_{10}$ ...), followed by ring closure and formation of monocyclic rings. These rings later arrange in aromatic structures, large PAHs, fullerene and amorphous carbon grains. The resulting dust is the same as the AGB dust, but the chemical pathways to dust formation are different.

**Linsky**: Please comment on the role that UV photons could play in the formation or destruction of PAHs and PAH dimers.

**Cherchneff**: The UV stellar radiation field is assumed to be small. However, UV photons could have two effects on PAH, PAH-dimer chemistry. They could either destroy some species via photodissociation or trigger the PAH formation via ionchemistry. In combustion chemistry, there exist chemical schemes describing PAH and soot formation that involve ion-molecule reactions.

# GRAIN FORMATION IN THE WINDS OF COOL RED GIANT STARS

E. SEDLMAYR AND J. M. WINTERS
*Institut für Astronomie und Astrophysik*
*Technische Universität, Berlin, Germany*

**Abstract.** The problem of dust formation in the circumstellar envelopes of Asymptotic Giant Branch stars is reviewed. Special emphasis is put on the consistent modelling of the dust-forming circumstellar shell, where due to a strong coupling the dust formation process governs the dynamical behavior of the object.

## 1. Introduction

Cosmic dust plays an important role in current astrophysics, manifesting itself as an important component of the interstellar medium and dominating the appearance of various objects like red giants and supergiants, and even episodic events such as R CrB stars, novae, and supernovae. For this reason it seems natural that cosmic grain physics, in particular the problem of grain formation, has attracted much interest in recent decades. In particular, red giants and supergiants, showing pronounced stellar outflows, have proven to be ideal objects to study in detail the process of grain nucleation and growth.

From a microscopic point of view, grain formation can be conceived as a sequence of chemical reactions, starting with suitable molecules and finally ending with macroscopic dust particles. In this respect, the grain formation process itself seems to be controlled only by proper pressure and temperature conditions and a certain chemical composition of the gas phase providing suitable molecules to condense. So far the problem appears to be rather simple: Adopting some nucleation theory, one has first to determine regions of high nucleation efficiency in the $p-T$ plane (i.e. regions of sufficiently high particle densities and sufficiently low temperatures), and second to determine those trajectories along the evolution of a material element which cross the dust-forming region for a time sufficiently long to enable effective grain formation in the element. Natural systems for this process are stellar outflows, starting from the initial plasma phase in the stellar photosphere. With decreasing temperature a complex chemistry evolves and finally, at some critical radius, favorable conditions for dust nucleation and growth occur as indicated by the observed steep increase in the optical depth, the high infrared excess and a significant increase of the expansion velocity in the shells of these objects (cf. Figure 1). In this respect, dust formation in the winds of red giant stars is a physically and mathematically well posed problem, the initial conditions of which can be uniquely inferred from the atmospheric analysis.

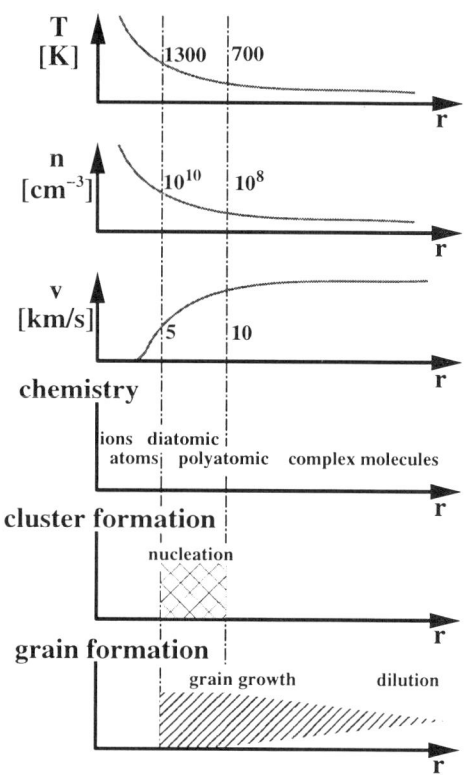

*Figure 1.* Schematic structure of a dust-forming circumstellar shell. Dust nucleation is well confined to a region approximately between 1300 K and 700 K while the region of dust growth extends much farther outward.

Despite this simple basic picture, astrophysical dust formation appears to be an extremely complicated non–linear process. The reason for this behavior is that the presence of even a small amount of dust, due to its huge absorption and scattering coefficient, has severe consequences for the local physical and chemical conditions in the dust-forming region. In particular the following three effects play an important role:

- *Dust is a very effective transmitter of energy* by efficiently absorbing and thermalizing short–wavelength photons and emitting their energy at infrared wavelengths. By this process, dust strongly affects the thermodynamical behavior of the system.
- *Dust is a very effective transmitter of momentum* by transferring the absorbed photon momentum by frictional coupling to the ambient gas. Therefore, dust strongly affects the hydrodynamical structure of the system.

- *Dust is a catalyst for chemical reactions:* gas particles absorbed at the grain's surface can react with the grain material or another admolecule to form new species which cannot be formed from the gaseous phase. Hence dust strongly determines the chemical behavior of the system.

As a consequence, any realistic treatment of circumstellar dust formation can only be performed by means of a self-consistent modelling of the dust-forming objects, with thermodynamics, chemistry, cluster physics, radiation/matter interaction, and nonlinear dynamics as essential ingredients (e.g. Sedlmayr & Winters 1991; Fleischer et al. 1992; Winters et al. 1994a).

## 2. Circumstellar Dust Formation

The consistent treatment of a dust-forming shell of a red giant requires the consideration of at least three closely coupled complexes: circumstellar chemistry, the dust nucleation and grain growth problem, and the hydrodynamical and thermodynamical wind structure.

### 2.1. CIRCUMSTELLAR CHEMISTRY

In principle, chemistry plays a three–fold role with regard to circumstellar shells:

*i*) for the interpretation of the observed molecular lines, i.e. the diagnostics of the objects, the detailed chemical structure and thus the local concentrations of the molecular species have to be known,

*ii*) the chemical species present determine the transport coefficients in the atmosphere and hence the local physics in the inner parts of the shells (e.g. temperature, density, ...), and

*iii*) the chemistry determines in particular the local abundance of the dust-forming species and, thereby, the primary (high–temperature) condensates to be expected.

To give a general picture of the complex processes to be involved in modelling the chemistry of the circumstellar shell, Figure 2 is a sketch of the various chemical processes which might be important.

#### 2.1.1. *Chemical equilibrium (CE)*

This most simple approach requires the chemical reaction timescales to be small compared to all other timescales determining the structure of the outflow (e.g. the hydrodynamical timescale). This condition only holds for sufficiently high particle densities, small velocity fields, and systems where photoreactions can be neglected, i.e. for radiation fields with a negligible UV component. This situation seems to be realized for the cool photospheres and inner shell regions of cool C stars and M stars without chromospheres. This CE condition is very severe and easily violated by the presence of even

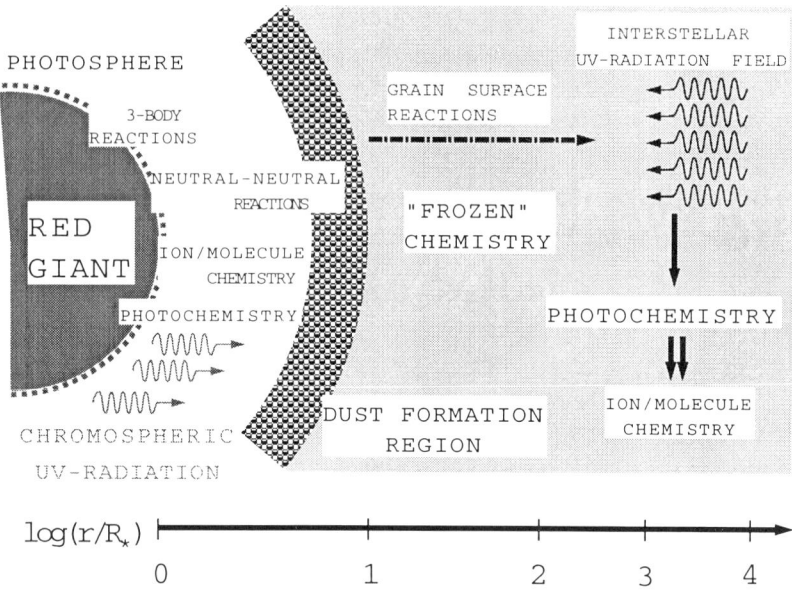

*Figure 2.* Chemical processes in different regions of the circumstellar shells around red giants (from Patzer 1998).

weak UV contributions and/or by moderate velocity fields (e.g. Goeres et al. 1988; Beck et al. 1992). Their model calculations for $\alpha$ Ori indicate dramatic deviations of particular species from CE caused by only a very moderate photospheric UV field ($T = 3000\,\mathrm{K}$) or by small temporal variations of the radiation field as observed for $\alpha$ Ori (Dupree et al. 1987).

### 2.1.2. *Chemical non–equilibrium*

This situation requires the solution of the time-dependent network of chemical rate equations (non–kinetic equilibrium) in the case that velocity fields and/or short time variations of the radiation field have to be considered. This is the usual case for the modelling of long-period variable stars (LPVs), where the dynamical timescales determine the local physics. Such a treatment is also necessary in the outer regions of stationary outflows, where the hydrodynamical expansion and acceleration timescale becomes comparable to or even shorter than the chemical timescales involved. For these systems often a frozen chemistry evolves due to the rapid decrease of the chemical reaction rates caused by dilution in the wind. As an example, Figure 3 shows a comparison between a pure CE and a non–CE calculation of the chemical structure for a stationary dust-driven wind model describing the dust shell of a C star.

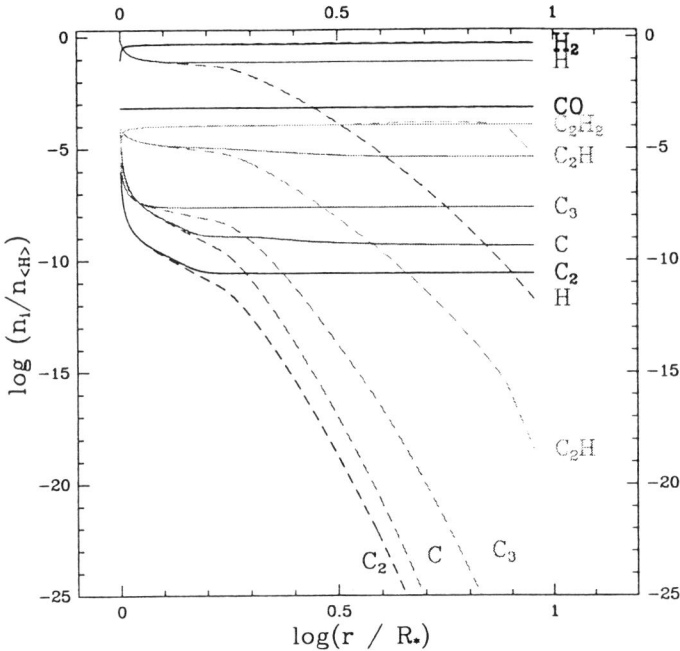

*Figure 3.* Equilibrium concentrations (dashed lines) and non–equilibrium concentrations (solid lines) for different molecules in a stationary model for a dust-driven wind. Due to fast dilution in the wind, a frozen chemistry develops in the non–CE calculation. The figure is from Patzer (1998).

## 2.2. DUST NUCLEATION AND GROWTH

Basically, dust formation can be thought of as a chain of chemical reactions starting with small molecules and ending with macroscopic specimens called dust grains. To follow this idea by a theoretical description requires, in principle, the construction and solution of a suitable chemical network, containing in the small regime molecules of increasing size, in the intermediate regime clusters (containing $\sim 10$ constituents) and in the large-particle regime grains having $\sim 10^{10}$ constituents. This approach, however, is much too elaborate not only due to the huge number of reaction equations (corresponding to the number of constituents), but also due to the lack of input data necessary to describe the reaction rates, structure, and thermodynamical properties of the particles involved. When applied to PAH formation in C-star outflows, this method turned out to be limited in practice to sizes of polyaromatic molecules having less than six aromatic rings (e.g. Cherchneff et al. 1992).

A detailed inspection of the dust formation process shows, however, that the problem can be split into two different phases which can be treated separately: (*i*) the nucleation process, i.e. the formation of so-called critical clusters out of the gas phase, and (*ii*) the growth of these critical clusters to macroscopic specimens. As all clusters smaller than the critical cluster have the tendency to evaporate and all clusters larger than the critical cluster tend to grow, the critical cluster plays the role of a 'bottleneck,' which any effective dust formation process has to surmount in order to arrive at macroscopic grains. The properties of the critical cluster, defined by an extremum of its Gibbs free enthalpy, are therefore decisive for the whole dust formation problem and its efficiency.

2.2.1. *Nucleation regime*

For the treatment of the first phase — the formation of the critical cluster — usually two different methods are adopted:

1) *Classical nucleation theory.* In this concept, the molecule–cluster transition is conceived as a thermodynamic phase transition governed by the thermodynamic functions of the emerging products and by the supersaturation ratio of the condensing gaseous species. Classical nucleation theory results in a critical cluster having maximum free enthalpy and in a critical nucleation rate, solely given by the thermodynamic properties of the critical cluster and the local supersaturation ratio of the condensing species which has to be larger than unity to allow the formation of the seed particle. This method for example yields excellent results if applied to the formation of water droplets in cloud chamber experiments or in the earth's atmosphere but is highly questionable in application to astrophysical problems. This is due to the fact that classical nucleation theory conceptually refers to critical clusters having already well defined surfaces (which is true for water droplets consisting of at least $10^9$ monomers) but not for astrophysical critical clusters usually having about 5–20 basic constituents. In this case, the surface tension concept cannot be applied and the thermodynamic functions of all individual small clusters involved either have to be provided by suitable laboratory experiments or calculated by molecular dynamics methods (e.g. Patzer et al. 1995, Köhler et al. 1997). So far, however, such an approach has been limited to chemically homogeneous clusters, whose composition and the corresponding molecular interaction potentials are known. A similar description of heterogeneous structures is still lacking. For this reason, all nucleation theories applied so far in astrophysics consider the formation of homogeneous critical clusters (Gail & Sedlmayr 1986; Kozasa & Hasegawa 1987; Kozasa et al. 1996; Köhler et al. 1997).

2) *Chemical pathway.* This approach has been applied to the formation of polyaromatic molecules in C stars (e.g. Keller 1987; Cherchneff et al.

1992; Goeres 1993; Goeres et al. 1996), polyaromatic carbons in R CrB stars (Goeres & Sedlmayr 1992), and fullerenes in the Krätschmer–Huffman experiment (Krätschmer et al. 1990; Goeres & Sedlmayr 1991). The basic idea of this method is that in the corresponding reaction networks a few dominant reaction pathways can always be isolated along which the grain formation process essentially proceeds. By a detailed discussion of the individual reactions a most efficient pathway can be constructed leading from the basic molecule (the monomer) into the regime beyond the critical cluster. On this path the critical cluster is defined as that step having the minimum net growth rate. Application of this method to PAH formation in C–star winds shows that the critical cluster is rather small at low temperatures (less than 20 C atoms for $T < 1000\,\mathrm{K}$; see Goeres et al. 1996), and therefore only few reaction steps have to be considered.

2.2.2. *Growth regime*
This particle size regime, which is essentially defined by the addition of growth species to the already existing clusters for both the classical nucleation theory and the pathway description, can be treated with sufficient accuracy by assuming thermal growth (Gail & Sedlmayr 1988; Gauger et al. 1990). This requires the knowledge of the corresponding reaction rates (i.e. the sticking coefficients) which in particular for inorganic grains are only poorly known or even completely lacking (e.g. Patzer et al. 1995).

## 3. Models of Dust–Forming Circumstellar Shells

In the context of circumstellar dust formation, three situations in particular have been modelled: (*i*) stationary dust-driven winds, (*ii*) shells of pulsating variables like LPVs, and (*iii*) episodic phenomena including R CrB stars, novae and supernovae.

3.1. STATIONARY DUST–DRIVEN WINDS

This phenomenon, which only occurs for objects showing negligible variability and extremely high mass-loss rates, is confined to the upper-right region of the HR diagram (cf. Dominik et al. 1990). The physical condition for a stationary dust-driven wind to exist is that the dust formation region is situated inside the sonic point of the wind in order for the radiation pressure on dust to be able to levitate the atmosphere beyond the escape point and to determine the mass-loss rate of the star (Gail & Sedlmayr 1987; Sedlmayr & Dominik 1995). A good example for such an object is IRC +10216, the self-consistent model of which (assuming spherical symmetry, radiative equilibrium and chemical equilibrium, and applying classical nucleation theory to describe the formation of amorphous carbon grains) provides not only the observed velocity field, the temperature structure, and the optical depths, but also the observed spectral appearance and the detailed shell structure leading to the observed fringe visibilities (Winters et al. 1994a).

### 3.2. PULSATING VARIABLES

Most red giants surrounded by a circumstellar dust shell are long-period variables (e.g. Jura 1986) with pulsation periods on the order of $P \approx 1$ yr. Modelling their circumstellar dust shells therefore requires a fully time–dependent treatment of the problem. The results of such calculations reveal a very complex structure of the circumstellar shell (e.g. a layered distribution of the dust component, dust-induced or dust-accelerated shocks) and the existence of an eigenperiod of the circumstellar shell leading to the effect of *multiperiodicity* which is caused by the very strong and highly non–linear coupling among hydrodynamics, thermodynamics, chemistry, and dust formation and growth (see Fleischer et al. 1991, 1992, 1995, 2000). These phenomena provide a physical explanation for the internal structures and the *multiperiodic* long–term behavior of the observed infrared lightcurves (e.g. Feast et al. 1984; Le Bertre 1992) and of the observed brightness profiles and corresponding fringe visibilities (e.g. Ridgway & Keady 1988; Danchi et al. 1994) of these objects as discussed in Winters et al. (1994b, 1995).

### 3.3. DUST FORMATION IN R CrB STARS

A third important astrophysical application of dust formation theories concerns the study of episodic phenomena like nova and supernova outflows and the R CrB occultation events. Whereas the first two cases can be treated simply by adopting a stationary flow, dust formation in R CrB environments turns out to be triggered by a critical phenomenon causing an instantaneous phase transition from molecules to carbon dust (cf. Goeres & Sedlmayr 1992). As R CrB even in the quiet state shows more or less periodic variations, the trigger mechanism might be some critical shock wave causing an extreme local increase in the temperature (up to $\sim 50\,000$ K), thereby providing an excellent initial condition for a subsequent cooling flow down to well below the radiative equilibrium temperature within only a few days (Woitke et al. 1996). This idea appears to be very appealing because it not only explains the observed large optical depths causing the obscuration but even seems to solve the long-standing debate between observers and theoreticians concerning the expected site of dust formation. Previous theories always claimed that dust formation takes place at a distance of about 10 stellar radii, where the temperature is low enough for grains to be able to form. This contradicts the observations which can only be explained if dust formation takes place very close to the star. This discrepancy could be resolved by the mechanism suggested above as it yields extreme supercooling and therefore high nucleation rates even at about 2 stellar radii (Woitke et al. 1996). However, a consistent modelling of this mechanism would require a 3-dimensional treatment of cloud formation which seems to be out of the scope of present-day capabilities. Moreover, atmospheres of R CrB stars seem to be close to their Eddington limit (Asplund 2000), suggesting sporadic "blob" outflows of matter which also could provide favorable conditions for dust formation along their cooling tracks.

## 4. Conclusions

The consistent modelling of red giants and supergiants, novae, and supernovae, including chemistry and dust formation, and the investigation of the obscuration phenomenon of R CrB stars, all yield results in excellent agreement with the observed spectral appearance and the timescales characterizing these objects. Nevertheless, all theories applied so far still suffer from rather restrictive assumptions either with regard to the microscopic physics (e.g. chemical equilibrium, classical nucleation theory, ...) or in order to reduce the numerical effort (e.g. spherical symmetry, ...). These limitations should be surmounted in future work in order to arrive at a realistic, consistent description and understanding of these interesting objects. This need is especially urgent in view of already existing and future high-precision instruments, which are collecting a wealth of very detailed observations, waiting to be interpreted on an equivalently high level.

We thank Beate Patzer and Axel Fleischer for useful comments on the manuscript. This work was supported by the BMBF (grant 05 3BT13A 6) and by the DFG (grant Se 420/8-1).

## References

Asplund, M. 2000, in IAU Symp. 177: *The Carbon Star Phenomenon*, ed. R. F. Wing (Kluwer), p. 521
Beck, H. K. B., Gail, H.-P., Henkel, R. & Sedlmayr, E. 1992, *A&A*, 265, 626
Cherchneff, I., Barker, J. R. & Tielens, A. G. G. M. 1992, *ApJ*, 401, 269
Danchi, W. C., Bester, M., Degiacomi, C. G., Greenhill, L. J. & Townes, C. H. 1994, *AJ*, 107, 1469
Dominik, C., Gail, H.-P., Sedlmayr, E. & Winters, J. M. 1990, *A&A*, 240, 365
Dupree, A. K., Baliunas, S. L., Guinan, E. F., Hartmann, L., Nassiopoulos, G. E. & Sonneborn, G. 1987, *ApJ*, 317, L85
Feast, M. W., Whitelock, P. A., Catchpole, R. M., Roberts, G. & Overbeek, M. D. 1984, *MNRAS*, 211, 331
Fleischer, A. J., Gauger, A. & Sedlmayr, E. 1991, *A&A*, 242, L1
Fleischer, A. J., Gauger, A. & Sedlmayr, E. 1992, *A&A*, 266, 321

Fleischer, A. J., Gauger, A. & Sedlmayr, E. 1995, A&A, 297, 543
Fleischer, A. J., Winters, J. M. & Sedlmayr, E. 2000, in IAU Symp. 177: *The Carbon Star Phenomenon*, ed. R. F. Wing (Kluwer), p. 377
Gail, H.-P. & Sedlmayr, E. 1986, A&A, 166, 225
Gail, H.-P. & Sedlmayr, E. 1987, A&A, 177, 186
Gail, H.-P. & Sedlmayr, E. 1988, A&A, 206, 153
Gauger, A., Gail, H.-P. & Sedlmayr, E. 1990, A&A, 235, 345
Goeres, A. 1993, *Rev. Mod. Astron.*, 6, 165
Goeres, A., Henkel, R., Sedlmayr, E. & Gail, H.-P. 1988, *Rev. Mod. Astron.*, 1, 231
Goeres, A., Keller, R., Sedlmayr, E. & Gail, H.-P. 1996, *Polycyclic Aromatic Compounds*, 8, 129
Goeres, A. & Sedlmayr, E. 1991, *Chem. Phys. Lett.*, 184, 310
Goeres, A. & Sedlmayr, E. 1992, A&A, 265, 216
Jura, M. 1986, *ApJ*, 309, 732
Keller, R. 1987, in *Polycyclic Aromatic Hydrocarbons and Astrophysics*, ed. A. Léger, L. d'Hendecourt and N. Boccara (Reidel), p. 387
Köhler, T. M., Gail, H.-P. & Sedlmayr, E. 1997, A&A, 320, 553
Kozasa, T., Dorschner, J., Henning, T. & Stognienko, R. 1996, A&A, 307, 551
Kozasa, T. & Hasegawa, H. 1987, *Prog. Theor. Phys.*, 77, 1402
Krätschmer, W., Fostiropoulos, K. & Huffman, D. R. 1990, *Chem. Phys. Lett.*, 170, 167
Le Bertre, T. 1992, *A&A Supp.*, 94, 377
Patzer, B. 1998, Ph.D. Dissertation, Technische Universität, Berlin, Germany
Patzer, A. B. C., Köhler, T. M. & Sedlmayr, E. 1995, *Planet. & Space Sci.*, 43, 1233
Ridgway, S. T. & Keady, J. J. 1988, *ApJ*, 326, 843
Sedlmayr, E. & Dominik, C. 1995, *Space Sci. Rev.*, 73, 211
Sedlmayr, E. & Winters, J. M. 1991, in *Stellar Atmospheres: Beyond Classical Models*, ed. L. Crivellari, I. Hubeny and D. G. Hummer (Kluwer), p. 397
Winters, J. M., Dominik, C. & Sedlmayr, E. 1994a, A&A, 288, 255
Winters, J. M., Fleischer, A. J., Gauger, A. & Sedlmayr, E. 1994b, A&A, 290, 623
Winters, J. M., Fleischer, A. J., Gauger, A. & Sedlmayr, E. 1995, A&A, 302, 483
Woitke, P., Goeres, A. & Sedlmayr, E. 1996, A&A, 313, 217

## Discussion

**Linsky**: Could you mention what critical laboratory or theoretical cross-sections and rates are needed to make major progress on this topic?

**Sedlmayr**: There is generally a need for cross-sections and thermodynamic data, in particular for calculating the silicon and sulfur chemistry in circumstellar envelopes. Also many neutral-neutral and three-body reactions — necessary for SiC–nucleation, for example — are unknown. Of great interest also are unimolecular reactions, like carbon chain to ring transitions, which play a role in the carbon grain formation in R CrB atmospheres.

**Gustafsson**: I do agree with your view that one should strive for detailed self-consistent theoretical modelling and avoid semi-empirical models with dubious or even meaningless free fitting parameters. However, observations now accumulate indicating the presence of inhomogeneities in circumstellar envelopes, and self-consistent models of these are probably still far off. For these phenomena we need to develop a physically reasonably sound semi-empirical approach.

**Sedlmayr**: This is just the point. On the one hand we have a great many excellent observations revealing both the overall appearance and also the whole variety of details. On the other hand we have the theoretical description based on our incomplete understanding of the relevant physical processes. The only way to connect the observations and theory is to develop consistent theoretical (yet highly simplified) models and hope to theoretically reproduce the structures observed. In this way an iteration procedure between theory and observation is defined which finally should lead to a real physical description and understanding of the observed phenomena.

# CARBON STAR DUST FROM METEORITES

UFFE GRÅE JØRGENSEN AND ANJA C. ANDERSEN
*Niels Bohr Institute, Copenhagen University Observatory*
*Copenhagen, Denmark*

**Abstract.** Inside carbonaceous chondrite meteorites are tiny dust particles which, when heated, release noble gases with an isotopic composition different from what is found anywhere else in the solar system. For this reason it is believed that these grains are (inter)stellar dust which survived the collapse of the interstellar cloud that became the solar system. We will describe here why we believe that the most abundant of these grains, microdiamonds, were formed in the atmospheres of carbon stars, and explain how this theory can be tested observationally.

## 1. Introduction

The discovery of meteoritic dust grains with origin outside the solar system has opened the possibility of studying presolar material in the laboratory, with all the advantages in details and accuracy such analyses allow for. Identification of the (possible) stellar origin of the meteoritic grains offers a unique opportunity to add important new constrains on models of stellar evolution (detailed elemental and isotopic abundances) and stellar atmospheric structure (elemental and mineralogical composition of the grains). There are several indications that a fraction, possibly the bulk, of the presolar meteoritic grains has its origin in carbon stars.

For the moment the amount of detailed information (such as isotopic ratios of tiny noble gas impurities) about the meteoritic grains is overwhelming (see e.g. Zinner 1995), whereas much of the fundamental data necessary in order to apply the meteoritic results to stellar modelling is entirely missing. For example, the necessary rate coefficients for formation of the most abundant presolar meteoritic grains (diamond dust) are lacking, and stellar wind models therefore do not predict diamond formation, but instead such models predict amorphous carbon (which has not been identified in meteorites) as the most abundant grain type in carbon-rich environments.

A combined self-consistent description of the full atmospheric region of a red giant star does not yet exist, but is slowly coming within reach. The meteoritic data combined with more fundamental laboratory data can be an important ingredient in constructing such a model for the first time. Successful construction of self-consistent models, followed possibly by verification of the formation place(s) of the most abundant stellar grains that contributed to the formation of the meteorites (and hence also the planets), would provide fundamental new knowledge about the sources of material for the solar system and the chemical evolution of the Galaxy.

The most common types of meteorites are fragments of larger protoplanetary bodies, which melted and chemically differentiated after their formation. Carbonaceous chondrites, on the other hand, are meteorites which have never been part of a larger body. They consist of spherical glass-like chondrules embedded in a fine-grained matrix. The matrix has had a gentle thermal history and is believed to be the (relatively unprocessed) original dust from which the planets formed. Therefore, the larger the amount of matrix in the chondrite, relative to the chondrule material, the more original solar nebula material is present, and the more primitive the chondrite is said to be.

When this matrix material was heated in the laboratory, it was realized already in the early 1960's (see Lewis & Anders 1983 for a review) that at certain temperatures the matrix released noble gases with an isotopic composition markedly different from everything else in the solar system. It was therefore concluded that the matrix contains one or more types of grains, formed before the solar system, in which non-solar-composition noble gases are trapped. After years of trials with different chemical purifications of the matrix material, and subsequent stepwise heating and isotopic noble gas measurements, the first presolar grains were finally isolated by Lewis et al. in 1987, and were identified as tiny diamonds.

Diamonds account for more than 99% of the identified presolar meteoritic material, with an abundance that can exceed 0.1% (1000 ppm) of the matrix (Huss & Lewis 1994b), corresponding to more than 3% of the total amount of carbon in the meteorite. The second and third most abundant types are SiC (6 ppm) and graphite (less than 1 ppm). They are all chemically quite resistant, which makes it possible to isolate them by dissolving the meteorite in acids. Further, a few of the SiC and graphite grains have been shown to contain tiny sub-grains of titanium and refractory carbides (Bernatowicz et al. 1991, 1992, 1994). Three isotopically anomalous, non-carbon-bearing grains have also been found. They are corundum ($Al_2O_3$), spinel ($MgAl_2O_4$), and silicon nitride ($Si_3N_4$) (Russel et al. 1995, Nittler et al. 1994, 1995). In the following sections we will discuss diamonds, SiC and graphite in some detail.

## 2. Diamonds

The individual diamond grains are very small, with a median diameter of less than 20 Å (Fraundorf et al. 1989). Since the diamond lattice distance is about 2 Å, a typical presolar diamond contains of the order $(\frac{20}{2})^3 = 1000$ carbon atoms, with $6\times10^2 \approx 50\%$ of these belonging to the surface. Since surface atoms have one unpaired bond, they will (in a hydrogen-rich atmosphere) resemble hydrogenated amorphous carbon (a-C:H). Only the $\sim 50\%$ "interior" atoms will sit in an actual diamond crystal structure. The presolar diamonds are therefore often called amorphous diamonds.

It is not obvious to what degree the extracted diamonds resemble the original diamond dust at its place of origin. Many alterations could have occurred in interstellar space or in the solar nebula, as well as during the chemical extraction process in the laboratory. However, the first step in an observational identification of their astronomical source of origin might be to model their synthetic spectrum. For this purpose we have measured the monochromatic absorption coefficient, described which of the features can be expected to be intrinsic to the diamonds (and which might be artifacts from the chemical processing in the laboratory), and computed synthetic carbon star spectra with the diamonds included (Andersen et al. 1998). The features which are most likely to be intrinsic are listed in Table 1 and compared with the results obtained by other groups.

As for the other grains, the strongest argument that the diamonds are formed outside the solar system is the peculiar, non-solar isotopic composition of their noble gas inclusions (and other trace element inclusions). There are several reasons why we believe the bulk of the diamonds formed in carbon stars, one of the most important being their $^{12}C/^{13}C$ ratio of $\sim 90$. This ratio is identical to what is observed in carbon stars with a large excess of carbon (i.e. with C/O $\gtrsim 1.5$, and strongly mass-losing), and it is not found in any other abundant astronomical objects. In contrast to this, SiC (the second most abundant presolar grain) has $^{12}C/^{13}C \approx 40$, which is typical (Lambert et al. 1986) for carbon stars with only small excesses of carbon (i.e. with C/O $\approx 1$). Hydrostatic MARCS photospheric models indicate that SiC grains will dominate the grain formation for C/O $\approx 1$ (where $^{12}C/^{13}C$ is as found in the meteoritic SiC grains), whereas pure carbon grains will dominate for the high C/O ratios (where $^{12}C/^{13}C$ is as in the meteoritic diamonds). This was the primary basis for our theory (Jørgensen 1988) that diamonds come from evolved carbon stars and SiC from less evolved carbon stars (actually, at the time the paper was written it was our prediction that SiC should exist in meteorites). The (radiative pressure driven) mass loss increases rapidly with increasing C/O (= increasing $^{12}C/^{13}C$) of the stars. If diamonds and SiC are formed in carbon stars, it is therefore a natural

TABLE 1. Spectral features, in cm$^{-1}$, detected in the spectra of presolar diamonds from the Allende, Murchison and Orgueil meteorites.

| ALLENDE | | | | MURCHISON | ORGUEIL | ASSIGNMENT |
|---|---|---|---|---|---|---|
| (1) | (2)$^a$ | (3) | (4) | (5) | (6)$^a$ | |
| | | 50 000 | | 50 000 | | paired N in diamond |
| | | 37 037 | | 37 037 | | paired N in diamond |
| 2919 | | 2954 | | 3000 | 2940 | aliphatic C–H stretch |
| 2849 | | 2854 | | 2800 | | |
| 1361 | | 1385 | 1399 | | 1380 | C–H deformation (CH$_3$) interstitial N |
| 1173 | 1143 | 1122 | 1084 | 1175 | 1042 | C–O/C–N stretch |
| 1028 | 1089 | 1054 | | 1090 | | CH$_2$ waging |
| | | 626 | | | 620 | CH out-of-plane |
| | | | | 396, 367 | | C=O=C or C=N=C |
| | | | | 310 | | |
| | | | | 130, 120 | | ?? |

(1) Lewis et al. 1989, (2) Koike et al. 1995, (3) Andersen et al. 1998,
(4) Lewis 1992, (5) Mutschke et al. 1996, (6) Wdowiak et al. 1988
$^a$ the spectra were obtained on diamond-like residues

consequence of this theory that the meteorites contain much more diamonds than SiC. A more quantitative simulation is still lacking because at present it isn't possible to include diamond formation in the model atmospheres (due to lack of basic input data).

A number of impurities have been identified in the presolar diamonds, including the noble gases (He, Ne, Ar, Kr, and Xe), Ba and Sr (which are slightly enriched in $r$-process isotopes; Lewis et al. 1991), H with $^1$H/$^2$D = 5193 (Virag et al. 1989; ($^1$H/$^2$D)$_{terrestial}$ = 6667) and N with $^{14}$N/$^{15}$N = 406 (Russel et al. 1991; ($^{14}$N/$^{15}$N)$_{terrestial}$ = 272). The most important of these, Xe, was actually known from stepwise heating techniques before the grains themselves were identified as diamonds. The Xe in the diamonds has a significant overabundance (compared to the solar isotopic ratios) of the very heavy isotopes (Xe-H, i.e. isotopes $^{134}$Xe and $^{136}$Xe) as well as the very light isotopes (Xe-L, i.e. $^{124}$Xe, $^{126}$Xe). This composition is often called Xe-HL to indicate that there is an excess of both heavy (H) and light (L) isotopes. There are no astronomical objects known (either from observation or from standard theories) which have both a solar $^{12}$C/$^{13}$C ratio and Xe-HL. An explanation therefore needs to involve either a non-standard model, not yet observationally verified, or an assumption of the diamonds being a mixture of populations from several different sources.

Heavy and light Xe isotopes are produced in supernovae (SN), and Clayton (1989) therefore proposed that the meteoritic diamond grains were formed in a supernova that also produced the Xe-HL measured in the diamonds. Since the progenitor star of a supernova in the standard theories has an oxygen-rich atmosphere (i.e., cannot produce carbon-rich grains) and a pure $^{12}$C interior shell (i.e., can only produce grains with $^{13}$C/$^{12}$C $\approx$ 0), a non-standard theory was necessary. In an extension of the model, Clayton et al. (1995) proposed a non-standard SN where mixing from a $^{13}$C-rich shell occurs in the right amount to give $^{12}$C/$^{13}$C $\approx$ 90. A non-standard $r$-process was assumed too, in order to avoid the production of $^{129}$I which decays to $^{129}$Xe and would therefore cause a very large excess of $^{129}$Xe, not observed in the meteorites. For a recent review of the standard $r$- and $s$- neutron capture processes, see Käppeler et al. (1989). Furthermore, the regular $r$-process cannot in itself produce the very large excess of $^{136}$Xe characteristic of the Xe-HL measured in presolar diamonds. Ott (1996), however, proposed that the standard $r$-process is active, but that a separation of xenon from iodine and tellurium precursors takes place in SN on a time scale of a few hours after termination of the neutron burst in the SN. Since $^{136}$Xe is formed minutes after the neutron burst, and the other $r$-process Xe isotopes are formed hours ($^{134}$Xe), days ($^{131}$Xe), or even years ($^{129}$Xe) later, a sufficiently early separation would allow almost infinite amounts of $^{136}$Xe relative to the other Xe isotopes which are produced. If a separation in the SN gas takes place two hours after the neutron burst, the meteoritic $^{136}$Xe/$^{134}$Xe ratio is established in the gas, and with a small amount of later mixing, the observed meteoritic Xe-H can be obtained.

Detailed supernova models supporting these isotopic arguments are lacking (as are simulations justifying, for example, the amount of $^{13}$C mixing or why only Xe from the separated gas is implanted in the grains when they form years after the neutron burst, etc.), but the success of fitting modified SN scenarios to the observed Xe-H makes it likely that part of the diamond grains originates in supernovae. However, there are several reasons why the bulk of the diamonds are unlikely to have formed in supernovae: (1) The hydrodynamical time scale of a supernova is short compared to the time scale for carbon grain formation (Sedlmayr 1994). (2) The mass loss is much stronger in carbon stars with high C/O ratios (where pure carbon grains will form) than in carbon stars with lower C/O ratios. If the SiC is formed in carbon stars (see next section), there will therefore have been expelled much more diamond dust from carbon stars (or other pure carbon grains, which are, however, not seen) into interstellar space than SiC. (3) Carbon stars were very abundant in the Galaxy prior to the formation of the solar system (due to their metallicity dependence). The resemblance of the carbon star $^{12}$C/$^{13}$C to the solar $^{12}$C/$^{13}$C is naturally explained if carbon stars were the source of the solar carbon (including carbon grains), whereas the standard SN not will produce this $^{12}C/^{13}C$ ratio.

The typical amount of Xe gas inclusion in the diamonds is $\sim 10^{-6}\,\text{cm}^3$ per gram of diamond. This corresponds to approximately one $^{132}$Xe atom and one $^{129}$Xe atom per $10^7$ diamonds, a bit less of the isotopes $^{131}$Xe, $^{134}$Xe, and $^{136}$Xe, one-tenth this amount of isotopes $^{128}$Xe and $^{130}$Xe, and only traces of $^{124}$Xe and $^{126}$Xe. A large number of diamonds is therefore necessary in order to perform an isotopic analysis (in practice $\sim 10^{10}$, Huss & Lewis 1994a), and attempts to separate diamonds into groups of different origin have so far not been successful (Huss & Lewis 1994b). If we assume that the trapping efficiency for Xe in the possible population of diamonds which originated in supernovae is sufficiently large compared to the trapping efficiency in the carbon star diamonds (perhaps because of the higher turbulent gas velocities in SN, higher densities, etc.), then the Xe-H can be explained as being connected with a small fraction of diamonds of pure $^{12}$C (which originated in SN), without altering the necessary bulk $^{12}C/^{13}C$ ratio of the carbon star diamonds. We therefore propose that the bulk of the presolar meteoritic diamonds originates in evolved carbon stars (as in our original theory) and are mixed with a smaller population (from SN II) which has a relatively high Xe content and is rich in the heavy isotopes. The content of light isotopes of Xe is very small (less than 1 atom per $10^9$ diamonds) and can be explained as coming from SN I (Lambert 1992) in binary systems where the low-mass component is an evolved carbon star as in our original theory (Jørgensen 1988), or as a by-product of the Xe production in SN II (Ott 1996).

## 3. Silicon Carbide

SiC is much less abundant (6 ppm) than diamonds (1000 ppm), but some of the SiC grains are large enough that isotopic ratios of Si and C (and the abundant impurities N, Mg-Al, Ti, Ca, He, Ne) can be measured in individual grains (Hoppe et al. 1994; Lewis et al. 1994; Anders & Zinner 1993, and references therein). The understanding of their stellar origin is therefore much better than in the case of the diamonds.

The SiC grain sizes have a large variety from less than 0.05 to 20 $\mu$m in equivalent spherical diameter, with about 95% (by mass) of the grains being between 0.3 and 3 $\mu$m (Amari et al. 1994). Ion micro-probe measurements can be performed on individual grains larger than 1 $\mu$m, and the results have made it possible to identify multiple stellar sources as their origin. The detailed match to the elemental and isotopic conditions in the He burning shell of AGB stars (Gallino et al. 1990) has made it generally believed that the bulk of the SiC originated in carbon stars.

In order to distinguish between various stellar origins, Hoppe et al.

(1994) have divided the coarse (2.1–5.9 $\mu$m) SiC grains into five subgroups:

1. The "mainstream" grains have $20 < {}^{12}C/{}^{13}C < 120$ and $200 < {}^{14}N/{}^{15}N < 10\,000$.
2. Grains A have ${}^{12}C/{}^{13}C < 3.5$.
3. Grains B have $3.5 < {}^{12}C/{}^{13}C < 10$.
4. Grains X have isotopically heavy N ($13 < {}^{14}N/{}^{15}N < 180$).
5. Grains Y have isotopically light C ($150 < {}^{12}C/{}^{13}C < 260$).

The mainstream, type A, and type B grains have comparable patterns of Si isotopes, distinctly different from type X and type Y grains. The mainstream grains constitute $\sim 94\,\%$ of all coarse-grained SiC, whereas grains from the groups A, B, X and Y account for only 2%, 2.5%, 1% and 1%, respectively. Based on this grouping, Amari et al. (1995a) find that grains X could originate from a supernova (SN II), and Lodders & Fegley (1995) find that grains A and B can be at least qualitatively understood if they originate from J-type carbon stars or carbon stars that have not experienced much dredge-up of He-shell material.

It is seen that the isotopic variations among the grains are very large. The ${}^{12}C/{}^{13}C$ ratio varies by a factor of more than 350, and ${}^{14}N/{}^{15}N$ varies by 300 times (and ${}^{30}Si/{}^{28}Si$ by a factor of 3). Variations in noble gases associated with SiC are large as well (Ott 1993). The two noble gas components, s-process Xe and neon-E (i.e., essentially pure ${}^{22}Ne$), show opposite correlations with grain size, s-process Xe being most abundant in fine-grained SiC and Ne-E in coarse-grained SiC (Lewis et al. 1994). Other elements which show s-process indications include s-process Kr, Ba, Sr, Ca, Ti, Nd and Sm.

The proportions of ${}^{80,86}Kr$ vary with release temperature of the gas. This variation reflects branching of the s-process at the radioactive progenitors, ${}^{79}Se$ and ${}^{85}Kr$ (Ott et al. 1988). These branchings depend sensitively on neutron density and temperatures in the s-process region, and the ${}^{80,86}Kr$ can therefore provide clues about in which stars the SiC formed, or if the stellar type is already known it can put constraints on the detailed modelling of these stars.

## 4. Graphite

The presolar graphite isolated from meteorites lies at the graphitic end of the continuum between kerogen, amorphous carbon, and graphite. It is not very abundant (less than 1 ppm), and it is much more complicated to extract than SiC and micro-diamonds (Amari et al. 1994). Presolar graphite occurs solely in the form of spherules, $0.8-8\mu$m in diameter, while graphite grains of other sizes and shapes have normal composition and are believed to have been formed in the solar nebula (Zinner et al. 1990). The presolar graphite has a very broad $^{12}C/^{13}C$ distribution, with $^{12}C/^{13}C$ ratios ranging from 7 to 4500, whereas the $^{14}N/^{15}N$ ratios range from 193 to 680 (Zinner et al. 1995).

The noble gases show systematic trends with sample density, suggesting more than one kind of graphite. Some have almost mono-isotopic $^{22}Ne$. Others contain neon with a somewhat higher $^{20}Ne/^{22}Ne$ ratio and are accompanied by s-process Kr, $^{4}He$, and other noble gases (Amari et al. 1995b).

The carbon and nitrogen isotopic ratios found in the grains indicate that they come from stellar sources dominated by H-burning rather than from sources dominated by He-burning (Amari et al. 1993). H-burning in the CNO cycle produces isotopically heavy carbon ($^{13}C$) and light nitrogen ($^{14}N$), in qualitative agreement with the measurements (Zinner et al. 1989; Hoppe et al. 1994). Systematic measurements of isotopic ratios of several other elements were recently done by Hoppe et al. (1995).

## References

Amari, S., Hoppe, P., Zinner, E. & Lewis, R.S. 1993, *Nature*, 365, 806
Amari, S., Hoppe, P., Zinner, E. & Lewis, R.S. 1995a, *Meteoritics*, 30, 679
Amari, S., Lewis, R.S. & Anders, E. 1994, *Geochim. Cosmochim. Acta*, 58, 459
Amari, S., Lewis, R.S. & Anders, E. 1995b, *Geochim. Cosmochim. Acta*, 59, 1411
Anders, E. & Zinner, E. 1993, *Meteoritics*, 28, 490
Andersen, A.C., Jørgensen, U.G., Nicolaisen, F.M., Sørensen, P.G. & Glejbøl, K. 1998, *A&A*, 330, 1080
Bernatowicz, T.J., Amari, S. & Lewis, R.S. 1992, *Lunar Planet. Sci.*, 23, 91
Bernatowicz, T.J., Amari, S. & Lewis, R.S. 1994, *Lunar Planet. Sci.*, 25, 103
Bernatowicz, T.J., Amari, S., Zinner, E. & Lewis, R.S. 1991, *ApJ*, 373, L73
Clayton, D.D. 1989, *ApJ*, 340, 613
Clayton, D.D., Mayer, B.S., Sanderson, C.I., Russel, S.S. & Phillinger, C.T. 1995, *ApJ*, 447, 894
Fraundorf, P., Fraundorf, G., Bernatowicz, T., Lewis, R.S. & Tang, M. 1989, *Ultramicroscopy*, 27, 401
Gallino, R., Busso, M., Picchio, G. & Raitari, C.M. 1990, *Nature*, 348, 298
Hoppe, P., Amari, S., Zinner, E., Ireland, T. & Lewis, R.S. 1994, *ApJ*, 430, 870
Hoppe, P., Amari, S., Zinner, E. & Lewis, R.S. 1995, *Geochim. Cosmochim. Acta*, 59, 4029
Huss, G.R. & Lewis, R.S. 1994a, *Meteoritics*, 29, 791
Huss, G.R. & Lewis, R.S. 1994b, *Meteoritics*, 29, 811
Jørgensen, U.G. 1988, *Nature*, 332, 702
Käppeler, F., Beer, H. & Wisshak, K. 1989, *Reports on Progress in Physics*, 52, 945
Koike, C., Wickramasinghe, C., Kano, N., Yamakoshi, K., Yamanoto, T., Kaito, C., Kimura, S. & Okuda, H. 1995, *MNRAS*, 277, 986
Lambert, D.L. 1992 *Astron. Astrophys. Rev.*, 3, 201
Lambert, D.L., Gustafsson, B., Eriksson, K. & Hinkle, K.H. 1986, *ApJ Supp.*, 62, 373
Lewis, R.S. 1992, published in Colangeli, L., Mennella, V., Stephens, J.R. & Bussoletti, E. 1994, *A&A*, 284, 583
Lewis, R.S., Amari, S. & Anders, E. 1994, *Geochim. Cosmochim. Acta*, 58, 471
Lewis, R.S. & Anders, E. 1983, *Sci. Am.*, 549, 54
Lewis, R.S., Anders, E. & Draine, B.T. 1989, *Nature* 339, 117
Lewis, R.S., Huss, G.R. & Lugmair, G.W. 1991, *Lunar Planet. Sci.*, 22, 807

Lewis, R. S., Tang, M., Wacker, J. F., Anders, E. & Steel, E. 1987, *Nature*, 326, 160
Lodders, K. & Fegley, B. Jr. 1995, *Meteoritics*, 30, 661
Mutschke, H., Corschner, J., Henning, Th. & Jäger, C. 1995, *ApJ*, 454, L157
Nittler, L. R., Alexander, C. M. O'D, Gao, X., Walker, R. M. & Zinner, E. 1994, *Nature*, 370, 443
Nittler, L. R. et al. 1995, *ApJ*, 453, L25
Ott, U. 1993, in *Protostars and Planets III*, ed. E. H. Levy and J. I. Lunine (Univ. Arizona Press), p. 883
Ott, U. 1996, *ApJ*, 463, 344
Ott, U., Begemann, F., Yang, J. & Epstein, S. 1988, *Nature*, 332, 700
Russel, S. S., Arden, J. W. & Pillinger, C. T. 1991, *Science*, 254, 1188
Russel, S. S., Lee, M. R., Arden, J. W. & Pillinger, C. T. 1995, *Meteoritics*, 30, 399
Sedlmayr, E. 1994, in IAU Coll. 146: *Molecules in the Stellar Environment*, ed. U. G. Jørgensen (Springer), p. 163
Virag, A., Zinner, E., Lewis, R. S. & Tang, M. 1989, *Lunar Planet. Sci.*, 20, 1158
Wdowiak, T. J., Flickinger, G. C. & Cronin, J. R. 1988, *ApJ*, 328, L75
Zinner, E. 1995, in *Nuclei in the Cosmos III*, ed. M. Busso, R. Gallino and C. M. Raiteri, AIP Conf. Proc., 327, p. 567
Zinner, E., Amari, S., Wopenka, B. & Lewis, R. S. 1995, *Meteoritics*, 30, 209
Zinner, E., Tang, M. & Anders, E. 1989, *Geochim. Cosmochim. Acta*, 53, 3273
Zinner, E., Wopenka, B., Amari, S. & Anders, E. 1990, *Lunar Planet. Sci.*, 21, 1379

# Session VIII

## Circumstellar Shells

Participants reclining on the 1850-year-old stone seats of the theater at Aspendos, watching nothing in particular.

# CIRCUMSTELLAR DUST AROUND M, S AND C STARS

IRENE R. LITTLE–MARENIN
*Wellesley College*
*Wellesley, MA 02181, U.S.A.*

**Abstract.** The circumstellar shells of M stars produce emission features due to amorphous silicates peaking around 10 and 18 $\mu$m, with additional emission at 11 $\mu$m due to crystalline olivine and at 13.1 $\mu$m (unknown carrier). C stars are associated with SiC dust emission at 11.2 $\mu$m with additional emission around 8.9 $\mu$m, whereas S stars have a relatively weak 10.5 $\mu$m emission feature which is due neither to silicates nor to SiC.

## 1. Introduction

With the advent of IRAS a new episode of infrared astronomy began with the identification of hundreds of thousands of infrared sources. Thousands of these sources were bright enough (>2 Jansky) to have their spectra recorded in the 8–22 $\mu$m region with the Low Resolution Spectrometer (LRS). This wavelength region is rich with different types of dust emission features. The existence of dust in circumstellar envelopes has been known for many decades. Stars on the AGB eject copious quantities of gas into space, part of which will condense into solid grains after the gas has cooled to T < 1500 K. The type of grain that is produced depends on the composition of the out-flowing material. In the oxygen-rich shells (C/O < 1) that are associated with M stars, the oxygen remaining after the formation of CO will condense into oxygen-rich solids such as amorphous silicates with emission features around 10 and 18 $\mu$m. On the other hand, in the carbon-rich shells (C/O > 1) associated with C stars, the extra carbon will condense into SiC grains with an emission feature around 11.2 $\mu$m.

IRAS was very good in characterizing large numbers of LRS by their major emission and absorption features. However, noisy spectra at times were assigned incorrect characterizations and previously unknown features could not be accommodated by the classification scheme.

## 2. The M stars

In 1990, Little-Marenin and Little (LML90) analyzed IRAS LRS spectra of Mira variables and showed that the 10 $\mu$m silicate emission feature (called class Sil) showed additional dust components at 11 $\mu$m (crystalline olivine; class Sil+) and at 13.1 $\mu$m (class 3C). The carrier of the 13.1 $\mu$m feature has not yet been unambiguously identified, but it does not appear to match the characteristics of corundum (Sloan et al. 1996). The strength of the 11 $\mu$m feature at times rivaled or surpassed the strength of the 10 $\mu$m feature (class Sil++). A weak, broad feature extending from 9 to 15 $\mu$m was also identified, possibly due to aluminum oxide. Besides the 10 $\mu$m feature, amorphous silicate has a long-wavelength feature that peaks at 18 $\mu$m. In stars which show Sil+ and 3C features, we found that the long-wavelength feature peaks at 19 $\mu$m rather than at 18 $\mu$m.

Two recent studies (Hron et al. 1997; Sloan & Price 1995) have done much to quantify the characteristics of the different types of dust emission seen in M stars. The two methods emphasize different aspects of the emission. Sloan and Price wanted to investigate correlations between the total amount of dust emission and the shape of the emission longward of 10 $\mu$m by using the flux ratios $F_{10}/F_{11}$ and $F_{11}/F_{12}$. They showed that the total dust emission in M stars is constrained to a narrow sequence of shapes ranging from the classic narrow feature peaking around 10 $\mu$m to the broad, low-contrast feature which peaks between 11 and 12 $\mu$m. They use this sequence to classify emission features of AGB stars. The LML90 classes more or less fall in different segments of their distribution. A limitation of their method is that they only used the longward portion of the emission features, whereas there are clear differences in the shape of the various classes that can be identified shortward of 10 $\mu$m.

Hron et al. on the other hand concentrated on investigating the shapes of the emission features by first subtracting a photospheric and dust "continuum" from the LRS. Using ratios of the remaining integrated flux in both the short- and long-wavelength portion of the feature, they found that they could recover the various LML90 classes. Their analysis showed that Miras on average tend to have thicker shells and stronger features than semiregular variables (SRb). It is unclear if their conclusion that Miras have lower stellar temperatures is not an artifact of their method. The 13.1 $\mu$m component appears to be associated primarily with SRb variables and with a fairly narrow range of shell optical depth. Sloan et al. (1996) also found that the 13.1 $\mu$m feature is associated primarily with SRb variables and occurs in approximately 75–90 % of these sources.

At large optical depths in the shell, the silicate features go into absorption. It is easy to confuse a partially self-absorbed silicate feature seen in

M stars with the SiC plus 8–9 μm feature seen in C stars, as has been done in several papers in the literature.

A few AGB stars show no dust emission features.

## 3. The C Stars

The circumstellar shells of many C stars are characterized by SiC dust emission at 11.2 μm. Unlike the 10 μm silicate feature, we found the SiC feature to be so uniform that it could be used for classification purposes (Little-Marenin et al. 1987). In a few stars (Y CVn, RY Dra, IRC+10216) the SiC feature is shifted so that it peaks at 11.4 μm.

We (Little-Marenin et al. 2000) and Goebel et al. (1995) have identified an emission feature in the 8–9 μm region possibly due to $a$:C–H, as well as absorption in the 13–15 μm region due to $C_2H_2$ and HCN. A few stars show a weak, broad emission feature which is clearly different in shape from the aluminum oxide feature seen in M stars. Its carrier has not as yet been identified.

After removing an estimated stellar contribution, we find that the majority of our sources fall into two categories: spectra with the classic SiC emission feature peaking around 11.2–11.5 μm (we will call this class SiC), and spectra where the SiC feature appears along with an additional component peaking around 8.5–9.0 μm (class SiC+). In a few stars the 8–9 μm feature rivals or exceeds the SiC feature in strength (class SiC++).

The classic SiC class contains mostly Mira variables, while the SiC+ and SiC++ classes contain mostly semi-regular and irregular variables. The periods of the classic SiC sources are longer than those of the SiC+ sources; that is true for both the Miras and the SRb variables. Classic SiC sources tend to have slightly redder [12]–[25] colors and correspondingly lower photospheric temperatures than the SiC+ and SiC++ sources. The classic SiC feature appears to be superimposed on a featureless continuum most likely due to amorphous carbon or graphitic material and is strongest for Miras. The C/O ratio increases along the sequence (SiC) → (SiC+) → (SiC++) from an average of 1.07 (SiC) to 1.2 (SiC+) to 1.3 (SiC++). Assuming that $a$:C–H is the carrier of the 8–9 μm feature (Goebel et al. 1995), we propose that this feature will strengthen with increasing C/O ratio. Support for this suggestion can be found in the increasing strength of the $C_2H_2$+HCN absorption feature seen in the 13–15 μm region and in the spectrum of VX And which has the largest C/O ratio (1.76) and the strongest contribution from the 8 – 9 μm feature.

Our sample of 99 stars was selected by cross-referencing the IRAS *Point Source Catalog* and the *General Catalogue of Variable Stars*. Four sources show no dust emission, 8 have a weak, broad feature that is difficult to classify, and 5 of these sources show emission from oxygen-rich dust, including two well-known silicate–carbon stars (BM Gem and V778 Cyg), and two S or SC stars which have been classified at various times as carbon stars (S Lyr and ST Cam). On the other hand, NP Pup appears to be related to the CS stars (Bidelman, private communication).

## 4. The S Stars

Pure S stars with C/O close to unity show an emission feature peaking around 10.5 $\mu$m which is subtly different from the 10 $\mu$m amorphous silicate or the 11.2 $\mu$m SiC feature (Little-Marenin & Little 1988). Neither Hron et al. (1997) nor Sloan & Price (1995) were able to clearly distinguish this S-star feature from other emission features by their methods. A direct superposition of the spectral shapes of the various emission features is needed.

In general S stars tend to have lower mass-loss rates and higher gas-to-dust ratios than M or C stars, implying less efficient dust formation in their circumstellar shells (Bieging & Latter 1994). Gas-to-dust ratios are estimated to be between 400 and 1000, at least a factor of two higher than for carbon stars, and hence strong dust emission features are not seen or expected.

I have identified seven S stars listed in the catalogue of Chen et al. (1995) which are listed as having very strong silicate features at 10 and 18 $\mu$m in their IRAS LRS spectra (LRS 25–29): IRAS 07197−1451 = TT CMa; 11169−6111; 15347−5555; 16490−4618; 19545−1122 = V1407 Aql; 21029+4917; and 22512+6100 = V386 Cep. However, all seven stars are actually either M or MS stars rather than pure S stars (Little-Marenin 2000), and hence they reflect the mass-loss rates and dust content associated with M stars.

I suggest that among true S stars we should find a few stars with strong 10 $\mu$m silicate features and enhanced $^{13}$C content with $^{12}$C/$^{13}$C ratios near the CNO equilibrium value of 3.4 rather than having values around 20, typical for S stars. If these stars are found they could be precursors to the Silicate–Carbon stars (Little-Marenin 1986), possibly produced by a helium core flash. However, no consensus about the evolutionary status of the Silicate–Carbon stars exists as yet. The MS stars with strong silicate features may fall into this category since all other MS stars I have analyzed have only weak silicate emission features. But no enhancement of $^{13}$C has yet been identified (or searched for) in these stars.

The spectra of carbon-rich proto-planetary nebulae show an emission feature around 21 $\mu$m.

## 5. Conclusion

Dust emission from circumstellar shells of AGB stars shows a variety of features in the 8–22 $\mu$m region. Besides the 10 and 18 $\mu$m amorphous silicate features, M stars have additional emission at 11 $\mu$m (crystalline olivine) and at 13.1 $\mu$m (unknown carrier), as well as a broad, low-contrast feature (9–15 $\mu$m) probably due to aluminum oxide. Most C stars show the classic SiC feature (11.2 $\mu$m), but many also show an 8–9 $\mu$m emission feature that strengthens with increasing C/O ratio. A featureless continuum (amorphous carbon or graphitic material) underlies the emission features. The S star feature peaking around 10.5 $\mu$m is difficult to identify except by directly comparing its shape to other emission features. A number of AGB stars show no dust emission. It has been suggested that the 11.3 $\mu$m feature superimposed on the amorphous silicate feature seen in some supergiants is due to PAHs rather than crystalline olivine. This finding is very interesting but needs further verification.

This work was supported by NASA grant NAG5-1667 and by Wellesley College.

## References

Bieging, J. H. & Latter, W. B. 1994, *ApJ*, 422, 765
Chen, P. S., Gao, H. & Jorissen, A. 1995, *A&A Supp.*, 113, 51
Goebel, J. H., Cheeseman, P. & Gerbault, F. 1995 *ApJ*, 449, 246
Hron, J., Aringer, B. & Kerschbaum, F. 1997, *A&A*, 322, 280
Little-Marenin, I. R. 1986, *ApJ*, 307, L15
Little-Marenin, I. R. 2000, in IAU Symp. 177: *The Carbon Star Phenomenon*, ed. R. F. Wing (Kluwer), p. 558
Little-Marenin, I. R. & Little, S. J. 1988, *ApJ*, 333, 305
Little-Marenin, I. R. & Little, S. J. 1990, *AJ*, 99, 1173 (LML90)
Little-Marenin, I. R., Ramsay, M. E., Stephenson, C. B., Little, S. J. & Price, S. D. 1987, *AJ*, 93, 663.
Little-Marenin, I. R., Sloan, G. C. & Price, S. D. 2000, in IAU Symp. 177: *The Carbon Star Phenomenon*, ed. R. F. Wing (Kluwer), p. 559
Sloan, G. C., LeVan, P. D. & Little-Marenin, I. R. 1996, *ApJ*, 463, 310
Sloan, G. C. & Price, S. D. 1995, *ApJ*, 451, 758

## Discussion

**Elitzur:** Extracting spectral features by subtracting a single-temperature blackbody is a dangerous procedure. The correct analysis must involve a full radiative transfer calculation in which the dust opacity, including features, is an input. Detailed results obtained otherwise cannot be trusted.

**Little-Marenin:** I agree that the best procedure for analyzing the low-resolution spectra is a full radiative transfer calculation, but my simple subtraction of a blackbody allows one to identify different dust components.

[**Unknown**]: Why would you expect less dust emission at maximum light than at minimum?

**Little-Marenin**: At higher luminosity and temperature, the condensation radius should move outward, evaporating grains, while at minimum the dust grain radius moves inward creating larger amounts of dust.

**Ivezić**: I would like to comment on the "discrepancy" between the behavior of the light curve and the strength of the 10 $\mu$m silicate feature. It is true that the optical depth should be larger during minimum light, but the strength of the feature can both decrease and increase with optical depth. Therefore, as the luminosity of the star changes, the feature strength can either increase or decrease.

**Little-Marenin**: Good.

**Speck**: The shift of the SiC peak from 11.2 to 11.4 $\mu$m was attributed to a change from $\alpha$–SiC to $\beta$–SiC. However, the $\beta$–SiC feature peaks at a shorter wavelength than $\alpha$–SiC, so this is unlikely. On the other hand, impurities in $\alpha$–SiC shift the peak to longer wavelength. My own work on the 8.6–8.9 $\mu$m feature suggests that there is no trend with optical depth. Maybe this suggests no evolutionary trend in the 8.6–8.9 $\mu$m feature.

**Little-Marenin**: Thank you for the information on $\alpha$– and $\beta$–SiC. We find trends in the strength of the 8.9 $\mu$m feature with varying C/O, but like you, we find no correlation with optical depth.

**van der Bliek**: The IRAS LRS spectra were calibrated using the spectrum of $\alpha$ Tau. However, the LRS spectrum of $\alpha$ Tau turned out to have an SiO absorption feature (7–10 $\mu$m), and this absorption then shows up as an "emission feature" in spectra without SiO absorption (Cohen et al. 1992, *AJ*, 104, 2030). Did you correct for this?

**Little-Marenin**: I corrected all the LRS spectra *à la* Cohen et al. (1992) and rescaled the spectra to the 12 $\mu$m PSC fluxes for that date.

# OBSERVATIONS OF MASS LOSS AND CIRCUMSTELLAR MATTER AROUND COOL CARBON STARS

T. LLOYD EVANS

*South African Astronomical Observatory*
*Observatory 7935, South Africa*

**Abstract.** Spectroscopy and infrared photometry of carbon stars show three distinct forms of circumstellar matter. IRAS 12311–3509 probably has an edge-on disk and the spectrum is dominated by resonance emission from atoms and molecules in the vicinity. The long-period variables V Hya and R Lep are undergoing deep fadings, apparently caused by dust formation around the star, while variable emission from circumstellar gas is seen. The semiregular variable T Mus showed absorption bands from very cool material during an unusual episode in 1994.

## 1. Introduction

Hoyle and Wickramasinghe (1962) predicted the formation of dust particles in the atmospheres of carbon stars. Excess infrared emission from some variable stars, including V Hya and R Lep which are discussed below, first revealed the presence of circumstellar dust (Woolf & Ney 1969; Gillett, Merrill & Stein 1971). More recent work on V Hya includes the discovery of an extended envelope with bipolar structure (Tsuji et al. 1988; Kahane, Maizels & Jura 1988). Another object discussed here, IRAS 12311–3509, probably has unresolved bipolar structure.

The fading and reddening of carbon Miras in the visible and the near infrared was found by Bessell & Wood (1983), Feast et al. (1984), and Le Bertre (1992). Percy et al. (1990) noted that such fading episodes were common in the visual light curves of carbon-rich Mira variables. Mattei & Foster (2000) report trends for some of these stars observed over 90 years, while Whitelock (2000) gives near-infrared light curves for several carbon variables and notes the occurrence of RCB-like dips in the light curve of R For.

## 2. IRAS 12311−3509

IRAS 12311−3509 has a large excess at $L$ and a unique optical spectrum characterised by emission of $SiC_2$ bands and of resonance lines of the alkali metals (Lloyd Evans 1991b). This was interpreted as an example of a star which is hidden by a dusty disk but seen by reflection from material out of the plane of the disk, as in the case of Herbig's (1970) model for the somewhat similar M star VY CMa. The emission lines and bands are produced by resonance emission from circumstellar gas, as in the case of cometary spectra. The star appears unresolved on the UK Schmidt Sky Survey photographs, whereas VY CMa appears in double-star catalogues as a complex object, but since IRAS 12311−3509 is 7 mag fainter, it is probably more distant and appears unresolved as a result.

Sarre, Hurst & Lloyd Evans (1996, 2000) have analysed the $SiC_2$ bands in the light of recent laboratory and theoretical work on the molecule. The particular suite of bands which appears strongly in emission is consistent with radiative excitation of cool material, while the band profiles indicate a temperature of a few hundred K.

Spectra taken over the period 1989–1996 show no significant change, nor do infrared observations taken near the extremes of this interval show a significant difference. This is consistent with the presence of a disk. The other stars discussed here show much more dynamic behaviour.

## 3. T Mus

T Mus is a semiregular variable with periods of 93 and 1082 days (Lysaght 1989). It is a J star with strong $SiC_2$ bands (Keenan 1993). T Mus was observed frequently from 1986 onwards to study spectral changes round the complex light cycle. The spectrum changes considerably, presumably as the result of temperature changes at the photospheric level; the spectrophotometric gradient in the 4700–5200 Å region and the $C_2$ bandhead at 4737 Å are especially variable. The $SiC_2$ bands also vary in intensity, though as with other carbon variables observed in the present programme the different bands seen in this spectral region maintain similar relative strengths. The spectrum showed a much greater change in appearance during the 1994 observing season when T Mus should have been on the declining part of the slow light variation (Lysaght 1989). The different $SiC_2$ bands no longer varied in step: several became very weak while the rest showed a narrower profile than usual.

The recent laboratory results for $SiC_2$ provide an explanation for this behaviour also (Sarre, Hurst & Lloyd Evans 1996, 2000). The missing bands are all "hot" bands, whereas those which remain strong are formed by absorption from the lowest vibrational level. The narrow profile indicates a rotational temperature below that typical of the stellar photosphere. Both observations show that the bands are formed in gas substantially cooler than that found even in the photospheres of Mira variables at any stage of their light cycle. This gas may be circumstellar; alternatively it may represent a substantial temporary extension of the normal photosphere.

## 4. Dust Minima in the Mira Variables

The early infrared and visual observations which led to the discovery of the dust fading events in carbon-rich long period variables were not supported by spectroscopic observations. We have spectroscopic observations covering a ten-year period for two Miras, R Lep and R For, the SRa star R Scl, and V Hya which is also an SRa of relatively small amplitude but in addition fades by 3 mag or so every 18 or 19 years. $JHKL$ photometry has been obtained during faint episodes of these stars. V Hya entered a deep minimum in 1992 and appears to be in the recovery stage now, while R Lep has been fading on average for some years and entered a deeper phase with spectacular spectroscopic consequences in 1994. R For has shown more modest phenomena of this type, while R Scl has been relatively inactive though it is believed to have suffered major mass loss in the past.

### 4.1. DUST FADINGS AND INFRARED COLOURS

The long runs of visual observations of Mira variables by the AAVSO (Mattei, Mayall & Waagen 1990) allow the detection of those stars with prominent dust fading events. The stars which undergo fadings with an amplitude greater than one magnitude are redder in the $J$-$K$, $K$-$L$ diagram than those with more constant mean light (Lloyd Evans 1997). The three stars noted above are redder during the fading in each case. Lloyd Evans (1997) used both $JHKL$ and IRAS data to show that the infrared colours from $1.2\,\mu$m to $12\,\mu$m are correlated with the amplitudes of the fadings in visual magnitude. Since these colours are often taken as a measure of the rate of mass loss, it seems that for carbon Miras, at least, the fading episodes represent individual epochs of mass loss. The visual fading as well as much of the reddening in the near infrared must result from absorption by newly formed circumstellar dust.

### 4.2. $C_2$ EMISSION DURING DUST FADING EPISODES

The most radical spectroscopic change seen in conjunction with the fading of visual light is the appearance of $C_2$ emission. This is seen first at the (0,0) bandhead at 5165 Å. This bandhead is normally a strong absorption feature without any emission, whereas the (1,0) 4737 Å bandhead is

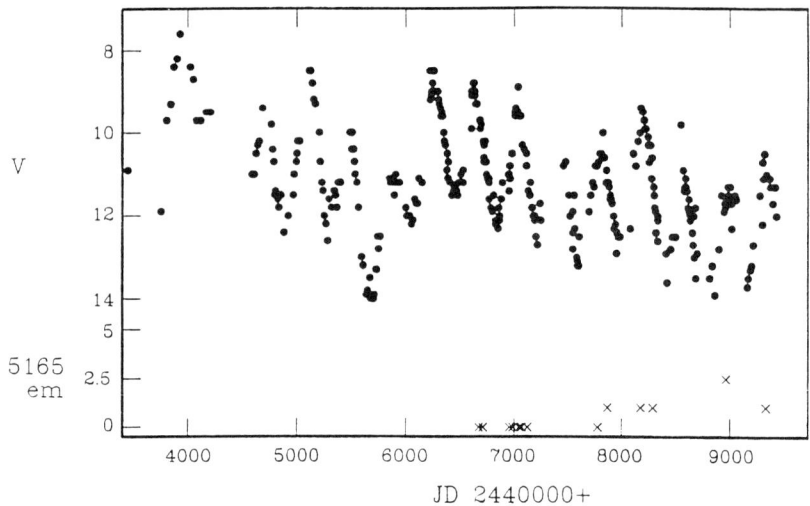

*Figure 1.* The visual light curve of R For from Southern African amateur observations (*above*), and the estimated emission intensities at the 5165 Å (0,0) bandhead of $C_2$ (*below*).

regularly seen in emission during part of the pulsation cycle (Lloyd Evans 1989). The emission at 5165 Å may be very intense so that the usual strong absorption band is reversed, when enhanced emission is also seen at 4737 Å (Lloyd Evans 1997). A truly quantitative estimate of the emission strength is hard to make because of the loss of the fiducial point provided by the now-obscured base of the absorption band, and qualitative estimates were made on a scale from 0 = absent to 5 = strong. Figure 1 shows a plot of emission against time for R For, with the visual light curve above it. The spectroscopic data are sparse because R For is too faint to be observed near minimum light. $C_2$ emission was absent when the star was bright early on, but as the mean light faded emission of moderate strength appeared. Lloyd Evans (1997) found that V Hya had no emission at 5165 Å despite almost invariably strong emission at the (1,0) bandhead until the deep minimum of 1992-95 when intense emission appeared. R Lep showed increasingly strong emission as the star faded slowly, especially from 1994 onwards. The mean light of R Lep varies almost continuously, with deep minima at intervals of about 40 years (Mayall 1963), and by 1995 the pulsation minima were as faint as had been observed at any time in the past.

### 4.3. ATOMIC LINE EMISSIONS

The ten-year observing programme also covered the Na D lines and H$\alpha$. V Hya showed no emission at Na D before the deep minimum. The farther red region of the resonance lines of K I and Rb I was only observed routinely after the fading was well established, although an early observation showed no emission a year before the start of the deep minimum. The absence of emission at K I before the fading began was established by Barnbaum, Morris & Kahane (1995) who reported only absorption at low velocity from May 1988 to January 1993. We found weakened absorption of Na D in July 1992 and emission in December 1992, weak emission of K I in July 1993 and strong emission in December 1993 and emission of Rb I from December 1993. Hα emission maintained the same cyclic behaviour that it exhibited before the fading started. The latter may indicate that the pulsation and the near-photospheric levels of the atmosphere were unaffected, which would be expected if the activity is a phenomenon of the circumstellar region as some theories predict (Fleischer, Gauger & Sedlmayr 1992; Winters et al. 1994). The Na and K lines showed a P Cygni structure. The blueshift of the absorption component corresponded to a velocity of 100 km s$^{-1}$ relative to the photosphere. This is in the range found previously for the bipolar outflow from V Hya (Tsuji et al. 1988; Kahane, Maizels & Jura 1988; Sahai & Wannier 1988; Lloyd Evans 1991a). The circumstellar gas may be accelerated outwards in the bipolar flow so that this velocity is not typical of mass-losing carbon stars in general.

R Lep differs in that the fadings are less discrete and more of an ongoing continuous variation in mean light, as can be seen by comparing its light curve (Mayall 1963) with that of V Hya (Mayall 1965). The Na D absorption varies greatly round the 427 d light cycle and becomes weak near maximum light. It was so weak near maximum light in 1994 that K I and Rb I were observed for the first time and found to be in emission. These lines varied little in the following two years, whereas Na D continued its normal variation between strong absorption and near disappearance; the weakness in 1994 was not highly abnormal when seen in perspective (see Figure 2). P Cygni structure has not been seen with certainty.

4.4. THE COMPANION OF V HYA

V Hya is atypical in having a companion. This was suspected from the unusual velocity-broadening of the photospheric absorption lines, attributed to rotation resulting from the star having been spun up by a companion (Kahane, Maizels & Jura 1988; Barnbaum, Morris & Kahane 1995). These authors favour a common envelope system, as opposed to the accretion disk about a detached companion postulated by Sahai & Wannier (1988) to drive the rapid bipolar flow. Continued observation of the violet spectral region in the six years since the observations of a blue continuum by Lloyd Evans (1991a) has shown that the continuum has both high and low inten-

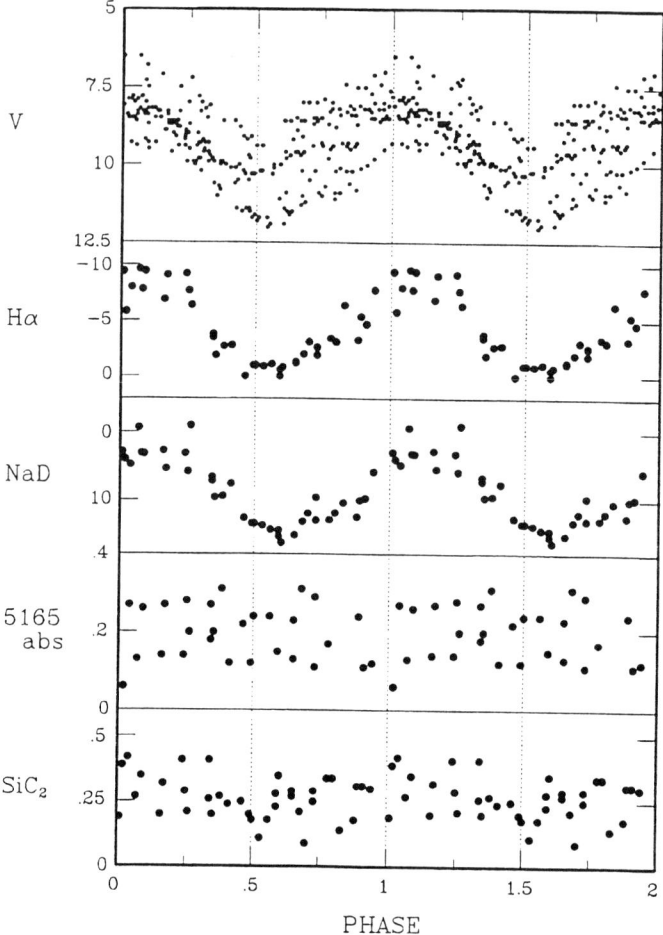

*Figure 2.* Phase dependence of magnitude and spectroscopic parameters for R Lep. The visual magnitude and the $C_2$ and $SiC_2$ absorptions do not repeat well as the star enters a dust fading, but H$\alpha$ emission and Na D absorption show little scatter. Note that there is weak net emission of Na D after light maximum on occasion.

sity states, each with its characteristic line spectrum. The spectrum is that of an accretion disk and does not match that of any normal star. Barnbaum, Morris & Kahane (1995) doubted the presence of an accretion disk because of the weakness of the UV continuum observed by IUE, a point also made by Luttermoser in discussion at this meeting. V Hya is surrounded by dust at all times and that may suffice to obscure the accretion disk in the UV. The Mg II emission seen in IUE spectra is likely to arise in the jet rather than to be of chromospheric origin. The continuum intensity at

4160 Å, which has been calibrated spectrophotometrically, also faded and brightened roughly in phase with the deep minimum of the carbon star although the amplitude was less. This is attributable to the same additional circumstellar absorption that affected the visual light of the carbon star. This observation excludes the possibility that the slow fading of the carbon star results from the extinction when it is on the near side of its orbit and is obscured by a dusty circumbinary disk. The accretion disk would be expected to be least obscured then, when it is on the far side of its orbit, unless it is indeed a close companion in which case a second more distant companion would have to be postulated to account for the size of the orbit.

### 4.5. AN ADDITIONAL PERIODICITY IN V HYA

The several atomic and molecular features which have been measured in R Scl, R For, R Lep and V Hya (avoiding the disturbed intervals for the last two stars) show time-dependent variations. The pulsation period of the carbon star is the principal determinant of the variations. V Hya completed nearly four cycles in its 530 d pulsation between the start of observation and the onset of the deep minimum. Most spectroscopic features followed this period, but the strength of the (0,0) band of $C_2$ at 5165 Å did not and may instead follow a period of double this length. This band is very strong in V Hya and it is probably formed at a high level in the atmosphere.

### 4.6. A HIGH-TEMPERATURE PHASE IN V HYA

The visual observations of V Hya showed the first departures below the normal light curve in the early months of 1992, and the spectrum had a unique appearance when observed in February and April. This was the only time when the pulsation-related emission of $C_2$ at 4737 Å disappeared and a strong absorption band appeared. The (0,0) band of $C_2$ at 5165 Å remained strong while the continuum in this spectral region was much flatter than on any other occasion. The whole appearance was that of a considerably hotter carbon star. It seems likely that the steep spectral continuum of a cool carbon star is related to the presence of a very distended atmosphere, and a possible explanation of this unusual observation is that the outer atmosphere of V Hya lifted off, becoming optically thin, to reveal a layer below, which would be hotter until it had time to adjust to the normal state of the star. The situation appeared normal in July 1992, by which time Na D absorption began to weaken.

The deep minima of R Lep and V Hya are still continuing. Long-term spectroscopic and photometric monitoring of these stars will be required to give a complete picture of these events. There is a need for spectroscopic observations of higher resolution and also for other types of observation such as spectropolarimetry.

I am indebted to the members of the Variable Star Section of the Astronomical Society of Southern Africa for data on these stars, and to Audrey van der Wielen for the diagrams.

## References

Barnbaum, C., Morris, M. & Kahane, C. 1995, *ApJ*, 450, 862
Beichman, C. A., Neugebauer, G., Habing, H. J., Clegg, P. E. & Chester, T. J. 1988, *Infrared Astronomical Satellite (IRAS) Catalogs and Atlases*, Vols. 1–6, National Aeronautics and Space Administration, Washington, D.C.
Bessell, M. S. & Wood, P. R. 1983, *MNRAS*, 202, 31P
Feast, M. W., Whitelock, P. A., Catchpole, R. M., Roberts, G. & Overbeek, M. D. 1984, *MNRAS*, 211, 331
Fleischer, A. J., Gauger, A. & Sedlmayr, E. 1992, *A&A*, 266, 321
Gillett, F. C., Merrill, K. M. & Stein, W. A. 1971, *ApJ*, 164, 83
Herbig, G. H. 1970, *ApJ*, 162, 557
Hoyle, F. & Wickramasinghe, N. C. 1962, *MNRAS*, 124, 417
Kahane, C., Maizels, C. & Jura, M. 1988, *ApJ*, 328, L25
Keenan, P. C. 1993, *PASP*, 105, 905
Le Bertre, T. 1992, *A&AS*, 94, 377
Lloyd Evans, T. 1989, in *Evolution of Peculiar Red Giant Stars*, ed. H. R. Johnson and B. Zuckerman (Cambridge Univ. Press), p. 241
Lloyd Evans, T. 1991a, *MNRAS*, 248, 479
Lloyd Evans, T. 1991b, *MNRAS*, 249, 409
Lloyd Evans, T. 1997, *MNRAS*, 286, 839
Lysaght, M. G. 1989, *JAAVSO*, 18, 17
Mattei, J. A. & Foster, G. 2000, in IAU Symp. 177: *The Carbon Star Phenomenon*, ed. R. F. Wing (Kluwer), p. 155
Mattei, J. A., Mayall, M. W. & Waagen, E. O. 1990, *Maxima and Minima of Long Period Variables* (Cambridge, Mass.: American Association of Variable Star Observers)
Mayall, M. W. 1963, *JRASC*, 57, 237
Mayall, M. W. 1965, *JRASC*, 59, 245
Percy, J. R., Colivas, T., Sloan, W. B. & Mattei, J. A. 1990, in *Confrontation between Stellar Pulsation and Evolution*, ed. C. Cacciari and G. Clementini, ASP Conf. Ser., 11, 446
Sahai, R. & Wannier, P. G. 1988, *A&A*, 201, L9
Sarre, P. J., Hurst, M. E. & Lloyd Evans, T. 1996, *ApJ*, 471, L107
Sarre, P. J., Hurst, M. E. & Lloyd Evans, T. 2000, in IAU Symp. 177: *The Carbon Star Phenomenon*, ed. R. F. Wing (Kluwer), p. 576
Tsuji, T., Unno, W., Kaifu, N., Izumiura, H., Ukita, N., Cho, S. & Koyama, K. 1988, *ApJ*, 327, L23
Whitelock, P. A. 2000, in IAU Symp. 177: *The Carbon Star Phenomenon*, ed. R. F. Wing (Kluwer), p. 179
Winters, J. M., Fleischer, A. J., Gauger, A. & Sedlmayr, E. 1994, *A&A*, 290, 623
Woolf, N. J. & Ney, E. P. 1969, *ApJ*, 155, L181

## Discussion

**Luttermoser**: I have obtained (as yet unpublished) low-resolution LWR IUE spectra of V Hya for an approximately 5–month duration in 1994. There was no evidence of an accretion disk (i.e. a hot continuum) in these spectra. Indeed, all of these spectra display "normal" N–type stellar chromospheric emission — Mg II and C II (2325 Å) emission lines. Can you comment as to why we don't see V Hya's accretion disk at UV wavelengths?

**Lloyd Evans**: The dust obscuration was strong then and may have hidden the accretion disk, which normally has the color temperature of an F star (7000 K). It is also possible, given the lack of velocity data, that these lines are formed in the 160 km s$^{-1}$ shock region which shows Ca II emission as well as [Fe II], [S II], etc.

**Giridhar**: For V Hya the periodicity from $C_2$ lines is twice that from optical features. A similar observation is found for the RV Tau star R Sct where infrared features give a period of 140 days whereas optical features give 70 days. The explanation offered was that optical spectral features formed near the photosphere experienced two shocks whereas infrared features formed in the outer layers experienced only one. Alternatively, non-sphericity such as a prolate or oblate shape could cause two different periods.

**Lloyd Evans**: It is interesting that in this respect, as well as that of the double periodicity, V Hya may have something in common with the RV Tauri stars.

**Kerschbaum**: When looking at your sketch of the geometry of IRAS 12311–3509, I wonder if it would not be helpful to have some polarimetric measurements to find out what comes from the star and what from its vicinity?

**Lloyd Evans**: Spectropolarimetry would be very valuable. We do not have the equipment to do this.

**Gustafsson**: Since R Scl shows extended resonance scattering (in Na D and K I $\lambda$ 7699) at least out to $\sim 10'$ from the star (see the poster by Gustafsson et al., these Proceedings), it might be rewarding to monitor it with a set of different slit lengths.

**Lloyd Evans**: I agree. Our spectrograph has only two channels (star, sky) so I have not attempted this.

# DYNAMICAL MODELS OF CIRCUMSTELLAR DUST SHELLS AROUND LONG–PERIOD VARIABLES

A. J. FLEISCHER, J. M. WINTERS AND E. SEDLMAYR
*Institut für Astronomie und Astrophysik*
*Technische Universität, Berlin, Germany*

**Abstract.** We present dynamical models of circumstellar dust shells around long-period variables which include time-dependent hydrodynamics and a detailed treatment of dust formation, growth and evaporation. Important effects due to the complex interaction between the dynamics of the pulsating atmosphere and the dust complex are demonstrated.

## 1. Introduction

We present results of our model calculations for circumstellar dust shells of long-period variable stars (LPVs). In particular, we will summarize the method and some results of the hydrodynamical calculations. What is our main aim and what can we learn from these models?

First, as we treat in detail the dynamics and physical processes of the dust shell, we can study its intrinsic structure, the distribution of the dust, the velocity field, the density structure, etc. which all are a result of the calculation.

Second, it turns out that it is absolutely necessary to take into account the relevant interactions among the physical processes. By doing so, we are able to study which processes contribute to the resulting structure and the dynamics of the model.

Third, as a result of the model we get the final outflow velocity, the mass loss rate, the dust–to–gas ratio, and the grain size distribution function. These values are a result of the calculation itself — they are *not* prescribed parameters — and therefore they can be directly compared with respective observations.

Finally, and perhaps most important, we can calculate the optical appearance of the model, the spectral energy distribution, light curves, brightness profiles, etc. and can compare these quantities with observations as well. If this approach is applied to a particular source and the theoretical spectral energy distribution, synthetic light curves, etc. match with the respective observations, we are able to determine the fundamental stellar parameters: stellar luminosity, stellar mass and temperature, and the elemental abundances. In this contribution we concentrate on the first three points and show typical results of the dynamics and intrinsic structure of the circumstellar dust shell.

The second part, the optical appearance, can be found elsewhere (Winters et al. 1994, 1995, 2000).

## 2. The Non–Linear Problem

In order to reach the objective described above we have to solve the following coupled problem:

**Time–dependent hydrodynamics.** The hydrodynamical description has to be time–dependent, to describe the variations of the velocity field and the occurrence of shock waves caused by the pulsation of the underlying star.

**Dust complex.** Since the formation of dust is a central phenomenon in these objects, it is necessary to use a theory which describes the formation of solid particles depending on the physical conditions, and which does not prescribe essential quantities such as the site where the dust forms.

**Chemistry.** Furthermore, we have to describe the chemistry of the gas phase in order to know at each instant of time and at each radial position how much condensable material is available.

**Radiative transfer.** Finally, we have to treat the radiative transfer problem in order to describe the influence of the radiation of the central object.

In contrast to the classical atmosphere problem, the dust complex now introduces a number of non-linear couplings and interactions which are of essential importance for the whole problem. A reliable theoretical modelling on the one hand requires a *physical* description of the various ingredients, and on the other hand, at least equally important, it requires taking into account all interactions among the different physical components.

It is evident from the list that the problem is non-linear and strongly coupled. Consider for instance the coupling between hydrodynamics and the dust complex: only if the density is high enough, is effective dust formation possible; but at the same time, since radiation pressure on dust enters into the equation of motion, dust formation immediately influences the velocity structure, which in turn alters the density stratification. The basic equations and the numerical method are described in detail in Fleischer et al. (1992) and Fleischer (1994).

## 3. A Typical Model Structure

The radial structure of a typical dynamical model is shown in Figure 1. It is evident that the dust is not distributed homogeneously across the shell but is concentrated in distinct layers such that the circumstellar dust shell exhibits an onion-like structure. Furthermore it can be seen that the dust quantities, e.g. the degree of condensation ($2^{nd}$ panel, dashed line), are intimately correlated with the hydrodynamical quantities, e.g. the velocity (upper panel, crosses). This suggests that there is a common mechanism which produces this structure. A more detailed analysis shows that the strong

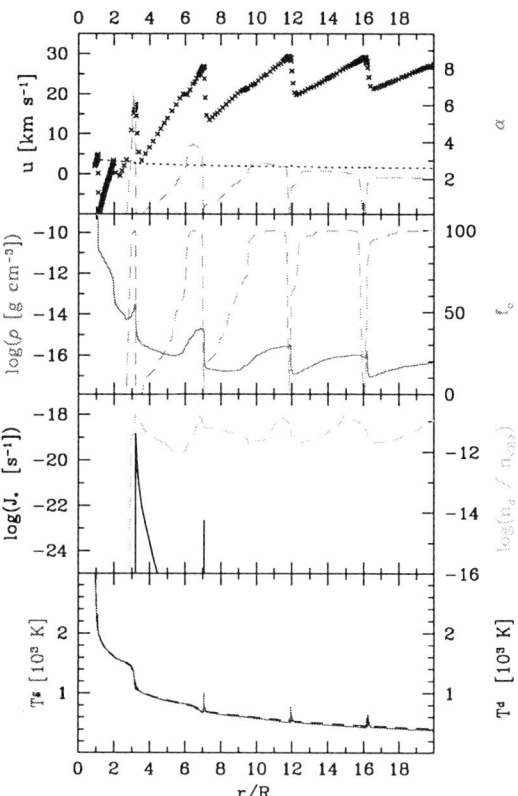

Figure 1. Typical radial structure of a dynamical model. The parameters of the model are $T_* = 2600\,\mathrm{K}$, $L_* = 10^4 L_\odot$, $M_* = 1.0 M_\odot$, $\epsilon_C/\epsilon_O = 1.80$, $P = 650\,\mathrm{d}$, $\Delta u = 2\,\mathrm{km\,s^{-1}}$.

shocks propagating through the atmosphere are not produced by the interior pulsation alone but are a product of the re-amplification of the pulsational shocks caused by the radiative acceleration $\alpha$ on dust grains in the discrete dust layers ($1^{st}$ panel, dashed line). Since $\alpha$ exceeds unity close to the star, the material is accelerated to velocities above the escape velocity $v_{\mathrm{esc}}$ already at radii around $4 R_*$. Due to its opacity, the dust also leads to a pronounced heating of the material inside the dust layers (backwarming), as is evident from the steps present in the temperature stratification around $3 R_*$ and $6 R_*$ (lower panel). In summary, we find that the dust completely determines the internal structure of the circumstellar shell, a result that is only revealed by the consistent treatment of the physical processes involved.

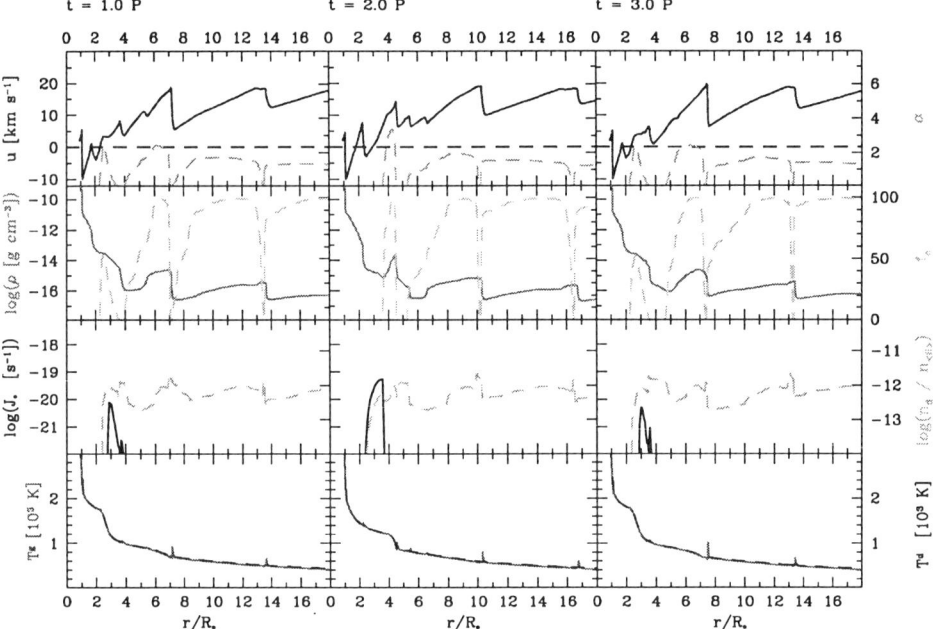

*Figure 2.* Same model as in Fig. 1 but with a reduced carbon-to-oxygen ratio of $\epsilon_C/\epsilon_O = 1.50$.

## 4. Multiperiodicity

An important property of the model shown in the preceding section is that its radial structure repeats after *one* pulsational cycle. Figure 2 shows three successive starting points of the hydrodynamical cycle of a model with a reduced overabundance of carbon to oxygen. All other parameters are the same as in Fig. 1. From Fig. 2 one can see that the radial structure of the model with a reduced carbon overabundance repeats only after *two* pulsational cycles. Even tiny details of the radial structure, e.g. the multiple shocks present below $4\,R_*$, are reproduced after this period of time. Around $2.5\,R_*$ at $t = 1.0P$ and $t = 3.0P$, respectively, it can be seen that the radiative acceleration $\alpha$ causes a pertubation in the velocity structure which later on turns into a dominant shock wave that sweeps up the preceding pulsational shock: cf. $t = 2.0P$ between 4 and 5 $R_*$. Due to the reduced amount of condensable material it takes two pulsational cycles to form a new dust layer. The effect of *multiperiodicity* strongly depends on the $\epsilon_C/\epsilon_O$ ratio. Lowering this number causes longer intervals between the formation of two dust layers; an increase causes the opposite effect. The time interval between the formation of two dust layers is of course not necessarily an integral number, e.g. models with $\epsilon_C/\epsilon_O$ values between 1.8 and 1.5 form a

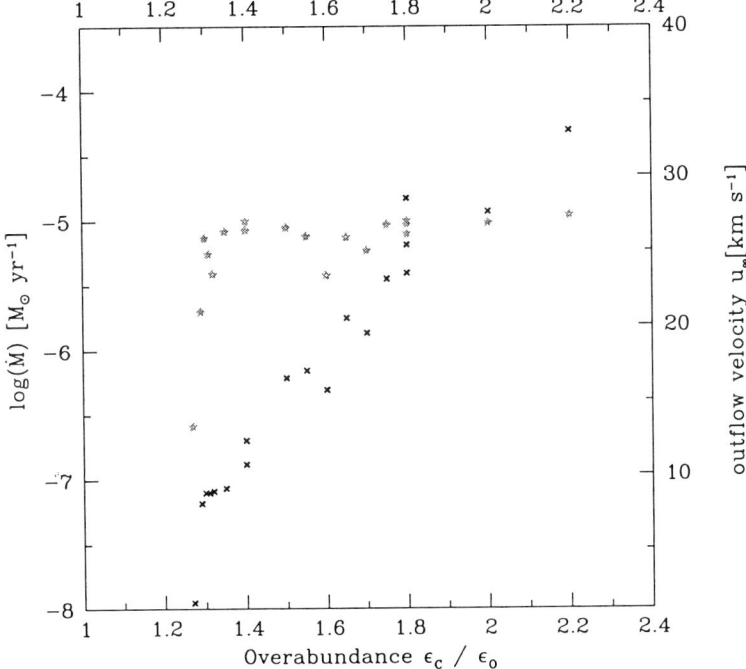

*Figure 3.* Dependence of the mass-loss rate $\dot{M}$ (asterisks, left axis) and the final outflow velocity $u_\infty$ (crosses, right axis) on the overabundance of carbon to oxygen $\epsilon_C/\epsilon_O$.

new dust layer on time scales larger than 1 and smaller than 2 pulsational periods. Depending on the remaining model parameters it is also possible that a dust layer forms on time scales of 3, 4 ... times the pulsational period, or even twice per pulsational cycle.

## 5. Dependence of $\dot{M}$ and $u_\infty$ on $\epsilon_C/\epsilon_O$

We have calculated a grid of models with the parameters given in the caption of Fig. 1 except for the overabundance of carbon to oxygen which is varied over a larger parameter range. The influence of this variation on the mass-loss rate and the final outflow velocity is shown in Fig. 3. Keeping all parameters except the $\epsilon_C/\epsilon_O$ ratio fixed results in a fairly constant mass-loss rate of $\sim 10^{-5} M_\odot \mathrm{yr}^{-1}$ for all models with $\epsilon_C/\epsilon_O > 1.30$. In contrast, $u_\infty$ is proportional to $\epsilon_C/\epsilon_O$. Below the value of $\sim 1.30$ for $\epsilon_C/\epsilon_O$ there is just enough dust formed to drive a wind which is, however, very slow and less massive. Going to even lower values of $\epsilon_C/\epsilon_O$, pulsation and radiative acceleration on dust grains alone no longer can support an outflow. A fitting equation which relates a given set of the six model parameters to the resulting mass-loss rate is derived in Arndt et al. (1997).

## 6. Shortcomings of the Models

The models suffer from two major shortcomings: the use of an equilibrium chemistry and the treatment of the cooling of the gas behind the shock fronts. As strong velocity changes and short-term variations in the radiation field are usually encountered in LPV atmospheres, the standard situation of the gas phase is chemical non-equilibrium (cf. Sedlmayr & Winters 2000). To consider these effects in the models would require solving a time-dependent reaction network which is extremely CPU-time-consuming. However, calculations of rate networks show that in the inner shell region, where the dust forms, the situation can be approximated by the equilibrium case if one is only interested in the densities of the dust-forming species (cf. Beck et al. 1992). In future work, the treatment of the post-shock cooling processes in the models also has to be improved. Calculations in the isothermal as well as the adiabatic limit case show that the structure of the circumstellar dust shell appears in the way described above. Using LTE cooling laws, as done in the models described in this paper and in Höfner et al. (1996), yields models very close to the isothermal limit case. Incorporating the laws proposed by Bowen (1988) results in models with an unrealistic temperature structure as the re-heating behind the shocks proceeds much too slowly. A solution to the post–shock cooling problem could be the incorporation of the scheme developed in Woitke et al. (1996).

This work has been supported by the DFG (grant Se 420/8-1) and by the BMBF (grant 05 3BT13A 6). The calculations were performed on the CRAY computers of the Konrad–Zuse–Zentrum für Informationstechnik Berlin (ZIB) and the Höchstleistungsrechenzentrum (HLRZ), Jülich.

## References

Arndt, T. U., Fleischer, A. J. & Sedlmayr, E. 1997, A&A, 327, 614
Beck, H. K. B., Gail, H.-P., Henkel, R. & Sedlmayr, E. 1992, A&A, 265, 626
Bowen, G. H. 1988, ApJ, 329, 299
Fleischer, A. J. 1994, *Hydrodynamics and Dust Formation in the Circumstellar Shells of Miras and Long–Period Variables*, Ph.D. thesis, Technische Universität, Berlin, Germany
Fleischer, A. J., Gauger A. & Sedlmayr, E. 1992, A&A, 266, 321
Höfner, S., Fleischer, A. J., Gauger, A., Feuchtinger, M. U., Dorfi, E. A., Winters, J. M. & Sedlmayr, E. 1996, A&A, 314, 204
Sedlmayr, E. & Winters, J. M. 2000, in IAU Symp. 177: *The Carbon Star Phenomenon*, ed. R. F. Wing (Kluwer), p. 337
Winters, J. M., Fleischer, A. J., Gauger, A. & Sedlmayr, E. 1994, A&A, 290, 623
Winters, J. M., Fleischer, A. J., Gauger, A. & Sedlmayr, E. 1995, A&A, 302, 483
Winters, J. M., Fleischer, A. J., Le Bertre, T. & Sedlmayr, E. 2000, in IAU Symp. 177: *The Carbon Star Phenomenon*, ed. R. F. Wing (Kluwer), p. 590
Woitke, P., Krüger, D. & Sedlmayr, E. 1996, A&A, 311, 927

# Discussion:

**Luttermoser**: This work you have just presented is phenomenal! I have one problem with it, however. We see UV emission lines — and, in the case of Miras, strong hydrogen Balmer lines — in the spectra of these stars. Yet the shocks in your model are not hot enough to produce these lines. Can you comment on this?

**Fleischer**: The models I just presented include an LTE cooling law, which yields a rather efficient cooling of the gas behind the shocks. Currently we are working on the incorporation of a more realistic treatment of the heating/cooling processes by using the approach given in Woitke, Krüger & Sedlmayr (in press) [A&A 311, 927, 1996 – Ed.]. I do not expect major changes in the overall structure of the circumstellar dust shell, as I already checked the influence of the cooling law by varying the C–parameter of Bowen's (1988) parameterized law.

**Linsky**: I encourage you to include an important physical process that is probably not in your calculations. Manfred Cuntz has computed models with a stochastic distribution of piston amplitudes and periods. He finds that models including a stochastic distribution of wave properties rather than monochromatic waves have supershocks due to coalescence of individual shocks and greatly increased mass-loss rates.

**Fleischer**: As far as I know, the work of Cuntz deals with sound waves generated by the outer convective zone. Our models try to simulate the large-amplitude pulsation of a Mira star. Nevertheless, I agree that we have to get rid of this piston approximation in favor of a physical model of the interior pulsation which, of course, could include the contribution of sound waves.

**Mowlavi**: What is, from your model calculations, the dependence of the mass-loss rate on the C/O ratio?

**Fleischer**: The mass-loss rate is essentially independent of the C/O ratio (Fleischer 1994, Dissertation, TU–Berlin).

**Mowlavi**: What are then the other parameters that influence the mass-loss rate? The mass-loss rates you obtain reach values as high as $10^{-5}$ or more. Could you comment on the development of "superwinds" as are often supposed to occur at some stage of the ascent of the AGB phase?

**Fleischer**: The mass-loss rate mainly depends on stellar temperature, stellar luminosity, and mass. The mass-loss rate increases with increasing luminosity, decreasing temperature, and decreasing mass. Since during the AGB evolution the stellar luminosity increases, the temperature (slightly) decreases, and due to mass loss the stellar mass also decreases, the mass-loss rate produced by our models easily reaches values as high as $\sim 10^{-4}$ $M_\odot/yr$. Therefore, dust-driven mass loss as it results from our models shows the typical characteristics of the so-called "superwind."

**Gustafsson**: A good criterion of very good theoretical modelling is that all good models enthusiastically suggest further complications. Here is another one: Your strong and very impressive coupling between pulsations, dust formation and wind suggests that non-radial perturbations could have severe consequences. Do you dare to comment on that?

**Fleischer**: Most likely this would yield a cloudy or patchy structure of the dust distribution. However, a quantitative answer can only be given by a more-dimensional treatment which within our approach is beyond today's computing power.

# MODELLING THE SPECTRAL ENERGY DISTRIBUTIONS OF AGB STARS IN THE LMC

M. A. T. GROENEWEGEN
*Max-Planck-Institut für Astrophysik, Garching, Germany*

J. TH. VAN LOON
*European Southern Observatory, Garching, Germany*

P. A. WHITELOCK
*South African Astronomical Observatory*
*Observatory, South Africa*

P. R. WOOD
*Mount Stromlo and Siding Spring Observatories*
*Canberra ACT, Australia*

AND

A. A. ZIJLSTRA
*European Southern Observatory, Garching, Germany*

**Abstract.** The spectral energy distributions of 16 AGB stars in the LMC are fitted. The 6 known oxygen-rich stars are well fitted with silicate dust. Due to the completely different absorption properties of silicate and carbonaceous dust we argue that of the 10 stars of unknown spectral type, 1 is probably O-rich while the other 9 are probably C-rich.

The Period-Luminosity relation is discussed. Both the O- and C-rich LPVs in this study appear to lie on extensions of $P-L$ relations derived for stars with much shorter periods.

## 1. Introduction

Modelling the spectral energy distributions (SEDs) of late-type stars provides information on their mass-loss rates and luminosities. Essentially all model fitting of this kind up to now has been done on AGB stars in our Galaxy (e.g. Groenewegen 1995 on carbon stars; Justtanont & Tielens 1992 on oxygen-rich stars).

The advantage of doing similar modelling for AGB stars in extra-galactic systems is that the distance is known (which is not usually the case for AGB stars in the Galaxy), and that one can then estimate the dependence of mass loss on metallicity.

The requirement to get reliable results from the model fitting is primarily a sufficiently large data set of photometric observations covering a broad range in wavelength. For AGB stars in our Galaxy these data either exist in the literature or are relatively easily obtained. In contrast, observations of AGB stars in the Magellanic Clouds, in particular in the near-infrared where most AGB stars emit most of their energy, have started to be performed only recently (Wood et al. 1992, Zijlstra et al. 1996).

In this paper we present the first results on the modelling of 16 AGB stars in the LMC, from which luminosities are derived. We also discuss the Period-Luminosity relation. A paper is in preparation which will include a discussion of the mass-loss rates and possible evolutionary consequences.

## 2. The Sample

From the available literature we selected all 16 AGB stars in the LMC which were detected by the IRAS satellite and for which pulsation periods have been determined. The periods and near-infrared data are taken from Wood et al. (1992), Zijlstra et al. (1996, 1998) and Wood (1998). One star (TRM 60) was previously fitted in Groenewegen et al. (1995), as one of only two AGB stars outside our Galaxy for which a silicate feature has been detected up to now. The near-infrared observations were supplemented by IRAS data. In order to obtain the best possible IRAS 12 and 25 $\mu$m fluxes we inspected two-dimensional images and one-dimensional co-adds of IRAS scans using the GIPSY software package (Assendorp et al. 1995). No useful 60 or 100 $\mu$m data could be obtained because of the high and variable background emission in the direction of the LMC.

The periods range between 1040 and 1400 days for the stars we believe to be O-rich (see below), and between 515 and 900 days for the stars we believe to be C-rich. These periods are significantly longer than those of LPVs selected from optical surveys (see Sect. 4). This is related to the fact that the stars were selected from the IRAS data-base which favors high mass-loss rates and high luminosities.

Six of the 16 stars are known oxygen-rich stars, either from the detection of OH maser emission (Wood et al. 1992) or the silicate dust feature (Groenewegen et al. 1995; TRM 60), or from optical spectroscopy (van Loon et al. 1998). The chemical type of the other stars is unknown, but we will argue later that the sample can be separated into O- and C-rich based on the modelling of the SEDs.

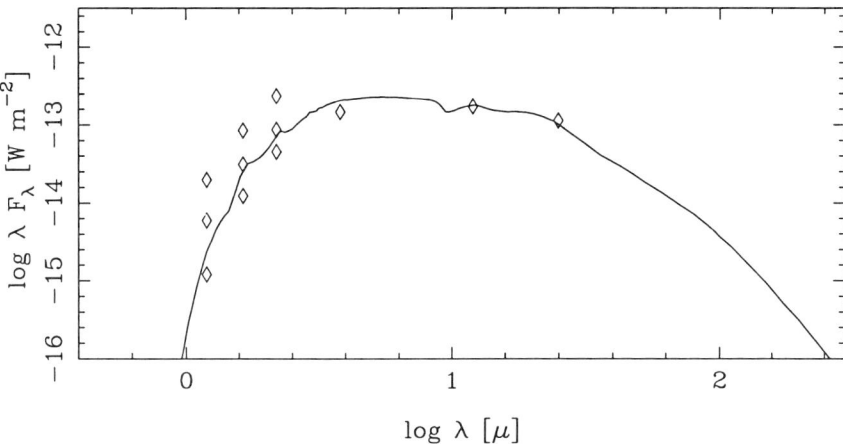

*Figure 1.* Fit to the spectral energy distribution of the oxygen-rich AGB star LMC 1506. Plotted are the IRAS data points at 12 and 25 $\mu$m and near-infrared data. The 3 values plotted for a given wavelength represent the fluxes at maximum, mean and minimum light.

## 3. Results

The SEDs are modelled with the radiative transfer model of Groenewegen (1993). Figure 1 shows the model fit to one of the known oxygen-rich stars. The dust opacity for silicates (Volk & Kwok 1988) is used. This type of dust gave the best results in fitting the 8–13 $\mu$m spectra of two AGB stars in the LMC and SMC (Groenewegen et al. 1995). The stellar photosphere is represented by an M5 model atmosphere (Fluks et al. 1994). This is the reason the model fit is not smooth in the NIR. The fitting is a two-step process. First the optical depth is determined by fitting the IRAS $S_{12}/S_{25}$ flux ratio. Then the final values of the mass-loss rate and luminosity are derived by adjusting the model to the observed 12 $\mu$m flux-density. Therefore the model predicts the near-infrared photometry and is not fitted to it. As the star is known to be oxygen-rich the model should lie between the observed fluxes at minimum and maximum light, as is indeed the case. In fact, all 6 known oxygen-rich stars are well fitted with silicate dust.

Figure 2 shows model fits to an AGB star of unknown spectral type. The upper panel is a fit to the IRAS data using silicate dust. This model clearly does not fit the observations. The lower panel is a fit using dust composed of 95% amorphous carbon (Rouleau & Martin 1991) with 5% silicon carbide (Pégourié 1988) and the stellar photosphere represented by a 2500 K blackbody. It is clear that the model with carbon-rich dust fits the data reasonably well, while the model with silicate dust completely fails. This is due to the completely different absorption properties of silicate and carbonaceous dust. We therefore argue that this star is a carbon star. Using

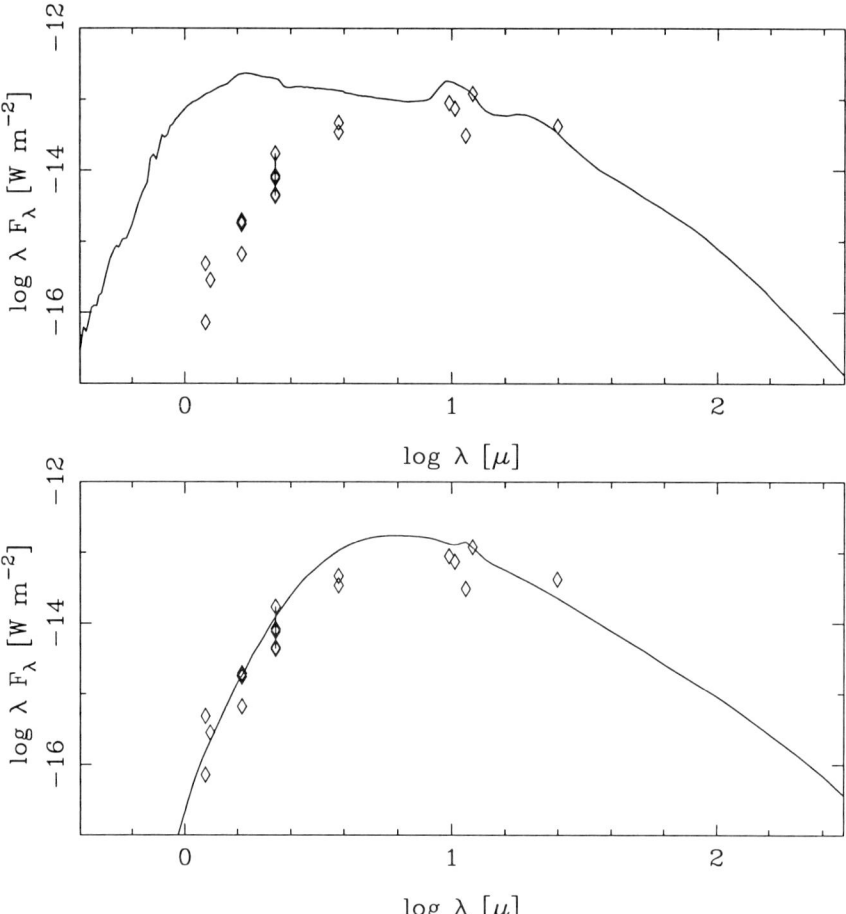

*Figure 2.* Fit to the spectral energy distribution of LMC 570, of unknown spectral type. In the upper panel silicate dust is used, in the lower panel amorphous carbon mixed with 5% silicon carbide.

this approach we find that of the 10 stars of unknown spectral type, 1 is probably O-rich, while the other 9 are probably C-rich.

## 4. Period-Luminosity Relations

The Period-Luminosity relation ($P$–$L$ relation) is a useful empirical tool. $P$–$L$ relations for O- and C-rich stars in the LMC were derived by Feast et al. (1989). Recently, an improved relation for C stars was presented by Groenewegen & Whitelock (1996). Both relations were limited to stars with periods less than ∼ 400–500 days because they were based on optically visible stars. The stars in the current sample have much longer periods compared to previous samples. It includes stars with periods of 700–1400 days compared to ≲500 days in the earlier LMC samples (Zijlstra et al. 1996;

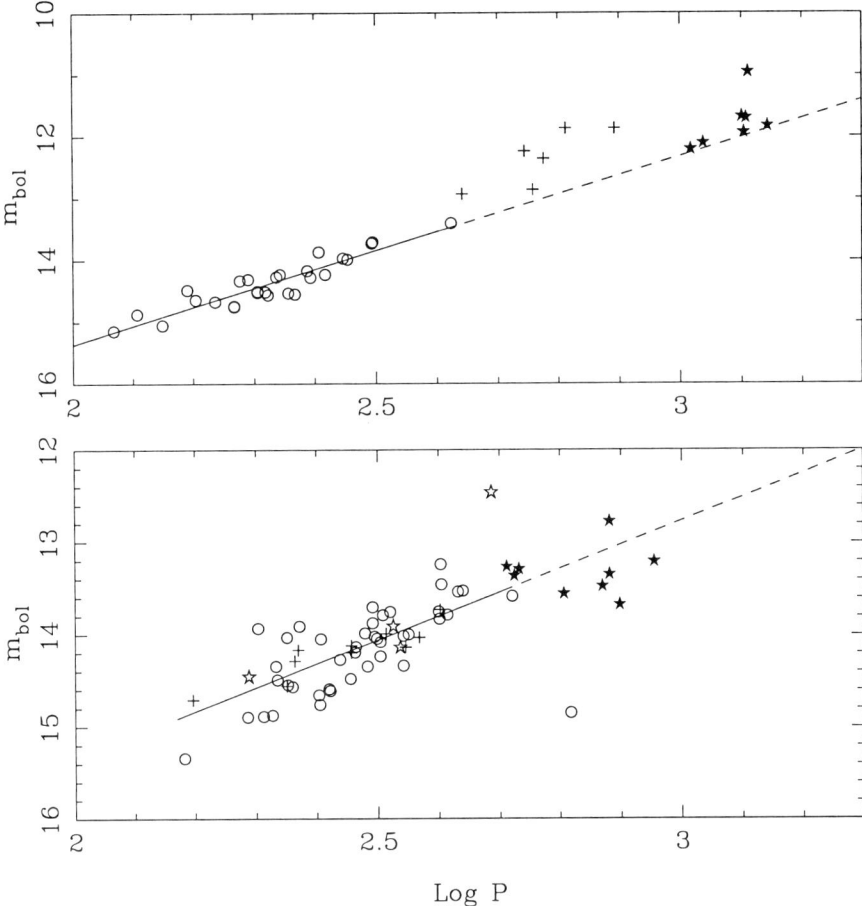

*Figure 3.* Period-Luminosity relation for O-rich (upper panel) and C-rich (lower panel) AGB stars in the LMC. The solid and dashed lines are least-squares fits (and the extrapolations, respectively) to the data points represented by the open symbols in the upper panel and open symbols and crosses in the lower one (except the two obvious outliers). The filled symbols represent the stars from the present sample. See text for further information.

Wood et al. 1992). The stars in the present study have better-determined luminosities and the separation between O- and C-rich is made more effectively than was previously possible.

In Fig. 3 the $P$–$L$ relations are shown. In the top panel the data used by Feast et al. (1989) for O-rich stars are shown (open circles and crosses). The solid line represents the $P$–$L$ relation derived by Feast et al. from a least-squares fit to the open circles; the dashed line is its extrapolation to longer periods. The filled symbols are data from the present paper. These points lie on the extrapolation of the Feast et al. $P$–$L$ relation derived for O-rich Miras with periods < 420 days.

In the lower panel the data from Groenewegen & Whitelock (1996) on C stars in the LMC are plotted, together with their $P$–$L$ relation and its extrapolation. This $P$–$L$ relation was derived from the crosses and open symbols which represent stars with periods less than 520 days. The C stars from the present sample (filled symbols) lie on the extrapolated $P$–$L$ relation.

In conclusion, the study of IRAS-detected AGB stars in the LMC allows one to investigate the most luminous AGB stars with high mass-loss rates and long pulsation periods. The model fitting enables a reliable determination of the carbon- or oxygen-rich nature of a star (due to the completely different absorption properties of silicate and carbonaceous dust). Both the O- and C-rich LPVs in this study appear to lie on extensions of $P$–$L$ relations derived for stars with much shorter periods.

## References

Assendorp, R., Bontekoe, T. R., de Jonge, A. R. W., Kester, D. J. M., Roelfsema, P. R. & Wesselius, P. R. 1995, *A&A Supp.*, 110, 395
Feast, M. W., Glass, I. S., Whitelock, P. A. & Catchpole, R. M. 1989, *MNRAS*, 241, 375
Fluks, M. A., Plez, B., Thé, P. S., de Winter, D., Westerlund, B. E. & Steenman, H. C. 1994, *A&A Supp.*, 105, 311
Groenewegen, M. A. T. 1993, Ph.D. Thesis, University of Amsterdam, Chapter 5
Groenewegen, M. A. T. 1995, *A&A*, 293, 463
Groenewegen, M. A. T., Smith, C. H., Wood, P. R., Omont, A. & Fujiyoshi, T. 1995, *ApJ*, 449, L119
Groenewegen, M. A. T. & Whitelock, P. A. 1996, *MNRAS*, 281, 1347
Justtanont, K. & Tielens, A. G. G. M. 1992, *ApJ*, 389, 400
Pégourié, B. 1988, *A&A*, 194, 335
Rouleau, F. & Martin, P. G. 1991, *ApJ*, 377, 526
van Loon, J. Th., Zijlstra, A. A., Whitelock, P. A., te Lintel Hekkert, P., Chapman, J. M., Loup, C., Groenewegen, M. A. T., Waters, L. B. F. M. & Trams, N. R. 1998, *A&A*, 329, 169
Volk, K. & Kwok, S. 1988, *ApJ*, 331, 435
Wood, P. R., Whiteoak, J. B., Hughes, S. M. G., Bessell, M. S., Gardner, F. F. & Hyland, A. R. 1992, *ApJ*, 397, 552
Wood, P. R. 1998, submitted
Zijlstra, A. A., Loup, C., Waters, L. B. F. M., Whitelock, P. A., van Loon, J. Th. & Guglielmo, F. 1996, *MNRAS*, 279, 32
Zijlstra, A. A. et al. 1998, in preparation

## Discussion

**Joyce**: I have completed a photometric monitoring program of long-period variables with dense circumstellar shells similar to that of Whitelock, but with a more extreme selection criterion, $K$–$[11\,\mu m] > 4$. I found one C-type star, AFGL 190, with $P = 1050$ d. This star is sufficiently faint and red ($K \approx 14$; $L \approx 7$) that an analog in the LMC may not be detectable in the near-infrared.

# IR EMISSION FROM DUSTY WINDS — SCALING AND SELF-SIMILARITY PROPERTIES

MOSHE ELITZUR AND ŽELJKO IVEZIĆ
*Department of Physics and Astronomy*
*University of Kentucky*
*Lexington, KY 40506-0055, U.S.A.*

**Abstract.** Infrared emission from radiatively heated dust possesses general scaling properties. The spectral shape is independent of overall luminosity when the inner boundary of the dusty region is controlled by dust sublimation; the only relevant property of the heating radiation is its frequency profile. For a given type of dust grains, the emission from red giants and supergiants is essentially controlled by a single parameter — the overall dust optical depth. This leads to tight correlations among different spectral properties that explain many available observations and enable systematic studies of large data bases.

## 1. Introduction

The radiation received on earth from late-type stars has undergone significant processing in the surrounding dust shell. Interpretation of the observations therefore necessitates considerable theoretical effort, involving detailed radiative transfer calculations. These calculations have traditionally required a large number of input parameters that fall into three categories:

*The Star.* Stellar input properties include the mass $M_*$, luminosity $L_*$, temperature $T_*$ and the mass-loss rate $\dot{M}$.

*The Shell.* This is described by a density profile $\rho(r)$, defined between some inner and outer radii $r_1$ and $r_2$, respectively.

*Dust Properties.* The dust abundance can be expressed via the dust-to-gas ratio $\rho_d/\rho$. The properties of individual grains must be specified, too, and quantities widely employed include the grain size $a$, the solid density $\rho_s$, the sublimation temperature $T_{sub}$ and the absorption and scattering efficiencies $Q_{abs}$ and $Q_{sca}$.

Once these input properties are prescribed, they are plugged into a detailed radiative transfer calculation whose output can be compared with the observations. The most widely used output quantity is the spectral shape

$$f_\lambda = \frac{F_\lambda}{F}, \qquad (1)$$

where $F_\lambda$ and $F$ are the observed flux density and bolometric flux, respectively. Another quantity that can be compared with observations when the angular resolution is sufficiently high is the surface brightness.

The rather large number of input parameters that had to be specified in traditional calculations creates two major practical problems. First, the volume of parameter space that must be searched to fit a given set of observations can become prohibitively large. Second, and more serious, even when a successful fit is accomplished, its uniqueness is questionable and the model parameters cannot be trusted as a reliable indication of the actual properties of the source.

## 2. Scaling

Much of the input required in past calculations is in fact redundant. In Ivezić & Elitzur (1995; IE95 hereafter) we were able to show that the IR emission from late-type stars obeys general scaling properties that greatly reduce the number of input parameters required to fully specify the radiative transfer problem. The primary input involves only the dust, for which we must specify two types of properties: (1) Optical properties, specified through the normalized spectral shape of the extinction efficiency $q_\lambda = Q_\lambda/Q_{\lambda_0}$, where $\lambda_0$ is some arbitrary fiducial wavelength; this spectral shape is controlled by the grain chemistry and size. (2) The overall optical depth $\tau_{\lambda_0}^T$; this is controlled primarily by the mass-loss rate. Additional input properties have only secondary significance and involve the dust sublimation temperature $T_{sub}$, the relative thickness of the dust shell $r_2/r_1$ and the stellar temperature $T_*$. Note that $T_*$ is the only stellar property that enters (and only in a secondary role). In particular, the stellar luminosity is irrelevant.

The proof of scaling is quite simple. The shell inner radius $r_1$ is controlled by dust sublimation, namely, $T_d(r_1) = T_{sub}$. Introduce the dimensionless radial distance $y = r/r_1$. Then the radiative transfer equation becomes

$$\frac{dI_\lambda}{dy} = \tau_\lambda^T \eta(y)(S_\lambda - I_\lambda) \qquad (2)$$

for $y \geq 1$. In this form, the equation contains no reference to either densities or dimensions. The only scale is determined by $\tau^T$; geometrical quantities enter only through the dimensionless radius $y$. The density enters only through $\eta = n(y)/\int n(y)dy$, its dimensionless, normalized radial profile. Because the outflows around late-type stars are controlled by radiation pressure on dust grains, the profile $\eta$ is uniquely determined by $\tau^T$ when the effects of gravity and dust drift are negligible (IE95). In fact, under these circumstances, the analytic expression

$$\eta \propto \frac{1}{y^2}\sqrt{\frac{y}{y-1+(v_1/v_\infty)^2}} \tag{3}$$

provides an excellent approximation to the actual density profiles we find in our detailed numerical calculations. This function requires as additional input the ratio of initial to final velocity, $v_1/v_\infty$. However, this parameter has a negligible effect on the result for its typical values, $\lesssim 0.1$.

Additional input required is the stellar radiation. This can be specified by its flux density $F_{*\lambda} = F_* \times f_{*\lambda}$, where $F_*$ is the bolometric flux. Each value of $F_1 = F_*(y=1)$ uniquely determines a corresponding value of $T_1$, the dust temperature at $y = 1$, so the reverse is also true. And since the inner boundary is controlled by dust sublimation, $T_{sub}$ fixes the value of $T_1$ and through it of $F_1$. When the stellar luminosity varies, the shell adjusts its inner boundary so that $F_1$ and $T_1$ ($= T_{sub}$) stay the same. The radiative transfer equation is oblivious to these changes because the inner boundary always corresponds to $y = 1$.

These scaling properties can be generalized to arbitrary geometries and density distributions (Ivezić & Elitzur 1997). The spectral shape of dust IR emission is independent of overall luminosity when the inner boundary of the dusty region is controlled by dust sublimation; the only relevant property of the heating radiation is its spectral shape. Densities and geometrical dimensions are likewise irrelevant; they enter only through one independent parameter, the overall optical depth. The geometry enters only through angles and aspect ratios. Dust properties enter only through dimensionless, normalized distributions that describe the spatial variation of density, and the wavelength dependence of scattering and absorption efficiencies.

## 3. Consequences

Scaling implies that for a given type of dust, the spectral energy distributions (SEDs) of late-type stars are controlled almost exclusively by overall optical depth and thus form a one-parameter family. Therefore, a single point on the normalized spectral shape $f_\lambda$ determines the entire function and any two spectral properties should be correlated with each other. Figure 1 displays in its left panel the distribution of normalized fluxes at 12 and 60 $\mu$m for a sample of 89 stars. These are all the IRAS objects identified

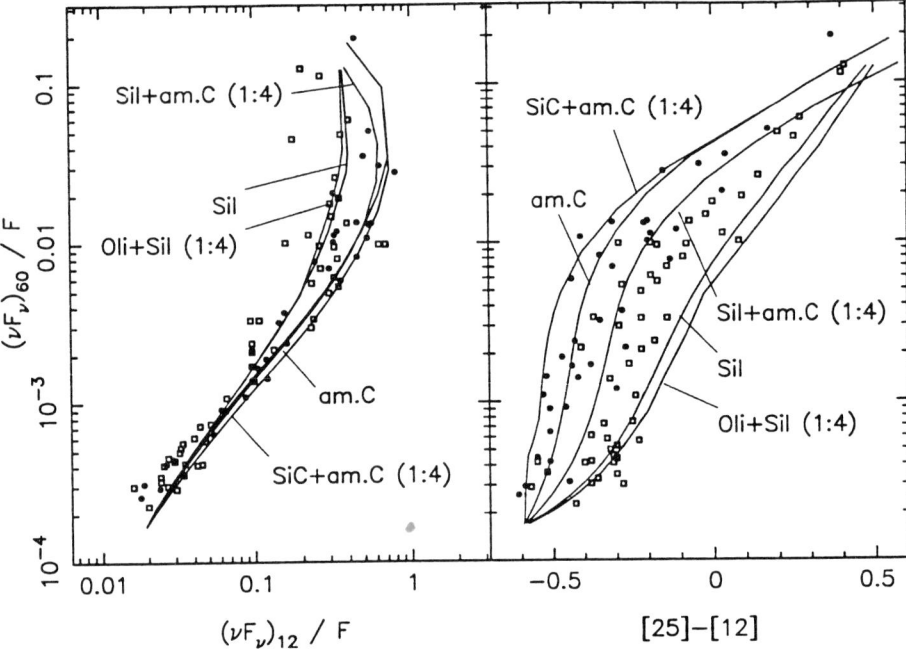

*Figure 1.* Test of correlations predicted by scaling. Open squares mark the data points for O-rich stars, solid circles C-rich stars. Curves describe the model predictions for various grain compositions, as marked.

as late-type stars for which we were able to find listings in the literature for total fluxes, luminosities, terminal velocities and mass-loss rates. As expected from scaling, the normalized fluxes do display a tight correlation that holds over the entire observed range, covering more than three orders of magnitude. The spread can be attributed to the differences in dust properties between carbon- and oxygen-rich stars. The right panel displays the corresponding distribution of $(\nu F_\nu)_{60}/F$ and [25]–[12] color. The expected correlation again is evident; the separation between C-rich and O-rich stars is more pronounced. The lines are our computed model spectra, displaying a close agreement with the data for various relevant chemical compositions ('Sil' stands for astronomical silicate; 'am.C' for amorphous carbon; 'Oli' for olivine). The agreement between model predictions and observations is better than a factor of 2 in almost all cases. In addition, O-rich and C-rich stars clearly separate and congregate in accordance with the trend of the model curves for corresponding chemical compositions. This behavior is particularly prominent in the right panel.

Location in IRAS color-color diagrams has become a widely used indicator of the nature of a source. Van der Veen & Habing (1988) identified the

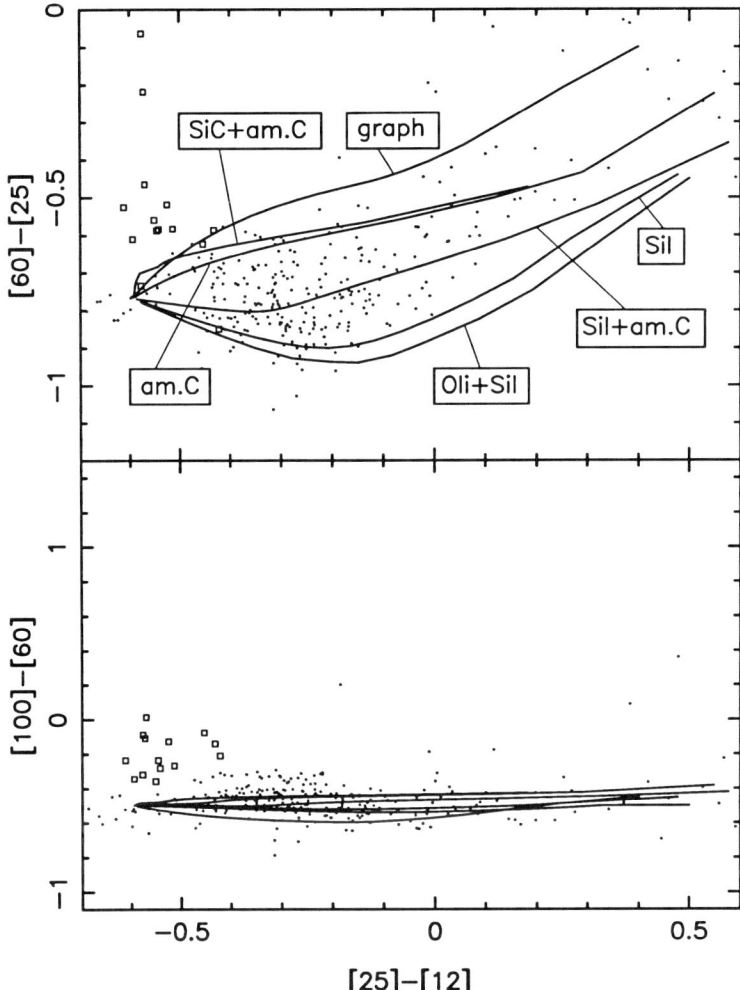

*Figure 2.* IRAS color-color diagrams of all sources with reliable data in the VH-window. Solid lines represent theoretical sequences for different grain compositions, as marked. In all mixtures, the abundance ratio of the first to second component is 1:4. In the lower panel, the order of the tracks from bottom to top around [25]–[12] ≈ 0 is graphite, Oli + Sil, Sil, Sil + am.C, am.C, and SiC + am.C.

appropriate IRAS region for late-type stars as [60]–[25] < 0 and [25]–[12] < 0.6, which we will refer to as the "VH-window." Figure 2 displays the color-color diagrams for all IRAS sources with reliable data whose colors fall in this window. Our model predictions for selected grain compositions are displayed as tracks. Position along each track is determined by $\tau^T$ and all tracks originate from the same spot, $\tau^T = 0$, corresponding to a blackbody spectrum at 2500 K convoluted with the IRAS instrumental profiles.

Distance from this common origin along each track increases with $\tau^T$. As $\tau^T$ increases, the dust emission becomes more prominent and the tracks of different grains branch out. The tracks for purely Si- and C-based grains outline the distribution boundaries of most IRAS sources in the VH-window, verifying the VH conjecture that this is the color-color location of late-type stars. The data points fill the entire region between these tracks and if this spread is real, it requires a similar spread in the optical properties of the grains. This could reflect chemical mixtures, as we proposed in IE95, or perhaps grain impurities.

It is important to note that we have removed from our sample sources whose IRAS fluxes are severely contaminated by cirrus emission. This contamination is measured by the cirrus contamination index

$$CCI \equiv \frac{\text{cirr3}}{F(60)}, \qquad (4)$$

where cirr3 is the 100 $\mu$m cirrus flux at the location of the source and $F(60)$ is its listed 60 $\mu$m flux. In IE95 we show that the [100]–[60] color and $CCI$ are perfectly correlated in sources with $CCI \gtrsim 2$, indicating that the IRAS 60 and 100 $\mu$m fluxes of these sources are unreliable. Sources in the upper-left corner of the color-color diagrams were thought to provide evidence for cool shells (Willems & de Jong 1986), but almost all of these sources turn out to be cirrus contaminated. Of the total of 292 sources displayed in Fig. 2, only 11 late-type stars (4%, marked with open squares) can be identified as a group whose colors are not well explained by steady-state radiatively-driven winds. Furthermore, among these 11 sources, 10 have borderline contamination with $1 < CCI < 2$. Based on new simultaneous observations in a wide spectral range, Miroshnichenko et al. (2000) present model fits for 4 carbon stars. The results are reproduced in Figure 3, showing excellent agreement with the model for steady-state winds without any additional shells. The only parameter that varies from model to model is the visual optical depth $\tau_v$. Two of the displayed stars, RY Dra and VY UMa, belong to the group marked with squares in Fig. 2. They are borderline contaminated and show a 100 $\mu$m excess proportional to $CCI$, so clearly they do not give any indication of detached shells. There is only one clearly uncontaminated C star, U Ant, in the "cool shell" zone. Indeed, this star has a thin CO shell (Olofsson et al. 1990). We conclude that, because of potential cirrus contamination, the SED is unfortunately not a good indicator of detached shells.

Thanks to scaling, the SEDs of large samples can be modeled systematically. We have embarked on a large-scale modeling project in which we are fitting all IR data of all the late-type stars in the IRAS catalogue ($\sim 5{,}000$ sources). The outcome of this project will be a supplemental catalogue,

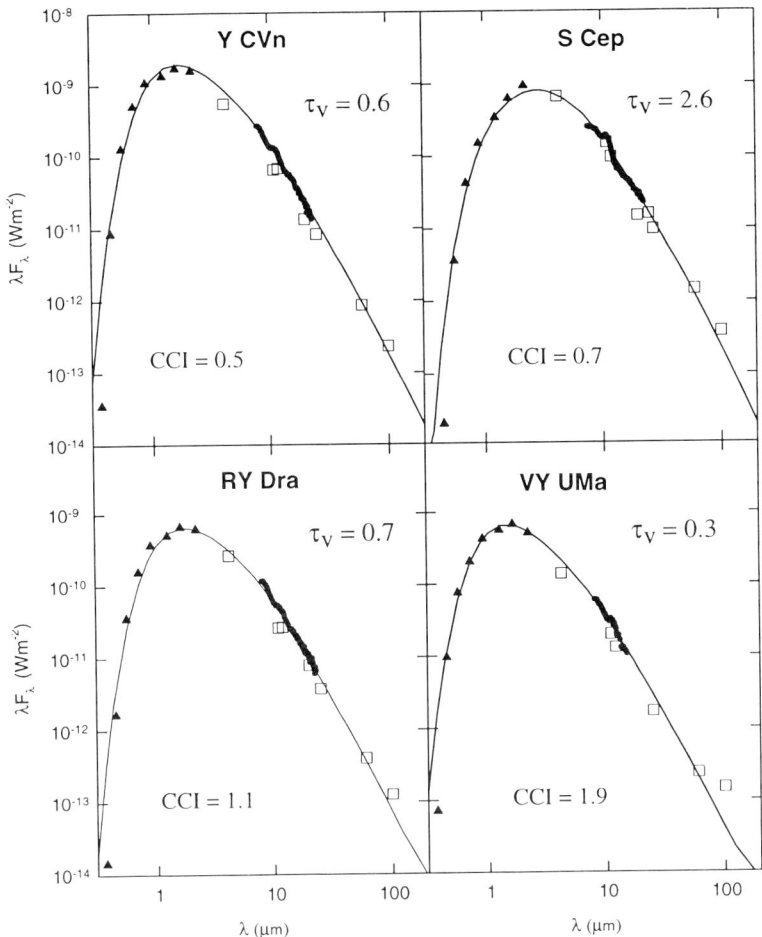

*Figure 3.* Model fits for four carbon stars with new ground based observations (triangles: Miroshnichenko et al. 2000). Circles denote LRS data, squares are IRAS PSC and RAFGL data. The only adjusted model parameter is the visual optical depth $\tau_v$.

listing the optical depths and dust properties of the largest sample yet modeled with a single, consistent approach.

## 4. Conclusions

The structure displayed in infrared color-color diagrams is a result of general scaling properties of radiatively heated dust. Sources segregate into families that differ from each other because of dust properties, not because of the central source. Within a family, the location of each source is controlled by its optical depth. Concerning late-type stars — properly selected IRAS data are almost fully explained by steady-state winds. The spectral shapes of sources free of cirrus contamination do not give evidence for detached shells in more than a handful of sources.

Support by NASA grant NAG 5-3010, NSF grant AST-9321847 and the Center for Computational Sciences at the University of Kentucky is gratefully acknowledged.

## References

Ivezić, Ž. & Elitzur, M. 1995, *ApJ*, 445, 415 (IE95)
Ivezić, Ž. & Elitzur, M. 1997, *MNRAS*, 287, 799
Miroshnichenko, A. S., Kuratov, K. S., Ivezić, Ž. & Elitzur, M. 2000, in IAU Symp. 177: *The Carbon Star Phenomenon*, ed. R. F. Wing (Kluwer), p. 566
Olofsson, H., Carlström, U., Eriksson, K., Gustafsson, B. & Willson, L. A. 1990, *A&A*, 230, L13
van der Veen, W.E.C.J. & Habing, H. J. 1988, *A&A*, 194, 125
Willems, F. J. & de Jong, T. 1986, *ApJ*, 309, L39

## Discussion

**Dorfi:** Concerning the 12 $\mu$m flux: How does the assumed $\rho \sim r^{-2}$ density law change your conclusions, and how does the 12 $\mu$m flux vary with phase?

**Elitzur:** We do not assume a $\rho \propto r^{-2}$ density. Rather, we solve fully the coupled equations of hydrodynamics and radiative transfer. The analytic expression we derive is an excellent approximation to the exact solution in most regions of interest. The variation of 12 $\mu$m flux with phase is not too large. These variations are noticeable mostly in the near IR. The exact variations of $F_{12}$ are dependent on the grain chemistry — silicate dust has a strong feature in this range. In general $F_{12}$ for C dust will be larger at maximum, while both trends are possible for Si dust.

**Hron:** In the [12]–[25]/[25]–[60] diagram, your pure silicate track runs below the bulk of M stars. The M stars are on the track with a mixture of silicate and carbon dust. How do you explain that?

**Elitzur:** For the tracks to go through the bulk of the data, there must be a spread in the absorption coefficients. In the original paper we suggested that implied mixed grain chemistry. Recently we learned that the literature includes a number of variations on the astronomical silicate absorption coefficient. The one we used was from Draine & Lee. It turns out that for this absorption coefficient, the ratio between the values at the 9.8 $\mu$m feature and short wavelengths is quite extreme. Recent suggestions of grain impurities appear to give a similar effect to that of adding C grains, providing a spread in absorption properties that goes in the right direction.

# DUST EMISSION FROM IRC +10216

ŽELJKO IVEZIĆ AND MOSHE ELITZUR
*University of Kentucky*
*Lexington, KY 40506-0055, U.S.A.*

**Abstract.** Infrared emission from the dust shell around IRC +10216 is analysed in detail, employing a self-consistent model for radiatively driven winds around late-type stars that couples the equations of motion and radiative transfer in the dust. The resulting model provides agreement with the wealth of available data, including the spectral energy distribution in the range 0.5–1000 $\mu$m, and visibility and array observations. Previous conclusions about two dust shells, derived from modelling the data with a few single-temperature components of different radii, are not supported by our results. The IR properties vary with the stellar phase, reflecting changes in both the dust condensation radius $r_1$ and the overall optical depth $\tau$ — as the luminosity increases from minimum to maximum, $r_1$ increases while $\tau$ decreases. We find that the angular size of the dust condensation zone varies from 0.″3 at minimum light to 0.″5 at maximum. The shortage of flux at short wavelengths encountered in previous studies is resolved by employing a grain size distribution that includes grains larger than $\sim 0.1$ $\mu$m, required also for the visibility fits. This distribution is in agreement with the one recently proposed by Jura in a study that probed the outer regions of the envelope. Since our constraints on the size distribution mostly reflect the envelope's inner regions, the agreement of these independent studies is evidence against significant changes in grain sizes through effects like sputtering or grain growth after the initial formation at the dust condensation zone.

## 1. Introduction

The purpose of this work is to perform a self-consistent case study of the bright infrared source IRC +10216 that employs a dust density distribution determined from the solution of the coupled system of radiative transfer and hydrodynamics equations for the wind. The equations are described elsewhere (Netzer & Elitzur 1993; Ivezić & Elitzur 1995, hereafter IE95).

As shown by Elitzur & Ivezić (2000, this Symposium), the solution of this system is essentially determined by a single quantity – the flux-averaged optical depth $\tau_F$. Once $\tau_F$ is determined, scaling relations listed in IE95 and in Ivezić & Elitzur (1996a, 1996b) can be used to constrain all other relevant quantities.

## 2. Spectral Energy Distribution

Our best-fitting model for the spectral shape is shown in Figure 1 together with the observations. The model is primarily determined by the overall optical depth and the dust composition. From previous work (e.g. Blanco et al. 1994), the dust grains around IRC +10216 are primarily composed of amorphous carbon with a minor inclusion of SiC to account for the 11.3 $\mu$m feature. With optical properties for amorphous carbon taken from Hanner (1988) and for SiC from Pégourié (1988), we find that the best fit to the 11.3 $\mu$m feature is obtained with a mixture of 95% amorphous carbon and 5% SiC (by mass), although varying the percentage of SiC in the range 3–8% still produces satisfactory agreement. In addition to the chemical composition, the distribution of grain radii $a$ also affects the optical properties. We employed two types of size distributions $n(a)$. Most often used is

$$n_{\mathrm{MRN}}(a) \propto a^{-3.5}, \qquad a \leq a_{\max} \qquad (1)$$

proposed by Mathis, Rumpl & Nordsieck (1977). Recently Jura (1994) proposed a similar form

$$n_{\mathrm{J}}(a) \propto a^{-3.5} e^{-a/a_0}, \qquad (2)$$

replacing the sharp cutoff with an exponential one. Both distributions can produce satisfactory fits: the MRN distribution requires $a_{\max} \approx 0.2$–$0.3$ $\mu$m, the Jura distribution $a_0 \approx 0.15$–$0.2$ $\mu$m. With these grain properties, the best-fit values for the visual optical depth are 20 at maximum luminosity and 24 at minimum.

We have thus determined the two major ingredients that affect the spectral shape, the grain optical properties and overall optical depth. In addition, the stellar temperature $T_*$ and dust condensation temperature $T_1$ have a discernible effect on the spectral shape, but only at short wavelengths. Our best fit gives $T_* = 2200 \pm 150$ K, $T_1 = 750 \pm 50$ K.

## 3. Spatially Resolved Observations

As the stellar luminosity varies from minimum to maximum, the dust condensation radius increases, and the overall optical depth decreases. Therefore in interpreting the spatially resolved observations, the source variability must be taken into account. From our best-fitting model for the SED, and

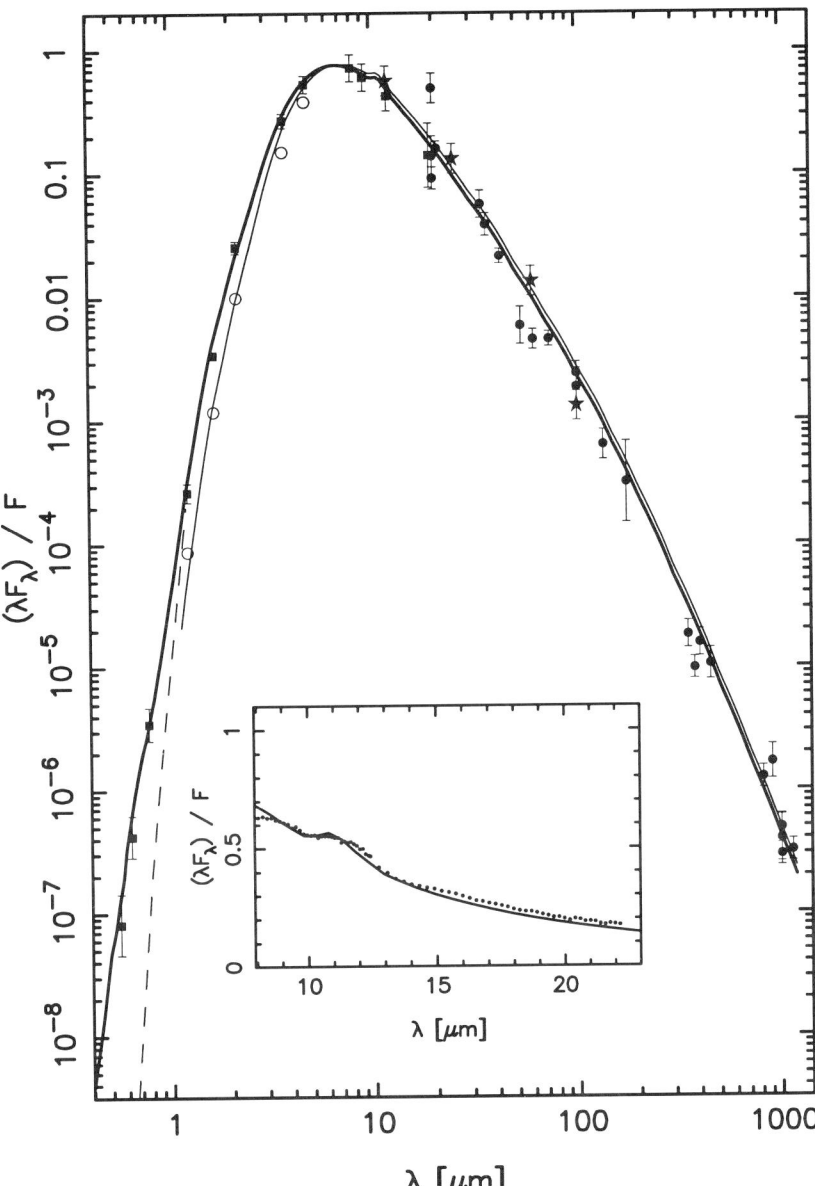

*Figure 1.* Spectral energy distribution for IRC +10216; lines represent model results, symbols the observations. Data are from the following sources: (■) Le Bertre (1987); (○) Le Bertre (1988); (●) Rengarajan et al. (1985); and (⋆) IRAS Point Source Catalogue. All observations are at maximum light except for those denoted by open circles, which were at minimum light. The thick solid line is the model result for maximum light, the thin solid line the result for minimum. The dashed line is the model result for maximum light and single-size (0.05 $\mu$m) grains. The inset shows an expanded view of the IRAS LRS spectral region: the dots are the data, taken close to maximum light, the solid line the model. From Ivezić & Elitzur (1996b).

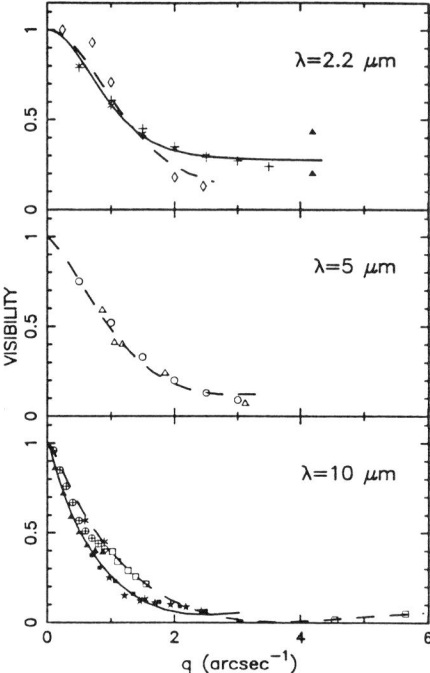

*Figure 2.* Visibility functions for IRC +10216. Lines represent model results, symbols the observations. Solid lines and full symbols (including + and *) correspond to phases close to maximum light, open symbols and dashed lines to phases close to minimum. Data are from the following sources: (⋆) Sutton, Betz & Storey (1979); (◇) Selby, Wade & Sanchez Magro (1979); (△) McCarthy, Howell & Low (1980); (o) Mariotti et al. (1983); (+) Dyck et al. (1984); (∗) Dyck et al. (1987); (⊕) Benson, Turner & Dyck (1989); and (□) Danchi et al. (1990, 1994). From Ivezić & Elitzur (1996b).

an expected bolometric amplitude of 1 mag, we find that the angular size of the dust condensation zone, $\theta_1$, varies from 0.″35 at minimum light to 0.″56 at maximum. These estimates for the angular scale must agree with high-resolution observations. We find this to be the case to within 15–20% for the visibility observations shown in Figure 2.

Bloemhof et al. (1988) obtained a single-scan image of IRC +10216 at 10 $\mu$m close to minimum light (phase $\simeq$ 0.4). We computed the profile expected in those observations from the model surface brightness determined for this phase from the spectral shape. Figure 3 shows the comparison between the observed profile (thick dashed line) and our convolved model result (outermost thin solid line, overlapping the observations). The innermost thin solid line is the model surface brightness distribution, the central peak corresponds to the stellar contribution, and the features at relative RA $\pm\theta_1/2$ correspond to the dust formation zone.

The close agreement between our models and the spatially resolved

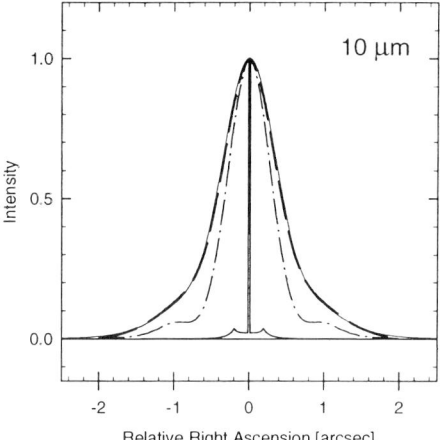

*Figure 3.* Single-scan (E–W) imaging of IRC +10216 at 10 $\mu$m. The thick dashed line represents the observations of Bloemhof et al. (1988). Superimposed on it is our model result drawn as a thin solid line, hardly distinguishable from the observations. It was obtained by a two-dimensional convolution of the surface brightness for $\theta_1 = 0.''35$ (innermost thin solid line) with the point-spread function (PSF, dot-dashed line; all profiles are normalized to unity at the peak).

observations shows that previous conclusions about two dust shells, derived from modeling the data with a few single-temperature components of different radii, are not supported by our results. The extended, continuous temperature and density distributions derived from our model obviate the need for such discrete shells.

Support by NSF grant AST–9321847, NASA grant NAG 5–3010, an AAS Travel Grant, and the Center for Computational Sciences of the University of Kentucky is gratefully acknowledged.

## References

Benson, J. A., Turner, N. H. & Dyck, H. M. 1989, *AJ*, 97, 1763
Blanco, A., Borghesi, A., Fonti, S. & Orofino, V. 1994, *A&A*, 283, 561
Bloemhof, E. E., Danchi, W. C., Townes, C. H. & McLaren, R. A. 1988, *ApJ*, 333, 300
Danchi, W. C., Bester, M., Degiacomi, C. G., McCullough, P. R. & Townes, C. H. 1990, *ApJ*, 359, L59
Danchi, W. C., Bester, M., Degiacomi, C. G., Greenhill, L. J. & Townes, C. H. 1994, *AJ*, 107, 1469
Dyck, H. M., Zuckerman, B., Leinert, Ch. & Beckwith, S. 1984, *ApJ*, 287, 801
Dyck, H. M., Zuckerman, B., Howell, R. R. & Beckwith, S. 1987, *PASP*, 99, 99
Elitzur, M. & Ivezić, Ž. 2000, in IAU Symp. 177: *The Carbon Star Phenomenon*, ed. R. F. Wing (Kluwer), p. 391
Hanner, M. S. 1988, in *Infrared Observations of Comets Halley and Wilson and Properties of the Grains*, ed. M. S. Hanner, NASA CP–3004, p. 22
Ivezić, Ž. & Elitzur, M. 1995, *ApJ*, 445, 415 (IE95)
Ivezić, Ž. & Elitzur, M. 1996a, *MNRAS*, 279, 1011

Ivezić, Ž. & Elitzur, M. 1996b, *MNRAS*, 279, 1019
Jura, M. 1994, *ApJ*, 434, 713
Le Bertre, T. 1987, *A&A*, 176, 107
Le Bertre, T. 1988, *A&A*, 203, 85
Mariotti, J. M., Chelli, A., Foy, R., Léna, P., Sibille, F. & Tchountonov, G. 1983, *A&A*, 120, 237
Mathis, J. S., Rumpl, W. & Nordsieck, K. H. 1977, *ApJ*, 217, 425
McCarthy, D. W., Howell, R. & Low, F. J. 1980, *ApJ*, 235, L27
Netzer, N. & Elitzur, M. 1993, *ApJ*, 410, 701
Pégourié, B. 1988, *A&A*, 194, 335
Rengarajan, T. N., Fazio, G. G., Maxson, C. W., McBreen, B., Serio, S. & Sciortino, S. 1985, *ApJ*, 289, 630
Selby, M. J., Wade, R. & Sanchez Magro, C. 1979, *MNRAS*, 187, 553
Sutton, E. C., Betz, A. L. & Storey, J. W. V. 1979, *ApJ*, 230, L105

## Discussion

**Wagenhuber**: Do you get information about the mass-loss rate out of your model, or is it an input parameter?

**Ivezić**: IR emission constrains optical depth and not $\dot{M}$. However, optical depth does depend on $\dot{M}$ and various other parameters. If there are independent estimates for these other parameters, then one can constrain $\dot{M}$, too.

**Bagnulo**: The value of 5 % for the SiC/AC ratio derived from your analysis is actually dependent on the assumption that both AC and SiC grains form at the same distance from the star.

**Ivezić**: Certainly. However, varying the difference in the condensation temperatures of amorphous carbon and SiC will have only a minor effect on the inferred SiC/AC ratio. Such differences in the SiC/AC ratio are on the order of the uncertainty in its determination, and I would say that it is probably in the range 3–10 %.

**Steffen**: What, in the context of your steady-state models, is the explanation for the existence of detached dust shells?

**Ivezić**: As discussed in our paper (*ApJ* 445, 415, 1995), one possibility is interaction between the wind and the interstellar medium. However, as recently shown by Schönberner and co-workers, the evolutionary scenario provides an equivalent, if not better, explanation.

**Kandemir**: There seems to be some uncertainty about the use of the term "envelope", e.g. where it starts — do some astronomers avoid using the term "envelope" and prefer "shell"?

**Ivezić**: We find that $r_2/r_1$ must be at least 700. Whether to call this an "envelope" or a "shell" seems to be more semantics than a real physical problem.

# Session IX

CIRCUMSTELLAR EMISSION
AND
ENVIRONMENT

Antalya astronomers Zeki Aslan and Orhan Gölbaşı arriving at the banquet with their wives and daughters.

Waiting for sushi: Takashi Iijima, Kunio Noguchi, Wako Aoki, and Takashi Tsuji posing before the banquet.

# MOLECULAR OPTICAL AND INFRARED EMISSION FROM THE RED RECTANGLE

*Carriers of Diffuse Circumstellar, Nebular and Interstellar Bands*

T. H. KERR, J. R. MILES, M. E. HURST,
R. E. HIBBINS AND P. J. SARRE

*Department of Chemistry, The University of Nottingham*
*University Park, Nottingham NG7 2RD, England*

**Abstract.** A link between stellar and interstellar molecules is discussed with particular reference to optical and infrared emission from the Red Rectangle.

The diffuse interstellar bands appear in spectra recorded towards stars which are reddened by interstellar dust. They extend across much of the near-UV, visible and near-infrared spectrum and now number over 150. The generally good correlation of band strength with extinction has led to the suggestion that the absorptions arise in or on the dust grains, but recent observations provide strong support for the alternative hypothesis that the carriers are free gas-phase molecules (for a review see Herbig, 1995).

In what way is the diffuse interstellar absorption band problem related to the carbon star phenomenon? First, it is likely that the diffuse interstellar band carriers are carbon-rich and so it is plausible that they are formed by reactions in carbon stars, although there is very little hard evidence that this is the case. It may be that the important chemical structures are built through the chemistry of carbon stars but that the spectroscopic absorptions appear only when the products are subjected to a strong UV radiation field in the interstellar medium, resulting in a change in chemical form. This could be through photoionization, photodissociation leading to hydrogen atom loss, photoisomerization, etc. Secondly, the diffuse band carriers probably bridge a gap in the size distribution of astronomical material, lying between diatomic molecules such as $C_2$ and dust grains, both of which are abundant in stellar atmospheres and in diffuse interstellar clouds.

The Red Rectangle is a unique object which has a strong unidentified optical emission spectrum together with the set of 'unidentified' infrared emission bands (UIBs). It is a biconical nebula with a binary star at the centre and it is probable that the optical emission is excited by the A0-type star which illuminates material emanating from a carbon-rich star. The 'UIBs' are excited by UV excitation of PAH molecules or material. The optical emission bands arise from a subset of the carriers of the diffuse interstellar absorption bands (Sarre et al. 1995b), and the form and evolution of the Red Rectangle spectra as a function of offset from the exciting star strongly suggest a molecular (rather than grain) origin. The emission bands have also been seen in spectra of the R CrB star V854 Cen during minimum light (Rao & Lambert 1993), and the fact that both the Red Rectangle and V854 Cen are carbon-rich gives support to the idea of carbon-rich molecules being responsible for the spectra.

Although not observed in the infrared spectra of circumstellar shells such as IRC+10216, the widespread appearance of the UIBs has led to suggestions that the diffuse interstellar bands also arise from PAHs, possibly in ionised, dehydrogenated or radical form. The unique geometry of the Red Rectangle allows comparison between the spatial distribution of the optical bands and the 3.3 $\mu$m emission, attributed to the C–H stretch of PAHs. Comparing our data obtained using CGS4 on UKIRT with those of Schmidt & Witt (1991) for the optical bands, shows that whereas the yellow/red bands are prominent along the interfaces of the bicone, the infrared emission is almost symmetrically distributed in the nebula. While a chemical link between PAHs and the optical bands may well exist, it appears that the optical emission arises from a significantly different chemical form in which the C–H component is not important. Recent ultra-high-resolution studies of diffuse absorption bands falling in the same subgroup have revealed fine structure in the spectra (Sarre et al. 1995a) which further supports a molecular origin. The spectra can be modeled in terms of molecular carriers, and pure carbon rings emerge as good candidates for the carriers of some of the diffuse bands (Kerr et al. 1996).

## References

Herbig, G. H. 1995, *Ann. Rev. Astron. Astrophys.*, 33, 19
Kerr, T. H., Hibbins, R. E., Miles, J. R., Fossey, S. J., Somerville, W. B. & Sarre, P. J. 1996, *MNRAS*, 283, L105
Rao, N. K. & Lambert, D. L. 1993, *MNRAS*, 263, L27
Sarre, P. J., Miles, J. R., Kerr, T. H., Hibbins, R. E., Fossey, S. J. & Somerville, W. B. 1995a, *MNRAS*, 277, L41
Sarre, P. J., Miles, J. R. & Scarrott, S. M. 1995b, *Science*, 269, 674
Schmidt, G. D. & Witt, A. N. 1991, *ApJ*, 383, 698

# SCATTERED LIGHT FROM ENVELOPES AROUND N-TYPE STARS

BENGT GUSTAFSSON AND KJELL ERIKSSON
*Uppsala Astronomical Observatory*
*Uppsala, Sweden*

DAN KISELMAN
*Royal Academy of Sciences, Stockholm Observatory*
*Saltsjöbaden, Sweden*

NILS OLANDER
*Dept. of Physics and Mathematics, MidSweden University*
*Sundsvall, Sweden*

HANS OLOFSSON
*Stockholm Observatory*
*Saltsjöbaden, Sweden*

AND

HUGO E. SCHWARZ
*Nordic Optical Telescope*
*Santa Cruz de La Palma, Canary Islands, Spain*

**Abstract.** Circumstellar emission in the Na I and K I resonance lines has been detected from three carbon stars using high-resolution spectroscopy. Some properties of the circumstellar envelopes are discussed.

Circumstellar envelopes around three bright N–type stars, R Scl, X TrA, and V Aql, have been detected in emissions in resonance lines from K I and Na I. This radiation, most probably scattered photospheric radiation, was discovered using high-resolution spectroscopy.

From the observations of the K I 769.9 nm emission we find systemic and expansion velocities in fair agreement with those obtained from the CO millimeter lines. We find a decline of the surface brightness of the scattered light as a function of the angular distance from the star, $\beta$, of approximately $\beta^{-3}$, in agreement with the assumption of optically thin emission and constant expansion velocity, mass-loss rate, and K I abundance.

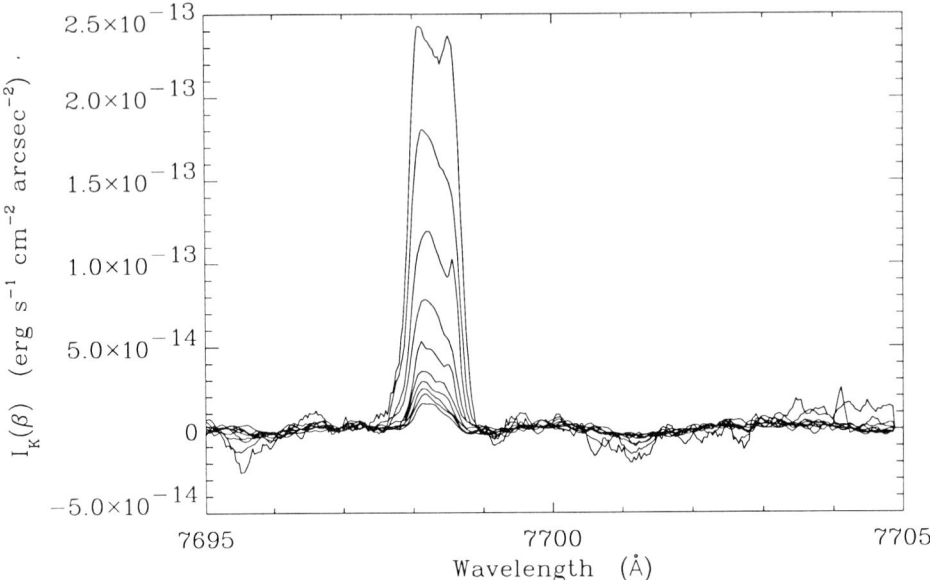

*Figure 1.* Subtracted K I spectra (off-star minus scaled on-star) for different distances from R Scl. The off-star spectra were obtained with the star approximately 4″ outside the slit, in the slit direction. The spacing between successive spectra is 0.″83.

Our mass-loss rate estimates from the K I line emissions agree rather well with those obtained from CO, which suggests that a considerable fraction of the potassium stays neutral throughout the envelope. If ionization of potassium from interstellar ultraviolet radition as well as from some assumed chromospheric fluxes is considered, the mass-loss rates increase by one order of magnitude. This puts strong upper limits on the photo-ionizing chromospheric UV emission from these stars. Details of this work are presented in Gustafsson et al. (1997).

Optical imaging in 5 nm wide Na and K filters shows stellar envelopes around the "detached CO shell stars" R Scl, U Ant, and S Sct, with envelope diameters between 20″ and 2′ and a flat brightness distribution. These diameters and the morphology are remarkably similar to those of maps of CO mm line emission of the objects. The optical images were obtained at the ESO 3.6-m telescope using a coronographic polarimetry technique to increase the contrast between faint envelopes and the stellar light scattered in the terrestrial atmosphere.

The optical images probably show general dust-scattered light, and the CO maps reflect the morphology of the molecular line emission, while the spectra represent the true resonance-scattering envelope.

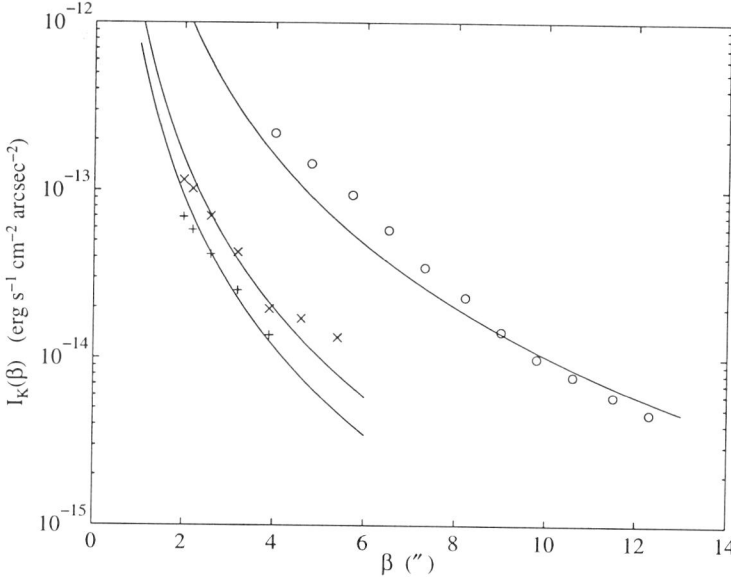

*Figure 2.* The wavelength-integrated circumstellar K I emission as a function of angular distance ($\beta$) to the stars R Scl (o), X TrA (x) and V Aql (+). The curves show $\beta^{-3}$ fits to the data.

## References

Gustafsson, B., Eriksson, K., Kiselman, D., Olander, N. & Olofsson, H. 1997, *A&A*, 318, 535

Marina at the site of the conference banquet.

# THE NEUTRAL ENVELOPES AROUND AGB AND POST–AGB OBJECTS

*Their Structure and Kinematics*

H. OLOFSSON
*Stockholm Observatory*
*S–133 36 Saltsjöbaden, Sweden*

**Abstract.** This review discusses the large–scale geometry and kinematics of envelopes around evolved stars, as inferred from radio line observations of circumstellar molecules, as a function of the evolutionary stage of the central object. In particular, the drastic change in morphology from largely spherical envelopes around asymptotic giant branch stars to distinctly non-spherical neutral envelopes around planetary nebulae is addressed. In addition, the small–scale structure of the envelopes, i.e., whether the circumstellar medium is smooth or clumpy, is discussed. Finally, a scenario for the morphological evolution, based on the assumption of a highly clumped medium, is presented.

## 1. Introduction

It is nowadays well established that stars on the Asymptotic Giant Branch (AGB) eventually lose considerable amounts of matter in a stellar wind, and that it is this process that significantly limits their lifetime on the AGB. The wind is slow enough that the envelope of gas and dust, formed by the mass loss, will remain in the vicinity as the star leaves the AGB and evolves towards its final stage as a white dwarf. Thus, the morphology and kinematics of the circumstellar envelopes (CSEs) provide us with information on the temporal evolution of the stellar mass loss on, as well as beyond, the AGB. The CSE also plays an important role in the formation of the planetary nebula (PN). The small–scale structure of the circumstellar medium gives information on the details of the mass-loss process. This is of crucial importance to our understanding of the mass-loss mechanism, which appears to involve physical processes that are critically dependent on density, temperature, etc. Due to lack of space, only the most recent, relevant publications are referred to.

## 2. The Probes

The small– and large–scale structure of the circumstellar medium can be probed by dust continuum emission and in atomic and molecular lines. Both types of probes have their advantages and disadvantages, and we will here concentrate on the molecular radio lines. In general, line emission has the advantage that it also carries kinematic information.

The present number of molecular species detected at radio wavelengths in AGB or post–AGB CSEs is 49 (Olofsson 1997). The majority of these are detected towards only one object (IRC+10216), or at most a few objects. Good statistics can be obtained essentially only from four molecules: SiO ($\geq 450$ objects), OH ($\geq 2000$ objects), $H_2O$ ($\geq 300$ objects), and CO ($\geq 450$ objects) [compilations of data can be found in Benson et al. (1990) and Loup et al. (1993)]. The first three, which are only seen in maser emission (except for SiO where there are also "thermal" lines), are restricted to oxygen-rich envelopes (i.e. C/O < 1, O–CSEs), but may be used to study the entire mass-loss phase of the AGB evolution, and to some extent also the post–AGB evolution (except for SiO). The CO emission is "thermal" in nature and probes both O–CSEs and C–CSEs (i.e. those with C/O > 1) in the AGB–phase and beyond, but the observational space is limited to about 10 kpc (Olofsson 1989). The masers are in general stronger, and in the LMC six objects have been detected in OH(1612 MHz) emission (Wood et al. 1992) and one object in SiO($v=1$, $J=2\rightarrow 1$) emission (van Loon et al. 1996).

## 3. The Deconvolution Problem

In the best case the observational data consist of two-dimensional brightness distributions at different line-of-sight velocities in a transition of a species X, $I(\alpha, \delta, v_z, X(u\rightarrow l))$. In order to get from this to the density distribution of species X, $\rho_X(r,\theta,\varphi)$, and to the kinematics for the CSE in the region probed by $X(u\rightarrow l)$, $v_{e,X}(r,\theta,\varphi)$, one has to correct for beam smearing (a severe problem since most emissions are unresolved by single–dish observations) and line-of-sight contamination (the $v_z$ information helps if the kinematics are simple), and finally calculate the excitation and the radiative transfer. For the latter a knowledge of the detailed density structure (i.e., a smooth or clumpy medium) as well as of the radiation field (which depends strongly on the evolutionary stage of the central star) is required. The density distribution $\rho_X$ is determined by the chemistry, i.e., one has to identify whether species X is of photospheric origin or if it is a product of processes in the CSE (see, e.g., Nercessian et al. 1989; Cherchneff & Glassgold 1993; Millar & Herbst 1994). In the latter case the molecule is either a direct photodissociation product or else it is produced in a photo-induced chemistry, and a knowledge of the relevant radiation fields (interstellar UV field, chromospheric radiation, hot central star radiation in the case of post–AGB objects) is required. It is the excitation and the chemistry that determine the region probed by the $X(u \to l)$ line. Observations of several species and transitions will eventually give the true total density distribution $\rho(r, \theta, \varphi)$ and the true kinematics $v_e(r, \theta, \varphi)$. The mass-loss geometry and history are finally obtained from the continuity equation, $\dot{M}(t - t_{\text{ret}}, \theta, \varphi) = 4\pi r^2 \rho v_e$ [where $t_{\text{ret}}$ is determined by $r$ and $v_e(r)$], provided that a number of conditions are fulfilled.

## 4. Morphology and Kinematics

### 4.1. AGB–CSEs

As outlined in the previous section, different probes probe different regions. Although this has obvious advantages, it also means that the connection between density structures at different scales is usually not easy to obtain. At the smallest scales we are restricted to maser emission, and unfortunately there are no useful probes for C–CSEs [two objects have been mapped in HCN maser emission, but there is no resolved structure (Carlstrom et al. 1990; Lucas & Guilloteau 1992)]: SiO(43 GHz) and $H_2O$(22 GHz, low–$\dot{M}$ CSEs) at $\leq 10^{14}$ cm, $H_2O$(22 GHz, higher–$\dot{M}$ CSEs) and OH(1665/1667 MHz) at $\leq 10^{15}$ cm. It is for various reasons very difficult to draw any definite conclusions on the geometry and kinematics of CSEs from the maser data. The most recent SiO observations show roughly circular ring structures (Diamond et al. 1994; Miyoshi et al. 1994) or elongated structures (Colomer et al. 1996). Important to note here is that at these small scales ($\leq 0''.1$) the position of the star has to be assumed. In only one case has the position and size of the star been measured simultaneously with the maser emission [$H_2O$(22 GHz) towards W Hya], and the result is an apparently spherical shell of emitting gas centred on the star (Reid & Menten 1990). In many other cases the $H_2O$ brightness distributions are highly asymmetric (see, e.g., Bowers & Johnston 1994; Yates & Cohen 1994). Also at the sligthly larger scales probed by the OH(1665/1667 MHz) masers it is not clear whether the density distribution is largely spherical, or if for instance there exists equatorial density enhancements (Bowers et al. 1989; Chapman et al. 1991, 1994). Kinematically it appears that a fair fraction of the gas has reached the terminal expansion velocity (as measured at large scales) already at about $10^{15}$ cm (Bowers & Johnston 1994).

At the larger scales, $\geq 10^{16}$ cm, the situation is better, from both a theoretical and an observational point of view. Here the OH(1612 MHz) maser emission outlines more or less overall spherically symmetric shells in about 15 cases (see e.g. Herman et al. 1985; Welty et al. 1987). The best examples are OH 26.5+0.6 and OH 127.8–0.0 (which may be a supergiant) (Bowers & Johnston 1990). The evidence for sphericity is further strengthened by good fits to the size–$v_z$ relation, $\theta(v_z) = \theta(0)[1 - (v_z/v_e)^2]^{1/2}$, where $\theta(v_z)$ is the measured size of the emission in a narrow velocity range around $v_z$, and $\theta(0)$ is the source size at the systemic velocity, expected to apply for a spherical shell expanding at the constant velocity $v_e$. At roughly the same size scale, maps of "thermal" SiO emission towards 10 objects show geometries consistent with overall spherical symmetry (Lucas et al. 1992; Sahai & Bieging 1993). At slightly larger scales, $\geq 5 \times 10^{16}$ cm, about 25 objects have now been mapped in various CO transitions using single telescopes, but since the majority of the objects are just barely resolved, and the dynamic range in the maps is limited, one should not put too much faith in the reported high frequency of elongations and asymmetries in the brightness distributions (Bujarrabal & Alcolea 1991; Stanek et al. 1995). In the four cases where the emission is clearly spatially resolved at scales of $10^{17} - 10^{18}$ cm, namely the C stars IRC +10216 (Truong-Bach et al. 1991), U Ant, S Sct, and TT Cyg (Olofsson et al. 1996), there is good evidence for overall spherical symmetry and isotropic expansion at a constant velocity.

Only in IRC +10216 has emission from a fair number of molecular species (16) been observed at high angular resolution (see, e.g., Bieging & Tafalla 1993; Guélin et al. 1993; Dayal & Bieging 1995; Gensheimer et al. 1995; Guélin et al. 1997). The combined picture is a CSE with an overall spherical symmetry (however, see below). Recent observations of six species towards the carbon star CIT 6, on the other hand, point to a marked elongation at PA $\approx 155°$ (Lindqvist et al. 1998).

Departures from "sphericity" certainly exist. A faster, bipolar wind and a (probably) largely spherical AGB–CSE (perhaps open at the poles) give a good explanation for the CO data on $o$ Ceti (Planesas et al. 1990), X Her (Kahane & Jura 1996), V Hya (Kahane et al. 1996), and $\pi^1$ Gru (Sahai 1992). Both the velocities of the AGB–CSEs (3–12 km s$^{-1}$) and the bipolar winds [(3–25)/cos $i$ km s$^{-1}$, where $i$ is the angle between the bipolar outflow axis and the line of sight) are fairly small in these cases (except for V Hya; see also Knapp et al. 1997). R Leo appears to have (possibly) a bipolar outflow ($\sim 3/\cos i$ km s$^{-1}$) close to the star ($\leq 2 \times 10^{14}$ cm) (Cernicharo et al. 1994). There is a symmetry axis in the IRC +10216 CSE in the sense that the brightness distributions appear open at PA $\approx 20°$ (see, e.g., Lucas et al. 1995). Furthermore, the star is not at the center of the extended envelope, possibly suggesting a binary nature for IRC +10216 (Guélin et al. 1993). The CSE of V Cyg has been convincingly shown to be elongated and not centered on the star (Bujarrabal & Alcolea 1991). Finally, in some stars there is evidence for low–intensity, higher–velocity wings (Bujarrabal & Alcolea 1991).

The data on IRC +10216 also show how difficult it is to infer the density structure from a single line. The $SiC_2(4_{23} \rightarrow 3_{22})$ line shows an opening in the CSE at PA $\approx 20°$ as do many other lines (Lucas et al. 1995), while both the $SiC_2(4_{04} \rightarrow 3_{03})$ (Takano et al. 1992) and the $SiC_2(4_{22} \rightarrow 3_{21})$ (Gensheimer et al. 1995) lines have their emission maxima in this direction.

## 4.2. EARLY POST–AGB CSEs

The early post–AGB objects can be crudely divided into three classes when it comes to the morphology and kinematics of the AGB–CSEs. It is not clear that these groups outline an evolutionary sequence for post–AGB objects, or whether the apparent differences are due to inherent differences (e.g. binarity). In general, the classification of post–AGB objects is a tricky business, and, in fact, many objects so classified may not even be post–AGB objects.

In the first class we have CRL 618 and CRL 2688, objects surrounded by large, essentially unperturbed and overall spherically symmetric AGB–CSEs that expand with a velocity of $\sim 20$ km s$^{-1}$ (Truong–Bach et al. 1990; Phillips et al. 1992; Yamamura et al. 1996). However, there is a distinct difference between the CO line profiles of these objects and those of the CSEs of AGB objects. The presence of low–intensity, high–velocity wings in the former suggests very high–velocity outflows [CRL 618, $\sim 200/\cos i \approx 280$ km s$^{-1}$, Gammie et al. (1989), Cernicharo et al. (1989); CRL 2688, $\sim 40/\cos i$ and $100/\cos i$ km s$^{-1}$, Jaminet et al. (1992), Young et al. (1992)]. The bipolar nature of these high–velocity winds is now established (Neri et al. 1992; Yamamura et al. 1995; Bieging & Nguyen–Q–Rieu 1996), and there is good evidence for elongated density enhancements at small scales (Nguyen–Q–Rieu et al. 1986; Hajian et al. 1995; Martin-Pintado et al. 1995). In CRL 2688 the HCN emission provides evidence for rotation in the molecular gas in the direction orthogonal to the bipolar axis (Bieging & Nguyen–Q–Rieu 1996).

In the second class the line profiles are dominated by the high–velocity outflow, and there seems to be little trace of a normal AGB–CSE, i.e. a fair fraction of the AGB–CSE gas has very likely been accelerated to high velocities. Examples of this class are M1-92 and OH 231.8+4.2. In the former, some gas remains in the vicinity of the star, but an elongated cavity has developed and the polar gas expands at $\sim 50/\cos i \approx 65$ km s$^{-1}$ (Bujarrabal et al. 1994). In the latter, the dominating CO emission comes from a very high–velocity ($\sim 250/\cos i \approx 330$ km s$^{-1}$), bipolar outflow that follows the optical outline (Alcolea et al. 1996).

In the third class high–velocity winds also exist, but spectra obtained towards the star are dominated by a strong, narrow (5–10 km s$^{-1}$) feature, possibly indicating the presence of a strong equatorial density enhancement. Examples of this class are IC 2220 (Nyman et al. 1993) and the Red Rectangle (Jura et al. 1995).

Finally, we mention the CO map data on an object whose post–AGB nature is not conclusively established, 89 Her. There appear to be two CSEs, an inner envelope and an outer shell as inferred from CO(1 → 0) data, with geometries and kinematics that are not easily interpreted (Alcolea & Bujarrabal 1995). On the other hand, when integrated over the entire velocity range of the emission, the brightness distribution is circularly symmetric.

It should be noted here that the external parts of thick AGB–CSEs may become very cold because of adiabatic cooling during the expansion (Sahai 1990), and hence the gas may escape detection. For the high–velocity gas the cooling may become even more severe, and the gas may in fact only be detectable in absorption against the microwave background.

4.3. LATE POST–AGB CSEs

There are presently about 45 PNe detected in CO emission (Huggins et al. 1996), and a fair fraction of these have been mapped in some detail. Only in the youngest PNe does one see an essentially unperturbed AGB–envelope. The best example is provided by NGC 7027 which is surrounded by a large AGB–CSE that appears overall spherically symmetric and that expands with $\sim 15$ km s$^{-1}$ (Bieging et al. 1991). At smaller scales there is a cavity, and the CO data suggest that it is surrounded by a prominent toroidal density distribution (Graham et al. 1993).

For the more evolved PNe the CO maps show brightness distributions that to a large extent follow the optical morphology, e.g. hour–glass structures (NGC 2346, Bachiller et al. 1989b) or rings (NGC 6720, Bachiller et al. 1989a; NGC 6781, Bachiller et al. 1993). In all cases this suggests equatorial density enhancements (probably of different strengths). The dominating CO emission has expansion velocities that lie at the high end of what is observed for AGB–CSEs ($\geq 20$ km s$^{-1}$), probably indicating that some acceleration of the gas has taken place. An extreme example is BD +30°3639 where the gas expands with $\sim 50$ km s$^{-1}$ (Bachiller et al. 1991). Bipolar winds are present but the expansion velocities are quite moderate ($\sim 30/\cos i$ km s$^{-1}$), i.e. much lower than for the early post–AGB CSEs, and they become significantly rarer among the elliptical PNe (Jaminet et al. 1991; Sahai et al. 1991; Sahai et al. 1994).

## 5. Small–Scale Structure of the Circumstellar Medium

It has traditionally been assumed that the mass loss occurs in the form of a steady and smooth stellar wind, mainly because this simplifies the models and our ability to derive quantitative results. However, evidence is gathering from a broad set of observational data that the medium is clumpy at some scale.

The SiO, $H_2O$, and OH maser maps of the inner regions consist of many, often unresolved, brightness spots, but it is essentially impossible to relate this to the small–scale density distribution, since velocity coherence along the line of sight plays, exponential amplification may play, and in the case of OH photodissociation plays, a major role. VLBI observations of SiO masers suggest spot sizes as small as a few $10^{12}$ cm (Colomer et al. 1992). Assuming a clump size of $10^{13}$ cm and $n_{H_2} \approx 10^{10}$ cm$^{-3}$ (the SiO masers are quenched at higher densities) leads to a clump mass of $\sim 10^{-7}$ $M_\odot$. In fact, for reasonable mass loss rates the gas within $10^{14}$ cm of the star must be substantially clumped to provide the densities required for SiO masers. Measured spot sizes for the $H_2O$ masers fall in the range $(0.5 - 2) \times 10^{13}$ cm at a typical distance of $5 \times 10^{14}$ cm from the star (Spencer et al. 1979), while the characteristic OH maser spot size is a few $10^{14}$ cm at a distance of a few $10^{15}$ cm from U Ori and U Her as measured in the detailed studies of Chapman et al. (1991, 1994).

The OH(1612 MHz) masers outline shells that are not uniformly filled with emission. Most of the spatially resolved brightness distributions towards IRC +10216 show a very patchy structure, and in many cases species with quite different excitation characteristics peak in the same region suggesting variations in column density rather than excitation (Guélin et al. 1997). The CN($N = 1 \to 0$) emissions towards U Cam (Lindqvist et al. 1996) and CIT 6 (Lindqvist et al. 1998) are very patchy, and even though CN is a photodissociation product of HCN and its emission may give an enhanced view of any density contrast, it is hard to reconcile these data with a smooth medium.

The CO emission is probably better related to the true density distribution than are any other molecular emissions. Therefore, the large, geometrically thin CO shells around the C stars U Ant, S Sct, and TT Cyg are probably the best objects for a study of the small–scale structure since there is very little circumstellar material along the line of sight, i.e. effects due to saturation or clump overlap are limited (Olofsson et al. 1996). The brightness maps have a clear patchy appearance ($\sim 10$–$20$ major clumps), but it has been shown that to infer the small–scale density distribution from these maps is not so easy. Emission from about $10^3$ identical, unresolved clumps that are randomly positioned within the shell gives a large–scale brightness distribution with a patchy appearance, at the scale set by the observational resolution, that resembles the observed ones (Bergman et al. 1993; Olofsson et al. 1996). Recent interferometer observations of a part of the shell of S Sct at an angular resolution of $\sim 5''$ show many unresolved clumps, and an extrapolation to the entire shell gives $\sim 500$ clumps (Bergman et al. 1998).

There is also evidence for clumpiness from the observations of post–AGB objects. In general, the patchiness of the CO brightness maps increases with the evolutionary stage of the central object (Bachiller et al. 1993). The most extreme example of this is the direct observation of CO emission from cometary globules in the Helix nebula (Huggins et al. 1992). The estimated clump properties are: $M = 5 \times 10^{-6}\ M_\odot$, $n_{H_2} \leq 10^5\,\mathrm{cm}^{-3}$, $T = 25\,\mathrm{K}$. Finally, several molecular species show anomalously strong lines in the more evolved objects, e.g. HCN, CN, and $HCO^+$, indicating that a photo-induced chemistry is initiated in high–density clumps (Deguchi et al. 1992; Cox et al. 1992; Howe et al. 1994).

## 6. A Scenario

In this final section we will try to give a possible explanation for the apparent morphology change of CSEs with evolutionary stage. We adopt the (perhaps somewhat extreme) view point that the medium is highly clumped already at its ejection from the central star. As a typical example, we suppose that one clump of mass $10^{-6}\ M_\odot$ is ejected per year (i.e., roughly one per pulsational period), i.e. $\dot{M} = 10^{-6}\ M_\odot\,\mathrm{yr}^{-1}$. The clump evolution as it recedes from the star is determined by the thermal motion, which in turn is determined by heating and cooling processes. We will follow the parametric treatment of this evolution that was introduced in Bergman et al. (1993), and also adopt the same parameters, which were shown by Olofsson et al. (1996) to give a reasonable explanation for the CO shells around the C stars R Scl, U Ant, S Sct, and TT Cyg (mainly the ages of the shells were changed). With an expansion velocity of $10\,\mathrm{km\,s}^{-1}$ we find: within $3 \times 10^{14}$ cm there are 10 clumps and $n_{H_2} > 10^9\,\mathrm{cm}^{-3}$, i.e. this is the regime of the SiO masers, and the $H_2O$ masers are quenched above this density; within $3 \times 10^{15}$ cm there are 100 clumps and $n_{H_2} > 4 \times 10^6\,\mathrm{cm}^{-3}$, i.e. the OH masers are found at the edge of this region (where OH is formed from the photodissociation of $H_2O$); and within $3 \times 10^{17}$ cm there are $10^4$ clumps and $n_{H_2} > 10^2\,\mathrm{cm}^{-3}$, i.e. at the outer edge CO is barely excited and it is, in fact, photodissociated before reaching this radius. Clump blocking along the line of sight occurs already at $\sim 2 \times 10^{15}$ cm. If one takes into account that the velocity dispersion within the clumps is much smaller than the expansion velocity, the blocking (in a narrow velocity range) occurs at much larger radii. The transition to a smooth medium occurs first at $\sim 10^{19}$ cm.

In this scenario the SiO masers are located inside the CSE as expected from theory (the SiO abundance is too low once dust starts to form), and as indicated by observations (see, e.g., Greenhill et al. 1995). In the region where the $H_2O$ masers are located there are too few clumps to give any reliable information on the morphology. The same applies partly to the OH(1665/1667 MHz) maser region for low $\dot{M}$, while for higher $\dot{M}$ accidental overlap of clumps combined with unsaturated amplification may give a highly perturbed view of the geometry. In the regions where the OH(1612 MHz) masers (for higher $\dot{M}$) and the CO line-emitting gas are located, the clump overlap is substantial and the brightness contrast is consequently decreased. One should also note that a clumpy medium leads to circular brightness distributions (if $v_e$ is isotropic) because the photodissociation, which normally limits the brightness distribution, is determined by the clump properties and not $\dot{M}(\theta, \varphi)$. This, combined with the geometrical projection effect, leads to brightness distributions from which an overall spherical symmetry is inferred, even though moderate equatorial density enhancements may be present. Any asymmetry, if existing, is most easily seen at small scales, and in the case of emission from geometrically thin shell regions.

Another interesting property of a clumpy medium is that since the photodissociation is determined by the clump properties, we would expect the sizes of molecular envelopes to be less dependent on the mass-loss rate than derived from "smooth density" models (unless the clump properties strongly depend on the mass-loss rate). Also the cut-offs in the molecular distributions will be much smoother than otherwise expected, and regions where the chemistry has developed differently will become mixed.

During the early post–AGB evolution an essentially isotropic, high–velocity, smooth wind starts to blow. Even if the equatorial density enhancement of the AGB–CSE is only moderate, the clumps in the polar region will be preferentially accelerated to higher velocities, while the clumps in the equatorial region start to pile up (in these clumps a completely different shock–induced chemistry may develop). A cavity develops, and the clumps at lower latitudes start to accelerate to intermediate velocities, while the material in the polar region gradually disappears. In this way an elliptical shell or an hour–glass morphology develops, depending on the strength of the equatorial density enhancement in the original AGB–CSE. At this point the UV–flux from the central star may drastically enhance the contrast in the brightness distribution as a result of increased photodissociation in directions of low optical depth (this may also lead to a new photo–induced chemistry). Finally, only the equatorial density enhancement remains, and the medium is very clumpy.

I am grateful to the Swedish Natural Sciences Research Council (NFR) and the IAU for travel support.

## References

Alcolea, J. & Bujarrabal, V. 1995, *A&A*, 303, L21
Alcolea, J., Bujarrabal, V. & Sánchez Contreras, C. 1996, *A&A*, 312, 560
Bachiller, R., Bujarrabal, V., Martin-Pintado, J.& Gómez–González, J. 1989a, *A&A*, 218, 252
Bachiller, R., Huggins, P. J., Cox, P. & Forveille, T. 1991, *A&A*, 247, 525
Bachiller, R., Huggins, P. J., Cox, P. & Forveille, T. 1993, *A&A*, 267, 177
Bachiller, R., Planesas, P., Martin-Pintado, J., Bujarrabal, V. & Tafalla, M. 1989b, *A&A*, 210, 366
Benson, P. J., Little-Marenin, I. R., Woods, T. C., et al. 1990, *ApJ Supp.*, 74, 911
Bergman, P., Carlström, U. & Olofsson, H. 1993, *A&A*, 268, 685
Bergman, P., Olofsson, H. & Bieging, J. H. 1998, in prep.
Bieging, J. H. & Nguyen–Q–Rieu 1996, *AJ*, 112, 706
Bieging, J. H. & Tafalla, M. 1993, *AJ*, 105, 576
Bieging, J. H., Wilner, D. & Thronson, H. A. Jr. 1991, *ApJ*, 379, 271
Bowers, P. F. & Johnston, K. J. 1990, *ApJ*, 354, 676
Bowers, P. F. & Johnston, K. J. 1994, *ApJ Supp.*, 92, 189
Bowers, P. F., Johnston, K. J. & de Vegt, C. 1989, *ApJ*, 340, 479
Bujarrabal, V. & Alcolea, J. 1991, *A&A*, 251, 536
Bujarrabal, V., Alcolea, J., Neri, R. & Grewing, M. 1994, *ApJ*, 436, L169
Carlstrom, J. E., Welch, W. J., Goldsmith, P. F. & Lis, D. C. 1990, *AJ*, 100, 213
Cernicharo, J., Brunswig, W., Pauber, G. & Liechti, S. 1994, *ApJ*, 423, L143
Cernicharo, J., Guélin, M., Martin-Pintado, J., Peñalver, J. & Mauersberger, R. 1989, *A&A*, 222, L1
Chapman, J. M., Cohen, R. J. & Saikia, D. J. 1991, *MNRAS*, 249, 227
Chapman, J. M., Sivagnanam, P., Cohen, R. J. & Le Squeren, A. M. 1994, *MNRAS*, 268, 475
Cherchneff, I. & Glassgold, A. E. 1993, *ApJ*, 419, L41
Colomer, F., Baudry, A., Graham, D. A., et al. 1996, *A&A*, 312, 950
Colomer, F., Graham, D. A., Krichbaum, T. P., et al. 1992, *A&A*, 254, L17
Cox, P., Omont, A., Huggins, P. J., Bachiller, R. & Forveille, T. 1992, *A&A*, 266, 420
Dayal, A. & Bieging, J. H. 1995, *ApJ*, 439, 996
Deguchi, S., Izumiura, H., Nguyen–Q–Rieu, Shibata, K. M., Ukita, N. & Yamamura, I. 1992, *ApJ*, 392, 597
Diamond, P. J., Kemball, A. J., Junor, W., Zensus, A., Benson, J. & Dhawan, V. 1994, *ApJ*, 430, L61
Gammie, C. F., Knapp, G. R., Young, K., Phillips, T. G. & Falgarone, E. 1989, *ApJ*, 345, L87
Gensheimer, P. D., Likkel, L. & Snyder, L. E. 1995, *ApJ*, 439, 445
Graham, J. R., Serabyn, E., Herbst, T. M., et al. 1993, *AJ*, 105, 250
Greenhill, L. J., Colomer, F., Moran, J. M., Backer, D. C., Danchi, W. C. & Bester, M. 1995, *ApJ*, 449, 365
Guélin, M., Lucas, R. & Cernicharo, J. 1993, *A&A*, 280, L19
Guélin, M., Lucas, R. & Neri, R. 1997, in IAU Symp. 170: *CO: Twenty-Five Years of Millimeter-Wave Spectroscopy*, ed. W. B. Latter et al. (Kluwer), p. 359
Hajian, A. R., Phillips, J. A. & Terzian, Y. 1995 *ApJ*, 446, 244
Herman, J., Baud, B., Habing, H. J. & Winnberg, A. 1985, *A&A*, 143, 122
Howe, D. A., Hartquist, T. W. & Williams, D. A. 1994, *MNRAS*, 271, 811
Huggins, P. J., Bachiller, R., Cox, P. & Forveille, T. 1992, *ApJ*, 401, L43
Huggins, P. J., Bachiller, R., Cox, P. & Forveille, T. 1996, *A&A*, 315, 284

Jaminet, P. A., Danchi, W. C., Sandell, G. & Sutton, E. C. 1992, *ApJ*, 400, 535
Jaminet, P. A., Danchi, W. C., Sutton, E. C., et al. 1991, *ApJ*, 380, 461
Jura, M., Balm, S. P. & Kahane, C. 1995, *ApJ*, 453, 721
Kahane, C., Audinos, P., Barnbaum, C. & Morris, M. 1996, *A&A*, 314, 871
Kahane, C. & Jura, M. 1996, *A&A*, 310, 952
Knapp, G. R., Jorissen, A. & Young, K. 1997, *A&A*, 326, 318
Lindqvist, M., Lucas, R., Olofsson, H., Omont, A., Eriksson, K.& Gustafsson, B. 1996, *A&A*, 305, L57
Lindqvist, M., Lucas, R., Olofsson, H., et al. 1998, in prep.
Loup, C., Forveille, T., Omont, A. & Paul, J. F. 1993, *A&A Supp.*, 99, 291
Lucas, R., Bujarrabal, V., Guilloteau, S., et al., 1992, *A&A*, 262, 491
Lucas, R., Guélin, M., Kahane, C., Audinos, P. & Cernicharo, J. 1995, in *Circumstellar Matter 1994*, ed. G. D. Watt and P. M. Williams (Kluwer), p. 293
Lucas, R. & Guilloteau, S. 1992, *A&A*, 259, L23
Martin-Pintado, J., Gaume, R. A., Johnston, K. J. & Bachiller, R. 1995, *ApJ*, 446, 687
Miyoshi, M., Matsumoto, K., Kameno, S., Takaba, H. & Iwata, T. 1994, *Nature*, 371, 395
Millar, T. J. & Herbst, E. 1994, *A&A*, 288, 561
Nercessian, E., Guilloteau, S., Omont, A. & Benayoun, J. J. 1989, *A&A*, 210, 225
Neri, R., García-Burillo, S., Guélin, M., Cernicharo, J., Guilloteau, S. & Lucas, R. 1992, *A&A*, 262, 544
Nguyen–Q–Rieu, Winnberg, A. & Bujarrabal, V. 1986, *A&A*, 165, 204
Nyman, L.-Å., Olofsson, H., Rogers, C., Heske, A. & Sahai, R. 1993, in *Mass Loss on the AGB and Beyond*, ed. H. E. Schwarz, ESO Conference and Workshop Proceedings No. 46, p. 451
Olofsson, H. 1989, in IAU Coll. 106: *Evolution of Peculiar Red Giant Stars*, ed. H. R. Johnson and B. Zuckerman (Cambridge Univ. Press), p. 321
Olofsson, H. 1997, in IAU Symp. 178: *Molecules in Astrophysics: Probes and Processes*, ed. E. F. van Dishoeck (Kluwer), p. 457
Olofsson, H., Bergman, P., Eriksson, K. & Gustafsson, B. 1996, *A&A*, 311, 587
Phillips, J. P., Williams, P. G., Mampaso, A. & Ukita, N. 1992, *A&A*, 260, 283
Planesas, P., Kenney, J. D. P. & Bachiller, R. 1990, *ApJ*, 364, L9
Reid, M. J. & Menten, K. M. 1990, *ApJ*, 360, L51
Sahai, R. 1990, *ApJ*, 362, 652
Sahai, R. 1992, *A&A*, 253, L33
Sahai, R. & Bieging, J. H. 1993, *AJ*, 105, 595
Sahai, R., Wootten, A., Schwarz, H. E. & Clegg, R. E. S. 1991, *A&A*, 251, 560
Sahai, R., Wootten, A., Schwarz, H. E. & Wild, W. 1994, *ApJ*, 428, 237
Spencer, J. H., Johnston, K. J., Moran, J. M., Reid, M. J. & Walker, R. C. 1979, *ApJ*, 230, 449
Stanek, K. Z., Knapp, G. R., Young, K. & Phillips, T. G. 1995, *ApJ Supp.*, 100, 169
Takano, S., Saito, S. & Tsuji, T. 1992, *PASJ*, 44, 469
Truong–Bach, Morris, D. & Nguyen–Q–Rieu 1991, *A&A*, 249, 435
Truong–Bach, Morris, D., Nguyen–Q–Rieu & Deguchi, S. 1990, *A&A*, 230, 431
van Loon, J. Th., Zijlstra, A. A., Bujarrabal, V. & Nyman, L.-Å. 1996, *A&A*, 306, L29
Welty, A. D., Fix, J. D. & Mutel, R. L. 1987, *ApJ*, 318, 852
Wood, P. R., Whiteoak, J. B., Hughes, S. M. G., Bessell, M. S., Gardner, F. & Hyland, A. R. 1992, *ApJ*, 397, 552
Yamamura, I., Onaka, T., Kamijo, F., Deguchi, S. & Ukita, N. 1995, *ApJ*, 439, L13
Yamamura, I., Onaka, T., Kamijo, F., Deguchi, S. & Ukita, N. 1996, *ApJ*, 465, 926
Yates, J. A. & Cohen, R. J. 1994, *MNRAS*, 270, 958
Young, K., Serabyn, G., Phillips, T. G., Knapp, G. R., Güsten, R. & Schulz, A. 1992, *ApJ*, 385, 265

## Discussion

**Elitzur**: Two comments: (1) SiO comes from inside the dust formation zone. Models that attempt to integrate the atmosphere and the wind should consider the SiO observations. Polarization shows that the magnetic field is a significant dynamic force. (2) Excitation may make a symmetric shell look asymmetrical and *vice versa*. In IRC +10216, the IR images are asymmetric at short wavelengths and spherically symmetric at long wavelengths.

**Olofsson**: I agree with both comments. Concerning the excitation, one has to use many transitions and many molecular species before one can draw any definite conclusions about the large-scale density structure.

**Whitelock**: I am worried about the effects of duplicity in the general interpretation of evolution from the AGB to PNe. Some of the stars you describe as early post–AGB are interacting binaries. Hans Van Winckel provided us with a clear indication that the so-called "post–AGB stars" with a near-IR excess are binaries which could not have undergone normal AGB evolution (their binary periods are too short). So I would like to make a general request that we stop calling these stars "post–AGB."

**Olofsson**: I agree that there are objects that should not be included in this discussion, and I have tried to avoid them. However, binarity is probably also common for true AGB and post–AGB objects, and it will probably have an effect on the mass-loss geometry on the AGB, in particular the strength of the equatorial density enhancement. This, in turn, will have an effect on the morphology of, for instance, planetary nebulae formed by the interaction of a fast wind and the slow AGB wind.

**Richards**: Just like your last model, bipolar outflow on a very small scale is seen in $H_2O$ masers from NML Cyg. Monitoring over years shows what clumps survive proper motion in the NML Cyg outflow, and random effects in more symmetric shells. Individual clumps are well resolved in $H_2O$ maser emission. So far supergiants have been analyzed; Miras will also be mapped.

# EXTENDED DUST SHELLS AROUND CARBON STARS IN THE INFRARED AND IN OPTICAL LIGHT

HIDEYUKI IZUMIURA
*Okayama Astrophysical Observatory, NAOJ*
*Kamogata, Asakuchi, Okayama 719-0232, Japan*

L. B. F. M. WATERS AND T. DE JONG
*Astronomical Institute "Anton Pannekoek",*
*University of Amsterdam*
*NL–1098 SJ Amsterdam, The Netherlands*

C. LOUP
*Institut d'Astrophysique de Paris,*
*CNRS, 75014 Paris, France*

AND

O. HASHIMOTO
*Gunma Astronomical Observatory,*
*1-18-7 Ohotomo, Maebashi, Gunma 371, Japan*

**Abstract.** We investigate the structure of extended dust shells around optical carbon stars in the far-infrared and in optical light. In the optical we have discovered that R Scl and U Ant are associated with circularly extended emission, the radii of which are about $20''$ and $58''$, respectively. The emission is probably scattered light of the central star by dust grains in their circumstellar shells. In the far-infrared we have discovered a double shell structure surrounding U Ant in high resolution IRAS images, which is direct evidence of a periodic change of mass-loss on a time-scale of the order of $10^4$ years in the AGB evolution. Relating the two shells to two consecutive thermal pulses allows for a self-consistent determination of the interpulse period, core-mass, luminosity, and distance. Direct mapping of Y CVn at 90 and 160 $\mu$m with ISOPHOT on board ISO has revealed a very extended detached dust shell around this star. The mass-loss rate is found to have decreased drastically by two orders of magnitude $1.4 \times 10^4$ years ago, which should be an important clue to the understanding of J–type stars.

## 1. Introduction

It is widely accepted that carbon stars are produced from low- and intermediate-mass stars by the third dredge-up at the thermally pulsing asymptotic giant branch (TP–AGB) phase (Iben & Renzini 1983). Such TP–AGB stars undergo intense mass-loss, which gives rise to a copious circumstellar shell of gas and dust. If very extended dust shells indeed exist around carbon stars, they must be one of the best sites for the study of the actual interplay between thermal pulses and mass-loss, which is now thought to be a key process in understanding the AGB evolution. Examining the structure of such a dust shell will allow us to obtain mass-loss history on time scales of 1,000 to 10,000 years. In the following sections we describe our recent results on the extended dust shells of some carbon stars.

## 2. Dust Shells of R Scl and U Ant in Optical Light

In Figure 1 we show images of R Scl and U Ant in blue light retrieved from the Digitized Sky Surveys. In the images we find that these stars are surrounded by an extended circular emission centered at the star. For R Scl the emission is smaller than a ghost pattern which is also seen for the star of similar magnitude at the right in the same panel. For U Ant the emission extends outside the ghost pattern.

The images were recorded on SERC-J (blue) plates taken with the UK Schmidt Telescope. For each star we examined images of other stars with similar brightness on both the same and different SERC-J survey plates. We did not find any similar extended emisson components. We therefore conclude that the extended components around R Scl and U Ant are true extended emission, probably the light from the central stars scattered by dust grains in their circumstellar dust shells.

The radii of the optical shells are $20''$ and $58''$ for R Scl and U Ant, respectively, which are in close agreement with those of their CO gas shells of $9'' + (5\text{--}10)''$ and $41'' + (5\text{--}10)''$, respectively (Olofsson et al. 1996). This excellent agreement between the shell sizes in the optical and CO gas emission strongly supports our idea that the extended emission in the blue optical light is due to light scattered by dust grains.

It is notable, however, that the dust emission is slightly but significantly more extended than the CO gas emission in both stars. The outer boundary of the CO gas shell may be truncated by the photodissociation process by the interstellar UV radiation field. Another possibility is the gas/dust drift (Gilman 1972; Goldreich & Scoville 1976; Kwan & Hill 1977), which has long been suggested but never measured directly. The drift velocity can be as large as 9 km s$^{-1}$ or 5 km s$^{-1}$ when the momentum transfer efficiency factor is 0.05 (Sopka et al. 1985) or 0.015 (Huggins, Olofsson & Johansson

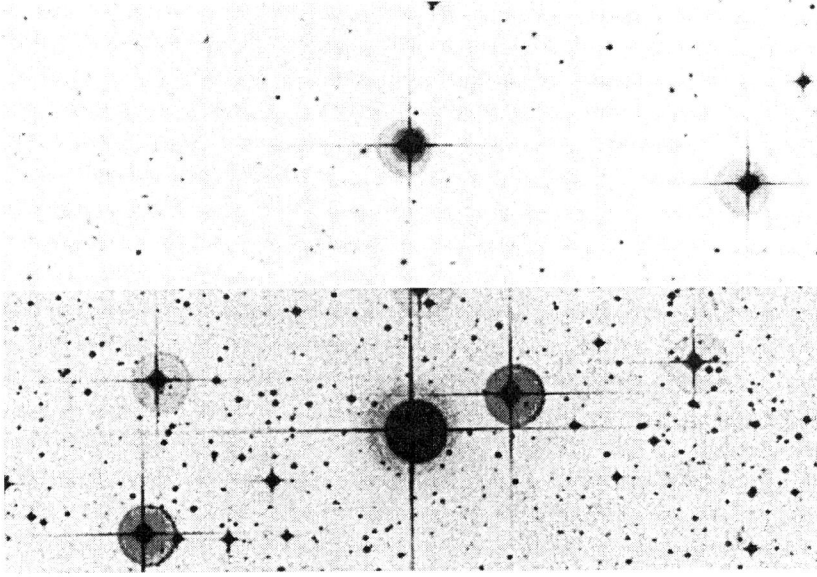

Figure 1. Top: Dust shell image of R Scl recorded on a blue photographic plate retrieved from the Digitized Sky Survey. Bottom: Same as the top panel, but for U Ant. The images measure $5'.12 \times 15'.0$ and the carbon stars are at the center.

1988), respectively, assuming a mass-loss rate of $5\times10^{-6}\,M_\odot\,yr^{-1}$, a gas expansion velocity of 21 km s$^{-1}$, and a luminosity of $1.7\times10^4\,L_\odot$ (Izumiura et al. 1997). Hence the size difference could be due to the gas/dust drift.

## 3. HIRAS Images of the Dust Shell of U Ant

High resolution IRAS (HIRAS) images in the far-infrared are now available with the use of Pyramid Maximum Entropy image reconstruction techniques (Bontekoe et al. 1994). We initiated a survey program of extended dust shells around optically bright carbon stars in the HIRAS images (cf. Izumiura et al. 1995). Here we show the most spectacular results, on U Ant, in Figure 2. The 60 $\mu$m image shows a considerably extended central component (FWHM 55") with an additional component which is nearly circularly distributed around the central component at about 3' in radius. In the 100 $\mu$m image the central component has a FWHM of about 90", which implies that it is somewhat resolved, and it shows a further slight extension in the lowest three contours. We have confirmed that these features are also discernible in the original IRAS survey scan data.

We analysed the brightness distribution obtained on the basis of a double detached dust-shell model (Izumiura et al. 1997). It is reasonable to assume that the inner dust shell is also detached considering the existence

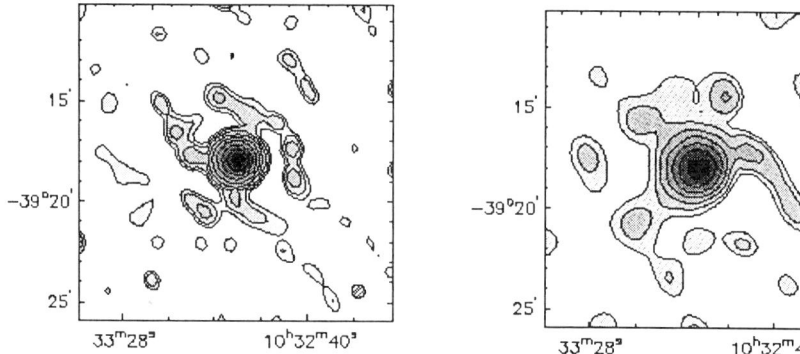

*Figure 2.* HIRAS images of U Ant in the 60 μm band (*left*) and the 100 μm band (*right*). Each image measures 16′ × 16′. The contour levels are given in steps of factors of 2 in MJy sr$^{-1}$ starting at 1 MJy sr$^{-1}$. The hatched circle at the bottom-right corner shows the resolution (FWHM): 39″ and 80″ in the 60 and 100 μm images, respectively.

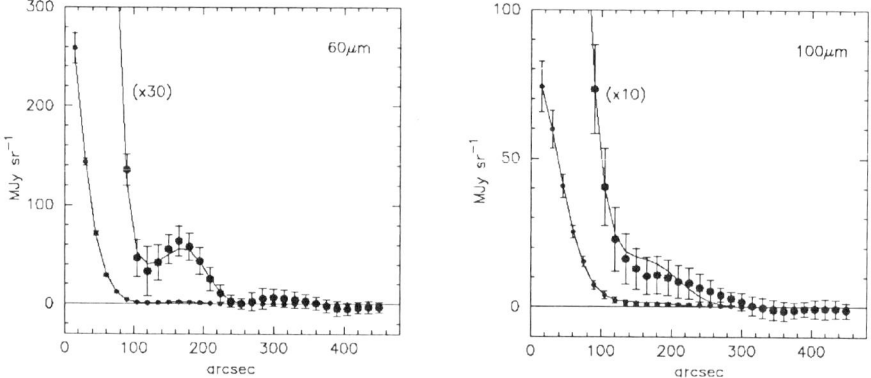

*Figure 3.* Simultaneous model fitting of a double dust shell to the brightness profiles of the dust shell of U Ant at 60 μm (*left*) and 100 μm (*right*). Filled circles are the data and the solid line shows the best fitting model for a shell thickness of 2.0×10$^{17}$cm (3000 years). The larger symbols show the same brightness profiles but thirty- and ten-times enlarged in the vertical direction in the left and right panels, respectively. Error bars show an estimated 1σ uncertainty range.

of a detached CO gas envelope with a mean radius of 40″ in this star (Olofsson et al. 1996). Simultaneous model fitting to the 60 and 100 μm data is shown in Figure 3, which gives a shell separation of 6.0×10$^{17}$cm and mass-loss rates in the inner and outer shells of 1.5× and 3.4×10$^{-6}$ M$_\odot$yr$^{-1}$, respectively, for a thickness of 2.0×10$^{17}$cm (i.e. a formation period of 3000 years with the expansion velocity of 21 km s$^{-1}$) for both shells and a distance of 280 pc. The fitting results suggest that this star experienced two cycles of a periodic mass-loss variation on a time-scale of some 10$^4$ years, the

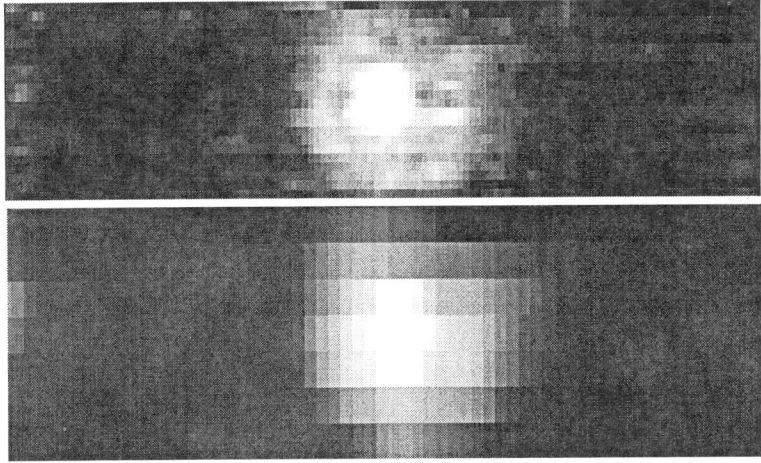

*Figure 4.* ISOPHOT images of Y CVn in the 90 μm band (*top*) and 160 μm band (*bottom*), where the pixel sizes are $15'' \times 23''$ and $30'' \times 92''$, respectively. The images measure $8'.4 \times 32'.0$ and $10'.7 \times 32'.0$

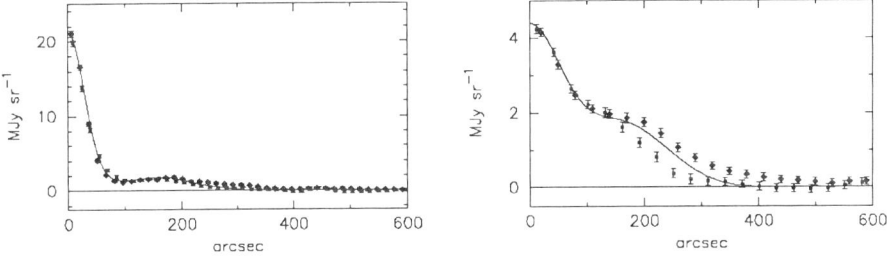

*Figure 5.* Brightness profiles of Y CVn at 90 μm (*left*) and 160 μm (*right*). Data points in the left half of the images are shown by squares and in the right half by diamonds. Error bars show only the statistical uncertainty. The solid line is a simultaneous model fit of a detached dust shell to the profiles in the two bands.

amplitude of which reached two orders of magnitude. Furthermore, relating the two shells directly to two consecutive thermal pulses allows us for the first time to determine the distance, interpulse period, core mass, and luminosity of the star self-consistently: they are 436 pc, $1.4 \times 10^4$ years, $0.77\,M_\odot$, and $1.7 \times 10^4\,L_\odot$, respectively (Paczyński 1975; Izumiura et al. 1997).

## 4. First ISO Images of the Dust Shell of Y CVn

We are also mapping the dust shells of selected AGB stars in the far-infrared using the ISOPHOT imaging photo-polarimeter (Lemke et al. 1996) on board the Infrared Space Observatory. Y CVn was observed with the C100 camera through the C90 filter ($\lambda_c = 95.1\,\mu$m) and with the C200 camera through the C160 filter ($\lambda_c = 174\,\mu$m) using the AOT 32 oversampling mapping mode on 25 April 1996. The maps obtained are shown in Figure 4. In the 90 $\mu$m map, sampling frequencies are 15″ and 23″ in the horizontal and vertical directions, respectively. In the 160 $\mu$m map they are 30″ and 92″. The maps demonstrate the first direct evidence of a hollow dust shell surrounding an AGB star (Izumiura et al. 1996).

We have made a simultaneous model fit of a detached dust shell to the brightness profiles shown in Figure 5. It is found definitely necessary to introduce a hollow dust shell to reproduce the local brightness minimum observed in the 90 $\mu$m data. The fit gives the mass-loss rate in the shell to be about $\sim 10^{-5}\,M_\odot\text{yr}^{-1}$, which is two orders of magnitude higher than the present-day mass-loss rate. The fitting results indicate that Y CVn underwent a sudden decline in mass-loss activity by two orders of magnitude $1.4 \times 10^4$ years ago (assuming a distace of 250 pc) and has been staying in that state.

With this picture of the mass-loss behavior, we conclude that the J–type carbon star Y CVn should be on the asymptotic giant branch. Our results will be important for understanding the evolutionary status of J–type carbon stars in the Galaxy.

The authors are indebted to the Digitized Sky Surveys that were produced at the Space Telescope Science Institute under U.S. Government grant NAG W-2166. The images of these surveys are based on photographic data obtained using the Oschin Schmidt Telescope on Palomar Mountain and the UK Schmidt Telescope.

### References

Bontekoe, Tj. R., Koper, E. & Kester, D. J. M. 1994, *A&Ap*, 284, 1037
Gilman, R. C. 1972, *ApJ*, 178, 423
Goldreich, P. & Scoville, N. 1976, *ApJ*, 205, 144
Huggins, P. J., Olofsson, H. & Johansson, L. E. B. 1988, *ApJ*, 332, 1009
Iben, I., Jr. & Renzini, A. 1983, *Ann. Rev. Astron. Astrophys.*, 21, 271
Izumiura, H., Hashimoto, O., Kawara, K., Yamamura, I. & Waters, L. B. F. M. 1996, *A&Ap*, 315, L221
Izumiura, H., Kester, D. J. M., de Jong, T., Loup, C., Waters, L. B. F. M. & Bontekoe, Tj. R. 1995, *Ap&SS*, 224, 495
Izumiura, H., Waters, L. B. F. M., de Jong, T., Loup, C., Bontekoe, Tj. R. & Kester, D. J. M. 1997, *A&Ap*, 323, 449
Kwan, J. & Hill, F. 1977, *ApJ*, 215, 781
Lemke, D., Klaas, U., Abolins, J., et al. 1996, *A&Ap*, 315, L64
Olofsson, H., Bergman, P., Eriksson, K. & Gustafsson, B. 1996, *A&Ap*, 311, 587
Paczyński, B. 1975, *ApJ*, 202, 558
Sopka, R. J., Hildebrand, R., Jaffe, D. T., Gatley, I., Roellig, T., Werner, M., Jura, M. & Zuckerman, B. 1985, *ApJ*, 294, 242

## Discussion

**Frogel**: For U Ant, why should the shell flashes have produced dust shells? You need something special about the two flashes you claim have produced dust shells.

**Izumiura**: Each shell flash may have produced a dust shell. Or each shell flash might have interrupted a high mass-loss phase. Neither can be ruled out at present. We need much higher spatial resolution to distinguish the two possibilities.

**Little–Marenin**: Y CVn has no Tc and hence is not likely to be an AGB star. Why do you think the two dust shells are produced by thermal pulses?

**Izumiura**: On the basis of the mass-loss behavior of Y CVn, it is almost indistinguishable from other AGB carbon stars. RGB stars usually show much lower mass-loss rates. Therefore we think the star is likely to be on the AGB in spite of the absence of Tc absorption lines. The star showing a double shell is not Y CVn but U Ant. U Ant has never been examined for Tc absorption lines.

Instructions in belly-dancing were provided after the banquet. Here, SOC Chairman Robert Wing checks out the instructor.

# ASYMMETRIES AROUND LUMINOUS RED VARIABLES

ANTONIO MÁRIO MAGALHÃES
*Instituto Astronômico e Geofísico, Universidade de São Paulo*
*São Paulo, Brazil*

AND

KENNETH H. NORDSIECK
*Space Astronomy Laboratory, University of Wisconsin*
*Madison, WI 53706, U.S.A.*

**Abstract.** Departures from spherical symmetry are known to exist around Luminous Red Variables (LRVs). One of the earliest pieces of evidence came from the discovery that light from these objects is polarized. Additional evidence comes from other techniques and observations of related objects. This talk briefly reviews these data and focuses on the mechanisms that produce polarization in LRVs. Emphasis is given on recent spectropolarimetric monitoring data of selected key objects and what we have learned about the LRV environment from such data.

## 1. Introduction

Mass loss plays a central role in the late stages of stellar evolution. In addition, mass loss from Asymptotic Giant Branch (AGB) stars, such as Mira variables, and from red supergiants may be responsible for up to 60%, relative to all stars, of the interstellar dust input into the interstellar medium (Gehrz 1989). Mass loss for low and intermediate mass stars occurs at rates of up to $10^{-4}$ solar masses/yr in the AGB stage (see, e.g., Chapman 1994 for a review). It has been proposed (e.g. Bedijn 1988; Bowen & Willson 1991) that the mass-loss rate actually increases with time so that a star loses a good share of its mass near the end of the AGB evolution, its envelope being finally stripped away.

Despite the success of this general picture, details of the physical processes occurring during mass loss are still not well understood. Models for pulsating red giant stars have been able to incorporate a great deal of physics in describing the atmospheric structure and stellar winds from Luminous Red Variables, or LRVs (Bowen & Willson 1991). However, these impressive models assume spherical symmetry, despite the fact that departures from spherical symmetry are known to exist around LRVs. One piece of evidence for this is that light from these objects typically shows some degree of linear polarization.

Intrinsic optical polarization in red long-period variables was discovered in the early 1960s (see Magalhães 1988 for a review) and indicated that non-spherically-symmetric structures do exist in the extended atmospheres of LRVs (see §2). Further evidence of asphericity in LRV envelopes comes from details of OH maser emission profiles (Collison & Fix 1992), K I resonant scattering (Plez & Lambert 1994), rings of SiO maser emission (Diamond et al. 1994; Greenhill et al. 1995) and OH radio images (Chapman et al. 1994). Aspherical symmetries such as found by Trammell, Dinerstein & Goodrich (1994) for post–AGB stars from spectropolarimetry may then be naturally understood, as they are already present in earlier stages. The observed, more obvious non-spherical symmetries in Protoplanetary and Planetary Nebulae (PPN) studied in the optical and IR (Kwok 1993) and also seen in optical polarimetry of evolved objects (Johnson & Jones 1991; Trammell et al. 1994) are also consistent with the origin of the asymmetries being early in the AGB phase.

## 2. Mechanisms that Produce Polarization in LRVs

Polarization observations yield otherwise unobtainable diagnostics related to the geometrical structure of unresolvable objects as well as the nature of the scattering physics. The degree of polarization depends on the nature and optical depth of the scatterers and on their geometrical distribution. The polarimetric wavelength dependence arises from the wavelength dependence of single-scattering polarization, of the scattering cross-section, of any competing opacity, and of any unpolarized, diluting stellar light. Time dependence arises when the asymmetry or the optical depth of the scatterers varies with time, as in a pulsating star or variable stellar wind.

The degree of linear polarization $P$, in percentage, and the position angle $\theta$ are usually transformed for analysis into two of the Stokes parameters, $Q = PI\cos(2\theta)$ and $U = PI\sin(2\theta)$, where $I$, the first Stokes parameter, is the beam intensity (the fourth Stokes parameter is $V$, which describes circular polarization). In the $Q - U$ plane, the polarization is represented by polar coordinates $\{[Q^2 + U^2]^{1/2}, 2\theta\}$. If the observed polarization results from multiple components (e.g. intrinsic plus interstellar), the $Q, U$ components of each polarization add like a vector at each wavelength. Since the polarization is usually small, this will also be true for the more often used normalized Stokes parameters, $q = Q/I$ and $u = U/I$.

In general, the optical linear polarization of LRVs is moderately large (a few to several percent); the polarization position angle and wavelength dependence typically vary with time, and the polarization typically increases with decreasing wavelength, which is taken as indicative of Rayleigh or Mie scattering. Two basic models have been invoked to explain the observations. One of them involves scattering in a non-spherically-symmetric dust cloud (Kruszewski, Gehrels & Serkowski 1968; Shawl 1975). Shawl (1975) has considered optically thin models whereas the Monte Carlo method can be used for arbitrary optical depths (Lefèvre & Daniel 1988).

In contrast, photospheric scattering, coupled to some asymmetry over the stellar disk, may cause the observed polarization (Harrington 1969). The strong gradient of the photospheric source function with optical depth at the shorter wavelengths will cause radiation at these wavelengths to come predominantly from deeper layers. Due to molecular Rayleigh scattering, the phase function of which has maximum polarization at a 90° degree scattering angle, this may cause strong limb polarization, with position angle usually tangent to the limb.

The needed asymmetries over the stellar disk may arise from non-radial pulsations, stellar spots, or other, more systematic variations in the temperature across the stellar surface. Temperature differences could indeed result from the formation of giant convective cells (Schwarzschild 1975). Origins for asymmetries in models for the atmospheric structure of LRVs are further discussed by Willson & Bowen (1988). The wavelength dependence of the polarization produced by an asymmetric photosphere will depend on how the ratio of absorption to scattering varies with optical depth and wavelength and the temperature gradient (Magalhães, Coyne & Benedetti 1986). A combination of stellar surface asymmetries with scattering by a dust shell could also give rise to a net polarization. Changes in the polarization across molecular bands may also result from dust scattering alone, as the size of the source as seen by the dust shell will vary with wavelength across the feature. A large stellar spot, with a different overall temperature than the remainder of the stellar disk, may provide the asymmetry and cause changes in the polarization across spectral features.

Finally, circular polarization may also appear in LRVs from multiple Mie scattering in an asymmetric dust envelope or magneto-emission from stellar spots (Hulenstein 1993). Multiple scattering by molecules *per se* will cause no circular polarization. Light leaving the photosphere linearly polarized may be circularly polarized when scattered off a dust envelope.

## 3. Spectropolarimetric Monitoring of LRVs

The accurate detection of changes in the polarization across photospheric features and with time should thus provide a wealth of information concerning the processes occurring in the LRV environment and help distinguish among different models (Magalhães 1988). Since 1989, a dedicated spectropolarimeter, HPOL, has been operating on every clear night at the University of Wisconsin's 90-cm telescope at Pine Bluff Observatory (PBO). Its main features are high polarimetric stability (0.002% in broad band work), high precision due to the large dynamic range of the CCD, and wide wavelength coverage (presently 3200 – 10500 Å, with 8 Å resolution). We have been monitoring a few specific red variables which earlier data have shown to be of particular interest. A survey of several red variables is also being conducted. Below we discuss some of the data for three red variables, namely $\alpha$ Ori, V CVn and $o$ Cet (Mira), which are among the ones with better phase coverage. These data are currently being prepared for publication and give further clues as to the processes involved in causing polarization and the diverse phenomena that may take place in the complex circumstellar environments of LRVs.

### 3.1. $\alpha$ Ori

Alpha Ori is a red supergiant of type M2 Iab, with probably around 10 solar masses, 1350 solar radii and 3500 K effective temperature. It is a semiregular variable of type SRc.

Alpha Ori has been monitored spectropolarimetrically at PBO since 1990. In December 1990 and February 1995 it was also observed from 1500 Å to 3000 Å by the Wisconsin Ultraviolet Photo Polarimeter Experiment, WUPPE, on board the Space Shuttle Astro–1 and Astro–2 observatories. In the PBO data, the different TiO band systems ($\gamma$, $\gamma'$ and $\alpha$) show distinct behavior in percent polarization and position angle. Further, "B filter" polarization measurements (obtained from the spectral data) and the TiO $\gamma$–system polarizations track each other in the Stokes $Q-U$ plane, while the TiO $\alpha$–system behaves as a separate, unmodulated component (Figure 1). The "B filter" measurements actually describe a seasonal loop in the $Q-U$ plane, with an amplitude of about 0.8% and a random position angle from year to year.

The WUPPE Astro–1 UV data show a dramatic rise in polarization to about 2% at 3000 Å, followed by a steep drop below 2900 Å (Nordsieck et al. 1994). The Mg II emission line at 2800 Å is unpolarized. The UV upturn has the same position angle as the $\alpha$ system in the optical region. There are then two geometrically separate components. One component is responsible for the visible blue continuum and TiO $\gamma$–system polarization

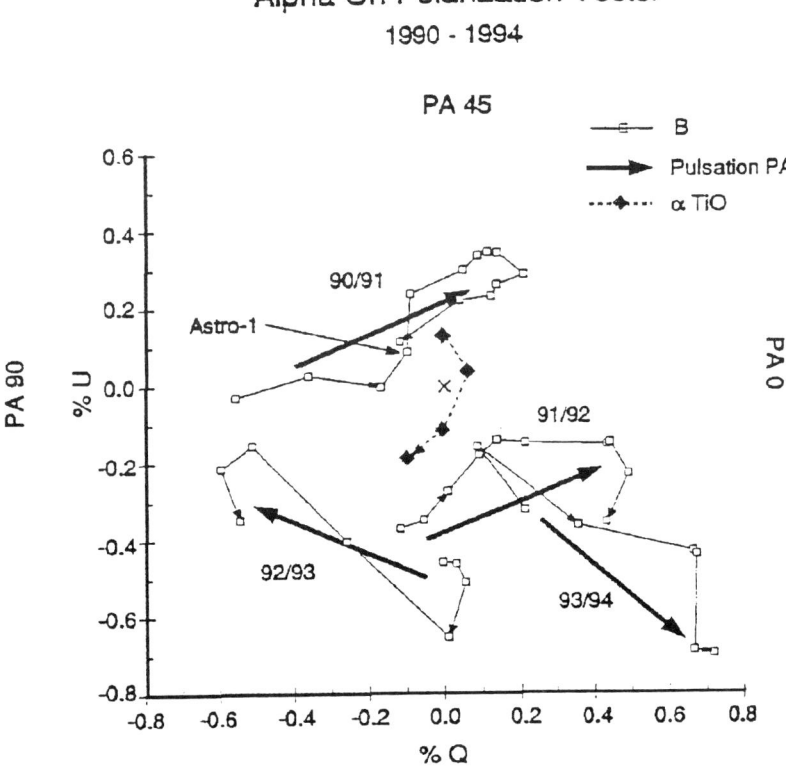

*Figure 1.* Normalized $Q-U$ plot of $\alpha$ Ori data. The cross and "PA" indicate, respectively, the origin and position angle along the four directions shown.

and is modulated by a yearly cycle, albeit with a random position angle. The TiO $\alpha$ system and UV upturn component seem to vary smoothly on a longer timescale. Other correlations confirm that the "B filter" component is indeed modulated by a radial pulsation.

The data indicate that the polarization is likely caused by photospheric Rayleigh scattering, with an asymmetry provided by a photospheric spot (Nordsieck et al. 1994). Non-radial pulsations (Holenstein 1993) could also originate in a hot region over the stellar disk. In the spot scenario, the polarization position angle would be parallel to the limb (see §2) and normal to the spot position angle. Remarkably, the UV Astro–2 WUPPE observations do indicate an average position angle very close to being 90° apart from the position angle of the hot spot detected by Gilliland & Dupree (1996) with the HST within a week of the Astro–2 flight in March 1995.

## 3.2. V CVn

The remarkable semiregular variable V CVn (M4–6 IIIe, $P = 192$ days) was one of the first objects noted to show large intrinsic polarization in broad-band data (Serkowski 1966). Most of the time V CVn shows a rather well behaved light curve and could be considered a small-amplitude Mira.

The PBO data for V CVn show that the polarization generally increases towards the violet, reaching at other times up to 10% (!) in that spectral region. The average polarization level is small when V CVn is bright and *vice versa*. PBO data, as well as the earlier data (Coyne & Magalhães 1977, 1979) and in contrast to $\alpha$ Ori, show that the position angle is always observed to be within rather limited bounds, $\pm 15°$ or so, over the years. Finally, the polarization generally decreases across TiO bands, sometimes with modest position angle changes.

The V CVn data are consistent with most of the optical polarization being produced in its photosphere. Some asymmetry exists already at the photospheric level, even though its nature is not completely clear. The stability of the position angle, as observed for decades now, raises the possibility that the asymmetry could be related to the rotation of the star. A modest rotation, for instance, could cause a noticeable pole-to-equator temperature variation over the stellar surface (Coyne & Magalhães 1979).

## 3.3. o Cet

The prototype of the Mira variable class is also known to be intrinsically polarized. The broad-band data of Shawl (1975) showed that the polarization of Mira in the blue increased sharply around phase 0.8 in the light curve, when the hydrogen lines from the periodic shock front first appear.

PBO data for Mira show that the overall polarization around maximum light (phases 0.9 through 0.3) is actually strongly related to the shock wave that progresses outward in each cycle. Also, the degree of polarization increases sharply through the Balmer emission lines (and Ca I 4226 Å; see Figure 2), as the latter appear and then fade, with little or no position angle rotation through the lines. H$\gamma$ and H$\delta$ reach their peak values earlier than H$\beta$, the latter falling within a TiO band. The overall degree of polarization also rises and falls through the cycle, being steeper into the blue when the polarization is higher, around phase 0.0. Changes in the polarization percentage and position angle across TiO bands are also seen.

These data suggest a non-spherically-symmetric shock front in Mira. The polarization then reflects the shock energy distribution, modified by the wavelength-dependent scattering and absorption opacities. Non-LTE effects, i.e. scattering in the lines (cf. Coyne & Magalhães 1977 on V CVn), may be present. The fact that line radiation may have such an additional

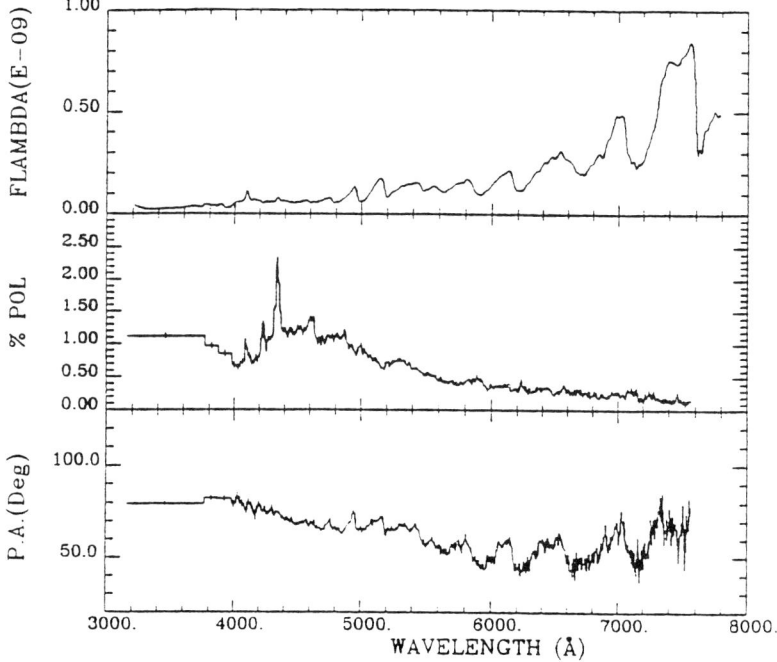

*Figure 2.* Flux, polarization and position angle for o Ceti on September 27, 1990, around maximum light. Pine Bluff Observatory observations.

component besides (unpolarized) thermal recombination radiation indicates that polarization through the Balmer emission lines can be useful as a probe through the atmosphere.

## 4. Final Remarks

Spectropolarimetric monitoring data for almost 40 LRVs now show that asymmetries in LRVs are indeed ubiquitous and that the phenomenon observed at later stages of stellar evolution (OH/IR stars, PPN, PN, etc.) appears early in the AGB and red supergiant phases.

Radiative transfer of polarized radiation will be useful to fully analyze the expected limb darkening and polarization from an LRV photosphere. Such an analysis could be done using the results from model atmospheres such as those by Plez, Brett & Nordlund (1992). Comparison with the observations should provide important feedback to the model atmospheres themselves and to details such as TiO band formation. Monte Carlo scattering models (Lefèvre & Daniel 1988; Holenstein 1993; Rodrigues & Magalhães 1995) have become sophisticated enough that modelling is already possible when scattering in an extended atmosphere and a combination of geometries and processes (e.g., molecular, dust and atomic resonant scattering) are involved.

Finally, future coordination between polarimetric observations and techniques such as optical interferometry and radio imaging, for selected objects, may prove instrumental for gaining a complete picture of the asymmetries around LRVs and improved knowledge of the mass-loss process.

AMM is greatly indebted to the Brazilian agency CNPq and the SOC for supporting his attendance at the conference. Research with HPOL is supported by NASA contract NAS5-26777 with the University of Wisconsin. Polarimetry at the University of São Paulo is supported by FAPESP.

## References

Bedijn, P. J. 1988, A&A, 205, 105
Bowen, G. H. & Willson, L. A. 1991, ApJ, 375, L53
Chapman, J. M. 1994, Southern Stars, 35, 241
Chapman, J. M., Sivagnanam, P., Cohen, R. J. & Le Squeren, A. M. 1994, MNRAS, 268, 475
Collison, A. J. & Fix, J. D. 1992, ApJ, 390, 191
Coyne, G. V. & Magalhães, A. M. 1977, AJ, 82, 90
Coyne, G. V. & Magalhães, A. M. 1979, AJ, 84, 1200
Diamond, P. J., Kemball, A. J., Junor, W., Zensus, A., Benson, J. & Dhawan, V. 1994, ApJ, 430, L61
Gehrz, R. D. 1989, in IAU Symp. 135: Interstellar Dust, ed. L. J. Allamandola and A.G.G.M. Tielens (Kluwer), p. 445.
Greenhill, L. J., Colomer, F., Moran, J. M., Backer, D. C., Danchi, W. C. & Bester, M. 1995, ApJ, 449, 365
Gilliland, R. L. & Dupree, A. K. 1996, ApJ, 463, L29
Harrington, J. P. 1969, Astrophys. Lett., 3, 165
Holenstein, B. D. 1993, PASP, 105, 322.
Johnson, J. J. & Jones, T. J. 1991, AJ, 101, 1735
Kruszewski, A., Gehrels, T. & Serkowski, K. 1968, AJ, 73, 677
Kwok, S. 1993, Ann. Rev. Astron. Astrophys., 31, 63
Lefèvre, J. & Daniel, J.-Y. 1988, in Polarized Radiation of Circumstellar Origin, ed. G. V. Coyne et al. (Vatican Obs.), p. 523
Magalhães, A. M. 1988, in Polarized Radiation of Circumstellar Origin, ed. G. V. Coyne et al. (Vatican Obs.), p. 461
Magalhães, A. M., Coyne, G. V. & Benedetti, E. K. 1986, AJ, 91, 919
Nordsieck K. H. et al. 1994, BAAS, 26, 864
Plez, B., Brett, J. M. & Nordlund, Å. 1992, A&A, 256, 551
Plez, B. & Lambert, D. L. 1994, ApJ, 425, L101
Rodrigues, C. V. & Magalhães, A. M. 1995, in IAU Symp. 163: Wolf-Rayet Stars: Binaries, Colliding Winds, Evolution, ed. K. A. van der Hucht and P. M. Williams (Kluwer), p. 260
Schwarzschild, M. 1975, ApJ, 195, 137
Serkowski, K. 1966, ApJ, 144, 857
Shawl, S. J. 1975, AJ, 80, 602
Trammell, S. R., Dinerstein, H. L. & Goodrich, R. W. 1994, AJ, 108, 984
Willson, L. A. & Bowen, G. H. 1988, in Polarized Radiation of Circumstellar Origin, ed. G. V. Coyne et al. (Vatican Obs.), p. 485

# Session X

NUCLEOSYNTHESIS AND EVOLUTION

Turkish students and their partners gather for a photo on the way to Bakırlıtepe, while the bus driver adds cool spring water to his radiator.

# HEAVY-ELEMENT NUCLEOSYNTHESIS IN AGB STARS

VERNE V. SMITH
Dept. of Physics, University of Texas at El Paso
El Paso, TX 79968, U.S.A.
and
McDonald Observatory, University of Texas at Austin
Austin, TX 78712, U.S.A.

**Abstract.** The production of certain neutron-rich elements heavier than iron occurs during He shell-burning on the asymptotic giant branch (AGB). These neutron captures occur at rather low neutron densities and, thus, the resulting heavy-element nucleosynthesis is characterisitic of the so-called s-process. Abundance analyses of the s-process elements in AGB stars can reveal details of the neutron densities and the stage of AGB evolution at which s-processing occurs, as well as the nature of the neutron source. These details derived from observations can constrain models of stars evolving along the AGB. Recent results concerning the nature of the s-process as a function of metallicity and the nature of the neutron source are reviewed.

## 1. Introduction

Many of the types of stars discussed at this symposium on "The Carbon Star Phenomenon" have played a key role in our understanding of the origins of the chemical elements. The idea that many (and as it has turned out, most) of the elements heavier than helium can be synthesized in the interiors of stars certainly received strong affirmation by Merrill's (1952) classic, and oft cited, spectroscopic detection of Tc I in certain S stars. With only relatively short-lived isotopes (half-lives of $\sim 10^5 - 10^6$ yrs), the observed technetium is "currently" synthesized in the interiors of these red giants and mixed to their surfaces. The production of Tc, and most of the other heavy elements (i.e. those heavier than Fe), is driven by successive neutron captures and $\beta$-decays which build up the range of elements beyond iron. Two neutron-capture processes are recognized: one happens at rapid capture rates which exceed the $\beta$-decay rates, and is referred to as the

$r$-process. The other process occurs at capture rates which are less than typical $\beta$-decay rates and is referred to as the $s$-process. The details of the $s$- and $r$-processes were elucidated in the classic work by Burbidge, Burbidge, Fowler & Hoyle (1957). The technetium observed by Merrill in the S stars was created at low neutron-capture rates (low neutron densities) and is a product of the $s$-process.

It has been known since the work of Schwarzschild & Harm (1967) and Sanders (1967) that the preferred astrophysical site for the production of much of the $s$-process nuclei was a region associated with the H- and $^4$He-burning shells found in stars evolving along the asymptotic giant branch (AGB). Of course, it is in these same $^4$He-burning shells that $^{12}$C is made, which is then mixed to the surface to produce the C-rich stars we are discussing at this meeting: the $s$-process heavy elements and $^{12}$C are thus intertwined. Much of the early work connecting stellar models evolving through the AGB to $s$-process nucleosynthesis was carried out by Iben and collaborators and is summarized nicely in the review by Iben & Renzini (1983).

This is a brief discussion of a few current topics concerning the $s$-process in AGB stars, but we list here for the reader some other recent reviews and papers which cover various aspects of this topic: Smith (1990) and Lambert (1991) cover observational topics, while Iben & Renzini (1983) and Lattanzio (1989) present excellent overall reviews of theory.

## 2. The Heavy-Element Rich AGB Stars

As quasi-periodic shell $^4$He-burning occurs during thermal pulses on the AGB, mixing episodes increase the atmospheric C/O ratio in these red giants, giving rise to the spectral sequence of M to MS to S to C. This increase in the C/O is accompanied, in general, by increasing $s$-process overabundances, so that the MS, S, and C stars are also heavy-element rich AGB stars. In addition, there are a number of warmer, lower-luminosity stars which exhibit $s$-process enhancements, such as the CH and barium stars (Keenan 1942; Bidelman & Keenan 1951). As first demonstrated convincingly by McClure, Fletcher & Nemec (1980) and most recently by McClure & Woodsworth (1990), these lower-luminosity heavy-element rich stars are members of binary systems whose companions, now white dwarfs, transferred heavy elements when they were AGB stars.

A number of detailed abundance analyses have been conducted on the various heavy-element rich stars over the last 15 years and can be summarized as follows: for the barium stars we have Tomkin & Lambert (1983, 1986), Smith (1984), Kovacs (1985), Smith & Lambert (1984), Malaney & Lambert (1988), and Busso et al. (1995). The CH stars have been studied by Vanture (1992a, 1992b). The MS and S stars have been studied by Smith & Lambert (1990), Plez, Smith & Lambert (1993), Busso et al. (1995), and Lambert et al. (1995), while the most detailed analysis of the s-process in the carbon stars is still the work of Utsumi (1985). Recently, Smith et al. (1996, 1997) have identified the yellow symbiotic binary systems AG Dra and BD −21°3873 as metal-poor members of the Ba star class.

Taken together, these abundance studies provide the observational results which can be used to probe our current understanding of the details of s-process nucleosynthesis in AGB stars and we now summarize these results.

## 3. The s-Process as a Function of Metallicity

One question that is basic to an understanding of the s-process in AGB stars is the nature of the neutron source. Since the realization that the slow capture of neutrons is a distinct stellar nuclear process, two candidate neutron producing reactions have been suggested: $^{13}C(\alpha,n)^{16}O$ and $^{22}Ne(\alpha,n)^{25}Mg$ (Cameron 1955; Greenstein 1954). These two separate sources have very different characters; the s-process in AGB stars is associated with $^4$He-burning temperatures of $T \geq 10^8$K, with the $^{13}C$ source being active near $1 \times 10^8$K, while the $^{22}$Ne source requires higher temperatures, nearer to 2–3 $\times 10^8$K. The different temperatures required to drive the s-process via the different neutron sources will occur in AGB stars of different mass, or at different points along the AGB. Historically, the $^{22}$Ne source was favored as the most likely s-process candidate because the production of $^{22}$Ne can occur quite naturally as the result of succesive $\alpha$-captures onto $^{14}$N in the He-burning shell of an AGB star ($^{14}N(\alpha,\gamma)^{18}F(e^+,\nu)^{18}O(\alpha,\gamma)^{22}Ne$), with the initial $^{14}$N itself coming from CNO-cycle H-burning. The $^{13}C$ neutron source required an extra mixing mechanism to move protons into the He-burning shell to produce $^{13}C$ via $^{12}C(p,\gamma)^{13}N(\beta^+,\nu)^{13}C$, and the physical nature of this extra mixing was not clear. The physics involved in any sort of extra mixing is both complicated and not easily constrained by modelling alone. However, the expected behavior of these neutron sources with metallicity is different and can be tested by observation. If the observations can determine the dominant neutron source, this alone can provide important constraints for stellar models.

A convenient parameter to characterize the relative abundance distribution of the s-process is the neutron exposure, $\tau$, defined as

$$\tau = \int_0^{t'} N_n(t)V(t)dt$$

where $N_n$ is the neutron density and $V(t)$ the relative velocity of neutrons and nuclei, with the integral taken over the interval of the s-process episode. A large neutron exposure leads to a larger amount of heavier nuclei produced (e.g. Ba, La, or Ce) relative to the lighter nuclei, such as Y or Zr.

As pointed out by Clayton (1988), the nature of the dominant s-process neutron source can be tested by using the heavy- to light-element abundance distribution of the s-process (a measure of $\tau$) as a function of metallicity. For example if, during the s-process episode, the neutrons are in local equilibrium between their production and destruction, then the neutron density will be

$$N_n = (\Sigma N_i N_j \langle \sigma \rangle_{i,j})/(\Sigma N_k \langle \sigma \rangle_{k,n})$$

where the numerator is summed over all neutron-producing reactions between species $i,j$, with Maxwellian-averaged cross sections of $\langle \sigma \rangle_{i,j}$, and the denominator has the neutron destruction (absorbing) reactions, summed over all neutron absorbers $k$, and again a Maxwellian-averaged neutron-absorption cross section of $\langle \sigma \rangle_{k,n}$. Making the simplifying assumption that the neutron-producing reactions come from $\alpha$-captures, and that the dominant neutron absorbers are the seed nuclei of Fe, the above expression can be approximated as

$$N_n \propto (N_{He} N_j)/(N_{Fe}),$$

where $N_j = N(^{13}C)$ for the $^{13}C$ source and $N(^{22}Ne)$ for the $^{22}Ne$ source. Since the $^{22}Ne$ in the He-burning layers comes from $^{14}N$ which, itself, comes from the initial $\Sigma$CNO which is $\propto N_{Fe}$, the expected neutron density (and neutron exposure) would be expected to be roughly independent of metallicity. $^{13}C$, on the other hand, must result from the mixing of protons into a $^{12}C$-rich region, where the $^{12}C$ comes from $^4$He, which is almost independent of metallicity. Thus, if the structures of the He-shells in AGB stars are not strong functions of metallicity, the numerator of the above expression is constant and $N_n$ (and $\tau$) is $\propto N_{Fe}^{-1}$.

The neutron exposure can be estimated using published abundance results and we show the results of such an exercise in Figure 1 on the next page. Here we have plotted the quantity [hs/ls] vs. [Fe/H] from a number of studies. The quantity [hs/ls] represents the heavy-s to light-s ratio and is an average overabundance of representative heavy s-process nuclei (Ba, La, and Ce) relative to an average overabundance of the light s-process species Y and Zr, measured in standard spectroscopic bracket notation. This ratio [hs/ls] is proportional to the neutron exposure defining the s-process abundance distribution. The scatter is rather large, although this is not surprising as we are comparing results from a large number of investigators using somewhat different techniques and data of differing quality. Nonetheless, there is a clear and significant trend of increasing [hs/ls] with decreasing metallicity. That is, there is a significant increase of neutron exposure with decreasing metallicity as predicted from Clayton's (1988)

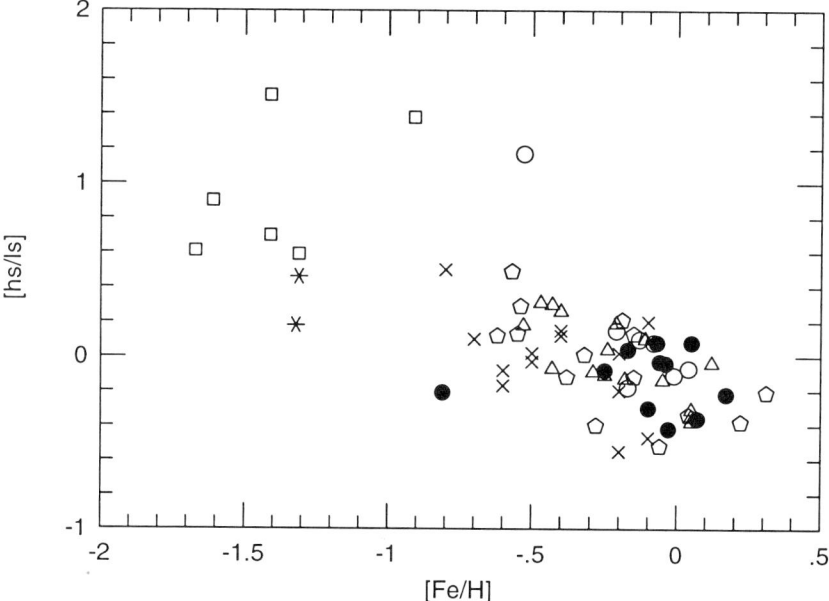

*Figure 1.* Published abundance results for the heavy *s*-process elements (Ba, La, and Ce) relative to the light *s*-process species (Y and Zr) for the MS/S stars (filled circles, with Tc, and open circles, without Tc), Ba giants (open triangles), Ba dwarfs (open pentagons), CH giants (open squares), CH subgiants (crosses), and yellow symbiotics (six-pointed stars). The increase of [hs/ls], which indicates an increasing neutron exposure, with decreasing metallicity [Fe/H] is evident. Such an increase is predicted by the $^{13}C(\alpha,n)^{16}O$ neutron source.

simple argument for the $^{13}C(\alpha,n)$ neutron source. Based on the observations, it would appear that the current best candidate for the *s*-process neutron source in AGB stars is $^{13}C$.

## 4. Summary

The topic of heavy-element nucleosynthesis in AGB stars is an exciting one. There is now emerging a unified picture which identifies the N-type carbon stars and the MS/S stars with technetium as AGB stars currently in the throes of active dredge-up, while the idea of mass transfer in a wide binary system adds the interpretation of dwarf, subgiant, and giant Ba and CH star abundance peculiarities as due to material processed in an AGB star being added to an "innocent companion." There is a wealth of observational and theoretical data being obtained concerning nucleosynthesis in AGB stars and it is fair to say that we are experiencing a real convergence of theory and observation which is leading us toward a much deeper understanding of this phase of stellar evolution.

## References

Bidelman, W. P. & Keenan, P. C. 1951, *ApJ*, 114, 473
Burbidge, E. M., Burbidge, G. R., Fowler, W. A. & Hoyle, F. 1957, *Rev. Mod. Phys.*, 29, 4
Busso, M., Gallino, R., Lambert, D. L., Raiteri, C. M. & Smith, V. V. 1992, *ApJ*, 399, 218
Busso, M., Lambert, D. L., Beglio, L., Gallino, R., Raiteri, C. M. & Smith, V. V. 1995, *ApJ*, 446, 775
Cameron, A. G. W. 1955, *ApJ*, 121, 144
Clayton, D. D. 1988, *MNRAS*, 234, 1
Gonzalez, G. & Wallerstein, G. 1994, *AJ*, 108, 1325
Greenstein, J. L. 1954, in *Modern Physics for Engineers*, ed. J. L. Ridenour (McGraw-Hill), p. 99
Iben, I. Jr. & Renzini, A. 1983, *Ann. Rev. Astron. Astrophys.*, 21, 271
Keenan, P. C. 1942, *ApJ*, 96, 101
Kovacs, N. 1985, *A&A*, 150, 232
Lambert, D. L. 1991, in IAU Symp. 145: *Evolution of Stars: The Photospheric Abundance Connection*, ed. G. Michaud and A. Tutukov (Kluwer), p. 299
Lambert, D. L., Smith, V. V., Busso, M., Gallino, R. & Straniero, O. 1995, *ApJ*, 450, 302
Lattanzio, J. 1989, in IAU Coll. 106: *Evolution of Peculiar Red Giant Stars*, ed. H. R. Johnson and B. Zuckerman (Cambridge Univ. Press), p. 161
Malaney, R. A. & Lambert, D. L. 1988, *MNRAS*, 235, 695
McClure, R. D., Fletcher, J. M. & Nemec, J. M. 1980, *ApJ*, 238, L35
McClure, R. D. & Woodsworth, A. W. 1990, *ApJ*, 352, 709
Merrill, P. W. 1952, *ApJ*, 116, 21
Norris, J. E. & Da Costa, G. S. 1995, *ApJ*, 447, 680
Plez, B., Smith, V. V. & Lambert, D. L. 1993, *ApJ*, 418, 812
Sanders, R. H. 1967, *ApJ*, 150, 971
Schwarzschild, M. & Harm, R. 1967, *ApJ*, 150, 961
Smith, V. V. 1984, *A&A*, 132, 326
Smith, V. V. 1990, in *Cool Stars, Stellar Systems and the Sun*, ed. G. Wallerstein, ASP Conf. Ser., 9, 340
Smith, V. V., Cunha, K., Jorissen, A. & Boffin, H. M. J. 1996, *A&A*, 315, 179
Smith, V. V., Cunha, K., Jorissen, A. & Boffin, H. M. J. 1997, *A&A*, 324, 97
Smith, V. V. & Lambert, D. L. 1984, *PASP*, 96, 226
Smith, V. V. & Lambert, D. L. 1990, *ApJ Supp.*, 72, 387
Tomkin, J. & Lambert, D. L. 1983, *ApJ*, 273, 722
Tomkin, J. & Lambert, D. L. 1986, *ApJ*, 311, 819
Utsumi, K. 1985, in *Cool Stars with Excesses of Heavy Elements*, ed. M. Jaschek and P. C. Keenan (Kluwer), p. 243
Vanture, A. D. 1992a, *AJ*, 103, 2035
Vanture, A. D. 1992b, *AJ*, 104, 1997
Vanture, A. D., Wallerstein, G. & Brown, J. A. 1994, *PASP*, 106, 835

## Discussion

**Plez:** Our observations of the Rb/Zr ratio in high-luminosity AGB stars in the SMC (Plez, Smith & Lambert 1993, *ApJ*, 418, 812) also fall on the general trend on Rb/Zr vs. Fe/H that you have shown.

**Smith:** Yes, and these SMC stars are some of the more luminous and, thus, fairly massive AGB stars.

# NUCLEOSYNTHESIS IN INTERMEDIATE-MASS STARS

JOHN C. LATTANZIO AND CHERYL A. FROST
*Department of Mathematics, Monash University*
*Clayton, Australia*

ROBERT C. CANNON
*School of Biological Sciences, Southampton University*
*Southampton, England*

AND

PETER R. WOOD
*Mount Stromlo and Siding Springs Observatories*
*Australian National University, Canberra, Australia*

**Abstract.** We discuss nucleosynthesis within 6 $M_\odot$ models with $Z = 0.02$, 0.008 and 0.004. The emphasis is on the AGB phase of evolution, with particular reference to thermal pulses and Hot Bottom Burning. We find strong CN cycling, with substantial Al production, especially at low metallicities.

## 1. Introduction and Method

Our knowledge of AGB stars has increased dramatically recently, driven primarily by observational studies, with theory lagging somewhat behind. Studies are desperately needed of the nucleosynthesis which occurs during thermally pulsing evolution and particularly "Hot Bottom Burning" (hereafter HBB), where the bottom of the convective envelope of a star reaches temperatures sufficiently high for significant nucleosynthesis to take place. Thus there is a thin layer at high temperatures, and envelope material is mixed through this region and then into the outer envelope. For nuclear reactions with time-scales no longer than the mixing time-scale, we cannot make the usual assumption of instantaneous mixing, and an average burning rate. The historical motivation for realising the importance of HBB has been briefly summarised in Lattanzio et al. (1996) and is not repeated here. We emphasise the importance of recent ion-probe studies of meteorites which reveal unusual isotopic compositions in some grains (such as SiC) which were almost certainly formed in the envelopes of AGB stars.

Our study requires both detailed nucleosynthesis and evolution models, with a time-dependent mixing algorithm. We have chosen to do this with two separate calculations. First we perform the evolutionary calculations with the Mt. Stromlo evolution code, modified to include the OPAL opacities (Rogers & Iglesias 1992). Following this, a nucleosynthesis code is used to calculate the detailed nuclear burning of 74 species with the stellar structure as previously calculated by the evolution code. Time-dependent mixing is included at this stage, in the way described by Cannon (1993). Nuclear reaction rates are primarily taken from the data base of Thielemann et al. (1987). There are some notable exceptions, such as the recent $\alpha$–particle capture rate for $^{18}$O (Käppeler et al. 1994). Details of both the code and the nuclear reaction rates may be found in Cannon et al. (1998).

## 2. Pre-AGB Evolution

The evolution of our models is qualitatively well understood, so we concentrate on composition changes. We will discuss the $M = 6\,M_\odot, Z = 0.02$ case (hereafter referred to as M6Z02) and then draw comparisons with other $Z$. (All masses given are in units of $M_\odot$).

### 2.1. CORE HYDROGEN BURNING AND FIRST DREDGE-UP

Following core hydrogen exhaustion, all models show similar composition profiles to the M6Z02 case seen in Figure 1. As we move inward from the surface, the increasing temperature starts the CN cycle, and we see an increase in $^{13}$C at the cost of $^{12}$C. The same region shows a decrease in $^{15}$N due to $^{15}$N(p,$\alpha$)$^{12}$C, and then a slight increase due to $^{18}$O(p,$\alpha$)$^{15}$N. Just slightly interior to $M = 3$ the $^{13}$C is itself destroyed in the production of $^{14}$N, which now starts to rise. From here inwards the carbon isotopes are in equilibrium. Interior to $M \simeq 2.5$ we see substantial destruction of $^{18}$O from the proton captures described above. At $M \simeq 1.8$ we see the production of $^{17}$O from proton captures on $^{16}$O, as the ON cycle starts to function more strongly. As the star ascends the first giant branch the convective envelope reaches down to $M = 1.256$ with consequent mixing from this point outward. This is known as the First Dredge-Up, and the key abundance changes are listed in Table 1.

### 2.2. CORE HELIUM BURNING AND SECOND DREDGE-UP

Within the helium core the star produces $^{12}$C and $^{16}$O. The abundant $^{14}$N left over from CNO cycling means that $^{14}$N($\alpha,\gamma$)$^{18}$F($\beta^+\nu$)$^{18}$O produces $^{18}$O

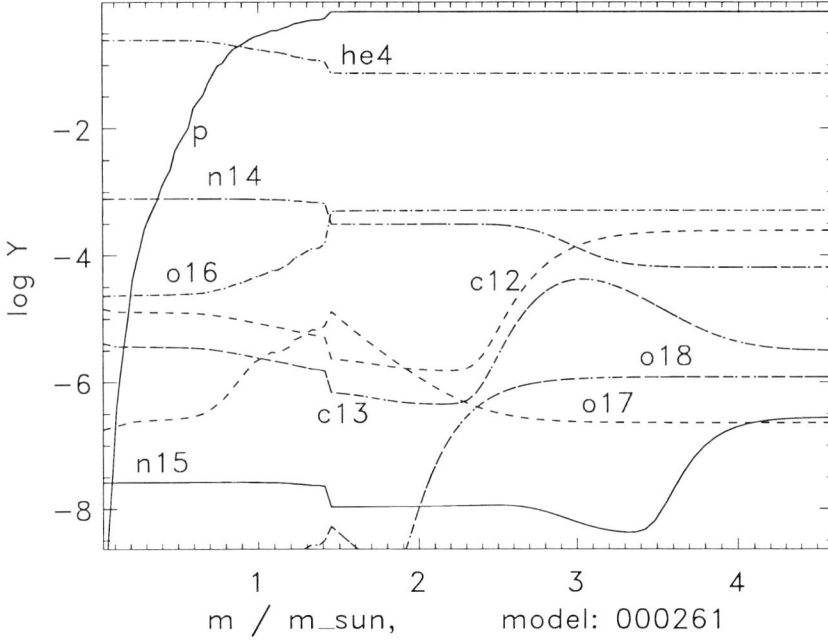

*Figure 1.* Composition profiles at the time of core hydrogen exhaustion for the M6Z02 model. Note that during first dredge-up the convective envelope will reach down to $M/M_\odot = 1.256$, thus homogenising the material from that point to the surface.

TABLE 1. Selected isotopic ratios after the first (FDU) and second (SDU) dredge-up events.

| Ratio | Initial | $Z = 0.02$ FDU,SDU | $Z = 0.008$ FDU,SDU | $Z = 0.004$ FDU,SDU |
|---|---|---|---|---|
| $^{12}C/^{13}C$ | 90 | 19.2, 18.0 | 18.8, 18.1 | 18.8, 17.8 |
| $^{14}N/^{15}N$ | 273 | 1297, 1931 | 1122, 2009 | 881, 2327 |
| $^{16}O/^{17}O$ | 2632 | 453, 408 | 708, 340 | 1746, 327 |
| $^{16}O/^{18}O$ | 500 | 562, 572 | 562, 590 | 501, 597 |

and then $^{18}O(\alpha,\gamma)^{22}Ne$ which produces some neutrons via $^{22}Ne(\alpha,n)^{25}Mg$. The hydrogen shell burning shows the usual signature of the CNO cycles, much as seen in Figure 1. The nucleosynthesis between the hydrogen "exhausted" region and the edge of the convective helium core mimics that which we will later see in the intershell region on the AGB. Outside the convective core the $^{13}C$ is destroyed by $(\alpha,n)$ and produces neutrons, which feed (n,p) reactions on $^{14}$N to produce $^{14}$C. Thus we see an increase in $^{14}$C and in the protons. Further inward the $^{14}$C has decayed into $^{14}$N.

## 3. Early AGB Evolution

The second dredge-up phase occurs early on the AGB. Table 1 gives some abundance ratios, but note that for $6\,M_\odot$ there is the complication that HBB begins during second dredge-up. After second dredge-up the star begins pulsing, which drives convection from the helium shell almost to the hydrogen shell (for details see Frost & Lattanzio 1996b). Later, envelope convection extends inward and penetrates the carbon-rich region after the $7^{th}$ pulse for M6Z02, the $6^{th}$ pulse for M6Z008 and the $5^{th}$ pulse for M6Z004. This periodic third dredge-up becomes very deep, and we soon find the dredge-up parameter $\lambda \simeq 0.9$ (see also Vassiliadis & Wood 1993). This is discussed further in Frost & Lattanzio (1996a).

With the onset of HBB the Cameron-Fowler beryllium transport mechanism produces large amounts of $^7$Li (see Boothroyd & Sackmann 1992). Our results are in agreement with these authors, to whom we defer for details.

The carbon-rich convective pocket mixes the products of helium burning throughout the intershell region, and at the next dredge-up the products of this nucleosynthesis are mixed to the surface. This pocket contains a relatively large amount of $^{22}$Ne at the cost of $^{14}$N and $^{18}$O. Also produced in the intershell zone is $^{19}$F from alpha captures on $^{15}$N, which is also dredged to the surface following each thermal pulse.

## 4. Advanced AGB Evolution

As the star ascends the AGB the temperature at the base of the convective envelope increases dramatically, reaching values of up to 80 or 90 million K. This causes CNO cycling as well as processing by the Ne-Na and Mg-Al cycles. Selected isotopic ratios are shown in Figure 2 for the thermally pulsing phase of the M6Z02 model. Mass loss (included here from Vassiliadis & Wood 1993) decreases the mass of the envelope, and eventually this shuts off the HBB. For M6Z02 this happens long before the third dredge-up ends (calculations for other models have not proceeded this far, yet) when the envelope mass is about $2.4\,M_\odot$. The details of the surface composition at this stage depend critically on the competition between these effects (e.g. Groenewegen & de Jong 1993). Thus, for most of its lifetime on the AGB, HBB prevents the M6Z02 model from becoming a carbon star. But when HBB ends, the third dredge-up may yet produce a carbon star (we will address this in a later paper). We now look in more detail at the nucleo-

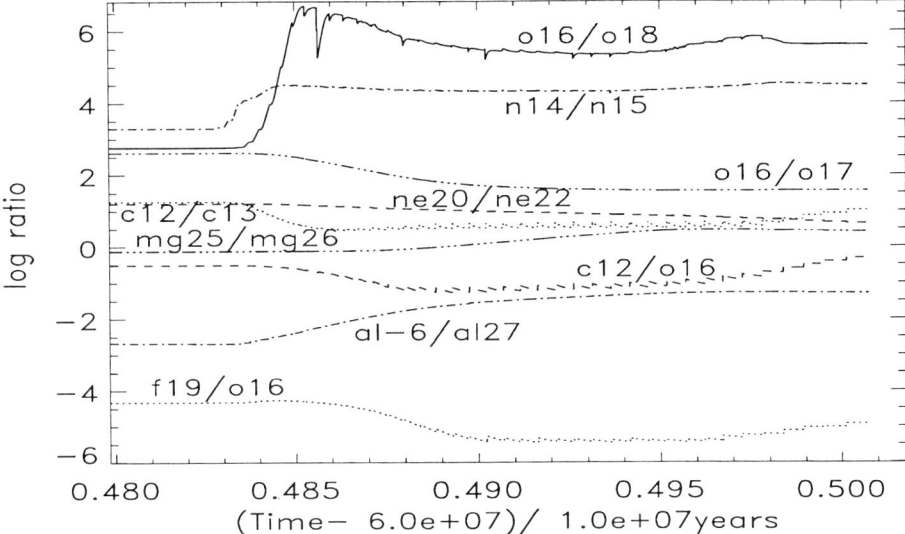

*Figure 2.* Selected isotopic ratios seen at the surface of the M6Z02 model during its thermally pulsing AGB evolution.

synthesis which occurs between the $30^{th}$ and $31^{st}$ pulses of M6Z02, being a typical interpulse phase showing HBB.

In the deepest layer of the convective envelope the temperatures are about 82 million K, and there is hydrogen burning via the CNO cycle. Initially the $^{12}C/^{13}C$ ratio is out of equilibrium because of the $^{12}C$ added to the envelope by the previous dredge-up episode. Thus the $^{13}C$ rises and $^{12}C$ falls as these come into equilibrium. The convective turnover timescale is such that the entire envelope passes through the hot bottom many times during the interpulse phase, producing the equilibrium abundances of $^{12}C$ and $^{13}C$. Then both C isotopes fall as they are processed into $^{14}N$. Further, at these high temperatures $^{26}Mg$ suffers substantial proton captures and produces $^{27}Al$. Also, the envelope $^{19}F$ is destroyed by (p,$\alpha$) reactions. Thus it would appear that stars experiencing HBB should not show increases of both $^{19}F$ and $^{7}Li$ simultaneously. This point has recently been made by Mowlavi et al. (1996), to whom we defer for details.

Figure 3 shows the region just below the convective envelope near the end of the interpulse phase — a region which has been processed by the hydrogen shell during the interpulse. First look at the $^{13}C$ profile. It leaves the hydrogen shell with $\log Y_{13} \simeq -4.9$, yet during the interpulse phase this burns totally via alpha captures, with a resulting increase in $^{16}O$. This was seen first in low-mass stars by Straniero et al. (1995) and later verified in more massive stars by Mowlavi et al. (1996). Some of the neutrons so

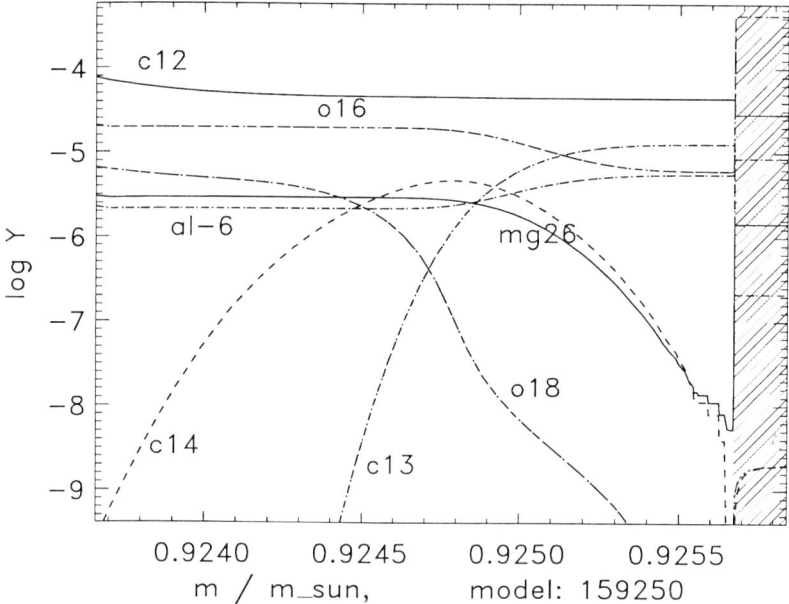

*Figure 3.* Compositions in a model near the end of the interpulse phase. The hatched region represents convection. "al-6" refers to the ground state of $^{26}$Al.

released are absorbed by the abundant $^{14}$N (not shown) to produce $^{14}$C via (n,p) reactions, as discussed above. This $^{14}$C also captures an alpha particle to produce the $^{18}$O rise seen in Figure 3. This will be absorbed by the next thermal pulse, to produce $^{22}$Ne and hence neutrons (the subject of a subsequent paper). The hydrogen shell also produces $^{27}$Al at the level $\log Y_{27} \simeq -5.3$ in the envelope, but it rises to $-4.7$ interior to the hydrogen shell, due to the proton captures on $^{26}$Mg which occur there. When this material is added to the envelope after the next dredge-up phase it will enhance the surface composition of $^{27}$Al.

We have seen the surface composition changes for the first 43 pulses of the M6Z02 model. The same calculation is shown in Figures 4 and 5 for $Z = 0.008$ and $0.004$, respectively. In each case the increase in $^{14}$N from CN cycling by HBB is quite apparent. As we have seen earlier, the $^{18}$O is almost completely destroyed within the envelope as soon as HBB begins. Likewise, $^{15}$N is seen in equilibrium with $^{14}$N for most of the evolution. We also see in the figures the production of $^{25}$Mg, $^{26}$Al and $^{27}$Al, all due to the operation of the Mg-Al cycle both in the hot-bottomed envelope as well as in the hydrogen shell, with subsequent dredge-up.

Having discussed the M6Z02 case we can see the contrasts with the lower metallicity cases. Decreasing $Z$ causes higher temperatures for HBB.

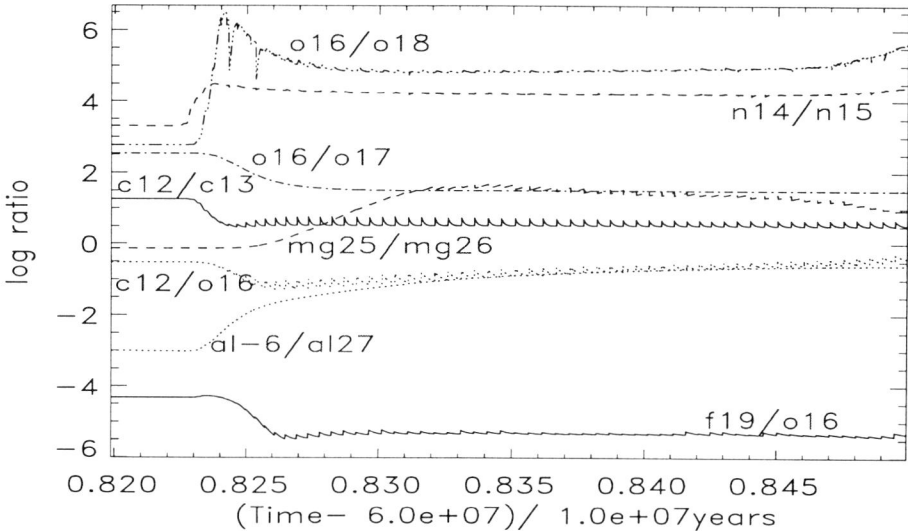

*Figure 4.* Surface composition for the M6Z008 model.

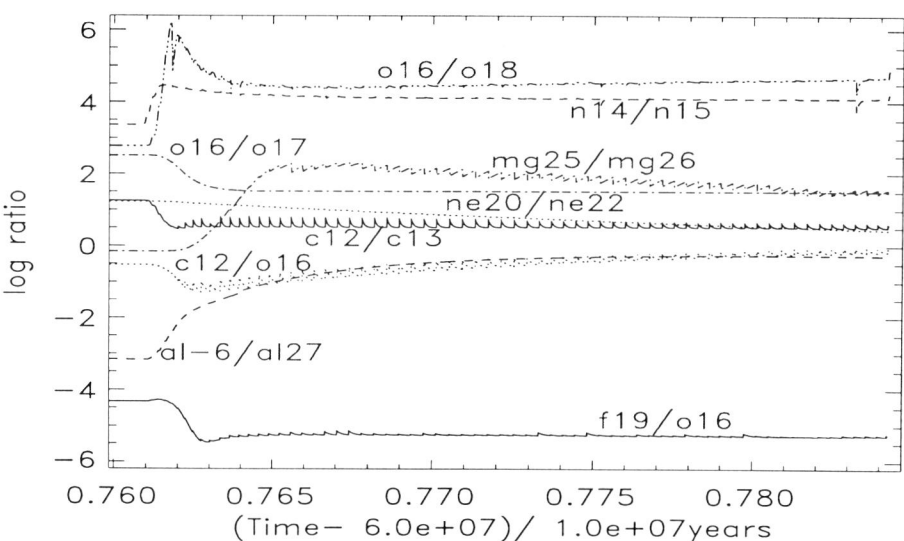

*Figure 5.* Surface composition for the M6Z004 model.

This is seen in the higher ratio of $^{26}\text{Al}/^{27}\text{Al}$ as $Z$ decreases. Note that with the most $^{26}\text{Al}$-rich meteorite grains (which reach values of $^{26}\text{Al}/^{27}\text{Al}$ up to 0.1; Zinner 1995), the current models are easily able to explain this measurement. A somewhat surprising result is the continued growth of $^{12}\text{C}/^{16}\text{O}$ during the HBB phase for the lower metallicity models. The dredge-up is so efficient at adding carbon to the envelope that despite the CN cycling, there is a continual increase in the $^{12}C/^{16}O$ ratio. The figures show that the $^{12}C/^{16}O$ ratios are close to unity (and the total C/O will be higher because of the $^{13}C$ present). This raises the possibility of a population of very high luminosity carbon stars. At the end of the calculations shown, the M6Z008 case has $M_{bol} = -6.6$ and for M6Z004 we have $M_{bol} = -7.2$. These stars may not be visible, however, because of the high mass-loss rates seen as they near the end of their lifetimes. Nevertheless, we note the identification of high luminosity, high mass-losing carbon stars by van Loon et al. (2000) at this meeting.

## 5. Conclusions

We have presented a brief summary of the nucleosynthesis in three model stars of 6 $M_\odot$ with $Z = 0.02$, 0.008 and 0.004. We find very deep dredge-up occurs with the dredge-up parameter $\lambda$ approaching unity. We have calculated the detailed processing during 43 ($Z = 0.02$) to 66 ($Z = 0.004$) pulses, and found that the models match quite well many known constraints (e.g. first and second dredge-up isotopic ratios). Further, during the AGB evolution a combination of dredge-up and HBB leads to many surface composition changes. A full analysis of these models, as well as six others of differing mass, will form the subject of later papers.

We thank Arnold Boothroyd and Roberto Gallino for many useful discussions. This work was supported by the Australian Research Council, and the British Council.

## References

Boothroyd, A. I. & Sackmann, I.-J. 1992, *ApJ*, 393, L21
Cannon, R. C. 1993, *MNRAS*, 263, 817
Cannon, R. C., Frost, C. A., Lattanzio, J. C. & Wood, P. R. 1998, in preparation
Frost, C. A. & Lattanzio, J. C. 1996a, *ApJ*, 473, 383
Frost, C. A. & Lattanzio, J. C. 1996b, in *Stellar Evolution: What Should Be Done?*, Proceedings of the 32nd Liège Colloquium, ed. A. Noels et al., p. 307
Groenewegen, M. A. T. & de Jong, T. 1993, *A&A*, 267, 410
Käppeler, F., Wiescher, M., Giesen, U., Görres, J., Baraffe, I., El Eid, M., Raiteri, C. M., Busso, M., Gallino, R., Limongi, M. & Chieffi, A. 1994, *ApJ*, 437, 396
Lattanzio, J. C., Frost, C. A., Cannon, R. C. & Wood, P. R. 1996, *Mem. Astron. Soc. Italia*, 67, 729
Mowlavi, N., Jorissen, A. & Arnould, M. 1996, *A&A*, 311, 803
Rogers, F. J. & Iglesias, C. A. 1992, *ApJ Supp.*, 79, 507
Straniero, O., Gallino, R., Busso, M., Chieffi, A., Raiteri, C. M., Limongi, M. & Salaris, M. 1995, *ApJ*, 440, L85
Thielemann, F.-K., Arnould, M. & Truran, J. W. 1987, in *Advances in Nuclear Astrophysics*, ed. E. Vangioni-Flam et al. (Editions Frontiers: Gif-sur-Yvette), p. 525
van Loon, J. Th., Zijlstra, A. A., Whitelock, P. A., Loup, C. & Waters, L. B. F. M. 2000,

in IAU Symp. 177: *The Carbon Star Phenomenon*, ed. R. F. Wing (Kluwer), p. 145
Vassiliadis, E. & Wood, P. R., 1993, *ApJ*, 413, 641
Zinner, E. K. 1995, in *Nuclei in the Cosmos III*, ed. M. Busso, R. Gallino and C. M. Raiteri, (AIP Press: New York), p. 567

## Discussion

**Frogel**: In my talk the other day I ruled out HBB as being able to turn a C star back into an M star for 1–4 $M_\odot$. For the more massive stars that you are talking about, the problem in testing your results is that there are too few LMC clusters in the appropriate age range to give good enough statistics.

**Lattanzio**: So it's the perfect theory....

**Frogel**: Almost certainly you can rule out high-mass C stars contributing 50 % of all AGB stars in the appropriate mass range, and maybe even at the 25 % level. Beyond that there are not enough real stars of known physical properties in the LMC.

**Lattanzio**: You are right, of course. There cannot be many of these stars, even if the models are correct. Also, as these "late conversion" C stars are experiencing rapid mass loss, they may not be visible at all!

**Ake**: Can you tell us what happens with Li?

**Lattanzio**: It should be very efficiently produced by stars in the mass range of about 4–6 $M_\odot$, as shown by Boothroyd & Sackmann (1992, *ApJ*, 393, L21). Once all the $^3$He has been used in producing this Li, then the HBB succeeds in destroying the Li, so that the Li-rich period is a temporary (but not particularly short) phase of evolution. Boothroyd has also been looking at the possibility that deep mixing on the AGB can produce Li in low-mass stars. Although this relies on parametrized calculations, the required values of the parameters are quite reasonable. Of course, we really need to understand properly what drives this proposed extra-mixing. But that is a very complicated problem.

**Gustafsson**: You did not mention sodium. There is some empirical evidence for an intermediate-mass sodium source. Do your models produce any?

**Lattanzio**: There is no significant change in the $^{23}$Na abundance on the AGB: a small increase is seen at second dredge-up and then a small decrease during HBB (at least, with the Caughlan & Fowler 1988 rates). There is significant production of $^{21}$Na, but it is always at a very low level, about log Y ≈ −10, which is unobservable, I believe!

**Jorissen**: Could $^{14}$N be used as a diagnostic of HBB?

**Lattanzio**: I think that is a question an observer could better answer. $^{14}$N is quite abundant, so you would need good quantitative analysis, I think. Maybe $^{14}$N/$^{15}$N is more useful. We know from measurements of this ratio in meteorite grains that there is a wide spread in the value found — wider than the models seem to give.

**Little-Marenin**: What is the minimum mass for HBB? And does this mean that this method only produces high-luminosity J stars?

**Lattanzio**: HBB seems to occur in stars of about 4–6 $M_\odot$, so the $^{13}$C production by this mechanism would only occur for relatively high luminosities, as you say. But if there is some deep mixing, and I think that the evidence now is that there probably is, then maybe this can produce enrichments of $^{13}$C just as found on the first giant branch in globular cluster stars.

# FLUORINE PRODUCTION IN ASYMPTOTIC GIANT BRANCH STARS

N. MOWLAVI

*Observatoire de Genève*
*CH-1290 Versoix, Switzerland*

AND

A. JORISSEN AND M. ARNOULD

*Institut d'Astronomie et d'Astrophysique*
*Université Libre de Bruxelles*
*B-1050 Bruxelles, Belgium*

**Abstract.** The present status of our understanding of fluorine production in asymptotic giant branch stars is reviewed, and future perspectives are presented.

## 1. Introduction

The question of the origin of the galactic $^{19}$F remained unanswered until the suggestion by Goriely et al. (1989) that it can be produced in He-burning environments by the $^{14}$N $(\alpha, \gamma)$ $^{18}$F $(\beta)$ $^{18}$O $(p, \alpha)$ $^{15}$N $(\alpha, \gamma)$ $^{19}$F reaction chain. The required protons are produced by $^{14}$N $(n, p)$ $^{14}$C, while the neutrons are supplied by $^{13}$C $(\alpha, n)$ $^{16}$O. Fluorine produced in such a way has been predicted to contaminate the surfaces of asymptotic giant branch (AGB) stars (Mowlavi et al. 1996, and references therein) or of Wolf-Rayet stars (Meynet & Arnould 1996), and consequently to contribute to the $^{19}$F galactic enrichment. These expectations have received an observational confirmation in the case of AGB stars (Jorissen et al. 1992). It has also been speculated that $^{19}$F could be a product of the $\nu$-process possibly accompanying Type II supernovae (Woosley et al. 1990 and references therein). However, the corresponding $^{19}$F yields remain very uncertain, as they depend sensitively on the neutrino temperature, which is a free parameter in the calculations. No observational test of this type of production is available to date.

This paper summarizes the status of $^{19}$F observations at the surface of AGB stars, and our present understanding of its production in these objects (§2). Perspectives for future studies are presented in §3.

## 2. Present Status

A definite proof that AGB stars can be net producers of $^{19}$F has been provided by the discovery that it is overabundant at the surface of S and C stars (Jorissen et al. 1992). So far, this is the only stellar site for which $^{19}$F production is directly confirmed by observation. In addition, these observations show that the surface $^{19}$F overabundance and the $^{12}$C/$^{16}$O ratio are correlated, confirming that fluorine is indeed produced in a He-burning environment.

Former AGB models including all reactions relevant to fluorine production (Mowlavi et al. 1996, and references therein) predicted that $^{19}$F is produced in abundances too low to match the observed values. In these models, the neutrons are supplied by $^{13}$C left over by the H-burning shell. More neutrons than those produced by this 'secondary' supply of $^{13}$C (since it scales as the primordial CNO content of the star) are needed to account for the observed fluorine abundances, thus calling for a primary $^{13}$C source.

Interestingly, the need for a primary $^{13}$C source in the He-burning layers of AGB stars also comes from the observations of s-process elements in S stars (Smith & Lambert 1990). Moreover, the s-process and $^{19}$F overabundances appear to be correlated in those stars (Jorissen et al. 1992). Parametric calculations of $^{19}$F and s-process nucleosynthesis mimicking the situation encountered in the He-burning shell of AGB stars (Mowlavi et al. 1998) indeed confirm that a primary $^{13}$C$(\alpha,n)^{16}$O neutron source is able to account for the observed abundances of both $^{19}$F and s-process elements in these stars.

## 3. Perspectives

The efficient production of fluorine along with s-nuclides in AGB stars is now well understood, at least qualitatively (the only puzzling fact is the large F overabundance reported for the super Li-rich star WZ Cas, since the Cameron-Fowler process invoked for the production of Li is expected to destroy F). However, reliable *quantitative* predictions of the surface $^{19}$F abundance in AGB stars are still hampered by two important shortcomings affecting current AGB models: (*a*) the *primary* $^{13}$C supply and the subsequent neutron production at the level required by both the $^{19}$F and s-process nucleosynthesis in the He-burning shell are not yet obtained in a self-consistent way by the current stellar models; (*b*) neither is the so-called third dredge-up (bringing material from the He-burning shell to the surface) predicted self-consistently by the current AGB models (but see Mowlavi 1998). These two problems *must* be solved before any *reliable s*-process and $^{19}$F yields can be obtained and compared with the observations.

Once reliable F yields become available for AGB stars with a large variety of masses and metallicities, their contribution to the solar system F may be estimated through the use of a galactic chemical evolution model. Preliminary estimates (Mowlavi & Meynet 1998) suggest that this contribution may be significant. On the other hand, Meynet & Arnould (1996) conclude that Wolf-Rayet stars alone might well account for the entire solar system F.

A complementary question of importance concerns the relative contribution of these two classes of stars to the F content of the Galaxy at different epochs. As AGB and Wolf-Rayet stars are expected to lead to different trends of [F/Fe] versus [O/Fe], observations of F abundances in low-metallicity stars are of utmost importance. They have in fact already been undertaken by V. Smith and colleagues. Such an analysis might also shed light on the question of the necessity (advocated by Timmes et al. 1995) of calling for the $\nu$-process in order to account for the F in the Galaxy at some point in its history.

As an extension of the study of the metallicity dependence of the F content of the Galaxy, one might also envision detecting $^{19}$F in high-redshift objects. The interest of such observations relates to the recent claim by Timmes et al. (1997) that "any positive detection of $^{19}$F at sufficiently high redshifts ($z > 1.5$) would suggest strongly a positive detection of the neutrino process." The possibility of a significant thermonuclear production of $^{19}$F by short-lived (non-exploding) Wolf-Rayet stars would evidently blur this picture.

## References

Goriely, S., Jorissen, A. & Arnould, M. 1989, in *Proceedings of the 5th Workshop on Nuclear Astrophysics*, ed. W. Hillebrandt and E. Müller (Max Planck Institut für Astrophysik report MPA/P1), p. 60
Jorissen, A., Smith, V. V. & Lambert, D. L. 1992, *A&A*, 261, 164
Meynet, G. & Arnould, M. 1996, in *Stellar Evolution – What Should Be Done*, 32nd Liège Internat. Astrophys. Coll., ed. A. Noels, D. Fraipont-Caro, M. Gabriel, N. Grevesse and P. Demarque (Liège Univ.), p. 89
Mowlavi, N. 1998, in preparation
Mowlavi, N., Jorissen, A. & Arnould, M. 1996, *A&A*, 311, 803
Mowlavi, N., Jorissen, A. & Arnould, M. 1998, *A&A*, 334, 153
Mowlavi, N. & Meynet, G. 1998, in IAU Symp. 187: *Cosmic Chemical Evolution*, in preparation
Smith, V. V. & Lambert, D. L. 1990, *ApJ Supp.*, 72, 387
Timmes, F. X., Woosley, S. E. & Weaver, T. A. 1995, *ApJ Supp.*, 98, 617
Timmes, F. X., Truran, J. W., Lauroesch, J. T. & York, D. G. 1997, *ApJ*, 476, 464
Woosley, S. E., Hartmann, D. H., Hoffman, R. D. & Haxton, W. C. 1990, *ApJ*, 356, 272

## Discussion

**Plez**: I would like to warn observers to be very careful in the derivation of abundances in stars with C/O ratios very close to 1. Due to the relative absence of strong blanketing, their atmospheric thermal structure is quite different from more O- or C-rich stars. Good models with the appropriate chemical composition, especially the C/O ratio, must be used in the analysis.

# CARBON STARS IN THE EARLY–AGB STAGE OF EVOLUTION

JU. L. FRANTSMAN*
*Radioastrophysical Observatory*
*Latvian Academy of Sciences*
*Riga, LV-1527, Latvia*

*Deceased 1998 July 20

**Abstract.** The assumption that faint carbon stars, as well as high-luminosity carbon stars whose effective temperatures are well above those of ordinary N–type stars in the Magellanic Clouds, are in the early asymptotic giant branch (E–AGB) evolutionary stage, is examined using population simulation techniques.

## 1. Introduction

The cool N–type carbon stars are generally associated with the thermally-pulsing asymptotic giant branch (TP–AGB) stage of evolution. But there are two groups of carbon stars whose luminosities and effective temperatures indicate that they are not in this stage:

1. Both theory and observation show that the minimum luminosity of carbon stars in the TP–AGB stage is about $M_{bol} = -3.0$ mag (see, for example, Vassiliadis & Wood 1993; Groenewegen & de Jong 1993, 1994). Nevertheless, according to *JHK* photometry (Westerlund et al. 1992, 1995) and a low-dispersion spectroscopic survey (Rebeirot et al. 1993) of carbon stars in the Small Magellanic Cloud, there are N–type stars fainter than $M_{bol} = -3.0$ mag, the least luminous having $M_{bol} = -1.8$ mag.

2. Suntzeff et al. (1993) have presented the results of photometry of carbon stars in the Large Magellanic Cloud which had been classified as CH stars (Hartwick & Cowley 1988; Cowley & Hartwick 1991). These carbon stars are significantly bluer than ordinary carbon stars in the LMC. Their mean bolometric magnitude is $M_{bol} = -5.3$ mag while the brightest members reach $M_{bol} = -6.2$ mag.

The main idea is that these stars are in the E–AGB evolutionary phase. This phase is the first part of the AGB evolution, when the hydrogen-

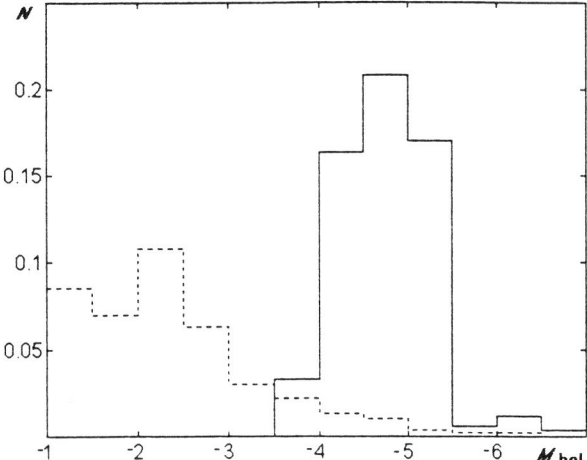

*Figure 1.* Theoretical luminosity functions of carbon stars with Z = 0.002. *Solid line*: TP–AGB stars; *dashed line*: E–AGB stars (the result of the evolution of close binaries).

burning shell is extinguished and helium burns steadily in a thin shell, providing most of the energy reaching the stellar surface. Towards the end of the E–AGB phase, H is re-ignited in a thin shell, and He burning continues in the form of the periodic thermal pulses. The TP–AGB stage then begins.

## 2. Simulation of Carbon Star Populations

The theory of two AGB stages (E–AGB and TP–AGB) was developed more than decade ago, but some investigators still do not take into account the E–AGB phase in spite of the fact that for some stars it lasts considerably longer than the TP–AGB stage.

Our scenario assumes that the above-mentioned two groups of stars are components of close binary systems. Mass was transferred from a TP–AGB carbon star by Roche lobe overflow to its companion which, as a result, becomes a carbon-enriched star. During its subsequent evolution, this star passes through all the later stages, including the E–AGB stage.

The results of a simulation of carbon star populations are plotted in Figures 1 and 2.

The luminosity functions are presented in Figure 1. The most significant result is a substantial extension of the carbon star luminosities towards fainter luminosities in comparison with TP–AGB carbon stars. Notice that a few high luminosity E–AGB carbon stars ($M_{bol} < -5$ mag) exist.

A comparison between the theoretical luminosity function for E–AGB carbon stars and the observational one for faint carbon stars in the SMC

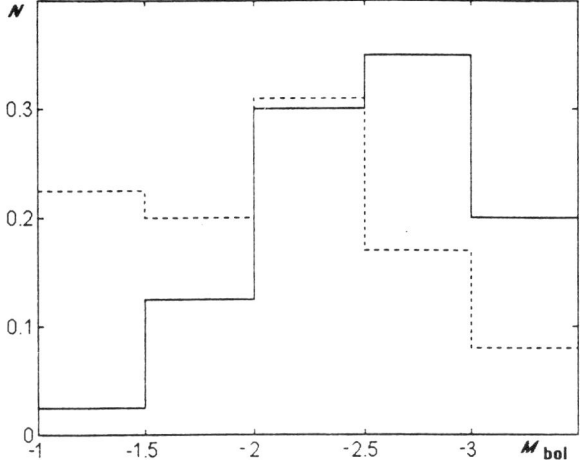

*Figure 2.* Luminosity functions of faint ($M_{bol} > -3.5$ mag) carbon stars. *Solid line*: the results of observations (Westerlund et al. 1995); *dashed line*: the results of calculations.

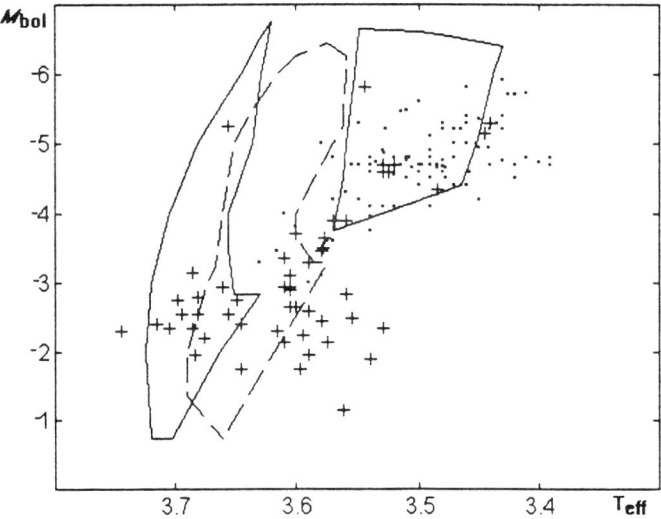

*Figure 3.* The positions of SMC carbon stars on the HR diagram. *Dots*: data from Westerlund et al. (1991); *crosses*: from Westerlund et al. (1995). The solid line contours outline the regions occupied by the E-AGB stars (left region) and by the TP-AGB stars (right region). The dashed line contour denotes the E-AGB region according to Fagotto et al. (1994).

(Westerlund et al. 1995) is presented in Figure 2. The two luminosity distributions are in qualitative agreement.

Figure 3 presents the HR diagram for SMC carbon stars, also showing the theoretically-calculated borders of regions occupied by the E-AGB and

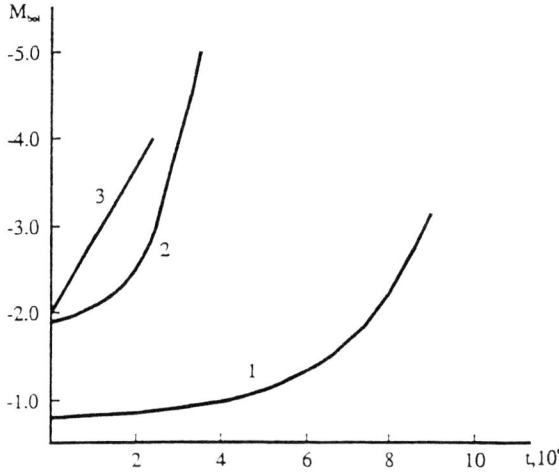

*Figure 4.* The dependence of the luminosity of a 3 $M_\odot$ E–AGB star on time, in units of $10^6$ yr.: (1) Z = 0.02; (2) Z = 0.001 (Lattanzio 1991); (3) our calculations for Z = 0.001, using the formulas of Iben & Renzini (1984).

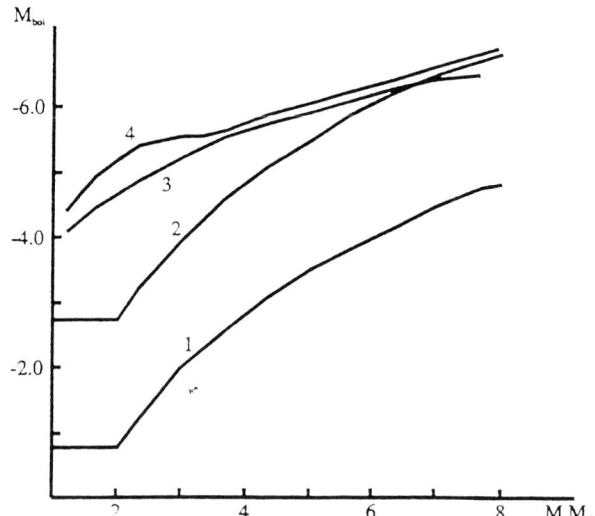

*Figure 5.* The dependence of luminosity on the initial mass for different stages of evolution: (1) the beginning of the E–AGB; (2) the end of the E–AGB; (3) the beginning of the TP–AGB; (4) the end of the TP–AGB.

TP–AGB models. All stars under consideration can be divided into two groups: (1) more luminous stars, with lower effective temperatures, and (2) fainter stars with higher temperatures. The former correspond closely to the TP–AGB model region, and the latter to the E–AGB region.

The dependence of the luminosity of a 3 $M_\odot$ E–AGB star on the time of evolution is presented in Figure 4. It is interesting to note that E–AGB stars can reach high luminosities only if the heavy-element abundance is low (typical for the Magellanic Clouds, but not for the Galaxy).

In Figure 5, the dependence of the luminosity of a theoretical model ($Z = 0.002$) on the initial mass is shown for different stages of evolution. Only massive stars reach a high luminosity, comparable to the TP–AGB star luminosity.

## 3. Conclusion

The main conclusion of this study is that the N–type carbon star population is not homogeneous but consists of objects which belong to two different stages of the evolution, the TP–AGB and the E–AGB.

## References

Cowley, A. P. & Hartwick, F. D. A. 1991, *ApJ*, 373, 80
Fagotto, F., Bressan, A., Bertelli, G. & Chiosi, C. 1994, *A&A Supp.*, 105, 29
Groenewegen, M. A. T. & de Jong, T. 1993, *A&A*, 267, 410
Groenewegen, M. A. T. & de Jong, T. 1994, *A&A*, 283, 463
Hartwick, F. D. A. & Cowley, A. P. 1988, *ApJ*, 334, 135
Iben, I. Jr. & Renzini, A. 1984, *Phys. Rep.*, 105, 329
Lattanzio, J. C. 1991, *ApJ Supp.*, 76, 215
Rebeirot, E., Azzopardi, M. & Westerlund, B. E. 1993, *A&A Supp.*, 97, 603
Suntzeff, N. B., Phillips, M. M., Elias, J. H., Cowley, A. P., Hartwick, F.D.A. & Bouchet, P. 1993, *PASP*, 105, 350
Vassiliadis, E. & Wood, P. R. 1993, *ApJ*, 413, 641
Westerlund, B. E., Azzopardi, M., Breysacher, J. & Rebeirot, E. 1991, *A&A Supp.*, 91 425
Westerlund, B. E., Azzopardi, M., Breysacher, J. & Rebeirot, E. 1992, *A&A*, 260, L4
Westerlund, B. E., Azzopardi, M., Breysacher, J. & Rebeirot, E. 1995, A&A, 303, 107

# FROM THE TIP OF THE AGB TOWARDS A PLANETARY: A HYDRODYNAMICAL SIMULATION

D. SCHÖNBERNER, M. STEFFEN, J. STAHLBERG,
K. KIFONIDIS AND T. BLÖCKER
*Astrophysikalisches Institut Potsdam*
*D-14482 Potsdam, Germany*

**Abstract.** We present a first exploratory investigation of the dynamical evolution of a dusty stellar wind envelope along the upper AGB and its transformation into a planetary nebula. We find the existence of AGB stars with detached shells to be a natural consequence of the mass-loss variations during a thermal pulse. It is also demonstrated that due to the large dynamical effects caused by the ionizing radiation field and the fast wind of the central star, it is impossible to deduce the AGB mass loss history from the planetary's density and velocity distribution. The structure of the halo, however, is still determined by the AGB mass loss history. The rapid decline of mass loss expected in the aftermath of thermal pulses leads to extended shells of low densities and explains halos with sharp boundaries.

## 1. Introduction

Recent attempts to model the development of planetary nebulae with evolving central stars by solving the coupled equations of radiation hydrodynamics have been remarkably successful, but so far all of these studies have started with initial structures which are exclusively based on guesses. This fact is, of course, just a reflection of our ignorance of the dynamical behavior of the dusty wind envelopes of AGB stars. Knowledge of the density, velocity, chemical, and ionization structures of stellar wind envelopes when the star leaves the AGB appears to us of paramount importance for understanding the evolution of proto- and well-developed planetaries as well as for the interpretation of observations. Important questions that have to be investigated are:

1. the short and long term behavior of AGB envelopes subject to stellar mass-loss variations on time scales smaller than the flight time of a gas parcel;

2. the final structures after the end of the AGB evolution when heavy mass loss ceases and the remnant starts to increase in $T_{\text{eff}}$;
3. whether a well-developed planetary-nebula shell still contains information about the mass loss history during the AGB evolution.

It is the purpose of this contribution to address these questions and to give first answers. We present here the very first attempt to compute the dynamical response of a stellar wind envelope for the last several hundred thousand years of AGB evolution and the following 5 000 years of post–AGB evolution. Based on self-consistent stellar evolution calculations, the adopted mass loss rate varies as a function of the stellar parameters. It should be emphasized that although the mass-loss law has not been derived from first principles (for details see Blöcker 1995), the stellar evolution calculations and the computations of the response of the envelope presented here are internally consistent once the physical details of the mass loss have been specified (see below).

## 2. Evolution towards the Tip of the AGB

Figure 1 illustrates the temporal evolution of luminosity and mass-loss rate for the last $\sim 350\,000$ years on the AGB and the following few thousand years of the post–AGB evolution of a solar composition stellar model computed by Blöcker (1995). Mass loss reduces the initial mass of 3 $M_\odot$ to a final mass of 0.605 $M_\odot$. The stellar outflow is assumed to be spherically symmetric, and the equations of hydrodynamics are solved for the gas and dust components, coupled by momentum exchange due to dust–gas collisions. The code is a modification of the one developed by Yorke & Krügel (1977) and makes use of the following simplifications:

1. Radiation transfer is considered only for the dust component, i.e. exchange of photons between dust grains and the gas is neglected.
2. The dust temperature is computed from radiative equilibrium, and the gas component (neutral hydrogen) is assumed to have the same (local) temperature. This is justified since we are not interested in computing line emissions, and since the influence of the gas-pressure gradient on the outflow velocity turns out to be quite small.
3. The dust component consists of single-sized grains, based on either oxygen or carbon chemistry, and a fixed dust-to-gas ratio is assumed at the condensation point.

A more detailed description of this fully implicit radiation hydrodynamics code, including comparisons with solutions for stationary outflows given in the literature, will be published in the near future.

The new approach applied here is to use time-dependent values of stellar mass, luminosity, effective temperature *and* the resulting variable mass loss

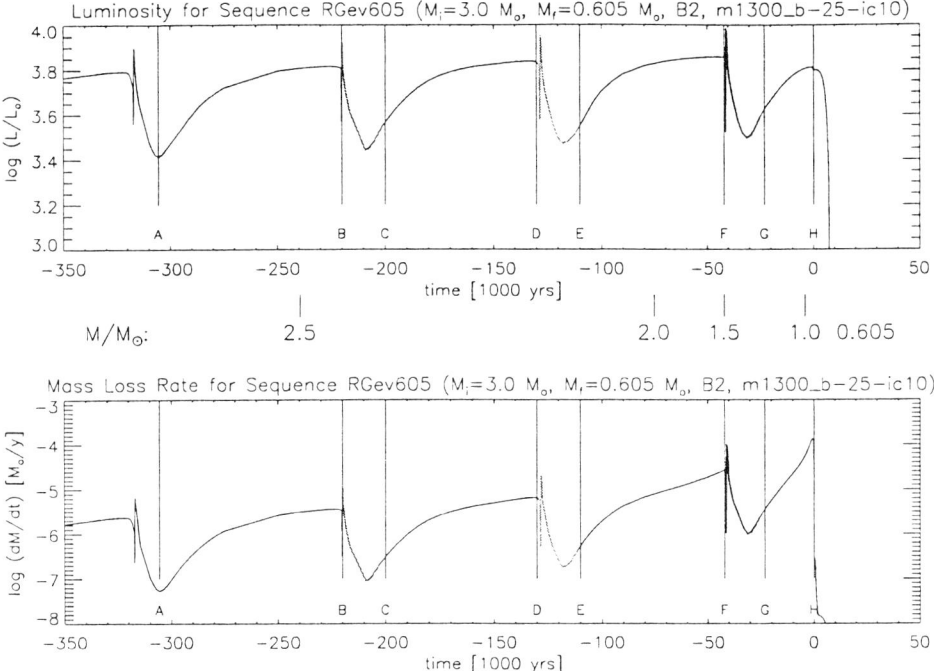

*Figure 1.* Luminosity vs. time (*top*) and mass loss rate vs. time (*bottom*) for the last four thermal-pulse cycles of a 3 $M_\odot$ model sequence, after Blöcker (1995). Time zero marks the beginning of the model's post–AGB evolution. The current stellar mass is indicated between the two panels. Labels A to H are referred to in the text.

(as shown in Fig. 1) at the inner boundary of our numerical grid, with a constant flow velocity equaling the local sound velocity ($\sim 3$ km s$^{-1}$). The radiation pressure on the grains and the momentum exchange with the gas leads to an acceleration of the material with typical final outflow velocities around 10 to 15 km s$^{-1}$, as are also observed. We started the computations at point A in Fig. 1 and continued through three consecutive thermal pulses until the end of the AGB phase is reached (H).

The pronounced variations of the mass loss rates during the thermal pulses (sections B–C, D–E and F–G in Fig. 1) lead to large deviations of density and velocity from the cases of stationary outflows. Correspondingly, the emerging spectral energy distributions (SEDs) change with time as demonstrated in Figure 2 for the first computed pulse (B–C). Dust grains with a radius of 0.05 $\mu$m are made of amorphous carbon and the dust-to-gas mass-density ratio is assumed to be $1.5 \times 10^{-3}$. The time-dependent models produce drastic changes of the spectral energy distributions on even rather short time scales (e.g. $\sim 300$ years between $t2$ and $t3$) which are solely due to the large variations of the mass-loss rate and the corresponding

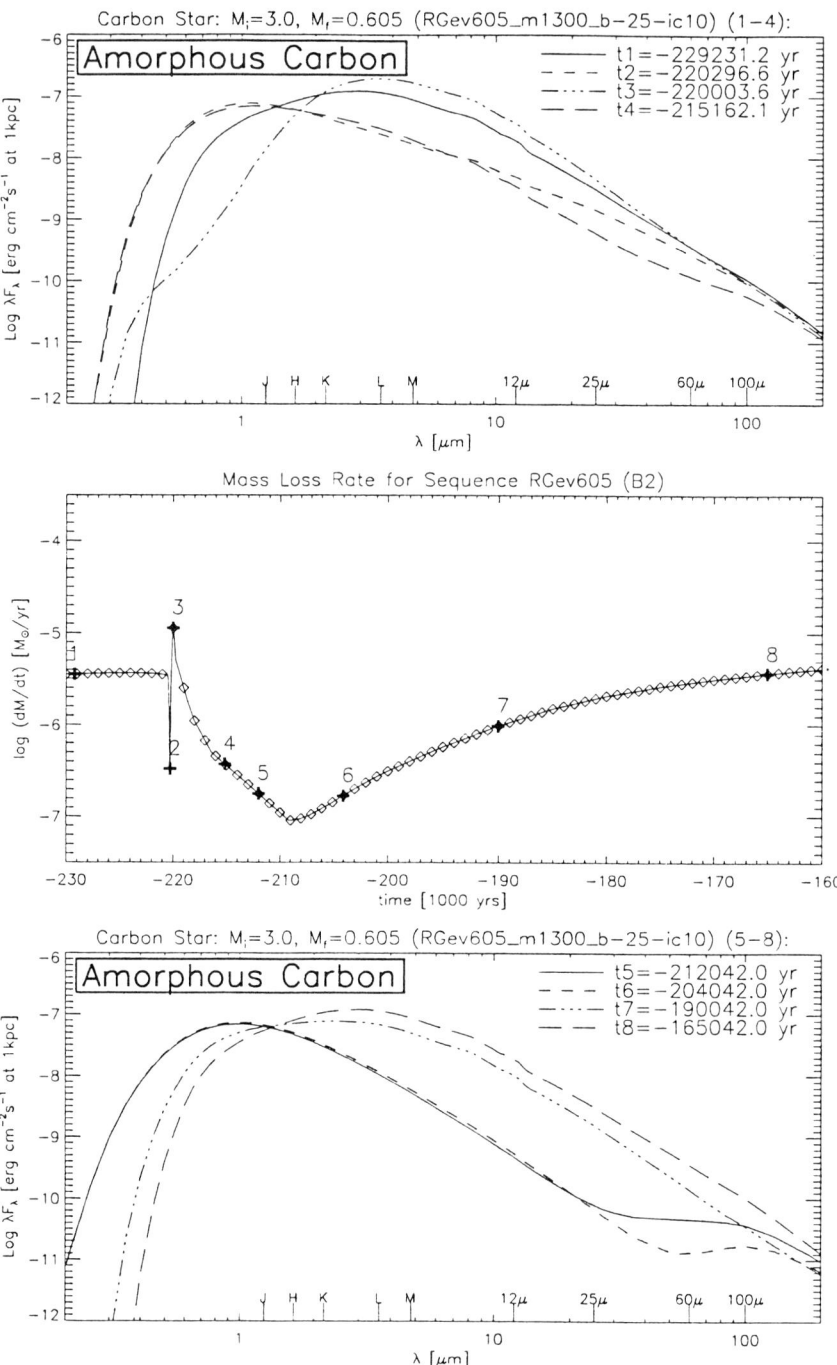

*Figure 2.* Spectral energy distributions (*upper* and *lower* panels) at selected times for a carbon star that experiences a thermal pulse (*middle* panel). The infrared and IRAS pass bands are indicated. For further details see text.

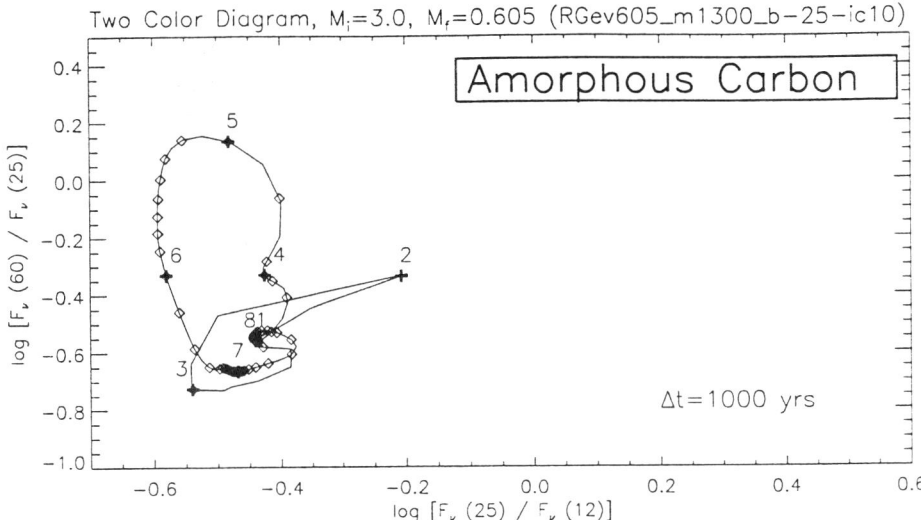

*Figure 3.* IRAS two-color diagram of the model sequence shown in Fig. 2, with numbered + signs indicating the same times as before. Diamonds outline the time evolution in steps of $\Delta t = 1\,000$ years. Colors at times 1 and 8 are identical.

deviations from stationarity. Although the mass-loss rates are practically equal at times $t2$ and $t4$, there are considerable differences in the resulting far-IR emission. The same holds for times $t5$ and $t6$.

During the whole mass loss minimum the dust shell is optically thin and the star is completely visible ($t2$, $t4$, $t5$, $t6$). The large far-IR excess is due to a cool detached dust shell containing matter which has been expelled well before the onset of the thermal pulse. Objects with far-IR excesses are known (e.g. Chan & Kwok 1988; Zijlstra et al. 1992; Kerschbaum & Hron 1996; Olofsson et al. 1996), and our computations give a quite natural explanation in terms of thermally pulsing AGB stars. Later the envelope becomes optically thick again ($t7$, $t8$) and can be approximated by stationary flow solutions.

The changes in spectral appearance shown in Fig. 2 are responsible for the object's movement in the two-color diagram as shown in Figure 3. Similar loops can be found by coupling envelopes with silicate grains to the stellar parameters.

Several remarks are in order. ($i$) In addition to the loops, our dynamical envelopes produce some intrinsic scatter in the two-color diagram, solely due to the deviations from stationary conditions. ($ii$) As the consequence of a thermal pulse, objects on the upper AGB can develop *detached dust shells* and spend about 10% of their time away from the main two-color relation of dusty AGB stars. Our computed SEDs are in good agreement with corresponding observations, which may, however, be disturbed by cirrus contamination in some cases. In contrast to Ivezić & Elitzur (1995), we believe that the modeling of dusty envelopes around AGB stars as stationary outflows is not justified *a priori* and may lead to erroneous interpretations!

## 3. Evolution towards a Planetary Nebula

In the mass-loss modeling by Blöcker (1995) it has been assumed that the mass loss rapidly decreases with the radial fundamental pulsational period of the remnant until the Reimers (1977) rate is reached at $P = 50$ days. This rate is then kept until the remnant becomes hot enough for the application of the theory of radiation-driven winds (Pauldrach et al. 1988) which provides mass-loss rate and wind speed. A more detailed description of the physical model used for the post–AGB evolution and how mass loss rate and wind speed vary in the course of the evolution can be found in Marten & Schönberner (1991).

The kinematic structure of the last model of the AGB evolution, computed at $t = 0$, is shown in the upper panel of Figure 4. The density structure is clearly different from the usual assumption of an $r^{-2}$ law: This model reflects the last 100 000 years of AGB evolution, showing a large density dip caused by the last thermal pulse about 30 000 years ago (cf. Fig. 1), but a rapid density increase inwards ($\propto r^{-3}$) due to the increasing mass-loss rate. Further inwards the density increase flattens somewhat ($\propto r^{-1}$). The outflow velocity is rather constant except for a slight decrease caused by the luminosity dip of the last pulse.

This model is then used as input for a newly developed, explicit radiation hydrodynamics code based on a second-order Godunov-type advection scheme (Walder, private communication) in order to follow the evolution into the domain of planetaries. The radiation part of this code considers time-dependent ionization, recombination, heating and cooling for up to nine elements (Marten & Szczerba 1997). Here we have used the elements H, He, C, N, O and Ne with all of their ionization stages. More details of our dynamical modeling of planetaries will be published in the near future.

The lower panel of Fig. 4 gives an example of a well-developed nebular structure of intermediate size. The fast but tenuous stellar wind terminates through a strong inner shock (near $r = 3 \times 10^{16}$ cm), and the high pressure of this shocked wind material compresses the inner parts of the planetary to a high-density, low-velocity shell (just beyond C in Fig. 4). The outer shell (just inside B) has much lower density, contains matter with larger velocities (up to 30 km s$^{-1}$) and snowplows through the halo which is still the original but ionized AGB wind. Please note that the outer shock front (B), which defines essentially the observed outer rim, expands with an even higher

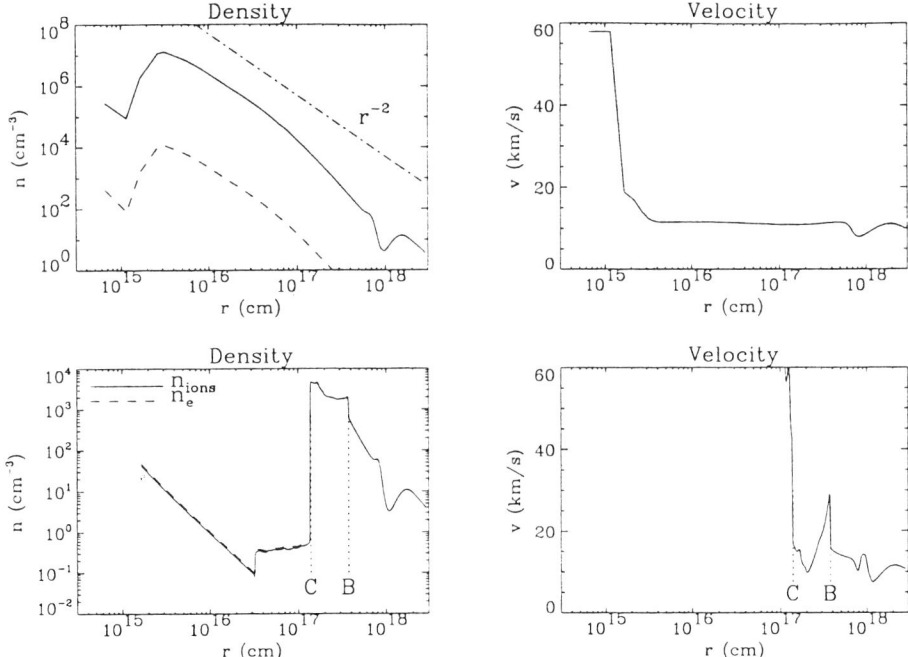

*Figure 4.* Upper panel: Kinematic structures of the initial model ($t = 0$) for post–AGB evolution, with central star parameters $L = 6310\,L_\odot$ and $T_{\text{eff}} = 6053$ K. Lower panel: 4603 years later: now $L = 5602\,L_\odot$ and $T_{\text{eff}} = 89\,851$ K. The planetary proper is the region between the contact discontinuity (C) and the outer shock (B).

velocity of about 40 km s$^{-1}$. Since the whole grid is completely ionized, such a planetary would be called "density bounded." Morphologically one would speak of a well-developed double-shell structure (Chu et al. 1987).

It is clear from Fig. 4 that the kinematic structure of the planetary proper (between C and B) has no resemblance whatsoever to the mass loss history that originally led to its existence. In particular, the double-shell structure has absolutely *nothing* to do with recurrent helium shell flashes, as has been proposed by Tuchman & Barkat (1980). Rather, this structure is a consequence of photoionization and wind interaction and appears also for simpler initial models, for instance for an $r^{-2}$ density law and also for the "superwind" model of Marten & Schönberner (1991).

The situation is different for the halo: The density dip (near $r = 10^{18}$ cm) caused by the last thermal pulse is still existent, and also the velocity field has not changed much. This low-density region has observational consequences for the surface-brightness distribution of, for instance, H$\beta$ (Figure 5). The halo appears to have a rather sharp boundary, and one may speak, in analogy, of a "density bounded" halo, which is typical for the few

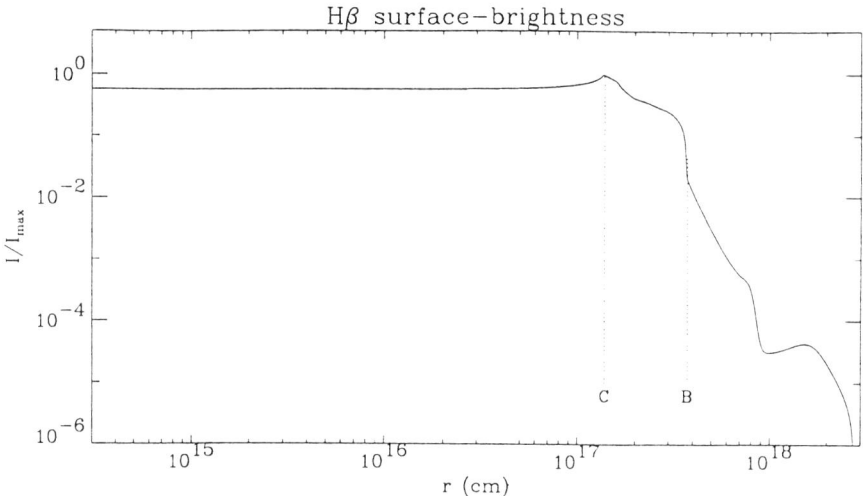

*Figure 5.* Surface-brightness distribution in Hβ of the nebular model shown in the lower panel of Fig. 4. Again, the positions of the outer shock (B) and the contact discontinuity (C) are indicated. Note the sharp outer rim of the halo near $r = 10^{18}$ cm.

known cases (Chu 1989). Our model shown in Fig. 5 resembles remarkably well the Hα image of NGC 6826 (cf. Plait & Soker 1990). Actually, NGC 6826 is classified as a triple-shell planetary (Chu et al. 1987), and our interpretation would be that these relatively rare objects are typical double-shell objects where the halo is visible as a third shell.

Part of this work was supported by the German Space Agency, DARA, under grant WE2 50 OR 9411.

## References

Blöcker, T. 1995, *A&A*, 297, 727
Chan, S. J. & Kwok, S. 1988, *ApJ*, 334, 362
Chu, Y.-H. 1989, in IAU Symp. 131: *Planetary Nebulae*, ed. S. Torres-Peimbert (Kluwer), p. 105
Chu, Y.-H., Jacoby, G. H. & Arendt, R. 1987, *ApJ Supp.*, 64, 529
Ivezić, Ž. & Elitzur, M. 1995, *ApJ*, 445, 415
Kerschbaum, F. & Hron, J. 1996, *A&A*, 308, 489
Marten, H. & Schönberner, D. 1991, *A&A*, 248, 590
Marten, H. & Szczerba, R. 1997, *A&A*, 325, 1132
Olofsson, H., Bergman, P., Eriksson, K. & Gustafsson, B. 1996, *A&A*, 311, 587
Pauldrach, A., Puls, J., Kudritzki, R. P., Méndez, R. H. & Heap, S. R. 1988, *A&A*, 207, 123
Plait, P. & Soker, N. 1990, *AJ*, 99, 1883
Reimers, D. 1977, in *Problems in Stellar Atmospheres and Envelopes*, ed. B. Baschek, W. H. Kegel and G. Traving (Springer), p. 229
Tuchman, Y. & Barkat, Z. 1980, *ApJ*, 242, 199
Yorke, H. W. & Krügel, E. 1977, *A&A*, 54, 183
Zijlstra, A., Loup, C., Waters, L.B.F.M. & de Jong, T. 1992, *A&A*, 265, L5

## Discussion

**Dorfi**: What is your argument for using the mass-loss rate of Blöcker also during the thermal pulse phases?

**Schönberner**: Blöcker developed his mass-loss formula because, at that time, Bowen's calculations were the only ones available that considered radial pulsations together with radiation pressure on grains. We used Blöcker's prescription in the dynamical calculations presented here in order to keep the internal consistency between mass loss and stellar evolution. A detailed treatment of dust formation together with a self-consistent determination of mass-loss rates along the AGB is a tremendous amount of work which, however, certainly has to be done.

**Wagenhuber**: Depending on the mass-loss rate dependence on the effective temperature, the time scales for the transition from the AGB to $\sim 10^4$ K can be very different, so that cores of higher mass may mimic the evolutionary timescales of lower-mass cores.

**Schönberner**: In the model of Blöcker, it is assumed that the high AGB mass-loss rate declines when the stellar pulsational period decreases from 100 to 50 days. This *always* ensures that the transition towards the PN region is rather fast. This treatment seems to be in rough agreement with the observations since none of the observed post–AGB stars appears to be cooler than 5000 K, as is also predicted by Blöcker's models.

# Session XI

## GALACTIC CHEMICAL EVOLUTION

"Duck Soup" – One of the stories of the 14th-century folk teacher/preacher Nasrettin Hoca, as envisaged by Bengt Gustafsson (see p. 494).

# CARBON STARS AND NUCLEOSYNTHESIS IN GALAXIES

BENGT GUSTAFSSON AND NILS RYDE
*Uppsala Astronomical Observatory, Uppsala, Sweden*

**Abstract.** The role of carbon stars in the build-up of chemical elements in galaxies is discussed on the basis of stellar evolution calculations and estimated stellar yields, abundance analyses of AGB stars, galactic-evolution models and abundance trends among solar-type disk stars. We conclude that the AGB stars in general, and carbon stars in particular, probably are main contributors of $s$-elements, that their contributions of flourine and carbon are quite significant, and that possibly their contributions of lithium, $^{13}$C and $^{22}$Ne are of some importance. Also contributions of N, Na and Al are discussed. The major uncertainties that characterize almost any statement concerning these issues are underlined.

## 1. Introduction

Any discussion of the role of carbon stars (C stars) in the chemical evolution of galaxies raises a number of important questions such as: For which metal abundances and initial stellar masses do stars become C stars? How much mass does a star lose during its C star phase? What are the elemental compositions of these ejecta?

C stars are brilliant and show easily recognizable spectral characteristics. Thus, their occurrence and frequency in different galaxies may be interpreted in terms of the properties (metallicity and age) of the stellar populations, if we know which stars become C stars. Current answers to this question are, however, not very precise, although stars with masses somewhere in the interval 1.2 to $4\,M_\odot$ may be a realistic answer. Studies of the distribution of C stars perpendicular to the Galactic plane suggest typical masses in the interval $1.2 - 1.6\,M_\odot$ (Claussen et al. 1987; cf. also Groenewegen et al. 1995). For stellar masses greater than about $4\,M_\odot$ hot-bottom burning (HBB) is thought to prevent C star formation (Boothroyd et al. 1993). In metal-poor populations, stars with even smaller masses may become sufficiently carbon-enriched to show C star spectra, while HBB may also be effective in inhibiting C star formation at lower masses than for Pop. I stars. An important issue in this respect is the frequency of C-rich planetary nebulae (PNe) in the Galaxy, which is so high (cf. Zuckerman & Aller 1986; Rola & Strasińska 1994) that a considerable fraction of all stars must become C stars in the end — this is a strong reason for not increasing the lower limit of the mass range discussed here too much above 1 $M_\odot$.

There is consensus that mass loss probably both sets the maximum luminosity that a star achieves and determines the time it spends in its evolution up along the AGB. The total mass returned to the interstellar medium is obviously the difference between the initial stellar mass and the final remnant mass (typically about 0.65 $M_\odot$ for the stars discussed here). In order to determine the total mass lost from the C stars we first need to know how much mass these stars have lost before becoming C stars. Most authors assume that the total mass loss on the RGB is on the order of 0.2 $M_\odot$ for solar-mass stars, following arguments by Renzini & Fusi Pecci (1988), while more massive stars — being more compact and not undergoing the He core flash — presumably lose less. Many studies also assume some continuous mass loss, specified by the Reimers relation (with a suitable fitting parameter), followed at the end of the AGB by a "superwind" with a steeper dependence of mass loss on stellar parameters. Detailed models of pulsating stars (Bowen & Willson 1991; Willson et al. 1996) demonstrate the inadequacy of simple parametrizations — the mass loss rate turns out to be quite sensitive to stellar parameters, including mass and composition, and this cannot be determined empirically from available observed mass loss rates because the stellar parameters are not well known.

Since the atmospheres of AGB stars, and especially C stars, are enriched in many elements by interior nucleosynthesis and various mixing mechanisms, they contribute substantial amounts of heavy elements to the interstellar medium (ISM). (Also, the physical and chemical properties of the ISM are affected by the grains produced by C stars.) The main yield to the ISM from the C stars depends, however, on the interplay between mass loss and dredge-up of processed material from the inner stellar layers. Neither of these processes is very well understood. In recent model calculations dredge-up, as well as mass loss, is incorporated with free parameters adjusted to fit some selected observational constraints, such as the luminosity distribution of C stars in the Large Magellanic Cloud (LMC). Major recent contributions of this type are due to Groenewegen & de Jong (1993), Groenewegen et al. (1995), Blöcker (1995) and Marigo et al. (1996). This may be a questionable procedure in view of the assumptions made concerning the dependence of the adjustable parameters on stellar properties.

Key issues in the present discussion are when, during the AGB evolution with its gradual dredge-up of enriched material, the most significant mass loss occurs and how this depends on the metallicity of the star. We are still far from definitive answers to these questions. Here, a rather empirical discussion will be given, based on the observed chemical compositions of stars and PNe. For a detailed and more theoretical approach, based on calculations of nuclear processing and mixing during the AGB with adjustable parameters describing the dredge-up as well as with semiempirical recipes for mass loss rate as a function of luminosity and pulsation period, see Marigo et al. (1996). These authors also contribute tables of predicted yields, partially superseding those of Renzini & Voli (1981).

Another, more indirect but very significant basis for estimating yields will also be used below: the trends in relative abundances of solar-type stars of different ages in the solar neighborhood. From such results (e.g. those of Edvardsson et al. 1993) one can estimate relative time scales of the build-up of different elements, and thus separate the contribution of elements from, say, slowly evolving low mass stars as compared with contributions from rapidly evolving high mass stars through supernova (SN) explosions.

We shall here review the current knowledge, and lack of knowledge, as regards the role of C star production of the nuclei of Li, C, N, O, F, Ne, Na, and Al as well as of the $s$-process elements. Our discussion will be based on what is known from stars in the galactic disk and in the LMC, and it is important to stress that the understanding concerning the mechanisms of the formation, nucleosynthesis and mass loss of C stars is so weak that extrapolations to stellar populations with other properties are highly uncertain. In fact, the mixture of arguments, partly based on galactic, partly on LMC C stars, in contemporary discussions is in itself dangerous, but necessary.

## 2. Nucleosynthesis

### 2.1. LITHIUM

As has been demonstrated by Reeves et al. (1990), spallation of CNO nuclei in the interstellar medium, although contributing $^6$Li, Be and B, cannot be the main process for raising the $^7$Li abundance from the Pop. II plateau value found by Spite & Spite (1982) to the more than 10 times higher value found in the solar system and in young stars. The processes by which Li has been built up are as yet not determined, but several processes have been proposed: Li-production in AGB stars, the $\nu$-process in SNe of Type II (SNeII), and production in novae. In a study based on models of galactic chemical evolution and observed abundances of light elements, Brown (1992) suggested intermediate-mass stars to be main contributors.

Some C stars show very strong Li I resonance lines (McKellar 1940) and these stars have been proposed to be a significant source of Li to the ISM (Scalo 1976). These rare stars, having logarithmic Li abundances, log $\epsilon$(Li), of about 10 times the solar system value or even greater ($\sim 4$ on a scale with log $\epsilon$(H) set to 12), are still not fully understood; attempts to explain them include HBB and the Be transport mechanism of Cameron & Fowler (1971) (Sackmann & Boothroyd 1992; Abia et al. 1993b).

Abia & Isern (1996) found that the $^{13}$C abundances of galactic C stars correlate with their Li abundances; this may give a further clue to the origin of the super Li-rich phenomenon. These authors suggest that HBB might produce the abundances found if one invokes a more efficient convection than is usually assumed. This could decrease the lower mass limit of HBB to below $2\,M_\odot$. Alternatively, the plume mixing model of Scalo & Ulrich (1973) might operate in these stars.

The significance of C stars in general for galactic Li production is entirely dependent on the role of the super Li-rich stars; the more normal N stars have log $\epsilon$(Li) ranging from –2 to 2 (Abia et al. 1993b) which is far too low for the stars to be of significance in this respect. Key questions concerning the super Li-rich stars are: (1) How common are they? (Note that the conclusion of Abia et al. that these stars may provide a significant fraction of the galactic Li is based on only one star!) (2) How does their frequency depend on mass, metallicity and other parameters? (3) What are their true Li abundances? (We note that the abundance estimates are severely dependent on model atmospheres, spectral synthesis and adopted C/O ratios for the models. Thus, Abia et al. (1993a) derive an abundance for WZ Cas of log $\epsilon$(Li) = 5, while Boesgaard et al. (1996) find a 3 times lower abundance with the same grid of model atmospheres.) (4) Are the super Li-rich stars overrepresented, or represented at all, among the dust-enshrouded stars with mass loss rates $> 10^{-5}\,M_\odot$/year? (5) Do the super Li-rich stars represent relatively short episodes, followed by Li burning, and certain limited mass intervals — as is suggested by the model calculations by Sackmann & Boothroyd (1992) and Frost & Lattanzio (1996) — or could they possibly mark the ending stage of AGB evolution in general?

A conclusion in several studies is that the C stars probably do not contribute significant amounts of Li (cf. Matteucci et al. 1995), even though contributions as high as 30% or more of the interstellar Li may be possible if some super Li-rich stars also have high mass-loss rates (cf. Abia et al. 1993b).

An interesting issue is whether this situation is different for low metallicities — i.e., in early phases of galactic evolution or in the present evolution of dwarf galaxies. A low metal abundance seems to lead to HBB at smaller stellar masses (cf. Sackmann & Boothroyd 1992). On the other hand, the low metallicity stars may lose their mass at lower rates, such that the Li produced may be burnt away before most of the mass loss occurs (Matteucci et al. 1995). Empirical support for this may be the Li abundance of log $\epsilon$(Li) $\approx 3$ derived for luminous AGB (S) stars in the SMC (with a metallicity 1/3 of solar) found by Plez et al. (1993), which, in spite of being significantly higher than expected from standard AGB evolution is still lower than values for the most Li-rich AGB stars in the Galaxy.

In conclusion it seems probable that another process, such as the $\nu$-process in SNeII (Woosley et al. 1990), is the main mechanism responsible for Li formation. However, until we understand the nature of the super-Li rich C stars better, more definitive statements must be avoided.

2.2. CARBON

For some time, intermediate or low mass stars have been considered as probably important for the production of carbon in the Galaxy (see, e.g., Tinsley 1978). Sarmiento & Peimbert (1985) estimated the carbon contribution from AGB stars to be on the order of 60–80% of all galactic carbon. Their argument was essentially based on the fact that models of SNe and novae suggested that these objects could not produce more than a minor part of the carbon in the interstellar medium. Sackmann & Boothroyd (1991) claimed, on the basis of calculations of dredge-up of carbon in AGB model sequences, that low and intermediate mass AGB stars are the dominant sources of carbon in the universe. The more recent supernova models of Woosley & Weaver (1995) and Thielemann et al. (1996) predict yields that are mutually different by about a factor of two, the Thielemann et al. yields being lower due to a high rate for the $^{12}C(\alpha,\gamma)^{16}O$ reaction. Adopting the higher yields of Woosley & Weaver (1995), Timmes et al. (1995) still find in their models of galactic evolution that SNe and novae cannot produce enough, so that the contributions from stars with a mass less than 11 $M_\odot$ must be dominant. Timmes et al. also adopt the (now partly obsolete) yields from intermediate and low mass stars of Renzini & Voli (1981) and find that the carbon abundance relative to iron, [C/Fe], was lower by about a factor of two than its present value in the early evolution of our galactic disk; then it increased above its present value and finally decreased again as a result of iron production in less massive SNeIa.

This particular behavior of [C/Fe] relative to [Fe/H] is not seen in the composition of solar-type disk stars of different ages, according to Tomkin et al. (1995). Instead, these authors find a steady decrease in [C/Fe] vs. [Fe/H], suggesting a less dramatic variation of contributing sites during the history of the Galaxy. In fact, the ratios of carbon relative to oxygen and $\alpha$-elements (produced in SNeII) only vary slowly and smoothly with [Fe/H]. This suggests that the carbon may well have been produced in high-mass stars, although carbon does not follow oxygen in the halo (Tomkin et al. 1992), nor in dwarf galaxies (Garnett et al. 1995). The decrease of [C/Fe] with increasing [Fe/H] in the galactic disk is very different from the variation of the $s$-element abundances with [Fe/H] (see also Edvardsson et al. 1993). This suggests that the $s$-elements were formed in stars with considerably longer characteristic lifetimes, i.e. smaller masses.

Prantzos et al. (1994) studied the formation of oxygen and carbon in the Galaxy and found that, if the duration of the halo phase was on the order of 1–2 Gyears as is currently believed, intermediate or low mass stars should not have been the main carbon sources. Instead, these authors suggest that massive stars contribute with metal-dependent yields (Maeder 1992). We note that the model calculations of Prantzos et al., with contributions added also from the intermediate and low mass stars according to Renzini & Voli (1981), fit the results of Tomkin et al. (1995) rather well, although the C/O ratios of comparatively metal-rich halo stars become too high. In that model, about 40% of the carbon in the disk stars was contributed by intermediate and low mass stars. Adoption of the yields of Marigo et al. (1996) would diminish the discrepancy for the metal-rich halo stars but may produce too much carbon at late stages.

Other data suggest that the intermediate and low mass stars may play a significant role in carbon synthesis. Adopting the findings by Zuckerman & Aller (1986) and Rola & Strasińska (1994) that about half of the PNe have C/O > 1, we find that these stars should contribute at least about 0.001 $M_\odot$/year of carbon to the Galaxy. (Here, we have used the PNe birthrate of about 1 PN per year in the Galaxy estimated by Pottasch (1992), and a characteristic PN mass of 0.3 $M_\odot$.) Assuming the yields recently derived from a semi-empirical modelling of the AGB phase by Marigo et al. (1996) would presently give contributions on the order of 0.003 $M_\odot$, depending on the star formation rate adopted. This contribution is of the same order of magnitude as the total present contribution by SNeII, assuming yields from Woosley & Weaver (1995) and a SN rate of 3 per century. In making this estimate we have assumed a closed box model; this may overestimate the significance of SNe relative to PNe, in view of the much higher expansion velocities of material ejected in SNe. One should note that the number of C-rich PNe is so high that they cannot only be produced by intermediate-mass stars; a significant fraction of the C-rich PNe must be formed by stars in the mass interval between 1 and 2 $M_\odot$.

If low-mass stars are assumed to produce significant amounts of carbon, further discrepancies between models of galactic chemical evolution and observed abundances may occur in the relative-abundance diagrams, such as those of Tomkin et al. (1995). Possibly, this conflict could be resolved by advocating more efficient carbon production by intermediate and low mass metal-poor stars, following the discussion of Boothroyd & Sackmann (1988) that suggests a more efficient dredge-up of carbon for models with low abundances of heavy elements. We note that the carbon yields by Marigo et al. (1996) are several times greater at a given initial stellar mass for their metal-poor (Z=0.008) models than for their Pop. I (Z=0.02) models. (This is, however, dependent on the assumptions made in these models that the two adjustable third-dredge-up parameters, the minimum core mass for dredge-up and the dredge-up efficiency, are independent of metallicity, and that the mass loss rate is dependent only on the metallicity through its effects on evolutionary tracks and pulsation period.) Another possibility, which should be further investigated, is that the spectrum analysis of the high-excitation C I lines in solar-type stars may give systematic errors, different for stars of different metallicities, e.g. due to errors in the effective temperature scale, or errors due to inhomogeneities and departures from LTE (we note, however, the small non-LTE effects found for carbon in the Sun by Stürenburg & Holweger 1990). The study of Andersson & Edvardsson (1994), using the low excitation [C I] line at 8727 Å, seems to verify the results of Tomkin et al. but the weakness of that line only admitted upper limits for most of the metal-poor disk stars in their sample. Further studies based on this line with higher S/N are underway.

Do the C stars contribute significantly to galactic carbon? Number densities and mass-loss observations of bright, as well as dust-enshrouded, C stars in the Galaxy (Olofsson et al. 1993a, 1996; Jura & Kleinmann 1989) suggest a total contribution of about $2 \times 10^{-4} \, M_\odot$/year. The indication, from the statistics of PNe, that C stars may contribute one order of magnitude more is the result of the fact that about every second PN is carbon rich. This illustrates the key significance, mentioned in the Introduction, of the question of the chemical composition of the stellar envelope just prior to the extensive mass loss that ends the AGB evolution.

We conclude that the significance of low and intermediate mass stars for carbon synthesis is still unclear, and deserves further exploration.

## 2.3. THE CARBON ISOTOPE $^{13}$C

Prantzos et al. (1996) have discussed the observations of carbon isotope ratios in the galactic disk and conclude that $^{13}$C has probably a mixed origin, being produced both as a primary element — presumably in intermediate-mass stars by HBB or by other processes where protons are mixed with $^{12}$C produced in the star by He burning — and as a secondary element by CN burning, occurring in stars of all masses. The $^{12}$C/$^{13}$C ratios observed in C stars range in the interval 30–100, with a pronouced peak around 50 (Lambert et al. 1986; a lower range, 15–40, has, however, been obtained by Ohnaka & Tsuji 1996, and these differences need further exploration). Excluded from this is a minority of J–type stars with considerably lower ratios. Values around 50 are consistent with the C/O ratios derived by Lambert et al.; that is, the result may be explained as the consequence of mixing $^{12}$C to the surface layers if $^{13}$C is left from the first dredge-up. These values agree reasonably well with those (30–40) derived for dusty C stars from mm-line observations by Kahane et al. (1992). These authors have also observed the C-rich PN NGC 7027 and find a $^{12}$C/$^{13}$C lower limit of 65. In fact, the difference in C/O ratio between bright N stars and C-rich PNe (see Lambert et al. 1986) suggests that further $^{12}$C enrichment has taken place in the latter, which would correspond to an increase from 50 to about 80 in $^{12}$C/$^{13}$C. Although this agrees with the local ISM value, that may well be fortuitous — little is known about the $^{12}$C/$^{13}$C ratios in PNe. One should also note that the $^{13}$C "produced" by normal C stars is, as presumed above, just the result of the CNO burning and the first dredge-up. That is, however, not necessarily the case; we note that Plez et al. (1993) find low $^{12}$C/$^{13}$C ratios for S stars in the Small Magellanic Cloud indicating H-burning in the envelope in late evolutionary phases.

One might ask what the role of the extremely $^{13}$C-rich J stars may be in providing $^{13}$C. The evolutionary history of these stars is still not understood (cf. Lambert et al. 1986 for some suggestions). They constitute about 1/5 of the bright N–type stars in the magnitude-limited Lambert et al. sample and have a $^{12}$C/$^{13}$C ratio of typically 5. They tend, however, to be systematically oxygen-poor; therefore, the $^{13}$C abundances relative to hydrogen are not more than typically a factor of 5 higher than those of other C stars. The J–type stars do not show any tendency of having higher present mass-loss rates than the rest of the N stars (Olofsson et al. 1993a). From this one would only conclude that the J stars at present contribute approximately the same total amount of $^{13}$C as the rest of the visible N stars. Similarly, if the J–type stars lose their envelopes without first burning their $^{13}$C, and if the fraction of J stars among C stars of 20% really is representative also of the population of immediate progenitors of C-rich PNe, which is highly questionable, the J stars could then deliver as much $^{13}$C as the sum of the rest of the C stars. There is at least one additional circumstance that might suggest that the J–type stars play a significant role in this respect: a great fraction of the most luminous red AGB stars in the LMC are $^{13}$C-rich C stars (Richer et al. 1979), possibly indicating that a greater fraction than 20% of the more massive C stars reach this stage close to the end of their AGB evolution, at least in metal-poor populations. However, there is also a population of low-luminosity J–type stars present (cf. Richer 1981; Bessell et al. 1983) which does not necessarily correspond to the latest phases of AGB evolution. Brewer (1996) recently discovered 7 J–type stars among 48 C stars in two M31 fields and found the J stars relatively faint.

2.4. NITROGEN

$^{14}$N is thought to be produced by equilibrium CNO–burning in the hydrogen burning shell, brought to the surface layers at the first dredge-up and even more at the second dredge-up, occurring in stars of 4–8 $M_\odot$ and expelled during later stellar evolution. Timmes et al. (1995), however, conclude, on the basis of abundance trends among galactic stars and radial gradients of CNO abundances in H II regions in galaxies, that $^{14}$N has a strong primary component which they ascribe to low metallicity massive stars. Woosley & Weaver (1995) show that it is possible to create this isotope in massive stars and claim that approximatively 25% of the solar abundance can be ascribed to these sites. Vila-Costas & Edmunds (1993) show evidence for a delayed primary and a secondary component relative to oxygen, the delay indicating that intermediate mass stars are responsible. The low N/O ratio observed for a high redshift gas cloud by Pettini et al. (1995) also supports an intermediate-mass primary component. The secondary component is dominant at high metallicities. The bulk of the solar abundance of $^{14}$N is attributed to low and intermediate mass stars (see e.g. Timmes et al. 1995).

Empirically, the evidence that C stars contribute a significant extra amount of N is weak. Lambert et al. (1986) find, from their analyses of CN lines in N–type star spectra, that these stars are not very N-rich, and this result is strengthened by recent values of the dissociation energy of CN. Olofsson et al. (1993b) find an N abundance higher by about a factor of 5 or so from their analysis of HCN mm lines from the circumstellar envelopes of the same sample of N stars, but this may be due to an underestimated mass loss rate (cf. also Olofsson et al. 1996). The C-rich PNe are, according to Zuckerman & Aller (1986), generally not very enriched in N — only very few in their list have N abundances in excess of twice the solar value. Pasquali & Perinotto (1993) list a mean nitrogen abundance of twice solar for their C-rich PNe of type II and III, which represent disk stars of intermediate and low mass, while the mean N abundance given for PNe of type I, that are more carbon-poor and represent younger, more massive objects, is three times greater. Kingsburgh & Barlow (1994) give a mean nitrogen abundance of only 1.4 times solar for their 36 non-Type I PNe. Similar tendencies may be traced in the yields calculated by Marigo et al. (1996). It seems probable that intermediate-mass stars that generally are prevented from becoming C stars by HBB may provide most of the nitrogen. We note that Brett (1991), in his analysis of luminous AGB stars in the SMC, finds strong CN bands, indicating C to N conversion by HBB.

## 2.5. FLUORINE

The galactic production of the single stable isotope of flourine, $^{19}$F, has been a matter of discussion and is still rather unclear. Jorissen et al. (1992) identified lines of HF in the 2 micron region for K, M, S and N giants and derived F abundances. They found that the F/O ratio in AGB stars increases with C/O, with maximum values for stars with C/O ratios not too much in excess of 1 (SC stars). Their result seems to imply that the thermal pulses produce fluorine, and they (cf. also Forestini et al. 1992) suggest as the most probable scenario that $^{13}$C$(\alpha,n)^{16}$O reactions produce neutrons, some of which are captured by $^{14}$N to produce $^{14}$C and protons, which in turn are captured by $^{18}$O: $^{18}$O$(p,\alpha)^{15}$N$(\alpha,\gamma)^{19}$F. The flourine is then brought to the surface by convection following the thermal pulse. This scenario is complicated by several circumstances, e.g. the new higher rate for the reaction $^{18}$O$(\alpha,\gamma)^{22}$Ne which may deplete $^{18}$O enough to decrease the significance of the second part of the reaction chain (cf. Frost & Lattanzio 1996). In any case, the over-abundances of F found in SC and N stars by Jorissen et al. (1992), ranging from 3 to 30 times the solar values, indicate that these stars may well be major contributors to the galactic flourine; an alternative is production in SNII (Woosley et al. 1990; Timmes et al. 1995).

An interesting way of clarifying this issue further may be to determine the flourine abundances in less evolved M stars with different metal abundances, since different scalings of, say, [F/O] with [Fe/H] may be expected for the different production sites.

## 2.6. NEON

The dominant neon isotope in the solar system, $^{20}$Ne, is probably mainly produced by SNeII (cf. Woosley & Weaver 1995; Timmes et al. 1995). The SNe yields of $^{22}$Ne are about one order of magnitude lower than those of $^{20}$Ne which is consistent with the isotopic ratios observed in the solar system. However, $^{22}$Ne may be formed at thermal pulses in AGB stars in the convective intershell through $\alpha$-capture on $^{14}$N, which after $\beta$-decay leads to $^{18}$O$(\alpha,\gamma)^{22}$Ne (Boothroyd & Sackman 1988; Gallino et al. 1990). In the AGB model calculations of Marigo et al. (1996), considerable amounts of $^{22}$Ne are produced and ejected, in particular for models with masses around 2.5 $M_\odot$ which get most carbon rich. The predicted $^{22}$Ne yields are fairly independent of metallicity (though higher for Z=0.008 than for Z=0.02), and so high that the AGB stars are suggested to be a major contributor of galactic $^{22}$Ne. In particular, the envelopes of C stars are predicted to have a $^{20}$Ne/$^{22}$Ne ratio considerably less than 1, even close to 0.1.

The Ne enrichment expected in C-rich PNe according to these results was not traced in the sample of southern PNe studied by Kingsburgh &

Barlow (1994), nor can it be seen in the sample of PNe from the Magellanic Clouds studied by Leisy & Dennefeld (1996). However, Corradi & Schwarz (1995) found a sample of bipolar PNe to be enriched in Ne and Marigo et al. (1996) suggest that the build-up of $^{22}$Ne could be the reason for this. We note that Lewis et al. (1990) and Nichols et al. (1993) find consistently low isotopic ratios in gas-rich meteoritic SiC grains.

## 2.7. SODIUM AND ALUMINUM

Na and Al are, as well as Mg, generally identified as products of Ne and C burning in massive stars while the production in SNeIa is probably small (Nomoto et al. 1992). The synthesis of Na and Al is controlled by the neutron flux during Ne and C burning which in turn is dependent on the initial metallicity and primarily on the initial O abundance ($^{16}$O being converted to $^{22}$Ne, via $^{14}$N, in He-burning and the extra neutrons in $^{22}$Ne, when being liberated, are essential to the formation of Na and Al). Therefore, one expects a rapid increase of Na/Mg ratios with Fe/H, and such a tendency is also apparent in the calculations of Timmes et al. (1995). For the most metal-rich solar-type stars the positive correlation between Na/Mg and Fe/H is well established (Feltzing & Gustafsson 1998). However, the observations of disk stars by Edvardsson et al. (1993), as well as observations of halo stars, do not show this tendency very clearly (Timmes et al. 1995). Edvardsson et al. find a tendency for the abundance ratio of Na/Mg to be smaller in the inner Galaxy than in the outer, at a given Mg/H, for disk-stars more metal-poor than the Sun. This suggests that Na and Mg, respectively, were formed in stars of different types. Thus, there might be additional sources of sodium and possibly of aluminum.

There is some additional evidence that intermediate-mass stars or even low-mass stars might contribute in these respects. In intermediate-mass stars sodium may be synthesized in the hydrogen-burning shell through the neon-sodium cycle, and by proton captures on $^{22}$Ne (Denisenkov 1989; Denisenkov & Denisenkova 1990; Langer et al. 1993). On longer time-scales, i.e. smaller masses, proton capture on $^{20}$Ne may also be significant, and the results may be brought to the surface if efficient mixing is at hand. Observationally, yellow field supergiants appear sodium rich (Boyarchuk & Lyubimkov 1985) but they hardly contribute significantly to the galactic Na. A fraction of low mass red giants in globular clusters show enhanced Na and Al and are correspondingly poor in O (Norris & Da Costa 1995 and references given therein). Recently Cavallo et al. (1996) have attempted to model this, by following the nucleosynthesis in low-mass red giant model sequences, and they find substantial Na-enrichments in the region above the hydrogen-burning shell throughout the giant branch, independent of metallicity. For the lowest metallicities Al is also produced. With suitable mixing processes these elements will be visible on the surface and may even, after mass loss, contribute to the global enrichment of the Galaxy.

There is, however, no strong argument for this low-mass star production of Na and Al to be linked to the existence of C stars. Conversely, the meager abundance analyses that exist do not support such a hypothesis; e.g. Lambert et al. (1986) find normal Na/Ca ratios on the basis of 1 line for each element in the 2 $\mu$m region.

There is an interesting possibility that AGB stars contribute radioactive $^{26}$Al (Kudryashov & Tutukov 1988), decaying to $^{26}$Mg with emission of 1809 keV photons, detected in space. Guélin et al. (1995) have recently found some evidence for $^{26}$AlF millimeter emission from the carbon-rich envelope of IRC +10216. Huss & Wasserburg (1996) comment on the finding of high $^{26}$Al/$^{27}$Al ratios in some SiC grains. The high temperatures needed to produce $^{26}$Al suggest rather high mass stars as sites if HBB is invoked. This is, however, not easily reconciled with other isotopic ratios.

## 2.8. THE s-ELEMENTS

The early discovery by Merrill (1952) of the unstable element technetium in S stars was a clear indication that these stars could be a source of the s-process elements in general (see also Cameron 1955). AGB stars are now generally found to be more or less s-element rich. C stars are believed to be the most enriched (see below).

Lambert (1992) gives several arguments in favor of the idea that s-elements are produced in AGB stars of low mass ($\lesssim 3 M_\odot$) rather than in stars of intermediate mass. In intermediate-mass AGB stars, the temperature in the He-burning shell is high enough to ignite the $^{22}$Ne($\alpha$,n)$^{25}$Mg neutron source. That would lead to an excess abundance of $^{25}$Mg (cf. Lambert 1991, and references therein). Smith & Lambert (1986) found no $^{25}$Mg enhancement in their seven randomly selected s-element enriched stars, which suggests that these were all low-mass AGB stars. In these stars the $^{13}$C($\alpha$,n)$^{16}$O reaction is the probable neutron source.

The neutron density at the s-process site can be determined from observations of isotope abundances at certain branching points of the s-path in the chart of nuclides. Two useful branching points are the ones at $^{95}$Zr and $^{85}$Kr. Lambert et al. (1995) found no evidence of $^{96}$Zr in ZrO bands in S star spectra, implying a neutron density of less than $5 \times 10^8$ cm$^{-3}$, much less than what is expected in regions where the $^{22}$Ne neutron source operates. This again suggests that intermediate-mass AGB stars are not the major contributors of s-processed elements in our Galaxy.

Until recently, it was believed that the $^{13}$C neutron source was in operation during the thermal pulses of low-mass stars. Straniero et al. (1995) have, however, found from an evolutionary model sequence (for 3 $M_\odot$) that the $^{13}$C neutron source is activated during the intervals *between* the thermal pulses. This suggests a lower neutron density ($10^7$ cm$^{-3}$), and that the $^{13}$C is consumed during the interpulse period. Semiconvection is essential for the s-process (see Sackmann & Boothroyd 1991, and references therein); this will allow mixing of protons into the carbon pocket of the thermal pulses leading to fresh $^{13}$C. Predictions from the models of Straniero et al. regarding one of the signatures of the low neutron density, the Rb/Sr ratio which depends on the s-path at the $^{85}$Kr branch, are in agreement with recent observations of MS and S stars by Lambert et al. (1995).

Mean neutron exposures deduced from observations of 34 MS and S stars by Smith et al. (1987) suggest that the solar system s-element distribution could arise from a mixture of MS, S and C stars which return s-element enriched material to the ISM. Their s-element excesses are for MS stars a factor of 2–3, for S stars 4–6 and for C stars 5–10. Note that the latter abundances are based on the uncertain abundance analysis of the crowded visual spectra by Utsumi (1985). These figures, in combination with estimated total mass loss, are high enough to explain the s-element abundance in the solar system. Also, Parthasarathy (1996) found s-element enhancement relative to the Sun of 2–40 for some post–AGB stars.

There is also other more indirect, though strong, empirical evidence for the s-elements being formed mainly in low-mass AGB stars. Edvardsson et al. (1993) showed, from studies of the composition of solar-type disk stars of different ages, that the major contributors to the enrichment of s-elements in the disk are stars with characteristic evolutionary times $> 3\times 10^9$ years, i.e. low-mass stars. This is significantly longer than the time scale of the iron production from SNeIa. The low-mass contribution to the s-elements in the galactic disk is also verified by Pagel & Tautvaišienė (1997). Moreover, from abundances in the halo they also trace a contribution of unknown origin but with a time scale characteristic of stars of $\sim 8\,M_\odot$.

The third dredge-up may be even more significant for populations with lower metallicity than solar (cf. Wood 1981) and provide strong s-element enrichment. We note that Kipper et al. (1996) found three presumably intrinsic C stars of Pop. II to have s-element enhancements of 1–3 dex.

We conclude that mass loss from precursors of C-rich PNe are most probably a main source of the s-elements in galaxies.

## 3. Conclusions

It seems most probable that the AGB stars in general, and C stars in particular, are the main contributors of s-elements; probable that their contributions of flourine and carbon are quite significant; and possible that their contributions of lithium (from the super Li-rich C stars), of $^{13}$C (from J stars) and of Ne (i.e. $^{22}$Ne) are of some importance. However, all of these statements are more or less uncertain, and the uncertainty is also great concerning the role of intermediate and low mass stars in the production of N, Na and Al. This is because a number of relatively fundamental questions concerning C stars still remain unanswered. In order to improve the situation new and more detailed and reliable abundance analyses for C stars and PNe are needed (also in nearby galaxies with different metallicities), as well as better models of nucleosynthesis and dredge-up in AGB stars. We also need to know more about mass loss — how much matter the stars eject, as a function of initial mass and composition, as well as when the stars lose mass relative to the time when their envelope enrichment occurs. Further studies of abundances in less evolved stars of different stellar populations are also of great significance. As regards the nucleosynthesis of two of the most important elements, carbon and nitrogen, more work on stars more massive than normal C stars seems also important.

The situation reminds us of one of the stories about the famous Nasrettin Hoca of Konya, a place not very far from Antalya. Nasrettin saw ducks on the lake shore and, thinking of his dinner, tried to catch one but it jumped into the lake and swam out of reach. The next one did the same. After several attempts he took a spoon and sat down at the shore and started eating water. Some friends passed by and asked him what he was doing. He answered: "I am having duck soup."

We have tried to "catch" the evasive secrets of the C stars with different, more or less sophisticated methods. However, in spite of much progress presented at this meeting, the solutions of the problems seem to be at some distance. This partly reflects the experience, so common in science, that phenomena tend to become more complex when studied. In any case, more systematic investigations are now needed. If we undertake them, it seems that we shall get the most important answers long before Nasrettin Hoca has been able to walk to the center of his lake and catch his ducks.

John Lattanzio, Bernard Pagel, Lee-Anne Willson and Bob Wing are thanked for valuable suggestions, and Martin Asplund and Kjell Eriksson for comments on the manuscript.

## References

Abia, C., Boffin, H. M. J., Isern, J. & Rebolo, R. 1993a, *A&A*, 272, 455
Abia, C. & Isern, J. 1996, *ApJ*, 460, 443
Abia, C., Isern, J. & Canal, R. 1993b, *A&A*, 275, 96
Andersson, H. & Edvardsson, B. 1994, *A&A*, 290, 590
Bessell, M. S., Wood, P. R. & Lloyd Evans, T. 1983, *MNRAS*, 202, 59

Blöcker, T. 1995, *A&A*, 297, 727
Boesgaard, A., Eriksson, K. & Gustafsson, B. 1996, in prep.
Boothroyd, A. I. & Sackmann, I.-J. 1988, *ApJ*, 328, 653
Boothroyd, A. I., Sackmann, I.-J. & Ahern, S. C. 1993, *ApJ*, 416, 762
Bowen, G. H. & Willson, L. A. 1991, *ApJ*, 375, L53
Boyarchuk, A. A. & Lyubimkov, L. S. 1985, *Bull. Crimean Astrophys. Obs.*, 66, 119
Brett, J. M. 1991, *MNRAS*, 249, 538
Brewer, J. P. 1996, *PASP*, 108, 379
Brown, L. E. 1992, *ApJ*, 389, 251
Cameron, A. G. W. 1955, *ApJ*, 121, 144
Cameron, A. G. W. & Fowler, W. A. 1971, *ApJ*, 164, 111
Cavallo, R. M., Sweigart, A. V. & Bell, R. A. 1996, *ApJ*, 464, L79
Claussen, M. J., Kleinmann, S. G., Joyce, R. R. & Jura, M. 1987, *ApJ Supp.*, 65, 385
Corradi, R. L. M. & Schwarz, H. E. 1995, *A&A*, 293, 871
Denisenkov, P. A. 1989, *Soviet Astr. Lett.*, 14, 435
Denisenkov, P. A. & Denisenkova, S. N. 1990, *Soviet Astr. Lett.*, 16, 275
Edvardsson, B., Andersen, J., Gustafsson, B., Lambert, D. L., Nissen, P. E. & Tomkin, J. 1993, *A&A*, 275, 101
Feltzing, S. & Gustafsson, B. 1998, *A&A Supp.*, 129, 237
Forestini, M., Goriely, S., Jorissen, A. & Arnould, M. 1992, *A&A*, 261, 157
Frost, C. A. & Lattanzio, J. C. 1996, in the $32^{nd}$ Liège Inst. Astrophys. Coll.: *Stellar Evolution: What should be done?*, ed. A. Noels et al., p. 307
Gallino, R., Busso, M., Picchio, G. & Raiteri, C. M. 1990, *Nature*, 348, 298
Garnett, D. R., Skillman, E. D., Dufour, R. J., Peimbert, M., Torres-Peimbert, S., Terlevich, R., Terlevich, E. & Shields, G. A. 1995, *ApJ*, 443, 64
Groenewegen, M. A. T. & de Jong, T. 1993, *A&A*, 267, 410
Groenewegen, M. A. T., van den Hoek, L. B. & de Jong, T. 1995, *A&A*, 293, 381
Guélin, M., Forestini, M., Valiron, P., Ziurys, L. M., Anderson, M. A., Chernicharo, J. & Kahane, C. 1995, *A&A*, 297, 183
Huss, G. R. & Wasserburg, G. J. 1996, *Lun. Planetary Sci.*, 27, 573
Jorissen, A., Smith, V. V. & Lambert, D. L. 1992, *A&A*, 261, 164
Jura, M. & Kleinmann, S. G. 1989, *ApJ*, 341, 359
Kahane, C., Cernicharo, J., Gómez-González, J. and Guélin, M. 1992, *A&A*, 256, 235
Kingsburgh, R. L. & Barlow, M. J. 1994, *MNRAS*, 271, 257
Kipper, T., Jørgensen, U. G., Klochkova, V. G. & Panchuk, V. E. 1996, *A&A*, 306, 489
Kudryashov, A. S. & Tutukov, A. 1988, *Astron. Zirk.*, 1525, 11
Lambert, D. L. 1991, in IAU Symp. 145: *Evolution of Stars: the Photospheric Abundance Connection*, ed. G. Michaud and A. Tutukov (Kluwer), p. 299
Lambert, D. L. 1992, in the $31^{st}$ Herstmonceux Conf.: *Elements and the Cosmos*, ed. M. G. Edmunds and R. Terlevich (Cambridge Univ. Press), p. 92
Lambert, D. L., Gustafsson, B., Eriksson, K. & Hinkle, K. H. 1986, *ApJ Supp.*, 62, 373
Lambert, D. L., Smith, V. V., Busso, M., Gallino, R. & Straniero, O. 1995, *ApJ*, 450, 302
Langer, G. E., Hoffman, R. & Sneden, C. 1993, *PASP*, 105, 301
Leisy, P. & Dennefeld, M. 1996, *A&A Supp.*, 116, 95
Lewis, R. S., Amari, S. & Anders, E. 1990, *Nature*, 348, 293
Maeder, A. 1992, *A&A*, 264, 105
Marigo, P., Bressan, A. & Chiosi, C. 1996, *A&A*, 313, 545
Matteucci, F., D'Antona, F. & Timmes, F. X. 1995, *A&A*, 303, 460
McKellar, A. 1940, *PASP*, 52, 407
Merrill, P. W. 1952, *ApJ*, 116, 21
Nichols, R. H. Jr., Amari, S., Hohenberg, C. M., Hoppe, P. & Lewis, R. S. 1993, *Meteoritics*, 28, 410
Nomoto, K., Tsujimoto, T., Yamaoka, H., Kumagai, S. & Shigeyama, T. 1992, in the $31^{st}$ Herstmonceux Conf.: *Elements and the Cosmos*, ed. M. G. Edmunds and R. Terlevich (Cambridge Univ. Press), p. 55

Norris, J. E. & Da Costa, G. S. 1995, *ApJ*, 441, L81
Ohnaka, K. & Tsuji, T. 1996, *A&A*, 310, 933
Olofsson, H., Bergman, P., Eriksson, K. & Gustafsson, B. 1996, *A&A*, 311, 587
Olofsson, H., Eriksson, K., Gustafsson, B. & Carlström, U. 1993a, *ApJ Supp.*, 87, 267
Olofsson, H., Eriksson, K., Gustafsson, B. & Carlström, U. 1993b, *ApJ Supp.*, 87, 305
Pagel, B. E. J. & Tautvaišienė, G. 1997, *MNRAS*, 288, 108
Parthasarathy, M. 1996, private communication
Pasquali, A. & Perinotto, M. 1993, *A&A*, 280, 581
Pettini, M., Lipman, K. & Hunstead, R. W. 1995, *ApJ*, 451, 100
Plez, B., Smith, V. V. & Lambert, D. L. 1993, *ApJ*, 418, 812
Pottasch, S. R. 1992, *Astron. Astrophys. Rev.*, 4, 215
Prantzos, N., Aubert, O. & Audouze, J. 1996, *A&A*, 309, 760
Prantzos, N., Vangioni-Flam, E. & Chauveau, S. 1994, *A&A*, 285, 132
Reeves, H., Richer, J., Sato, K. & Terasawa, N. 1990, *ApJ*, 355, 18
Renzini, A. & Fusi Pecci, F. 1988, *Ann. Rev. Astron. Astrophys.*, 26, 199
Renzini, A. & Voli, M. 1981, *A&A*, 94, 175
Richer, H. B. 1981, *ApJ*, 243, 744
Richer, H. B., Olander, N. & Westerlund, B. E. 1979, *ApJ*, 230, 724
Rola, C. & Strasińska, G. 1994, *A&A*, 282, 199
Sackmann, I.-J. & Boothroyd, A. I. 1991, in IAU Symp. 145: *Evolution of Stars: the Photospheric Abundance Connection*, ed. G. Michaud and A. Tutukov (Kluwer), p. 275
Sackmann, I.-J. & Boothroyd, A. I. 1992, *ApJ*, 392, L71
Sarmiento, A. & Peimbert, M. 1985, *Rev. Mex. Aston. Astrofís.*, 11, 73
Scalo, J. M. 1976, *ApJ*, 206, 795
Scalo, J. M. & Ulrich, R. K. 1973, *ApJ*, 183, 151
Smith, V. V. & Lambert, D. L. 1986, *ApJ*, 311, 843
Smith, V. V., Lambert, D. L. & McWilliam, A. 1987, *ApJ*, 320, 862
Spite, F. & Spite, M. 1982, *A&A*, 115, 357
Straniero, O., Gallino, R., Busso, M., Chieffi, A., Raiteri, C. M., Limongi, M. & Salaris, M. 1995, *ApJ*, 440, L85
Stürenburg, S. & Holweger, H. 1990, *A&A*, 237, 125
Thielemann, F.-K., Nomoto, K. & Hashimoto, M.-A. 1996, *ApJ*, 460, 408
Timmes, F. X., Woosley, S. E. & Weaver, T. A. 1995, *ApJ Supp.*, 98, 617
Tinsley, B. M. 1978, in IAU Symp. 76: *Planetary Nebulae*, ed. Y. Terzian (Reidel), p. 341
Tomkin, J., Lemke, M., Lambert, D. L. & Sneden, C. 1992, *AJ*, 104, 1568
Tomkin, J., Woolf, V. M., Lambert, D. L. & Lemke, M. 1995, *AJ*, 109, 2204
Utsumi, K. 1985, in *Cool Stars with Excesses of Heavy Elements*, ed. M. Jaschek and P. C. Keenan (Reidel), p. 243
Vila-Costas, M. B. & Edmunds, M. G. 1993, *MNRAS*, 265, 199
Willson, L. A., Bowen, G. H. & Struck, C. 1996, in *From Stars to Galaxies*, ed. C. Leitherer, U. Fritze-von Alvensleben and J. Huchra, ASP Conf. Ser., 98, 197
Wood, P. R. 1981, in *Physical Processes in Red Giants*, ed. I. Iben Jr. and A. Renzini (Reidel), p. 135
Woosley, S. E., Hartmann, D. H., Hoffman, R. D. & Haxton, W. C. 1990, *ApJ*, 356, 272
Woosley, S. E. & Weaver, T. A. 1995, *ApJ Supp.*, 101, 181
Zuckerman, B. & Aller, L. H. 1986, *ApJ*, 301, 772

# Session XII

## Observing Facilities

Bus laden with symposium participants struggles up the winding road to nearly the top of Bakırlıtepe.

Robert Wing easily scales the last 100 m to the summit.

# VARIABILITY STUDIES WITH INTERNATIONAL NETWORKS

FRANÇOIS R. QUERCI AND MONIQUE QUERCI
*Observatoire Midi-Pyrénées*
*31400 Toulouse, France*

**Abstract.** Variable stars can be monitored through observing campaigns which coordinate multi-site telescopes at various longitudes. A new practice is now being developed, involving devoted networks of robotic telescopes. We will review these two technologies and will emphasize the NORT (Network of Oriental Robotic Telescopes) project, which we are promoting in Middle Eastern countries.

## 1. Introduction

For over a decade, coordinated international campaigns have taken place from sites with good longitude and latitude coverage and/or with various instruments working at complementary wavelengths. The aim is to monitor variable stars as continuously as possible to reduce the aliasing problems bound to observations at a single site.

The final goal is to study and to understand the pulsational behavior of the variables, given that the pulsation on the star's surface probes its interior structure and composition in the framework of asteroseismology.

## 2. Advantages of Campaigns

Coordinated campaigns with existing telescopes give better tools to solve scientific problems by allowing one (*a*) to compare individual observational technologies such as CCD cameras, (*b*) to compare data reduction procedures, (*c*) to develop methods for filling gaps in data (*e.g.* Horne & Baliunas 1986; Serre et al. 1992) due to local bad weather conditions, and (*d*) to develop period-finding techniques such as Fourier analysis or least squares of brightness residuals (LSR), adapted to the observed objects with single and multiple periods or in more complex situations such as variable periods, flares, semi-regularity, and so on.

Multi-site campaigns also lead to a better international organization of science, since they are based on cooperative programs among many experts in various scientific and technical fields: theoreticians, observers, engineers. They are also valuable for the training of young observers, PhD students and young research associates, who follow the scientific evolution of the project from program definition to the interpretation of the results, thereby enabling them to choose their own speciality.

Some campaigns give access to multi-wavelength techniques. They are dedicated to various wavelength regions by using space telescopes (*e.g.* IUE, ROSAT, HST), as well as many existing small- and medium-sized ground-based telescopes around the world.

In a word, they produce better science.

## 3. Drawbacks of Campaigns

The main drawbacks of multi-site campaigns can be summarized as follows:

Many nights are lost to clouds as some existing telescopes are not located at the best sites in the world.

The costs are significant due to overhead and associated expenses such as travel and accommodation costs for observers going to faraway observatories, and for equipment maintenance or replacement. Also there are considerable handling requirements to mount and dismount instrumentation, to transport equipment to distant observatories, and to adapt it to various existing telescopes. In fact, these costs and handling are so high that they seriously restrict the number of campaigns that can be carried out per year. As examples, we cite the WET (*Whole Earth Telescope*) campaigns (Sullivan 1995) with white dwarfs as prime targets, and the $\delta$ Scuti STEPHI (*Stellar Photometry International project*) network campaigns (Belmonte et al. 1994). These and other campaigns are run only once or twice per year or even once in two years for about a one week duration. In consequence, only one or a few stars are monitored during a campaign, and the stars analysed are necessarily short-period variables such as white dwarfs, $\delta$ Scuti stars, RoAp stars, post-novae, cataclysmic variables, etc., having periods of some seconds to a few hours or days.

Some technical inconveniences could be added if the scientific organization is badly coordinated, as for example when data from different sites are reduced by different techniques, or obtained by not sufficiently similar instruments at different sites, or even at the same site in long-term monitoring programs, as emphasized by Young (1994).

Problems associated with coordinated observing campaigns are reviewed for example by Sterken (1988) and Breger (1992, 1994).

## 4. Observational Status of Variability in Red Giants

As the coordinated international campaigns operate one or two times a year for about a week at a time, they are not well adapted to the slowly varying red giants. So, it is not surprising that this subject does not appear in the literature on cool stars. Even variables such as Miras can appear "unfavorable for long-term projects" as stressed by Szabados (1994). However, it is encouraging that a first step was recently taken for the solar-type stars with the presentation of a new project, the SONG (*Stellar Oscillations Network Group*) (see the Madison Meeting of the AAS, 9-13 June, 1996).

Roughly speaking, what do we know about red-giant variability?

Red giants have long periods of about one year. Superimposed on the dominant long period there may also be shorter periods, of say a month to a few months. Moreover, they can show rapid or slow period changes, a critical behavior around phase 0.7–0.8 in the case of Miras, and flares and very rapid variations over a day or less (for example, see a review by Querci, 1986). This knowledge was mainly obtained by single ground-based sites dedicated to long-term monitoring programs. Let us cite (*a*) the pioneering photographic observations made by Campbell, Cannon, Hetzler, Merrill, Payne, etc., (*b*) the contributions to visual photometry made by associations of amateur astronomers such as AAVSO, AFOEV, GEOS, BAN, etc., and (*c*) the photoelectric photometry by Gow, Lockwood, Wing, etc.

Nowadays, the AAVSO includes red giants in its photoelectric photometry survey (*e.g.* Percy et al. 1994), and other societies also make a contribution to the observation of cool stars (see GEOS Circulars, IBVS, etc.). Photographic monitoring by Latvian astronomers led to the discovery of light-curve anomalies. In 1982, red giants were targets in the ESO–LTPV project (Sterken 1994): international teams collaborated for more than a decade (see Jorissen 1994 for results on Barium and S stars).

For the past 10 years, automatic photometric telescopes (APT) have operated on Mt. Hopkins for long-term monitoring of semi-regular variables (Baliunas et al. 1987; Cristian et al. 1995).

More recently, MERLIN, the Multi Element Radio-Linked Interferometer, mapped out $H_2O$ emission in the inner circumstellar envelopes of Miras, SR variables, and red supergiants, giving information about mass loss (Yates & Cohen 1994), and it has also observed OH/IR stars (Migenes et al. 1995). A radio survey could be organized in cooperation with optical networks.

It appears evident that we need a world-wide coordinated monitoring of long-period and semi-regular variables which will permit the detection of important unknown features in their light curves and, consequently, will lay the observational foundation for red-giant asteroseismology.

## 5. An International Network: The Network of Oriental Robotic Telescopes (NORT)

Global networks of automated telescopes (GNAT) have been proposed by Budding (1993, 1995), Crawford (1992, 1993, 1995), Querci & Querci (1992), Querci et al. (1993, 1995a,b), and Querci (1995). Here we would like to describe the philosophy of the NORT project, which is a network of 1.3-m diameter automated telescopes for photometric and spectrographic studies, to be installed in desert sites from Morocco to China.

### 5.1. FRAMEWORK OF THE PROJECT

The NORT will deal with variable stars (mainly red giants), planetary nebulae, and post-novae, to stimulate asteroseismology of long-term variables. As various characteristic times of variation have to be considered, we need a continuous follow-up of some typical stars. This constraint can only be satisfied through using the "best" sites, *i.e.* semi-desert countries, around the world. Some such countries are along the Tropic of Cancer from Morocco to the Chinese deserts, with the further advantage that the longitude interval they cover is complementary to that covered by the U.S.A. where robotic telescopes already exist. It is to be stressed that these countries had great astronomers in the past, but nowadays few of them perform research in astrophysics. However, their universities, their sites (high mountains in semi-desert climate) and their wishes for development can allow them to make progress in astrophysics.

In the NORT project, we propose (*a*) to collaborate in education in astronomy and astrophysics, (*b*) to help in the development of laboratories by giving advice for building well-equipped 60-cm telescopes for student training, and (*c*) to offer engineer and technician training in the French observatories such as Haute-Provence Observatory (OHP), Midi-Pyrénées Observatory (OMP), etc., to collaborate in setting up the network and for making the scientific choice of objects to be observed.

There will be a direct transmission of all the observations collected each day to all universities that are members of the network, via the Internet (or ARABSAT, METEOSAT, ...). The whole network will be fully robotized. Data reduction and interpretation could be made in common by sharing scientific and technical results.

This network is supported by:
- INSU/CNRS (Institut National des Sciences de l'Univers),
- OHP and OMP Observatories in France,
- UN–Space Affairs Division,

and several of the developing countries.

### 5.2. WHY SITES IN THESE COUNTRIES ?

The oriental countries are suitable because:
- they have high mountains (3000 to 4000 m) in semi-desert areas (north-tropical latitude from 15° to 35°), and consequently a large annual number of nights with a clear sky and low telluric absorption,
- they are in a longitude interval (about 10° West to 110° East) complementary to that of some automated stations already devoted to variable star research, including the Hawaii volcano, Arizona mountains, Chilean cordillera, Etna volcano, and South African desert.

### 5.3. HOW TO SELECT GOOD SITES FOR OBSERVATIONS?

By using 14-year archives of METEOSAT data and the METEOSAT and NOAA on-line data from the Dax Observatory Science Club, we discovered a set of meteorologically very good sites from Morocco to the Takla Makan and Gobi deserts in China. The final site selection will be made from local astronomical tests such as seeing measurements. It ought to give a list of sites not subjected to the same airstreams. Also, the local access facilities will presumably be a non-negligible factor in the selection.

A minimum number of 9 to 12 stations should be able to follow the variable stars each night, without interruption, throughout the year.

### 5.4. PROPOSED OBSERVATIONAL TECHNIQUES

We propose the following basis for a decade of collaboration on variable stars:
- First, by photometry, which measures the stellar flux variation itself at several wavelengths and helps to disentangle the evolution and the internal structure of the stars,
- Later, by spectroscopy, which in its low-resolution modes gives the stellar abundances, and in its high-resolution modes is able to give the physical parameters of the stars and their dynamical behavior, and
- Finally, by interferometry at visible frequencies, to obtain diameter variations or the shapes of stars, a detailed description of their external layers, and the eventual discovery of planets.

### 5.5. PRESENT STATUS OF THE PROJECT

#### 5.5.1. *Training and Scientific Aspects*
A French Committee within CNRS was created in October 1994. Its mission is to take care of the educational problems and the supervising of PhD theses in cooperation. It has eight members, among them two professors in

Astrophysics, two astrophysicist specialists in equipment, two theoretician astrophysicists, and two engineers in electronics and computer science.

The technical training courses have begun at OHP. PhD theses are in progress in France on variable stars of various spectral types.

### 5.5.2. *Technical Aspects*

Several 60-cm diameter telescopes dedicated to student use, built by a private company in collaboration with the OHP and OMP observatories, are already at work. The mechanical structure calculations for 1.0-m and 1.3-m diameter telescopes are underway at OMP observatory in Toulouse, as well as the optical drawings of their focal-plane instrumentation.

### 5.5.3. *International Relationships*

We had this idea of a Network of Oriental Robotic Telescopes some years ago. We told our Moroccan colleagues about it and one of us (F.Q.) visited some oriental countries.

During the "First International Conference on Space and Astronomy," held in Amman, Hachemite Kingdom of Jordan, in September 1994, an international committee was created to promote the project. Its members represent the following countries (by longitude): Morocco, France, Libya, Egypt, Lebanon, Iraq, Jordan, and Yemen. Contacts are in progress with the following countries (by longitude): Mauretania, Algeria, Tunisia, Syria, Iran, Saudi Arabia, Bahrain, Pakistan, India, Malaysia, Indonesia, and Brunei. Contacts still have to be developed with Kuwait, Qatar, United Arab Emirates, Oman, Tadjikistan, Uzbekistan, and China.

At the General Assembly of the International Astronomical Union (IAU) held in The Hague (Holland) in August 1994, the Arabic astronomers had a meeting about the NORT. Contacts have been developed (or are to be developed) with the following organizations: IAU, ALECSO, TWNSO, IFSTAD, UNESCO, AUPELF, IMA, etc.

## 6. Concluding Remarks

We should like to conclude by noting the realistic fact that long-term monitoring of variable stars through global automated dedicated networks leads to a "new field of astronomy – astroeconomics," according to Budding (1995) and Crawford (1995). Today, this fact is still more opportune with respect to the multi-site campaigns and some space missions.

If the project succeeds, as we hope, it should allow several oriental countries from Machreq to Maghrib to make a rapid jump into contemporary astrophysics.

# References

Baliunas, S. L., Donahue, R. A., Loeser, J. G., Guinan, E. F., Genet, R. M. & Boyd, L. J. 1987, in *New Generation Small Telescopes*, ed. D. S. Hayes, R. M. Genet and D. R. Genet (Fairborn Press), p. 97

Belmonte, J. A. *et al.* (17 authors) 1994, *A&A*, 283, 121

Breger, M. 1992, in *Variable Star Research: An International Perspective*, ed. J. R. Percy, J. A. Mattei and C. Sterken (Cambridge), p. 171

Breger, M. 1994, in *The Impact of Long-Term Monitoring on Variable Star Research*, ed. C. Sterken and M. de Groot (Kluwer), p. 393

Budding, E. 1993, in IAU Colloquium 136: *Stellar Photometry – Current Techniques and Future Developments*, ed. C. J. Butler and I. Elliot (Cambridge), p. 257

Budding, E. 1995, *Astrophys. & Space Sci.*, 228, 299

Crawford, D. L. 1992, in *Automated Telescopes for Photometry and Imaging*, ed. S. J. Adelman, R. J. Dukes Jr. and C. J. Adelman, ASP Conf. Ser., 28, 123

Crawford, D. L. 1993, in IAU Colloquium 136: *Stellar Photometry – Current Techniques and Future Developments*, ed. C. J. Butler and I. Elliott (Cambridge), p. 244

Crawford, D. L. 1995, in *Robotic Observatories*, ed. M. F. Bode (Wiley–Praxis), p. 77

Cristian, V. C., Donahue, R. A., Soon, W. H., Baliunas, S. L. & Henry, G. W. 1995, *PASP*, 107, 411

Horne, J. H. & Baliunas, S. L. 1986, *ApJ*, 302, 757

Jorissen, A. 1994, in *The Impact of Long-Term Monitoring on Variable Star Research*, ed. C. Sterken and M. de Groot (Kluwer), p. 143

Migenes, V., Lüdke, E., Cohen, R. J., Shepherd, M. & Bowers, P. F. 1995, *Astrophys. & Space Sci.*, 224, 515

Percy, J. R. *et al.* (18 authors) 1994, *PASP*, 106, 611

Querci, F. R. 1986, in *The M-Type Stars*, ed. H. R. Johnson and F. R. Querci, NASA SP-492, p. 1

Querci, F. R. 1995, in *Problems of Astronomy in Africa*, ed. A. H. Batten, *Highlights of Astronomy*, 10, 677

Querci, F. R. & Querci, M. 1992, in *Automated Telescopes for Photometry and Imaging*, ed. S. J. Adelman, R. J. Dukes Jr. and C. J. Adelman, ASP Conf. Ser., 28, 67

Querci, F. R., Querci, M., Kadiri, S. & de Rancourt, L. 1993, in IAU Colloquium 136: *Poster Papers on Stellar Photometry*, ed. I. Elliott and C. J. Butler (Dublin Institute for Advanced Studies), p. 122

Querci, F. R., Querci, M., Kadiri, S. & Benkhaldoun, Z. 1995a, in *Robotic Telescopes: Current Capabilities, Present Developments, and Future Prospects for Automated Astronomy*, ed. G. W. Henry and J. A. Eaton, ASP Conf. Ser., 79, 239

Querci, F. R., Querci, M., Kadiri, S. & de Rancourt, L. 1995b, in *Robotic Observatories*, ed. M. F. Bode (Wiley–Praxis), p. 85

Serre, T., Auvergne, M. & Goupil, M. J. 1992, *A&A*, 259, 404

Sterken, C. 1988, in *Coordination of Observational Projects in Astronomy*, ed. C. Jaschek and C. Sterken (Cambridge), p. 3

Sterken, C. 1994, in *The Impact of Long-Term Monitoring on Variable Star Research*, ed. C. Sterken and M. de Groot (Kluwer), p. 1

Sullivan, D. J. 1995, *Baltic Astronomy*, 4, 467

Szabados, L. 1994, in *The Impact of Long-Term Monitoring on Variable Star Research*, ed. C. Sterken and M. de Groot (Kluwer), p. 213

Yates, J. A. & Cohen, R. J. 1994, *MNRAS*, 270, 958

Young, A. T. 1994, in *The Impact of Long-Term Monitoring on Variable Star Research*, ed. C. Sterken and M. de Groot (Kluwer), p. 421

## Discussion

**Mattei**: Will there be a database created with NORT observations, and will this database be accessible by the astronomical community?

**Querci, F.**: Yes. However, we have not yet decided where the database will be located.

**Gustafsson**: Could you give a typical price for one of these units in the NORT?

**Querci**: At this time we cannot, because not all the elements are manufactured and fully tested at l'Observatoire de Haute-Provence. We can give the price of the equipped telescopes (60-cm and 1.3-m), i.e. the mechanics, optics, electronics, and driving computer. The prototypes for the meteorological station, i.e. the photometer, the spectrographs, and the data transmission system, are respectively undergoing tests, under construction, and in development.

**Bakker**: Besides astroeconomics, there might also be astropolitics. Most of the countries you've selected are politically unstable. Did you consider that maybe the telescopes are destroyed every two years?

**Querci**: If astronomy is part of the country's society, they are proud of it. Battling parties may agree that astronomy should be above their struggle. Therefore the telescopes will not be destroyed.

# A NEW OPTICAL OBSERVATORY IN TURKEY

ZEKİ ASLAN

*Akdeniz University & Turkish National Observatory*
*Physics Dept., Topçular, TR-07200 Antalya, Turkey*

SELİM O. SELAM

*Ankara University Observatory*
*Science Faculty, Tandoğan, TR-06100 Ankara, Turkey*

AND

AKİF ESENDEMİR

*Middle East Technical University*
*Physics Dept., Emek, TR-06100 Ankara, Turkey*

**Abstract.** Site-testing observations carried out between 1982 and 1986 have shown that Southwest and Southeast Turkey contain good potential observatory sites. The mountain known as Bakırlıtepe near the Mediterranean coast was found to be a good site with a high percentage of nights with clear skies and good seeing.

Here we summarize the results of the site-test observations and give general information about the site.

## 1. Introduction

This paper gives some brief information about the new optical observatory site which some of the participants visited after the conclusion of IAU Symposium 177. The site is located on a mountain known as Bakırlıtepe about 50 km (35 km as the crow flies) west of the city of Antalya at a height of 2485 m at latitude 36°51′ N and longitude 30°20′ E (Figure 1). It was selected after site-testing observations carried out between 1982 and 1986. Among the sites tested, Bakırlıtepe was found to be the best, with a high percentage of nights with clear skies and good seeing.

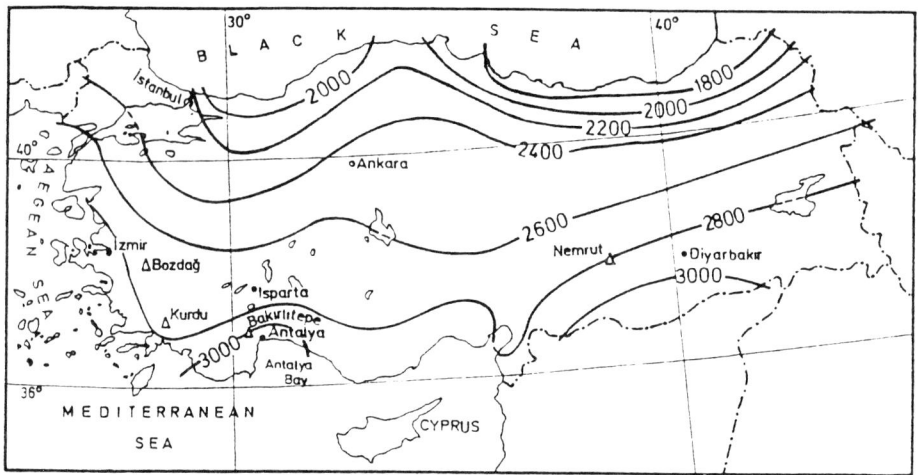

*Figure 1.* Map of Turkey showing the positions of sites tested (triangles), radio sonde stations (filled circles), and contour lines of sunshine hours per year.

## 2. Summary of Site-Testing Observations

Meteorological observations and astronomical seeing measurements made during two seasons are summarized in Figures 2 – 5, and in Tables 1 and 2. The details can be found in Aslan et al. (1989). As seen from Tables 1 and 2, Bakırlıtepe compares favorably with the Roque de los Muchachos Observatory on La Palma, one of the world's best observatories. It should be noted that the seeing measurements on Bakırlıtepe and La Palma were made by the same method (Walker 1984) using the same Polaris trail telescopes (borrowed from the Royal Greenwich Observatory) and Walker's set of calibration trails. It is now known that the actual seeing on La Palma is much better than indicated in Table 1, because the calibration of the Polaris trails contained the dome seeing of the 120-inch Lick telescope, against which the calibration was made. What this means is that the actual seeing on Bakırlıtepe might also be better than that indicated in Table 1.

## 3. The Facilities

The observatory on Bakırlıtepe is presently under construction and should be operational by the end of 1997. The site will be operated as the Turkish National Observatory (TUG) under the administration of the TUG Institute of TÜBİTAK (Scientific and Technical Research Council of Turkey). Initially there will be two facilities: telescopes of 0.4-m and 1.5-m aperture. The first telescope will be used for photometry, the second mainly for spectroscopy. The 0.4-m telescope was donated by the University of Utrecht. The 1.5-m telescope is a joint project between IKI-RAN (Space Research

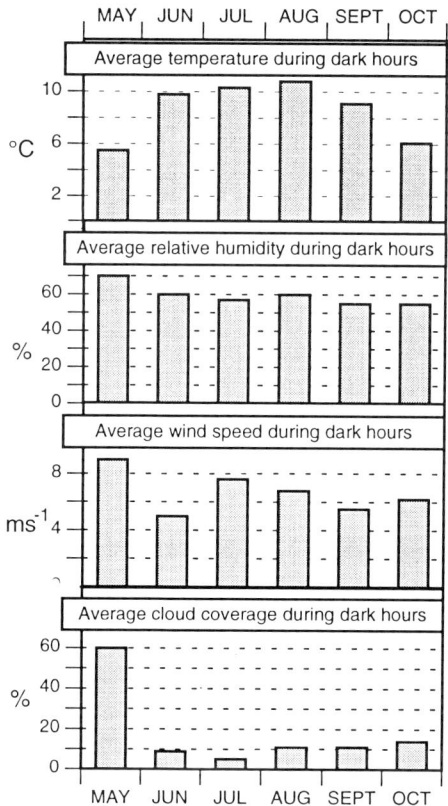

Figure 2. Average values of meteorological observations made in 1984 and 1985.

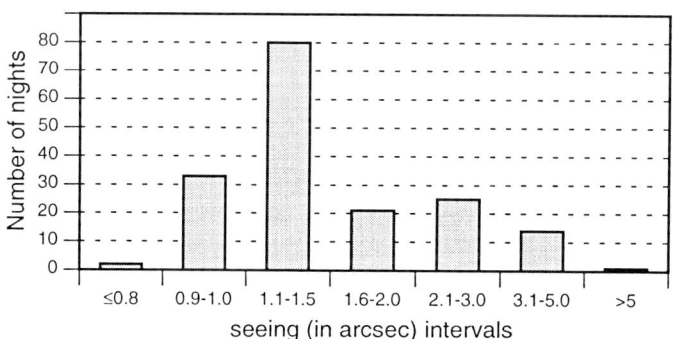

Figure 3. Distribution of average seeing values.

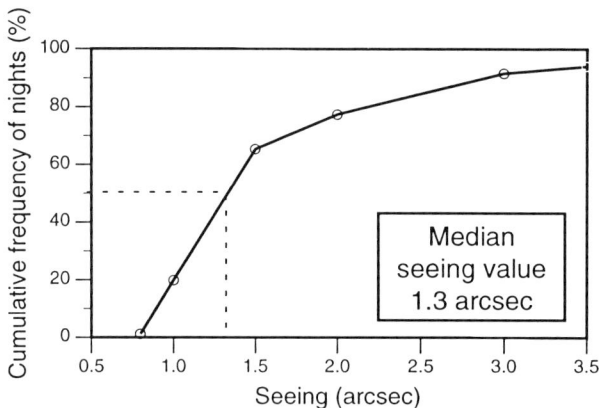

*Figure 4.* Percentage of nights with average seeing smaller than a given value (May to October).

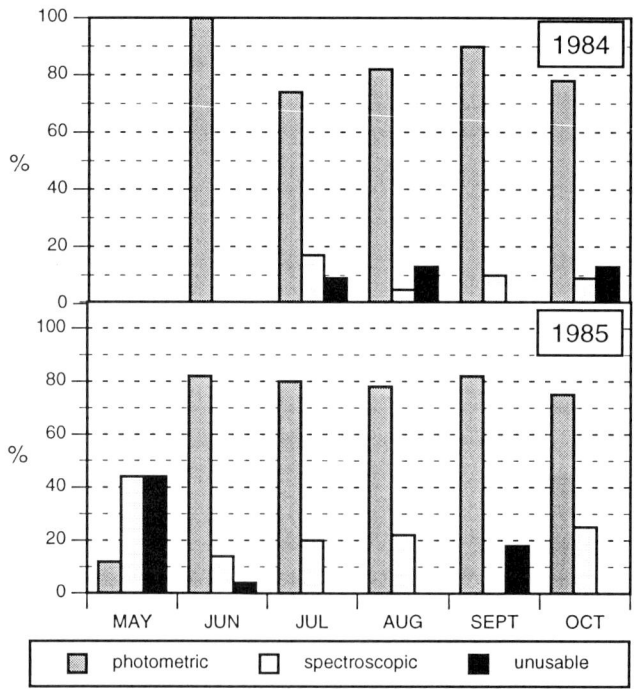

*Figure 5.* Monthly percentages of night quality.

TABLE 1. Comparison of atmospheric seeing on Bakırlıtepe and Roque de los Muchachos Observatory (RMO)

| RMO | | Bakırlıtepe | | |
|---|---|---|---|---|
| Interval | Median seeing | Interval | Median seeing | |
| | | | Hourly | Nightly |
| 1984 Summer | $1''\!.1$ | 1984 Summer | $1''\!.1$ | $1''\!.3$ |
| 1984/5 Winter | $1''\!.4$ | 1985 Summer | $1''\!.3$ | $1''\!.4$ |
| 1975 Winter+Summer | $1''\!.3$ | | | |

TABLE 2. Comparison of night quality on Bakırlıtepe and Roque de los Muchachos

| | Interval | No. of nights | Percentage of | | Ref. |
|---|---|---|---|---|---|
| | | | Photometric nights | Usable nights | |
| RMO | 1982+1983 (Winter+Summer) | 221 | 59 | 78 | Ardeberg, 1984 |
| | Jun84-Feb85 | 178 | | 80 | Murdin, 1985 |
| | May84-Dec84 | 194 | 50 | 81 | Murdin, 1985 |
| Bakırlıtepe | 1984+1985 (May to Oct) | 226 | 72 | 90 | Aslan et al., 1989 |
| | 1985+1986 (Nov to Apr) | 284 | 54 | | Aslan et al., 1989 |

Institute of the Russian Academy of Sciences), EAO-KSU (Engelhardt Astronomical Observatory–Kazan State University) and the TUG Institute of TÜBİTAK. It is a Ritchey-Chretien telescope with three cassegrain foci of focal ratios f/3, f/7.7, and f/17.3, and a coudé focus of f/48. The focal plane instrumentation is under study.

## References

Ardeberg, A. 1984, in *Site Testing for Future Large Telescopes*, ed. A. Ardeberg and L. Woltjer, ESO Conference and Workshop Proceedings No. 18, p. 73

Aslan, Z., Aydın, C., Tunca, Z., Demircan, O., Derman, E., Gölbaşı, O. & Marşoğlu, A. 1989, A&A, 208, 385

Murdin, P. G. 1985, in *Royal Greenwich Observatory: Telescopes, Instruments, Research and Services, October 1 1980 – September 30 1985*, ed. P. Murdin and J. Wall (Roy. Greenwich Obs.), p. 48

Walker, M. F. 1984, in *Site Testing for Future Large Telescopes*, ed. A. Ardeberg and L. Woltjer, ESO Conference and Workshop Proceedings No. 18, p. 3

# Abstracts of Posters

Jacco van Loon, known in some circles as *der Fliegende Holländer*, prepares to test his wings.

Tsitsino and Irakli Simonia, from the Republic of Georgia, feel right at home in the impressive scenery of the Toros mountains.

POSTERS

# V and R Observations of Two Carbon Stars: UX Dra and RY Dra

HASAN AK[1], BERAHİTDİN ALBAYRAK[1], ZEKİ ASLAN[2], OSMAN DEMİRCAN[1], ZEKERİYA MÜYESSEROĞLU[1], SACIT ÖZDEMİR[1], and KUTLUAY YÜCE[1]

[1] Ankara University Observatory, Ankara, Turkey
[2] Akdeniz University, Antalya, Turkey

The carbon stars UX Dra and RY Dra, classified as semiregular variables, were observed photometrically in the $V$ and $R$ filters at the Ankara University Observatory during the period between April 1995 and March 1996. Altogether ~19 hours of observations on 7 nights for UX Dra and ~11 hours of observations on 4 nights for RY Dra were used in search of both short-term and long-term light variations.

UX Dra was at maximum light in January 1996 and its brightness decreased by $0.16 \pm 0.02$ mag in both filters in about 100 days. The minimum light occurred in April 1996. The light variation is not symmetrical. The period of light variation cannot be less than about 200 days. This means the lengthening of the period is about 1.5 day per year, which confirms Vetesnik's (1983, IBVS No. 2329) finding. The $V-R$ color curve follows a trend similar to the light curves, i.e. the star is hotter at maximum. Two 6.5 hour continuous observations show that UX Dra exhibits short-term light variations of up to 0.03 mag in both filters.

RY Dra was also at maximum light in January 1996 and it faded by about $0.16 \pm 0.02$ mag in $V$, and by $0.10 \pm 0.02$ mag in $R$, in about 20 days. Due to gaps in our observations, the period of light variation cannot be estimated. As in the case of UX Dra, the $V-R$ color curve follows a trend similar to the light variation. No significant short-term light variation was detected in our observations.

# Unusual Light Curves of Some Carbon Stars

## ANDREJS ALKSNIS

*Radioastrophysical Observatory, Riga, Latvia*

Among more than 200 carbon stars monitored photographically with the Baldone Schmidt telescope for time intervals from several years to nearly 3 decades in two to five passbands, three large-amplitude variable stars with exceptional properties of light variation were noticed.

The cyclic component of the light variation of RW LMi = CIT 6 in blue light (cycle length between 350 and 600 d) is entirely different from that in the red and infrared (long-period variation typical for carbon stars: $P = 605$ d before 1988, $P = 628$ d afterwards). A secondary variation with a probable cycle length of about $8P$, however, is shared by both passbands, as are the shorter-timescale (3–30 d) fluctuations. What causes the cyclic blue light variation of RW LMi?

For DY Per, both a long-period (however rather irregular) variation with a cycle length of 792 d and deep light declines similar to those observed in R Coronae Borealis variables have been found. The last five decline events of the RCB type occurred with time intervals of 930, 770, 730 and 680 d, not very different from the cycle length. Is DY Per a long-period variable or an RCB variable, or both?

In AFGL 2881 = V366 Lac, a long-period variation of the Mira type ($P = 562.5$ d) is superimposed upon a secondary variation of extraordinarily large amplitude, unprecedented for a carbon star (5 mag in red light). From an optically recognizable carbon star during the years 1979–1996, the star has changed to an infrared carbon star.

More information on these stars is published in or submitted to *Baltic Astronomy* as well as submitted to *Astronomical and Astrophysical Transactions*.

# Modelling the M-S-C Giants Spectral Sequence

### FRANCE ALLARD[1], PETER H. HAUSCHILDT[2], DAVID R. ALEXANDER[1], MARTIN COHEN[3], and GORDON C. AUGASON[4]

[1] *Wichita State University, Wichita KS, U.S.A.*
[2] *Arizona State University, Tempe AZ, U.S.A.*
[3] *Radio Astronomy Laboratory, University of California Berkeley CA, U.S.A.*
[4] *NASA/Ames Research Center, Moffett Field CA, U.S.A.*

We present pressure-dependent line-by-line LTE model atmospheres for cool red giants ($T_{\rm eff}$ < 4000 K) in spherical geometry. The models are computed using the atmospheric code PHOENIX, and they constitute an extension of the existing cool dwarf grids of Allard & Hauschildt (1995, *ApJ*, 445, 433) to the pressure regimes of red giants. The grid covers C/O ratios ranging from 0.27 to 1.02 with otherwise solar metallicity. We find our models comparable to those of Kurucz in regimes where the plane-parallel approximation is valid, with the exception of the predicted strength of TiO bands as expected from the use of different TiO opacity sources. Departures from LTE in the Ti I lines are investigated for some selected models across the grid, but only modest NLTE effects are found in the abundance of the important absorber TiO. The models are used to construct a spectral sequence of M, S and C type giants for which both optical and infrared spectra are available. Colors of the combined dwarf and giant model grids are presented in the Wing eight-color system which reveals a clear separation of dwarf and giant stars, and of giants in carbon abundance and gravity, providing ideal grounds for the study of the chemical evolution of giants.

Model atmospheres, synthetic spectra and colors presented here will be made available upon request. This research is partially supported by grant AST-9217946 from the National Science Foundation.

# Carbon Isotope Ratios in Carbon Stars of the Galactic Halo

## WAKO AOKI[1,2] and TAKASHI TSUJI[1]

[1] *Institute of Astronomy, The University of Tokyo, Mitaka, Japan*
[2] *Department of Astronomy, The University of Tokyo, Tokyo, Japan*

We have analysed the CN red system ($\sim 8000$ Å) and the $C_2$ Swan system ($\sim 4700$ Å) to obtain carbon isotope ratios ($^{12}C/^{13}C$) for carbon stars in the Galactic halo, known as CH stars. Isotope ratios are obtained for 6 CH stars by a curve-of-growth analysis of isolated $^{12}CN$ and $^{13}CN$ lines. In this analysis, we directly compare $^{12}CN$ and $^{13}CN$ lines of similar intensity (iso-intensity method), and the resulting $^{12}C/^{13}C$ ratios are almost independent of the model atmosphere and its parameters. The $^{13}CN$ lines were too weak to measure in some CH stars, for which we applied the spectral synthesis method to the stronger $C_2$ Swan band and obtained $^{12}C/^{13}C$ ratios for two stars and estimated a lower limit to the $^{12}C/^{13}C$ ratio for two others. In this case, however, the results depend on the model atmosphere and its parameters. Results from our present and previous work show that most CH stars (12 stars) have values distributed around $^{12}C/^{13}C \approx 10$ while two stars have very high values ($^{12}C/^{13}C \geq 500$). The distribution of the $^{12}C/^{13}C$ ratio in CH stars is different from those of the Population I carbon stars and Population II oxygen-rich giants (G – K types). The CH stars of very high $^{12}C/^{13}C$ ratio can be explained by dredge-up of $^{12}C$ due to the $3\alpha$-process as in Population I carbon stars (N–type). On the other hand, the formation of CH stars with low $^{12}C/^{13}C$ ratios requires a large supply of $^{12}C$ followed by a process for decreasing the $^{12}C/^{13}C$ ratio.

# Carbon Stars in Open Clusters

**BERNHARD ARINGER**

*Institut für Astronomie der Universität Wien, Vienna, Austria*

We have performed a systematic search for carbon stars associated with open clusters. We have taken optical carbon stars from Stephenson's *General Catalog of Cool Galactic Carbon Stars* and infrared carbon stars from the IRAS LRS Catalog (class 4 n) and correlated their positions with the open clusters included in the Lyngå catalog. Subsequently all objects that can not be cluster members have been ruled out by investigating their radial velocities, magnitudes, and the probability of an incidental correlation of positions. In the end we came up with several candidates for AGB carbon stars. When it was possible, we calculated absolute bolometric luminosities based on the cluster distances.

# The SiO Molecule in the Atmospheres of Cool AGB Stars

BERNHARD ARINGER[1,2], UFFE G. JØRGENSEN[2],
STEPHEN R. LANGHOFF[3], and JOSEF HRON[1]

[1] Institut für Astronomie der Universität Wien, Vienna, Austria
[2] Niels Bohr Institute, Copenhagen, Denmark
[3] NASA/Ames Research Center, Moffett Field CA, U.S.A.

We have computed a large grid of synthetic SiO spectra based on hydrostatic atmospheres calculated with the MARCS code (Gustafsson et al. 1975, A&A, 42, 407). The results are compared to observations of the SiO features in the photometric $L$ band, which have been obtained using the cooled grating spectrograph IRSPEC at the ESO NTT. They cover a sample of 23 AGB stars. Since the observed SiO bands are much weaker than the calculated ones, we discuss different explanations for this behavior focusing on the effects of dynamic phenomena such as pulsation, mass loss and dust formation. We also present exploratory synthetic SiO spectra based on dynamic atmospheres.

Details of our work are described by Aringer et al. (1997, A&A, 323, 202) and by Aringer et al. (1998, A&A, submitted). This work is supported by the *Fonds zur Förderung der wissenschaftlichen Forschung* under project number S7308-AST.

# The Eddington Limit, Radiative Instabilities and the Declines of R Coronae Borealis Stars

## MARTIN ASPLUND

*Astronomiska observatoriet, Uppsala, Sweden*

The trigger mechanism for the famous declines of R Coronae Borealis stars may be found *in* the photospheres of the stars. Recently constructed model atmospheres show gas pressure inversions in deep layers ($\tau_{\mathrm{Ross}} \approx 10$), which is equivalent to the stellar luminosity locally exceeding the opacity-modified Eddington limit (Asplund et al. 1997, *A & A*, 318, 521). Observed parameters of R CrB stars fall very close to and along the computed Eddington limit, which suggests that there is indeed a relationship between the limit and the declines of R CrB stars. The similar hydrogen-deficient carbon stars, which do not show any declines, have therefore probably not yet reached the Eddington limit. In the present poster a radiative instability is presented which may eject gas clouds from the stellar surface due to the super-Eddington luminosities and the opacity dependence on temperature and density. The ejected gas will cool very rapidly both radiatively and adiabatically to conditions possibly favoring dust condensation, much the same way as described in Woitke et al. (1996, *A & A*, 313, 217), and thereby possibly causing a decline. Typical timescales for the instability are discussed, and the conditions under which the gas pressure inversion disappears and instead a continuous stellar wind is initiated are investigated. Observational consequences, such as high-velocity absorption components even at maximum light, are outlined.

# Modelling of Carbon-Rich Stars with Far Infrared Flux Excess

STEFANO BAGNULO[1], GERRY DOYLE[1],
CHRIS SKINNER[2],* and VINCENZO ANDRETTA[3]

[1] *Armagh Observatory, Armagh, N. Ireland*
[2] *Space Telescope Science Institute, Baltimore MD, U.S.A.*
[3] *Laboratory for Astronomy and Solar Physics*
   *NASA/Goddard Space Flight Center, Greenbelt MD, U.S.A.*
* *Deceased 1997 October 21*

It is now well-known that many carbon-rich stars — especially those with optically thin dust shells — show a large infrared (IR) excess at 60 and 100 $\mu$m. It is common opinion that such a phenomenon can be explained by assuming that the star is surrounded by a cool detached dust shell, placed far away from it. However, there is no agreement in the literature about the chemical composition or typical size of such detached shells, or their distance from the star. Here we present a set of coeval broadband photometric and spectrophotometric measurements for a sample of carbon stars which show large flux excess in the far–IR. We also present the preliminary results of a spectral analysis carried out considering both oxygen-rich and carbon-rich detached shells:

1. An inner shell composed of amorphous carbon and silicon carbide is required to account for a feature seen in the mid–IR spectrum at low resolution.

2. The inner radius of the detached shell is a few hundred stellar radii (that is, $< 0.01$ pc).

3. The far–IR flux excess can be explained by assuming either a detached C-rich shell and an O-rich one; however, with O-rich grains, the dust-to-gas ratio required to fit the spectral energy distribution would be too high with respect to what one expects from the cosmic abundances of Mg, Fe, Si, and O.

We suggest that the stars we have modelled undergo a short-time-scale modulation in the mass-loss rate, and that the observed detached shell, rather than representing a remnant of the star's former O-rich phase, instead represents a past episode of higher mass-loss rate as a C-rich star.

# Selective Depletion of Elements in Stellar Atmospheres: A Unified Picture?

## ERIC J. BAKKER, GUILLERMO GONZALEZ, and DAVID L. LAMBERT

*University of Texas, Austin TX, U.S.A.*

We have investigated stars which show abundance patterns resembling that of gas in the interstellar medium: the abundances of the elements in the stars' atmospheres correlate with condensation temperature. Critical to the detection of this pattern is the measurement of S and Zn, which are only slightly depleted in the interstellar environment and are not likely to be altered significantly by dredge-up episodes in low-mass stars.

The three groups of stars showing this pattern are:

| | | | | |
|---|---|---|---|---|
| 1: | binary post-AGB stars | 4 objects | <[Fe/H]> | = −3.6 |
| 2: | field Type II Cepheids: | | | |
| | a) RV Tau stars | 5 objects | <[Fe/H]> | = −1.3 |
| | b) W Vir stars | 1 object | [Fe/H] | = −1.5 |
| 3: | λ Boötes stars | 4 objects | <[Fe/H]> | = −1.7 |

where [Fe/H] is the iron abundance relative to the solar value.

The first two groups of stars have two things in common: (*a*) their evolutionary time scale is short, and (*b*) almost all stars (there are two exceptions) show a pronounced infrared excess with temperatures $< 1000$ K. We propose that in ALL these metal-depleted stars (the three groups identified) the fractionation process takes place in a circumbinary disk, implying that the RV Tauri stars with metal-depleted abundance pattern may be binaries with periods of the same order as the post–AGB binaries: 1 to 2 years.

For the third group of stars, the λ Boötes stars, a disk may be a remnant of the star formation process and binarity is not needed to explain the observed phenomena.

# How to Make Carbon Stars: A New Approach to Model Boundaries of Convective Regions

## T. BLÖCKER[1], F. HERWIG[1], D. SCHÖNBERNER[1], and M. EL EID[2]

[1] Astrophysikalisches Institut Potsdam, Germany
[2] American University of Beirut, Lebanon

Observationally, most of the carbon and $s$-process enriched AGB stars are found at rather low luminosities indicating that they are *low-mass* stars ($1...3 M_\odot$). Theoretically, we then meet two problems: ($i$) The $3^{\rm rd}$ dredge-up, i.e. the mixing of interior carbon to the surface, is not found self-consistently for low initial masses, and ($ii$) temperatures are too low to activate the $^{22}{\rm Ne}(\alpha,{\rm n})^{25}{\rm Mg}$ reaction. Instead, the $s$-process is most likely driven by the $^{13}{\rm C}(\alpha,{\rm n})^{16}{\rm O}$ neutron source, raising the question how to mix protons into carbon-rich layers in order to produce sufficient amounts of $^{13}{\rm C}$.

We present stellar evolution calculations for Pop. I models which incorporate recent results of hydrodynamical simulations for stellar convection. The hydrodynamical models of Freytag, Ludwig & Steffen (1996, *A & A*, 313, 497) show that convective motions extend well beyond the Schwarzschild boundary of convective instability. The velocity field of the convective elements continues beyond that boundary and declines exponentially. Its scale height $H_{\rm v}$ is proportional to the pressure scale height, $H_{\rm v} = f \cdot H_{\rm P}$, with $f$ depending on the stellar parameters considered. We treated convective mixing by solving a diffusion equation with diffusion coefficients given either by the mixing length theory for the "classical" convective regions or by the hydrodynamical simulations for the overshoot layers.

With this mixing treatment we found the $3^{\rm rd}$ dredge-up not only for massive models (e.g. $7\,M_\odot$) but also for a low-mass model of $3\,M_\odot$ after 10 thermal pulses. The diffusive tail of the hydrogen profile leads to the formation of a $^{13}{\rm C}$ pocket. Our calculations show that this pocket is burnt under radiative conditions before the onset of the next flash, confirming the findings of Straniero et al. (1995, *ApJ*, 440, L87).

# Opacities for Carbon Dwarfs and M Dwarfs

ALEKSANDRA BORYSOW[1,2] and UFFE GRÅE JØRGENSEN[1]

[1] *Niels Bohr Institute, University Observatory, Copenhagen, Denmark*
[2] *Physics Dept., Michigan Technological University, Houghton MI, U.S.A.*

We have computed new opacity data for collision-induced absorption (CIA) of $H_2-H_2$ and $H_2-He$, applicable to oxygen-rich and carbon-rich compositions, and updated existing line data. We have analyzed the combined effect of such data on the model atmospheric structure and the synthetic spectra of dwarfs and giants of various compositions.

The influence of molecular data on the models increases with decreasing effective temperature, and whereas CIA dominates at low metallicities and large gravities, "normal" line transitions dominate at solar metallicity. At $Z = 0.1\,Z_\odot$, collision-induced absorption has observable effects on the synthetic spectrum, and for $Z = 10^{-2}$ and lower, CIA absorption is the dominant feature in the infrared spectrum and has major effect on the over-all flux distribution. For low metallicities, the strong depression of the infrared continuum by CIA makes the stars look more metal-deficient than they really are. Spectra of M dwarfs in globular clusters will be totally dominated by CIA absorption, and we predict that their infrared spectra will look very "smooth" – with almost no trace of the usual bands of water and CO – because of the strong CIA continuum depression.

In metal-rich carbon stars the diatomic features (mainly CN and $C_2$) are much more pronounced in giant stars than in dwarfs, whereas the polyatomics ($C_2H_2$, $C_3$, and HCN) show smaller difference between dwarfs and giants. For oxygen-rich stars there is a corresponding balance between the intensities of $H_2O$, TiO and CIA spectral features, TiO being relatively stronger in the giants, and $H_2O$ being relatively stronger in the dwarfs.

Details of our CIA data and applications to oxygen-rich model atmospheres have been described by Borysow et al. (1997, *A&A*, 324, 185). The support of NATO Collaborative Research Grant CRG.941197 and of the NASA Astrophysics Theory Program is gratefully acknowledged.

# A Kinematic Survey of Carbon Stars in the Small Magellanic Cloud

RUSSELL CANNON[1], BARRY CROKE[2],
DESPINA HATZIDIMITRIOU[2], and DAVID MORGAN[3]

[1] *Anglo-Australian Observatory, Sydney, Australia*
[2] *University of Crete, Heraklion, Greece*
[3] *Royal Observatory, Edinburgh, Scotland*

We are carrying out a survey of carbon stars in the outer parts of the Small Magellanic Cloud (SMC), using the double-beam spectrograph on the ANU 2.3-m telescope at Siding Spring Observatory. These stars can be used as tracers of the older stellar populations (i.e. with ages of a few billion years). The stars were chosen from the large sample of new carbon stars found by Morgan & Hatzidimitriou (1995, *A & A Supp.*, 113, 539) on objective prism photographs taken with the UK 1.2-m Schmidt Telescope. The blue spectra cover the $C_2$ Swan bands and confirm that virtually all the stars are indeed carbon stars, while the higher dispersion red spectra yield velocities accurate to better than 5 $km\,s^{-1}$.

For 71 carbon stars surrounding the SMC, at radii of between 3° and 6°, we obtain a mean radial velocity of 146 ± 3 $km\,s^{-1}$ and a dispersion of 21 ± 2 $km\,s^{-1}$, similar to that found in other intermediate-age and old populations in the SMC. This implies a mass of about $1.2 \times 10^9$ solar masses under the somewhat unrealistic assumption of a spherical SMC halo. There is no evidence for any bifurcation of radial velocities as seen in some samples of younger objects, nor are there any significant velocity trends across the face of the SMC.

Three carbon stars have been found mid-way between the two Clouds; their velocities are typical of those in the main SMC sample. For a small sample of 16 stars in the SW corner of the LMC we obtain a higher mean velocity of 226 ± 3 $km\,s^{-1}$ with a dispersion of 12 ± 2 $km\,s^{-1}$.

A full account of this work has been published by Hatzidimitriou et al. (1997, *A & A Supp.*, 122, 507).

# Mid-Infrared Silicate Variation in Long-Period, Oxygen-Rich Variable Stars

## M. J. CREECH-EAKMAN and R. E. STENCEL

*University of Denver, Denver CO, U.S.A.*

We present preliminary results from our on-going monitoring campaign of a selected group of more than 30 Long-Period Variable (LPV) stars at 10, 11 and 18 $\mu$m. Our stars were chosen from a list by Little-Marenin & Little (1990, *AJ*, 99, 1173) based upon a classification scheme of silicate features of oxygen-rich LPV stars. We are monitoring these LPVs for changes in their silicate features at 10 and 18 $\mu$m with respect to IR continuum and optical phase. We are attempting to ascertain the relationship of dust formation to optical period, and any shell-shock interactions from the acoustic shocks originating in the photosphere and later impinging on these dust-forming areas. The ultimate goal is to determine what conditions lead to dust formation and destruction in these environments, and whether or not an evolutionary sequence can be inferred for AGB objects based on their spectra and dust formation. The instrument being used for data acquisition is Denver University's (DU) TNTCAM (Ten aNd Twenty micron CAMera), a liquid-helium-cooled, mid-IR camera using a Rockwell 128 $\times$ 128 Si:As hybrid array and housing 7 filters on an externally driven filter wheel (Klebe et al. 1996, in *Polarimetry of the Interstellar Medium*). A portion of this list of stars was chosen for our initial campaign at the Wyoming Infrared Observatory (WIRO) in 1993, including the brightest of our objects with a range of periods and LML types (Creech-Eakman et al. 1997, *ApJ*, 477, 825). The rest of our list consists of a sample of LML types with periods of 300–400 days and fluxes of 25 $\pm$ 5 Jy at 8 $\mu$m chosen for one of our ISO proposals. Ancillary mid-IR spectra exist from LRS on IRAS in 1983, our CAESR data from 1993, and CGS–3 data from UKIRT service time in 1995. We hope to obtain photometric data with TNTCAM and spectral data using ISO's SWS, and TGIRS, DU's new Two Grating mid-IR Spectrometer.

We would like to acknowledge partial support under NASA grant NGT-51290.

# LMC/SMC Outer Halo Carbon Star Survey: Radial Velocities of 500 Newly Identified Stars

SERGE DEMERS[1], W. E. KUNKEL[2],
and M. J. IRWIN[3]

[1] *Université de Montréal, Montréal, Canada*
[2] *Las Campanas Observatory, La Serena, Chile*
[3] *Royal Greenwich Observatory, Cambridge, U.K.*

Making use of the available UKST sky survey plates scanned by the APM, we have produced color-magnitude diagrams and identified carbon star candidates among the very red stars ($B-V > 2.0$) within the magnitude range of AGB stars at $\sim 55$ kpc in 19 fields representing 600 deg$^2$ around the Magellanic Clouds. Follow-up slit spectroscopy at a resolution of 2.3 Å has resulted in the identification of more than 500 previously unknown outer halo carbon stars extending up to 10° from the LMC and 8° from the SMC. Radial velocities of these stars are valuable for the kinematic study of the periphery of the Clouds.

We are investigating the use of spectral features in the 7000–9000 Å interval to establish criteria to discriminate between the two Cloud populations. Examples of observed spectra are shown.

The data discussed here have been published by Kunkel, Irwin & Demers (1997, *A&A Supp.*, 122, 463). The radial-velocity data are available, via Internet, at the CDS.

# The Backwarming Effect and Carbon Stars

## DIMITRI N. DOIKOV and EKATERINA M. DOIKOVA

*Odessa State University, Odessa, Ukraine*

The abundant molecules are an important component in the opacity source function. This importance increases for cooler carbon stars. All stellar spectra contain "transparent windows" with a pseudo-continuous flux. We have calculated the contributions from a mixture of molecules in these spectral regions and estimated the changes in effective temperatures due to the backwarming effect.

# Proposal for a Photometric System for the Classification of Carbon Stars

## ULDIS DZERVITIS

*Radioastrophysical Observatory, Riga, Latvia*

Because of the presence of molecular bands in the spectra of carbon stars resulting from their peculiar atmospheric compositions, most photometric systems are of limited value in determining the physical and chemical parameters of their atmospheres. Therefore, for the photometric classification of carbon stars in the visual and near infrared regions, a medium bandwidth (10–25 nm) photometric system is proposed in which the locations of pass-bands are in conformity with the spectral peculiarities, and which is useful for small and moderate-sized telescopes and performable with interference filters and standard photomultipliers.

Integrating spectrophotometric records of about 60 N and late R stars with a systematic, step-by-step variation of the location and width of the proposed bands, a search was made for the highest resolution color-index diagrams. Five primary and two supplementary filters were selected, as follows:

Characteristics of the proposed system

| Filter | Width (nm) | Measured feature |
|---|---|---|
| $\Phi_{51}$ | 505–520 | $C_2$ Swan bands $\Delta v = 0$ |
| $\Phi_{53}$ | 520–540 | Quasicontinuum |
| $\Phi_{65}$ | 640–655 | Quasicontinuum |
| $\Phi_{78}$ | 770–790 | Quasicontinuum |
| $\Phi_{80}$ | 790–820 | CN red system $\Delta v = 2$ |
| Supplementary filters | | |
| $\Phi_{59}$ | 585–595 | Na D resonance doublet |
| $\Phi_{89}$ | 870–900 | Quasicontinuum |

The five primary bands of the system provide two color-indices representing effective temperature and two molecular indices ($C_2$ and CN) which — as one can see from simplified equations of the thermochemical equilibrium in the very limited effective temperature interval of cool carbon stars (2500–3000 K) — can be treated as indicators of C/O and N/H. The supplementary filters provide the possibility of measuring one more pseudocontinuum color-index and the Na D resonance doublet intensity — a classical temperature indicator in carbon-star spectra.

# Carbon to Oxygen Abundance Ratios in the Atmospheres of Carbon Stars of the Orion and Perseus Galactic Arms

### I. EGLITIS and M. EGLITE

*Radioastrophysical Observatory, Riga, Latvia*

The abundance ratio C/O in the atmospheres of carbon stars of the Orion and Perseus galactic arms is determined on the basis of a great number of homogeneous spectra obtained with the 2.6-m telescope of the Byurakan Astrophysical Observatory (Armenia) in 1990–1991. The reciprocal dispersion for observations of faint carbon stars was 101–136 Å/mm and the wavelength range was 4000–6800 Å. We used the empirically-obtained correlation between the intensities of the CN Red System and $C_2$ Swan bands, on the one hand, and C/O ratios determined from model calculations of synthetic spectra (which were compared to 30 high-resolution spectra of carbon stars [Lambert et al. 1987, *ApJS*, 62, 273]), on the other hand, and so obtained the individual C/O abundances in the atmospheres of 342 faint carbon stars. Twenty-eight high-carbon-abundance C stars were found. They all have similar spatial distribution and many common spectral peculiarities.

The distributions of carbon stars by C/O abundance ratio interval were studied in the Orion and Perseus galactic arms. In Orion most carbon-star atmospheres have carbon to oxygen ratios in the interval $1.0 \leq C/O \leq 1.1$, but in the Perseus Arm most are in the interval $1.1 \leq C/O \leq 1.2$.

A detailed discussion of these results has been submitted to Astrophysics and Space Science.

# Condensation of SiC in Circumstellar Dust Shells of C–Rich Red Giants

## ANDREAS GAUGER[1], JOHN J. KEADY[1] and ERWIN SEDLMAYR[2]

[1] *Los Alamos National Laboratory, Theoretical Division*
*Los Alamos NM, U.S.A.*
[2] *Institut für Astronomie und Astrophysik, TU Berlin, Germany*

Since the late 1960's silicon carbide has been known to be one of the dust components in circumstellar dust shells around carbon-rich red giants due to the attribution of a prominent feature at 11.3 $\mu$m to solid SiC. From radiative transfer models the fraction of SiC is inferred to be a few percent by mass, and the shape of the feature can be fitted with the optical properties of $\alpha$–SiC powders. Although there exist a few theoretical investigations (e.g. Kozasa & Sogawa, this conference), the details of the condensation process of SiC in circumstellar shells are still unknown.

Our investigations show that homogeneous formation of SiC from the nominal molecule can be ruled out for typical conditions in circumstellar dust shells, due to the low abundance of the SiC molecule. Condensation of SiC has to proceed via processes involving more abundant species such as monoatomic silicon, $SiC_2$ and $Si_2C$. We have considered a condensation scheme in which hydrocarbons ($C_2H$, $C_2H_2$) and silicon carbide molecules ($SiC_2$, $Si_2C$) condense onto carbon seeds to form two distinct coexisting solid phases of carbon and SiC. In model calculations for dust-forming outflows this process yields a fraction of solid SiC of a few percent, in good agreement with the values derived from observations. In the present models nucleation of a separate population of SiC particles is suppressed because carbon seeds nucleate first, but SiC nucleation may become important under conditions where carbon condensation is less effective, e.g. in S stars where the carbon-to-oxygen ratio is close to one.

The calculations were performed on the SGI Cluster at the Institut für Astronomie und Astrophysik. The research at LANL was performed under the auspices of the U.S. Department of Energy.

# New Results from the Modeling of the Shell around IRC +10216

## M. A. T. GROENEWEGEN

*Max-Planck Institut für Astrophysik, Garching, Germany*

A spherically symmetric dust radiative transfer code is used to model the circumstellar dust shell around IRC +10216. Compared to numerous previous models a much larger body of observational data is used as constraints. The spectral energy distribution between 0.5 and 60000 $\mu$m, 2–4 $\mu$m and 8–23 $\mu$m spectra, optical, far-infrared and centimeter sizes, and interferometric visibility curves between 1.6 and 11.2 $\mu$m are used to constrain the model.

Key results are:

- In order to fit the visibility curve at 2.2 $\mu$m and the size of the shell in the optical, scattering has to be invoked. The strong dependence of the scattering coefficient on grain size allows one to derive a mean grain size of $0.16 \pm 0.01$ $\mu$m.

- Previous suggestions that the mass loss rate was higher in the past are confirmed. The principal argument is that with an $r^{-2}$ model the calculated far-infrared sizes are smaller than observed.

- Regarding the cm emission it is found that in small apertures dust emission is negligible for wavelengths $\gtrsim 2$ cm. Free-free emission is negligible for wavelengths $\lesssim 0.5$ cm. The free-free emission is found to be optically thin even at 6 cm. An ionization fraction of $7.8 \times 10^{-5}$ is derived which, according to the Saha equation, corresponds to an electron temperature of about 2400 K. Although there are uncertainties in the free-free emission model, this suggests that the free-free emission does not come from a chromosphere.

This research is discussed further in Groenewegen (1996, *A & A*, 305, L61) and Groenewegen (1997, *A & A*, 317, 503).

# JHK Photometry of AGB Stars in the SMC

M. A. T. GROENEWEGEN[1] and J. A. D. L. BLOMMAERT[2]

[1] *Max-Planck-Institut für Astrophysik, Garching, Germany*
[2] *ISO Science Operations Centre, Astrophysics Division of ESA Villafranca, Spain*

We have obtained *JHK(L)* photometry of 29 objects in the direction of the SMC that have IRAS 12 and 25 $\mu$m colors that suggest that they are AGB stars. Some of the objects are extremely red with $J-K$ up to 5.6. Two objects are certain foreground objects, one is the VV Cep object N55 (A. R. Walker 1983, *MNRAS*, 203, 25), three are known carbon stars and three are known oxygen-rich AGB stars.

We have modeled the spectral energy distributions of a few stars using a radiative transfer model (Groenewegen 1993, Ph.D. Thesis, University of Amsterdam, Chapter 5). In quite a few cases the IRAS photometry does not fit the near-infrared data. Variability may be the reason in a few cases. In others this seems unlikely and there may be actually more than one red source in the field of view: one that is so red that we did not pick it up down to $K \approx 13$, and one serendipitously discovered red AGB object for which we obtained the near-infrared photometry and which initially was believed to be the IRAS counterpart.

By modeling the energy distributions, carbon stars can be distinguished from oxygen-rich stars based on the completely different absorption efficiencies of carbonaceous and silicate dust.

# Frequency Sampling for Radiative Transfer Calculations in Cool Stars

CHRISTIANE HELLING[1,2] and UFFE GRÅE JØRGENSEN[1]

[1] Niels Bohr Institute, University Observatory, Copenhagen, Denmark
[2] Institut für Astronomie und Astrophysik, Technische Universität Berlin, Germany

A central problem in combining existing hydrostatic computations for photospheres of cool stars with existing hydrodynamic codes for computation of the wind and circumstellar environment is the thorough optimisation of the frequency selection for the radiative transfer problem. Most hydrostatic models are today based on the Opacity Sampling (OS) method, where the monochromatic opacity in a few thousand frequency points are used for solving the radiative transfer problem. In contrast, solution of the hydrodynamic problem is so much more CPU-demanding that existing codes have to rely on a much more simplified treatment of the radiative transfer. Typically, the radiative transfer in such codes is based on a single value of the opacity — for example the Rosseland mean opacity, or even simpler approximations.

We have analysed the effect of different choices of Opacity Sampling frequencies on hydrostatic photospheric models and have compared the results with those obtained by use of the Rosseland mean or constant-value opacities frequently applied in hydrostatic models. We find that considerable improvement over the results from such mean values can be obtained with Opacity Sampling in as few as 20 frequency points. In the upper photospheric layers of cool carbon-rich giants, typical temperature deviations between a detailed OS-treated model (with 5300 frequency points) and corresponding models based on absorption coefficients approximated by 1) OS with 500 points, 2) OS with 50 points, 3) OS with 20 points, 4) a constant-value of $10^{-4}$ cm$^2$/g, and 5) the Rosseland mean, were found to be 50 K, 100 K, 200 K, 700 K and 700 K, respectively. The Rosseland mean models approach the detailed models deep in the atmosphere, whereas the constant-value models deviate very much from the detailed models in deep layers.

Details of our analysis, and a corresponding analysis of oxygen-rich stars, are described in the Diploma Thesis of C. Helling (July 1996) and by Helling & Jørgensen (1998, *A&A*, in press). We acknowledge support from the Danish Natural Science Research Council.

# Macro-Molecules in Model Atmospheres

CHRISTIANE HELLING[1,2], UFFE GRÅE JØRGENSEN[1],
BERTRAND PLEZ[1], and HOLLIS R. JOHNSON[1,3]

[1] *Niels Bohr Institute, University Observatory, Copenhagen, Denmark*
[2] *Institut für Astronomie und Astrophysik, Technische Universität Berlin, Germany*
[3] *Indiana University, Bloomington IN, U.S.A.*

We present calculations of partial pressures of macro-molecules in stellar photospheric MARCS (Jørgensen et al. 1992, *A&A*, 261, 263) models of carbon stars. The molecular equilibrium data included were taken from Cherchneff & Barker (1992, *ApJ*, 394, 703). Our aim is to study the molecular route to the formation of carbonaceous grains. In our hydrostatic models the largest molecule that appears in a significant amount is $C_6H_2$. Open-shell species (radicals) have comparable abundances but all of them are found to be too rare to support grain formation, or to affect the photospheric structure by levitation.

A hydrodynamic model atmosphere (kindly provided to us by A. Fleischer) has higher pressure in the outer atmosphere, and much larger density, than our corresponding hydrostatic model. Although the dynamical model extends to much lower gas pressures, the $T - P_{\text{gas}}$ structures of the dynamical and hydrostatic models are quite similar in the region of overlap. The same molecular species are therefore formed in about the same amount in the overlap-region of the two models. Chemical equilibrium calculations for $(T, P_{\text{gas}})$ values corresponding to a hydrodynamic (wind) model show large concentrations of PAHs (polycyclic aromatic hydrocarbons) in the layers with $T \leq 850$ K (i.e., cooler than the photospheric models). Larger PAH molecules such as $C_{22}H_{14}$-benzo[c]crysene dominate over smaller PAHs and other hydrocarbon molecules. An exploration of the $T - P_{\text{gas}}$ plane, $T \epsilon$ [750 K, 1300 K] and $\log P_{\text{gas}} \epsilon$ [−5,3] (dyn/cm$^2$), shows that macro-molecules bind a large fraction of the available C not bound in CO and can bind even more C than CO for C/O $\geq 5.0$. The time scale of a realistic wind model is, however, much shorter than the time required for PAH formation. Hence, the hydrostatic as well as the hydrodynamic models have problems with forming carbon grains via the PAH route.

Details of our analysis are described by Helling et al. (1996, *A&A*, 315, 194). We acknowledge support from the Danish Natural Science Research Council.

# Molecular Hydrogen in the Circumstellar Shells of Carbon Stars

## KENNETH HINKLE[1] and JOHN KEADY[2]

[1] *NOAO/Kitt Peak National Observatory, Tucson AZ, U.S.A.*
[2] *Los Alamos National Laboratory, Los Alamos NM, U.S.A.*

The S-branch lines of molecular hydrogen were detected in 2.0–2.4 $\mu$m carbon-star spectra by Johnson et al. (1983, *ApJ*, 270, L63). We subsequently detected strong S-branch lines in the 2 $\mu$m infrared spectra of M-, S-, and C-type long-period variables. In long-period variables, the $H_2$ lines show very strong phase-dependent changes in velocity, strength, and line profile, with the $H_2$ line strength greatest near minimum light. The strength of the $H_2$ lines also depends on the visual amplitude of the variable. Analysis of the $H_2$ line profiles shows that the $H_2$ line is formed both in the photosphere and in an extended, non-photospheric (non-pulsating) region (or regions) of the stellar atmosphere.

In the case of the obscured carbon stars, the photospheric 2 $\mu$m spectrum has been nearly completely filled in by thermal emission from circumstellar dust. Our modeling of the circumstellar envelope of IRC +10216 indicated that the expanding circumstellar $H_2$ S(1) line should be detectable with a depth of a percent or so. We also felt that observations of this line would shed light on the origin of the non-photospheric $H_2$ profiles seen in non-obscured stars. We undertook very high signal-to-noise 2 $\mu$m spectroscopy of IRC +10216 to observe the $H_2$ S(1) line. However, the observations reveal that at the few percent level, the photospheric spectrum of IRC +10216 is present in the 2.12 $\mu$m spectrum. This greatly complicates the analysis. Limits on the mass-loss rate will be discussed.

# Atmospheric Dynamics and Dust-Driven Winds of Carbon Stars

## SUSANNE HÖFNER and ERNST A. DORFI

*Institut für Astronomie der Universität Wien, Vienna, Austria*

We have calculated radiation–hydrodynamical models of the atmospheres and circumstellar dust shells of C-rich AGB stars which include a time-dependent description of the dust formation process. Simulating the stellar pulsation by a variable inner boundary, we have investigated the time-dependent dynamics of the atmosphere and circumstellar envelope and its interaction with dust formation. We have varied the physical parameters of the models systematically to study their influence on the time-dependent behavior and time-averaged mass-loss characteristics of our models. The results can be summarized as follows: The dependence of the mass-loss rate on stellar parameters predicts a strong increase of mass loss as stars evolve along the AGB. As the stellar luminosity increases, while simultaneously the mass and effective temperature decrease, the atmosphere becomes more extended and conditions become more favorable for dust formation and higher densities in the acceleration region of the wind. The models agree nicely with mean mass loss – period relations deduced from observations of Mira stars, and the observed scatter of mass-loss rates for a given period can be understood in terms of pulsation amplitude or non-linearities of the dust condensation and wind mechanism. As expected from radiation-driven outflows, the wind velocities show a good correlation with a quantity characterizing the strength of radiation pressure on dust relative to gravitation.

This paper has appeared in *A & A* 319, 648, 1997. Synthetic spectra and IRAS colors based on these models have been calculated by Loidl et al. and by Windsteig et al. (this volume). This work is supported by the *Fonds zur Förderung der wissenschaftlichen Forschung* (FWF) under project number S7305–AST.

# The Shape of Silicate Features in Semiregular and Mira Variables

JOSEF HRON, BERNHARD ARINGER,
and FRANZ KERSCHBAUM

*Institut für Astronomie der Universität Wien, Vienna, Austria*

We have analyzed the mid-IR silicate emission features from O-rich semiregular (SR) and Mira variables by fitting blackbodies to the underlying photospheric and dust continua and computing ratios of the excess flux in selected wavelength bands. These ratios were then related to the photospheric temperatures and the optical depths of the dust shells. The ratio of the strengths of the 10 $\mu$m and 18 $\mu$m features decreases from the SRs to the Miras. Narrow 10 $\mu$m features are found for the 'hottest' stars (mostly SR) and for cool stars with thick envelopes (mostly Miras). Thus optical depth effects can almost certainly be ruled out as the main factor influencing the feature shapes. We discuss possible explanations in terms of current ideas on the formation of dust in oxygen-rich stars.

This work is supported by the *Fonds zur Förderung der wissenschaftlichen Forschung* under project number S7308–AST.

# FG Sge as a New-Born Carbon Star

## TAKASHI IIJIMA

*Astronomical Observatory of Padova, Asiago, Italy*

The spectral type of the post–AGB star FG Sagittae rapidly changed from B4 I in the 1950s to K2 Ib in the 1980s. The Swan bands of the $C_2$ molecule have been detected in some spectra taken in 1981 and later. This star seems to have become a carbon star as expected in theories of the evolution of post–AGB stars. The recent spectra, however, are much different from those of normal carbon stars. The absorption lines of Ba II, Sr II, La II *etc.* are unusually deep, while those of the iron group are weak. In most spectra the CN bands are absent even when the $C_2$ bands are easily visible. Since a weak CN band at 4215 Å was detected in a spectrum taken on 23 April 1994, the absence of CN bands in the other spectra may not have been due to a lack of nitrogen atoms. There might have been a peculiar atmospheric condition in which the $C_2$ molecule was formed, but not the CN molecule. Large spectral variations have been observed during the photometric decline which started in August 1992.

The full text has appeared in *MNRAS* 283, 141, 1996.

# Experimental Gas-Phase Spectroscopy of Carbon Molecular Structures and Diffuse Interstellar Bands

## J. JANČA[1], M. PLONKA[2], M. ŠOLC[3] and M. VETEŠNÍK[2]

[1] *Plasma-Chemistry Laboratory, Masaryk University, Brno, Czech Republic*
[2] *Department of Theoretical Physics and Astrophysics*
   *Masaryk University, Brno, Czech Republic*
[3] *Astronomical Institute of Charles University, Prague, Czech Republic*

Preliminary results of an experimental study of ultraviolet to near-infrared gas-phase spectra of possible candidate materials for Diffuse Interstellar Bands (DIBs) are presented. The experimental technique enabling the study of potential carriers of DIBs in conditions that mimic the natural environment of carbon-star atmospheres is described. Conclusions concerning some carriers to be searched for (including molecular features at $\lambda 2150$, $\lambda 2600$, $\lambda 5844$, $\lambda 5850$, $\lambda 9577$, and $\lambda 9632$) are also discussed.

We have searched for the spectral features of fullerene $C_{60}$ in the ultraviolet at wavelengths corresponding to the measurements of Leach et al. (1992, *Chem. Phys.*, 160, 451). Following a detailed analysis, no fullerene spectral features have been detected, although they could be fully masked by molecular absorptions that cannot be reduced. Such a conclusion holds true also for the observed feature at $\lambda 2150$ which can be identified with the rovibrational spectra of various types of molecules like $C_2N_2$, CO, $CO^+$, and $N_2$, while the true fullerene profile should be much broader.

Our attempt to throw light upon the DIBs at $\lambda 5844$ and $\lambda 5850$, recently studied by Jenniskens et al. (1996, *A&A*, 313, 649), has resulted in the conclusion that two spectral features observed in a helium atmosphere enriched with hydrogen atoms could be, under suitable conditions, considered as counterparts of the DIBs. This statement is supported by the structure seen in the feature at $\lambda 5844$.

We have also considered two new DIBs at $\lambda 9577$ and $\lambda 9632$, theoretically attributed to interstellar $C_{60}^+$ (Foing & Ehrenfreund 1995, in *The Diffuse Interstellar Bands*, ed. Tielens and Snow, p. 65). Unfortunately, as both of these DIBs coincide with the (2,1) band of the red system ($^2\Pi - {}^2\Sigma$) of CN, it is hard to say that the spectral features observed by us are due to $C_{60}^+$.

# The Population of Red Giant Stars in Globular Clusters of the Fornax Dwarf Galaxy

## UFFE GRÅE JØRGENSEN[1] and RAUL JIMENEZ[2]

[1] *Niels Bohr Institute, University Observatory, Copenhagen, Denmark*
[2] *Royal Observatory, Edinburgh, U.K.*

Fornax dSph is the only one of the Milky Way dwarf galaxies known to contain star clusters of its own, and the variety of its 6 clusters is as large as the variety among the many globular clusters in our own Galaxy. We have obtained CCD photometry of clusters 1 and 3 in $V$ and $I$ and through two narrow-band filters of the Wing system, designed to distinguish between carbon and M-type giants. The clusters contain a large population of carbon stars, in strong contrast to the (near) absence of carbon stars in the Galactic globular clusters. The carbon stars in the Fornax clusters populate only the lower 30% of the luminosity interval spanned by the Fornax field carbon stars. All the cluster carbon stars are fainter than the theoretical first helium shell flash luminosity. The large number of carbon stars could be explained by assuming the Fornax dSph to be about 6 billion years younger than the Galaxy, but this would still not explain their low luminosity. The $V$–$I$ colors of the giant branch stars define a much broader range than the corresponding very narrow Galactic globular cluster giant branches. The broad morphology can only be simulated by assuming a large metallicity spread among the Fornax giants, and we estimate that the individual stars along the giant branches span a range of metallicities all the way from [Fe/H] $= -2.4$ to [Fe/H] $= -1.6$.

The definite existence of carbon stars on the AGB in both of the observed Fornax clusters is a challenge to the theoretical models of the third dredge-up episode and to galactic chemical evolution models. It raises the central question of why the carbon stars in Fornax form already well below the theoretical carbon star luminosity limit, and also why they are present in relatively large numbers in the Fornax clusters and not at all in the Galactic globular clusters of the same metallicities.

Further details of this project are described by Jørgensen & Jimenez (1997, A&A, 317, 54).

# Synthetic JHK Colors for M Dwarfs, M Giants, and Carbon Stars

UFFE GRÆE JØRGENSEN[1] and ROBERT F. WING[2]

[1] *Niels Bohr Institute, University Observatory, Copenhagen, Denmark*
[2] *Ohio State University, Columbus OH, U.S.A.*

Using model atmospheres computed with the MARCS code and filter functions for the near-infrared $J$, $H$, and $K$ filters from Bessell & Brett (*PASP*, 100, 1134, 1988), we have computed synthetic $J-H$ and $H-K$ colors for giants ($\log g = 0$) and dwarfs ($\log g = +5$) of effective temperature 3000 – 5000 K, both for solar composition and for a carbon-enriched composition (C/O = 1.07). The three molecules which have the greatest effects on near-infrared spectra — CO, CN, and $H_2O$ — were individualy turned on and off when computing the synthetic spectra in order to investigate their separate effects.

For solar-composition models, we reproduce the observed bifurcation of the giant and dwarf sequences in the $J-H, H-K$ color-color diagram. Absorption by $H_2O$ has an important effect on the colors of dwarfs, starting at temperatures as high as 4000 K; CO has a noticeable effect on giant colors. However, a major portion (more than half) of the observed bifurcation is present in the model sequences even when no molecules at all are included in the synthetic spectra; this effect is attributed to the differing effects of $H^-$ opacity on models having different temperature structures.

In carbon-rich giants the combined effect of CO and $C_2$ is to change the $J-H$ and $H-K$ colors considerably from the corresponding continuum colors. For the coolest models $C_2H_2$ and $C_3$ have a strong effect in the $K$ filter. In carbon dwarfs CO and $C_2$ have small and opposite effects on the $JHK$ colors, and the colors of the dwarfs are therefore similar to those of the continuum alone.

# Near-Infrared Spectroscopy of High Galactic Latitude Carbon Stars – They Might Be Giants?

## RICHARD R. JOYCE

*NOAO/Kitt Peak National Observatory, Tucson AZ, U.S.A.*

Because of their high luminosity and distinctive optical spectra, carbon stars can be used as kinematic probes of the Galaxy to great distances (Bothun et al. 1991, *AJ*, 101, 2220). However, a number of faint carbon stars at high galactic latitudes (including three from the above work) show significant proper motion, suggesting a main-sequence luminosity. These carbon dwarf (dC) stars are generally believed to be binary systems in which the main-sequence dwarf has received processed material from a now invisible companion during the companion's ascent of the AGB. The optical spectra of dwarf and giant carbon stars are similar, and unambiguous discriminants such as proper motion can be problematic. Infrared spectra of a selection of faint carbon stars in the $J$, $H$, and $K$ bands indicate that the known dC stars have weak first-overtone CO bands for their $H-K$ color in comparison to the other stars in the sample and a selection of bright carbon giants. A similar segregation in the $J-H$, $H-K$ plane was noted by Green et al. (1991, *ApJ*, 380, L31). The spectroscopic results are consistent with the suggestion of Green et al. that $JHK$ colors, which are more easily obtained than proper motion or IR spectra, may provide a useful luminosity discriminant.

# A Simulation for a Carbon Star

## GÜLÇİN KANDEMİR and CEM GÜÇLÜ

*Physics Department, Istanbul Technical University*
*Istanbul, Turkey*

A simple one-dimensional electrostatic plasma simulation code has been applied at a carbon star to represent possible instabilities. Electrons are assumed to move against a stationary ion background. The temperature in the vicinity of the carbon star is adopted as 5,000–10,000 K and some plasma input parameters are calculated accordingly for this cold plasma. Here, the Maxwell and Vlasov equations are employed and the fast Fourier transform is applied in the program. Linear and non-linear effects resulting from plasma instabilities in carbon-star atmospheres have been investigated. Among the several types of plasma instability, electrostatic instability is the most destructive. Recently, the role of ionization is adopted as 10 times more in the carbon stars. The electrostatic instability grows by accumulation of charge and the perturbations grow exponentially with time. At a carbon-star plasma where large changes occur, the electrostatic instability should not be ignored. The present simulation shows that even small perturbations may cause instability in the carbon-star envelope. The calculated drift velocity found by this simulation is comparable to the velocity of 20 km s$^{-1}$ indicated by recent observations of the expanding envelopes of carbon stars. In addition to the expected linear behavior, the non-linear behavior in phase space found by this simulation agrees with the theory.

# On the Nature of Irregular Variables of Type Lb

## FRANZ KERSCHBAUM[1], PETER HABISON[1,2], and JOSEF HRON[1]

[1] *Institut für Astronomie der Universität Wien, Vienna, Austria*
[2] *Kuffner-Sternwarte, Vienna, Austria*

The study deals with one group of stars at the low-mass-loss part of the asymptotic giant branch, namely the Irregular variables of type Lb. Their distinctly different pulsational behaviour, when compared with the well studied Miras, combined with the presumed importance of pulsation for the stellar mass loss, make them important scientific targets. O-rich Lb variables were selected from the GCVS and the IRAS Catalogue. Additional near infrared data was used, mainly from our own observations. Comparison samples of Miras, SRa-, and SRb-type variables were selected in the same way.

For the four groups of O-rich variables, pulsational properties like the amplitude or the period (not for Lbs) are correlated with atmospheric properties like the effective temperature and can be used to separate at least Lbs, SRas and SRbs from the Miras. By means of NIR as well as IRAS two-color diagrams, O-rich Lbs are indistinguishable from SRas or SRbs. Hence, it seems likely that they have similar effective temperatures, atmospheric structure and also mass-loss rates. The Lbs fit nicely in the areas of "blue" and "red" SRVs. But additionally one finds a few objects displaying NIR properties similar to M giants. These stars may be not on the AGB at all. All three groups are well separated from the Miras. Also the blackbody-fit results turn out to be comparable for the Lb, SRa, and SRb stars and different for the Miras (for SRVs see *A&A* 308, 489, 1996). Whereas the first three have relatively high photospheric temperatures $T^*$, medium dust temperatures $T^d$ and a wide spread in the relative sizes of the dust blackbody, the later tend towards lower $T^*$ and somewhat higher $T^d$. This can be interpreted in the sense that Lbs, SRas and SRbs contain some stars with $60\,\mu$m excess originating from colder dust and that Miras suffer higher mass loss with perhaps hotter dust temperatures. IRAS $60\,\mu$m mass-loss rates are generally much higher for Miras and well correlated with some results of the BB fits, namely $T^*$ and $L^d/L^*$. The scale heights of Lbs and SRVs are quite similar, confirming our assumption of similar luminosities for these two groups.

This work is supported by the *Fonds zur Förderung der wissenschaftlichen Forschung* under project number S7308–AST.

# Comparison of C-Rich Mira, Semiregular, and Irregular Variables

## FRANZ KERSCHBAUM[1], PETER HABISON[1,2], RITA LOIDL[1], HANS OLOFSSON[3], and JOSEF HRON[1]

[1] *Institut für Astronomie der Universität Wien, Vienna, Austria*
[2] *Kuffner-Sternwarte, Vienna, Austria*
[3] *Stockholms Observatorium, Saltsjöbaden, Sweden*

Carbon-rich Mira variables and semiregular variables of type SRa and SRb as well as irregular variables of type Lb are compared with each other in various aspects. These groups of AGB stars having different pulsational properties are selected from the GCVS and the IRAS Catalogue. Additional near infrared and mm-CO data are used, partly from our own observations. Special emphasis is put on the stars studied by Olofsson et al. (1993, *ApJS*, 87, 267).

For the four groups of variables, pulsational properties like the amplitude or the period (not for Lbs) are correlated with atmospheric properties like the effective temperature and can be used to separate at least Lbs, SRas and SRbs from the Miras. Whereas the first three groups have relatively high photospheric temperatures $T^*$, medium to low dust temperatures $T^d$ and relatively high relative sizes of the dust blackbody, the later tend towards lower $T^*$ and somewhat higher $T^d$. This can be interpreted in the sense that Lbs, SRas and SRbs contain a significant fraction of stars with $60\,\mu m$ excess originating from cold, fossile dust shells and that Miras suffer higher mass loss with perhaps hotter dust temperatures. CO mass-loss rates are well correlated with some results of the BB fits, namely $T^*$, $T^d$ and $L^d/L^*$. The scale heights and number densities of Lbs would make these stars more massive than the Miras, however this result is based on rather small samples. The DENIS survey will provide much larger volume limited samples of AGB variables and thus allow a check of our results concerning the evolutionary status of Lbs. The main result of that study is that C-rich Lb variables are similar to semiregulars of the same chemistry in many aspects. For the future a more detailed analysis of the individual objects is forseen. This could help to get rid of deviating objects which influence the conclusions from our relatively small samples.

This work is supported by the *Fonds zur Förderung der wissenschaftlichen Forschung* under project number S7308–AST.

# The Chemical Composition of the Halo Mira V CrB

## TÔNU KIPPER[1] and UFFE G. JØRGENSEN[2]

[1] *Tartu Observatory, Toravere, Estonia*
[2] *Niels Bohr Institute, Copenhagen, Denmark*

We have derived CNO and metal abundances for the metal-deficient carbon Mira V CrB using the high-resolution spectra obtained by C. Barnbaum with the Hamilton Echelle Spectrograph of the 3-m telescope at Lick Observatory and an FTS near-IR spectrum from KPNO archives. These spectra were analysed by the synthetic spectrum method. New model atmospheres were calculated for this highly metal-poor carbon Mira. The atmospheric model used was calculated with continuum opacity sources and molecular opacity due to CO, CN, $C_2$, HCN, $C_2H_2$ and $C_3$. The important simplification is that this is a static atmosphere.

V CrB was found to be metal-deficient with [Fe/H] $= -2.12$. The CNO abundances are $\log A(C) = 7.17$, $\log A(N) = 5.3$, and $\log A(O) = 7.13$. The carbon isotopic abundance ratio is $^{12}C/^{13}C = 10.5 \pm 5.0$. The abundances of s-process elements are enhanced. For details see Kipper (1998, *Baltic Astronomy*, in press). The abundance pattern is similar to that of other late-type CH stars. The low C/O ratio of late-type CH stars cannot be explained by the mass transfer scenario, and we suppose that these stars have formed as intrinsic carbon stars. For more details of the proposed evolution scenario see Kipper & Jørgensen (1994, *A&A*, 290, 148).

The comparison of observed and computed near-IR spectra indicates the need of improving the $C_2H_2$ opacity data and using dynamical model atmospheres for such cool Miras.

# A 200 km/s Molecular Wind in the Carbon Star V Hya

## GILLIAN KNAPP[1], ALAIN JORISSEN[2], and KEN YOUNG[3]

[1] *Princeton University, Princeton NJ, U.S.A.*
[2] *Institut d'Astronomie, Université Libre de Bruxelles, Belgium*
[3] *Caltech Submillimeter Observatory, Hilo HI, U.S.A.*

The carbon star V Hya is remarkable for having a bipolar outflow resolved in observations of the CO (1–0) and (2–1) rotational lines (e.g. Kahane, Maizels & Jura 1988, *ApJ*, 328, L25). Observations of the CO (2–1) and (3–2) lines in this star using the 10.4-m telescope of the Caltech Submillimeter Observatory (Mauna Kea, Hawaii) now reveal a fast wind with an outflow velocity of about 200 km s$^{-1}$. Several distinct kinematic components can be identified in the V Hya wind, all of them having the same central velocity as the star: ($i$) a strong central component with an outflow velocity $V_e$ of about 45 km s$^{-1}$, ($ii$) a bipolar flow with $V_e = 7$ km s$^{-1}$, and ($iii$) a fast wind ($V_e = 200$ km s$^{-1}$), possibly bipolar as well. These features are unique among C stars, though they are reminiscent of proto-planetary nebulae like CRL 618 or CRL 2688. The C star V Hya may therefore be in the first stages of its transition to a planetary nebula.

# Formation of Core–Mantle Type Grains Consisting of a SiC Core and a Carbon Mantle in Circumstellar Envelopes of Carbon Stars

TAKASHI KOZASA[1] and HISATO SOGAWA[2]

[1] *Department of Earth and Planetary Sciences, Kobe University, Japan*
[2] *Department of Physics, Kyoto University, Japan*

We have proposed that the so–called 11.3 $\mu$m feature observed towards carbon stars can be attributed to small spherical core-mantle type grains consisting of a SiC core and a carbon mantle, using the optical constants of SiC tabulated by Choyke & Palik (1985, in *Handbook of Optical Constants of Solids*). Also we demonstrated qualitatively the possibility of formation of such a grain on the basis of a theory of nucleation and grain growth with a simple model of the circumstellar envelopes (Kozasa et al. 1996, *A&A*, 307, 551). Radiative transfer calculations have shown that the proposed dust grains reasonably reproduce the observed spectral features if the volume fraction of the carbon mantle increases with increasing distance in the outflow. In order to realize the formation of the proposed core–mantle type grains quantitatively, we have investigated the formation of SiC and carbon grains as well as the accretion of a carbon mantle on pre-condensed SiC grains together with the gas flow caused by the radiation pressure forces acting on the newly formed grains in the circumstellar envelope. The result of calculations shows that, after the avalanche of formation of SiC grains, accretion of carbon mantles on SiC and formation of carbon grains proceed in competition with each other. The volume fraction of the carbon mantle as well as the amount of isolated carbon grains increases as the mass-loss rate of carbon stars increases. We discuss the dependence of the structure and size of dust grains formed in the circumstellar envelopes on the mass-loss rate of carbon stars.

# Open and Globular Cluster Ages Using Theoretical Isochrones

## İBRAHİM KÜÇÜK

*Erciyes University, Kayseri, Turkey*

Clusters are key objects for knowing galactic structure as well as for checking stellar evolutionary models. They play an important role in linking the theories of stellar and galactic evolution.

We have obtained the photometric parameters of some selected galactic and globular clusters including their distance modulus, metallicity and reddening values in UBV from the literature. By using our theoretical evolutionary results we have compared the color – magnitude diagrams of selected clusters to the ones obtained from theoretical calculations.

# Tien Shan Astronomical Observatory

## KENESKEN KURATOV

*Tien Shan Astronomical Observatory*
*Fesenkov Astrophysical Institute, Almaty, Kazakhstan*

The instruments, astroclimate and research perspectives of the Tien Shan Astronomical Observatory are reviewed.

Tien Shan Astronomical Observatory (TSAO) was formed in 1994 by fusing the Solar Observatory of Fesenkov Astrophysical Institute (Republic of Kazakhstan) with the former Tien Shan Mountain Expedition of Shternberg Astronomical Institute (SAI), Moscow State University (Russia). At present work is being carried out to organize a joint Kazakh–Russian Astronomical Observatory on the basis of the structures and astronomical instruments of TSAO. We are interested in the participation of astronomical organizations from other countries, and any such proposals will be considered.

Tien Shan Astronomical Observatory is located 20 km from the southern boundary of Almaty city at altitudes of 2750 to 2780 m above sea level. Thus the Observatory combines the advantages of considerable altitude and the proximity of a large city.

The main astronomical telescopes available at the Observatory are:
  Two 1-m Ritchey-Chrétien/Coudé telescopes (Zeiss-1000)
  Two 48-cm Cassegrain telescopes
  The HSFA horizontal solar telescope-spectrograph (Carl Zeiss Jena), with a 60-cm mirror.

Besides the instruments mentioned, the Observatory is equipped with a 20-cm Coudé refractor (Opton), an ACU-5 horizontal solar telescope with an ASP-20 spectrograph, a large Nikolsky coronograph (diameter of main objective 53 cm), and a super illumination Schmidt astrograph (Schmidt plate diameter 40 cm, illumination 1:2). An 80-cm Carl Zeiss automated reflector is being mounted at present.

# Statistics of Carbon Stars in the Galaxy

## OMAR M. KURTANIDZE

*Abastumani Astrophysical Observatory*
*383762 Abastumani, Republic of Georgia*

Numerous low-dispersion objective prism spectral surveys have been conducted in recent decades in the yellow-red and near-infrared spectral regions for identification of carbon stars. The data from these spectral surveys have been merged by Stephenson into *A General Catalog of Cool Galactic Carbon Stars*, containing 5987 entries, the faintest of which are as faint as $I = 15.5$. A ten-degree equatorial belt of the Milky Way is now uniformly covered to $I = 13.0$. On the basis of these data, characteristics of the surface distribution of carbon stars have been determined. To study the latitude distribution, the whole galactic longitude range was divided into the strips 20° wide, and then by applying different statistical criteria the uniformity of the C-star distribution was checked. It is shown that in the longitude range $60° - 120°$ most carbon stars are located above the galactic plane, while in the range $260° - 320°$ most are below the plane. In other strips the distribution is uniform, except $40° - 60°$, where a strong concentration of C stars to the galactic plane is observed. The latitude distribution of carbon stars in the whole galactic longitude range is uniform. To study the longitude distribution, the numbers of C stars were counted in 5° bins. The longitude distribution is very non-uniform. A striking increase of the surface density of C stars is observed away from the galactic center. Their number in the region $310° - 50°$ is two times lower than in the region $130° - 230°$. This could not be fully attributed to galactic absorption, since M giants show quite the opposite distribution. The surface density of C stars reaches its maximmum values of 2.3–2.7 $\deg^{-2}$ in the Cas–Cyg and Carina directions. The numbers $N(50° - 130°)$ and $N(230° - 310°)$ are equal to 1281 and 1271, respectively. It may probably be concluded that most of the carbon stars are located outside the solar circle. By the nearest-neighbor method it is shown that the numbers of close associations of carbon stars with each other and with open clusters are not statistically significant. The connection of carbon stars with bright and dark clouds was also studied.

# A Close Association of Three Carbon Stars in the Direction of M92

## O. M. KURTANIDZE and M. G. NIKOLASHVILI

*Abastumani Astrophysical Observatory*
*383762 Abastumani, Republic of Georgia*

The discovery of the first faint ($V > 15.0$) high-latitude carbon star (FHLCS) was announced by Sanduleak (1980, *PASP*, 92, 246). It is located in the direction of the Magellanic Stream. On low-dispersion spectral plates (1250 Å mm$^{-1}$ at H$_\gamma$, IIIa–J,F) taken with the 70-cm meniscus telescope for identification of blue horizontal-branch stars in the globular cluster M92, two FHLCS were accidentally discovered (Kurtanidze 1980, *Astron. Tsirk.*, 1109, 3) near the known bright carbon star HD 156074 = Ste 3795. The data on these stars are tabulated below along with data for two other faint carbon stars found in the field of the cluster of galaxies A2199. Slit spectra of the two new C stars were obtained by P. Green (1995, private communication). The

| $\alpha_{2000}$ | $\delta_{2000}$ | $V$ | $B-V$ | $V-R$ | Notes |
|---|---|---|---|---|---|
| 16 18 21.3 | 39 23 20 | | | | New |
| 16 39 20.2 | 34 31 23 | | | | SP 105 |
| 17 13 31.2 | 42 06 23 | 7.80 | | | Ste 3795 |
| 17 14 11.8 | 42 00 31 | 14.60 | 1.14 | 1.16 | New |
| 17 14 57.5 | 42 10 24 | 15.98 | 2.22 | 1.22 | Ste 3801 |

surface density of C stars in this region is equal to 30 per sq. deg., although their mean surface density is only about 0.03 C star per sq. deg. at $|b| > 30°$. The proper motions of the faint carbon stars have been determined from the digitized POSS–I, II surveys. None of these four faint objects shows detectable proper motion (MacConnell 1996, private communication). The astrometric and photometric data were obtained from plates taken with 2-m Tautenburg Schmidt (1961–1990) and the 0.7-m Abastumani meniscus telescope (1978–1989).

# Near IR (J H K) Observations of Selected Carbon Stars

## VLADIMIR P. KUZ'KOV

*Main Astronomical Observatory, Kiev, Ukraine*

Infrared observations have great importance for the CH and Ba stars. Observations of selected carbon stars (HD 188650, HD 214714, HD 148897) were obtained with a small transportable IR photometer (see Kuz'kov 1989, *Kinem. Phys. Celestial Bodies*, v. 5, n. 2; First Eurasian Symposium on Space Science and Technology, Gebze, Turkey, 1993).

Observations were performed at the telescopes of Maidanak Observatory and Terscol Peak Observatory (Caucasus, altitude 3100 m) in collaboration with Ya. Shleivite of Lithuania.

The IR observations of HD 188650 indicate that this star has an IR excess and is a carbon CH star. On the diagram $J-H$, $H-K$ it was found that CH carbon stars are separated from other stars. Multicolor optical and IR photometry can be used to distinguish this type of carbon star from other stars.

# A Comparison between the Mira–OH/IR and the Carbon-Rich AGB Sequences

## JACQUES R. D. LÉPINE

*Departamento de Astronomia, Universidade de São Paulo, Brasil*

Carbon-rich and oxygen-rich AGB stars are known to separate clearly into two parallel groups in color-color diagrams like $K-L$ versus [12]–[25] (e.g. Epchtein et al. 1990, *A&AS*, 71, 39). We used this criterion to select a sample of 580 carbon-rich stars out of a larger sample observed in the $JHKLM$ bands with the 1-m ESO telescope. We investigate the energy distributions of these carbon-rich stars with the help of a simple model of a star represented by a blackbody, surrounded by a circumstellar dust shell. We show that the separation between carbon-rich and oxygen-rich stars in the color-color diagram can be explained by a difference in the condensation temperatures of dust grains, which implies different internal radius of the envelopes for the two flavors of AGB stars. A study of the oxygen-rich sequence from Miras to heavily obscured OH/IR stars (Lépine et al. 1995, *A&A*, 299, 453) showed that the main parameter which defines the position of a star along this sequence is the mass of the progenitor star. A single parameter like for instance the $K-L$ color index determines almost completely other parameters like optical depth, mass-loss rate, luminosity, and the absolute magnitudes in each band. A similar study is now presented for carbon-rich AGB stars. The main parameters of carbon-rich stars, derived by fitting their infrared energy distribution with a model of a star plus a circumstellar envelope, are given as a function of the index $K-L$. Among other parameters, we obtain luminosities, effective temperatures and mass-loss rates. Our results allow more precise determination of the distances of carbon-rich AGB stars, which have usually been estimated assuming a constant luminosity.

# Interferometric Molecular Line Observations of RW LMi

## MICHAEL LINDQVIST[1,2], ROBERT LUCAS[3], HANS OLOFSSON[4], FREDRICK LARSEN[4], ALAIN OMONT[5], KJELL ERIKSSON[6], and BENGT GUSTAFSSON[6]

[1] *Onsala Space Observatory, Onsala, Sweden*
[2] *Sterrewacht Leiden, Leiden, The Netherlands*
[3] *IRAM, Grenoble, France*
[4] *Stockholm Observatory, Saltsjöbaden, Sweden*
[5] *Institut d'Astrophysique de Paris, Paris, France*
[6] *Astronomical Observatory, Uppsala, Sweden*

We have observed the carbon star RW LMi (CIT 6) in the HCN($J=1\to 0$), CN($N=1\to 0$), HC$_5$N($J=34\to 33$), HNC($J=1\to 0$), SiS($J=5\to 4$) and HC$_3$N($J=10\to 9$) lines with the IRAM interferometer on Plateau de Bure. The SiS emission is clearly confined to regions close to the star. We see the expected structure of a hollow CN brightness distribution outside that of the HCN emitting region, but the CN brightness distribution appears to deviate significantly from spherical symmetry. The HNC molecules appear to be distributed in a shell, and so do the HC$_3$N and HC$_5$N molecules. In all cases, the results are qualitatively in accord with current models of photospheric and circumstellar chemistry. A model for the circumstellar molecular line emission will be used in the interpretation of the data.

# Are There Silicate – S Stars?

## IRENE R. LITTLE-MARENIN

*Wellesley College, Wellesley MA, U.S.A.*

I have identified seven S stars with very strong silicate features in their IRAS LRS spectra at 10 and 18 $\mu$m (LRS 25–29) (Chen et al. 1995, *A&AS*, 113, 51). This is highly unusual since S stars tend to have lower mass-loss rates and higher gas-to-dust ratios than M or C stars, implying less efficient dust formation in their circumstellar shells. Gas-to-dust ratios are estimated to be between 400 and 1000, at least a factor of two lower than for carbon stars, and hence strong dust emission features are not seen or expected. Also, pure S stars have a relatively weak emission feature which peaks in the 10–11 $\mu$m region and is subtly different from the 10 $\mu$m silicate or the 11.2 $\mu$m SiC feature. However, all seven stars have been found to be either M or MS stars rather than pure S stars, and hence they reflect the mass-loss rates and dust content associated with M stars. The stars and their characteristics are listed below.

Table 1: S? Stars with Strong Silicate Emission Features

| IRAS | Name | GCSS | LRS | $l$ | $b$ | Sp. Class |
|---|---|---|---|---|---|---|
| 07197 – 1451 | TT CMa | 341 | 27 | 230 | −0.2 | not S, maybe MS (a) |
| 11169 – 6111 |  | 738 | 29 | 292 | −0.5 | M5.5, not S (b) |
| 15347 – 5555 |  | 897 | 26 | 325 | −0.5 | M3 (b) |
| 16490 – 4618 |  | 944 | 25 | 340 | −1.5 | M1.5 (b) |
| 19545 – 1122 | V1407 Aql | 1175 | 29 | 30 | −19.6 | M6S (c) |
| 21029 + 4917 |  | 1259 | 28 | 90 | +1.7 | M3 (d) |
| 22512 + 6100 | V386 Cep | 1314 | 28 | 109 | +1.6 | M6 (c), M3 (a) |

(a) Stephenson, private communication; (b) Lloyd Evans & Catchpole 1989, *MNRAS*, 237, 219; (c) Bidelman, private communication; (d) Cohen et al. 1989, *AJ*, 97, 1759

# Classification of Dust Emission Features in Carbon Stars

IRENE R. LITTLE–MARENIN[1], GREGORY C. SLOAN[2], and STEPHAN D. PRICE[3]

[1] *Wellesley College, Wellesley MA, U.S.A.*
[2] *NASA/Ames Research Center, Moffett Field CA, U.S.A.*
[3] *Phillips Lab., Hanscom AFB, Hanscom MA, U.S.A.*

We have cross-referenced the IRAS PSC with the GCVS, searching for AGB carbon stars, and found 99 sources brighter than 28 Jy at 12 $\mu$m. We have classified their LRS spectra after removing an estimated stellar contribution. The majority of our sources fall into two categories: spectra with the classic SiC emission feature peaking around 11.2–11.5 $\mu$m (class SiC), and spectra where the SiC feature appears along with an additional component peaking around 8.5–9.0 $\mu$m (class SiC+). In a few stars the 8.5–9 $\mu$m feature rivals or exceeds the SiC feature in strength (class SiC++). Our sample also contains several unusual and low-contrast dust spectra which are difficult to classify. The classic SiC class contains mostly Mira variables, while the SiC+ and SiC++ classes contain mostly SRs and Lbs. Classic SiC sources tend to have redder [12]–[25] colors, correspondingly lower photospheric temperatures, and longer periods than SiC+ and SiC++ sources. The SiC feature appears to be superimposed on a featureless continuum most likely due to amorphous carbon or graphitic material. The C/O ratio increases along the sequence SiC $\rightarrow$ SiC+ $\rightarrow$ SiC++ from an average of 1.07 (SiC) to 1.2 (SiC+) to 1.3 (SiC++). If $a$:C–H is the carrier of the 8–9 $\mu$m feature (Goebel et al. 1995, ApJ, 449, 246), we suggest that this feature will strengthen with increasing C/O ratio. We find support for this suggestion in the increasing strength of the $C_2H_2$+HCN absorption feature seen in the 13–15 $\mu$m region and in the spectrum of VX And, which has the strongest 8–9 $\mu$m feature and the largest C/O ratio (1.76).

This work was supported by my friends in Washington and at Wellesley College.

# Synthetic Spectra for Carbon-Rich Long-Period Variables

RITA LOIDL[1], BERNHARD ARINGER[1], UFFE JØRGENSEN[2], SUSANNE HÖFNER[1] and JOSEF HRON[1]

[1] *Institut für Astronomie der Universität Wien, Vienna, Austria*
[2] *Niels Bohr Institute, Copenhagen, Denmark*

We have computed exploratory synthetic spectra for carbon-rich long-period variables by using the dynamical model atmospheres of Höfner et al. (see their abstract in this volume) as input for the spectral synthesis code of Jørgensen et al. (e.g. *A&A*, 261, 263, 1992). We compare the atmospheric structure of an extended, hydrostatic model atmosphere computed with the MARCS code with a corresponding hydrostatic initial model of Höfner and a few selected dynamical models. We find an overall agreement between the models in the range where the spectra are formed. Synthetic opacity-sampling spectra of two different models are shown for the three molecules $C_2$, CN and HCN for different phases of the light curve.

Since the symposium, we have compared the synthetic spectra presented here with ISO SWS spectra of the semi-regular carbon star R Scl (Hron et al. 1998, *A&A*, 335, L69) and the carbon Mira T Dra (Loidl et al. 1997, *Astrophys. Space Sci.*, 251, 243).

This work is supported by the *Fonds zur Förderung der wissenschaftlichen Forschung* under project numbers S7308–AST and S7305–AST.

# Circumstellar Envelopes of Peculiar and Normal J–Type Stars

## S. LORENZ MARTINS

*Observatorio Nacional-DAGE, Rio de Janeiro, Brazil*

Late-type stars present circumstellar envelopes which contain gas and grains. In the case of carbon stars, stellar flux is absorbed by the envelope and re-emitted by carbon-rich grains in the infrared. Several compounds have been predicted to condense in these environments: graphite grains, amorphous carbon grains, silicon carbide (SiC), and also MgS grains. J–type stars are identified in the visible range as carbon-rich and they can be separated into two distinct groups, according to their dust envelope properties: (1) normal J–type stars, which have carbon-rich envelopes, and (2) peculiar J–type stars, which have oxygen-rich envelopes. The modelling of normal and peculiar J–type envelopes has been performed by us using a Monte-Carlo type code for solving the radiative tranfer. For normal J–type stars two kinds of grains were considered simultaneously: amorphous carbon (A.C.), and silicon carbide (SiC). Silicate grains were used in modelling the peculiar ones. The temperatures of central stars and some characteristics of the circumstellar shell such as extinction opacities and its extension were determined by fitting the flux curves. The SiC/A.C. ratio as well as the energy distributions and temperature law have been studied. The results show that normal J–type carbon stars have thin envelopes (with extinction opacities about 0.02) and intermediate SiC/A.C. ratios. Peculiar J–type stars have thicker envelopes: their extinction opacities are about 0.60. Based on these results, two alternative scenarios are discussed: (1) Normal J–type carbon stars may be an intermediate group between the Groups I and II introduced by Lorenz-Martins & Lefèvre (1994, *A&A*, 291, 831). (2) J–type stars (normal and peculiar) could form an independent evolutionary sequence apart from that proposed for ordinary carbon stars.

This work was supported by CNPq–Brazil.

# Modelling of OH/IR Dust Envelopes

## S. LORENZ MARTINS and F. X. DE ARAÚJO

*Observatorio Nacional-DAGE, Rio de Janeiro, Brazil*

One of the least understood phases of stellar evolution occurs at the tip of the AGB and immediatly following the AGB. The stellar population at the tip of the AGB consists mainly of OH/IR stars and carbon stars. These classes are distinguished by the oxygen-rich envelope of OH/IR stars and the carbon-rich envelope of carbon stars. It is thought that both kinds of stars evolve from Mira variables and both can be progenitors of planetary nebulae. OH/IR objects can be classified in two groups. Type I OH/IR stars have lower mass-loss rates and they are frequently identified optically as Mira variables.

We have performed calculations of radiative transfer in circumstellar dust shells and the modelling of 19 OH/IR objects (types I and II). The Monte Carlo method was employed for representing the propagation of radiative energy, photon by photon, and solving the radiative transfer. We have modified previous versions of the numerical code in order to allow choosing the distribution of matter in the envelope, say the density law, while maintaining spherical symmetry. Then, we have modelled our sample considering $\rho \propto r^{-2.0}$, $\rho \propto r^{-2.5}$, and $\rho \propto r^{-1.3}$. Spherical grains with different radii in the range 900–8000 Å have been considered. The absorption and scattering efficiencies, as well as the albedo of the grains, were calculated using the Mie theory and the optical constants of David & Pégourié (1995, A&A, 293, 833). These constants are essential for building the models. The temperature of the central star and some characteristics of the circumstellar envelopes were determined by fitting the observed flux curves. Preliminary results show that type I OH/IR stars have thinner dust envelopes than type II ones. The temperatures also differ according to the types: type I OH/IR stars have higher temperatures than type II OH/IR stars. These results could indicate that type I OH/IR stars are less evolved than type II OH/IR stars.

This work was supported by CNPq–Brazil.

# Southern Carbon Stars Found on Near-IR Objective-Prism Plates

## D. JACK MacCONNELL

*Computer Sciences Corp., Space Telescope Science Institute  
Baltimore MD, U.S.A.*

A large set of deep objective-prism plates taken with the Curtis Schmidt telescope of The University of Michigan at Cerro Tololo, Chile is being used to search for cool supergiants and carbon stars along the southern galactic plane. The exposures are on hypersensitized I–N plates covering the 6800–8800 Å region at a dispersion of 3400 Å mm$^{-1}$ at the telluric A–band and reach a limiting magnitude of I $\approx$ 13.5. The sky coverage extends to 6°.5 on either side of the southern galactic plane, but not all fields have been searched to date. When not overexposed, cool carbon stars are readily distinguished by their prominent CN bands around 7900 Å and can be detected to the plate limit. A few hundred new C stars found on these plates were published in 1988, and here we report additional C stars found in recently-surveyed regions covering about 600 square degrees; about one-third of those found are unpublished.

# Absolute Magnitude and Kinematics of Barium Stars

MARIE-ODILE MENNESSIER[1], ANA GÓMEZ[2],
XAVIER LURI[3], SUZANNE GRENIER[2], LOUIS PRÉVOT[4],
JORDI TORRA[3], and FRANCESCA FIGUERAS[3]

[1] *Université Montpellier II, Montpellier, France*
[2] *Observatoire de Paris, Meudon, France*
[3] *Universitat de Barcelona, Barcelona, Spain*
[4] *Observatoire de Marseille, Marseille, France*

We apply to Ba stars a maximum-likelihood method able to distinguish several groups in a given sample and to provide estimates of the mean absolute magnitude and kinematic parameters for each of them. Three group can be distinguished:
- the first contains normal giants,
- the second contains more intrinsically luminous stars,
- the third is smaller and consists of subgiants.

This paper has been published by Gómez et al., *A&A*, 319, 881, 1997.

# The Overtone Spectrum of Molecular Hydrogen and Methane in the Visible: Recent Measurements

## MICHAEL MICKELSON, LEE LARSON, LARS ENGLISH, and DAVID FERGUSON

*Denison University, Granville OH, U.S.A.*

The increasing precision and accuracy of modern high-resolution spectroscopic observations of cosmic sources has to some extent driven the need for improved laboratory data and refined theories of atoms and molecules. This paper describes spectroscopic facilities in the Department of Physics and Astronomy at Denison University and recent measurements of some of the overtone spectra of molecular hydrogen and methane in the visible portion of the spectrum. For molecular hydrogen, we have recently completed the measurement of pressure shifts, line strengths and shapes, self-broadening coefficients and line positions for the S(0) and S(1) quadrupole transitions in the 4-0 vibration-rotation band. The data were obtained using a high-resolution dye laser system coupled to a White-type cell of unique optical design which is capable of obtaining optical paths of over 6 km. Using this system we also were able to detect — for the first time in the laboratory — the S(1) line of 5-0 band and to estimate its line strength and pressure-broadening coefficient. For methane, we have initiated a program to measure the strong visible and near-infrared absorption bands at moderate and eventually high resolution. Due to the complex and unresolved nature of these bands, empirical measurements of the absorption coefficient as a function of temperature have been carried out. These measurements were made using a lower-resolution dye laser system interfaced with a coolable three-meter basepath Chernin cell adjusted for optical paths up to one kilometer and at temperatures down to 123 K. It is hoped that these measurements of the overtone spectrum of hydrogen and methane will be of use in modeling the extended atmospheres of cool stars.

This work was supported by NASA Grant NAGW 1765 and the W. M. Keck Foundation.

# Multiwavelength Photometric Observations of Northern Carbon Stars

ANATOLY S. MIROSHNICHENKO[1],
KENESKEN S. KURATOV[2],
ŽELJKO IVEZIĆ[3], and MOSHE ELITZUR[3]

[1] *Central Astronomical (Pulkovo) Observatory, St. Petersburg, Russia*
[2] *Fesenkov Astrophysical Institute, Almaty, Kazakhstan*
[3] *University of Kentucky, Lexington KY, U.S.A.*

We discuss and justify the importance of simultaneous, multiwavelength observations of late-type stars. Such observations have a crucial role in constraining theoretical models of these objects and determining properties of the circumstellar medium. Performed in the visual and near-infrared wavelength range, they are also a valuable complement to the high-quality Low Resolution Spectra obtained by the *Infrared Astronomical Satellite*.

Here we report first results of our *UBVRIJHK* simultaneous observations of four northern carbon stars (S Cep, VY UMa, RY Dra, and Y CVn). The observations were performed at the Tien Shan Astronomical Observatory (Kazakhstan) in August 1995 and January 1996. Simultaneous multiwavelength photometry for S Cep was obtained for the first time. Comparison with previous observations for RY Dra, VY UMa, and Y CVn shows that their color indices have changed, and that RY Dra and Y CVn were observed at fainter brightness levels than previously. The monitoring of these stars will be continued and the sample size will be extended.

# AFGL 4106: Proto-Planetary Nebula or Post-Red Supergiant?

FRANK MOLSTER[1], JACCO VAN LOON[2],
RENS WATERS[1], and HANS VAN WINCKEL[3]

[1] *Astronomical Institute "Anton Pannekoek"*
  *Amsterdam, The Netherlands*
[2] *European Southern Observatory, Garching, Germany*
[3] *Instituut voor Sterrenkunde, Heverlee, Belgium*

Superficially, proto–planetary nebulae (PPNe) and post-red-supergiant (post–RSG) stars look alike: typically, they both have detached dust shells, A–G type supergiant classifications, and on-going mass loss. Usually a star with such properties is classified as post–AGB. However, there are two stars known in the literature, IRC +10420 (e.g. Jones et al. 1993, *ApJ*, 411, 323) and AFGL 2343 (Hawkins et al. 1995, *ApJ*, 452, 314), which were previously identified as post–AGB stars, but appeared to be post–RSG stars. AFGL 4106 might be the next candidate to join this illustrious couple.

AFGL 4106 is located in the direction of the Carina arm. It is heavily reddened and at IRAS wavelengths shows strong dust emission from a cool detached envelope. It has been classified in the literature as a G–type post–AGB star. However, new observations give rise to a skeptical approach to this classification. We show new and surprising observations: a 10 $\mu$m image, showing an asymmetric dust distribution; a spectacular H$\alpha$ image, indicating the presence of a bow shock; and spectra for different orientations of the slit.

# The Detection of Low-Luminosity Carbon Stars in the Leo II and Fornax Dwarf Galaxies

## GÉRARD MURATORIO[1] and MARC AZZOPARDI[1,2]

[1] *Observatoire de Marseille, Marseille, France*
[2] *Canada-France-Hawaii Telescope Corporation, Kamuela HI, U.S.A.*

Our very deep low-resolution spectroscopic surveys of the Leo II and Fornax dwarf spheroidal galaxies resulted in the identification of several new faint carbon star candidates. For this purpose we used, in the slitless spectroscopic mode, various spectrographs attached to the ESO 4-m class telescopes and CFHT. All observations were carried out through an interference filter ($\lambda_0 =$ 4880 Å; $\Delta\lambda \simeq$ 1000 Å) to narrow the instrumental spectral domain, thus keeping the number of overlapping images as low as possible. In this spectral region (4300–5300 Å), carbon stars can be identified through their pronounced Swan $C_2$ bands at 4737 and 5165 Å. A semi-automatic procedure in the MIDAS environment led us to select promising candidates whose carbon-star nature was then confirmed by medium-resolution slit spectroscopy. Data reduction techniques and results are shown.

# High-Resolution Coudé-Echelle Spectrometer for the 1.5-m Kazan University Telescope at the Turkish National Observatory

## FAIG MUSAEV[1] and ILFAN BIKMAEV[2]

[1] *Special Astrophysical Observatory, Russian Academy of Sciences*
  *Nizhnij Arkhyz, Russia 357147*
[2] *Engelhardt Astronomical Observatory, Kazan State University*
  *Kazan, Russia 420008*

Modern tasks of stellar spectroscopy demand the registration of spectra with high spectral resolution and high signal-to-noise ratio over a broad spectral region. The only way to realize these requirements simultaneously is with an echelle spectrometer working with a large-format CCD. For 1–3-m class telescopes such spectrometers would be installed in separate coudé rooms. We propose a project of modern coudé instrumentation for the 1.5-m Kazan University telescope being installed at the Turkish National Observatory in 1997. The use of two echelle gratings (R2 and R4) with prisms as cross-dispersers and two optical cameras will provide a spectral resolution up to $R = 200,000$ in a wide spectral range from 3500 to 9500 Å. As a basic model, the optical scheme of the coudé-echelle spectrometer of the 1-m SAO RAS telescope has been adopted. This spectrometer was designed and constructed during 1993–95 and is now providing high-quality spectral data (Musaev 1996, *Astronomy Letters*, v. 22, n. 10).

To obtain high-quality spectra during long observing periods, there are some important technical requirements for the telescope tower: (*a*) the coudé room should be well temperature-controlled and mechanically stabilized, (*b*) there should be no (or minimal) heat sources inside the tower, and (*c*) the telescope's automatic control system should be situated in a separate room at the 1st floor, or even outside the tower. The telescope tower is now being constructed by our Turkish collaborators according to these requirements (http://astroa.physics.metu.edu.tr).

Some of the spectrometer's optical elements have been manufactured in Russia during 1995–96 and we expect that observations will start in 1998. The main scientific goals for the coudé instrumentation of this 1.5-m telescope are: (*a*) detailed chemical composition determination of the atmospheres of hot and cool stars down to 8th magnitude, (*b*) study of spectral line profiles and possible time variations of line profiles with a resolution $R = 40,000 - 200,000$ and S/N ratio $> 100$, (*c*) the production of stellar atlases of a sample of bright stars of different spectral types, (*d*) a search for low-amplitude (50–500 m s$^{-1}$) variations of radial velocities of cool stars, (*e*) study of the spectra of magnetic A–F stars, and (*f*) study of the local interstellar medium.

# Spectroscopic Analysis of Single-Lined Spectroscopic Binaries with Unseen Companions

## FAIG MUSAEV[1], ILFAN BIKMAEV[2], and LAIMONS ZAČS[3]

[1] *Special Astrophysical Observatory, Nizhnij Arkhyz, Russia*
[2] *Engelhardt Astronomical Observatory, Kazan State University Kazan, Russia*
[3] *Radioastrophysical Observatory, Riga, Latvia*

Detailed abundance analyses have been carried out for 20 single-lined binaries with barium-star-like orbital elements using high-dispersion echelle spectra and the model atmosphere method. No substantial differences in the atmospheric abundances of 15 program stars relative to standard stars were found. Therefore, as a group, the single-lined spectroscopic binaries show atmospheric abundances similar to single stars, and the unseen companions did not have an influence on the atmospheric abundances of the primary stars studied.

The full version of this report has appeared in *A&A Supp.*, 122, 31, 1997 as a paper on "An abundance analysis of the single-lined spectroscopic binaries with barium stars–like orbital elements. I. Analysis and results" by L. Začs, F. A. Musaev, I. F. Bikmaev, and O. Alksnis.

# Radiative Transfer and Dynamics of Stellar Outflows

### NATHAN NETZER

*ORT Braude College, Karmiel, Israel*

The equation of radiative transfer is solved for spherically symmetric outflows. This is done by expanding the intensity function into Legendre series. The radiative transfer is coupled with the equation of motion of the outflow, where the driving force is radiation pressure on dust. It is found that there is a correlation between the outflow velocity of the gas and the mass-loss rate. There is a maximum possible mass-loss rate for carbon stars, which is of the order of $10^{-4} M_\odot \, \text{yr}^{-1}$. There is probably such a limit for oxygen stars as well, but it is much higher, as attempts to carry out the calculation to above $10^{-3} M_\odot \, \text{yr}^{-1}$ do not show any decrease in their outflow velocity.

# A Determination of the C/M5$^+$ Ratio in the Galactic Plane

### MARIA G. NIKOLASHVILI

*Abastumani Astrophysical Observatory*
*383762 Abastumani, Republic of Georgia*

Two deep low-dispersion objective-prism spectral surveys, in the yellow-red and near-infrared spectral regions, have been carried out at Abastumani Astrophysical Observatory for the identification of late-type stars, especially carbon stars, with the 70-cm meniscus telescope equipped with a 2° prism (1250 Å mm$^{-1}$ at $H_\gamma$ and 7000 Å mm$^{-1}$ at the A-band). As a result of both surveys about 2200 carbon stars have been identified, among them more than 1400 new ones. On the basis of these spectral surveys the C/M5$^+$ ratios have been determined in 100 survey fields located at latitudes $b = 0°$, ±3.6. The numbers of M stars were counted in $2° \times 2°$ squares. The results are given in

| Region | New | All | N/S | C/M5$^+$ | Region | New | All | N/S | C/M5$^+$ |
|---|---|---|---|---|---|---|---|---|---|
| 30–50 | 85 | 111 | 0.56 | 0.02 | 115–130 | 115 | 181 | 1.12 | 0.09 |
| 50–70 | 145 | 217 | 1.09 | 0.04 | 130–145 | 79 | 122 | 0.81 | 0.14 |
| 70–90 | 156 | 264 | 1.32 | 0.06 | 145–165 | 83 | 172 | 0.86 | 0.25 |
| 90–115 | 146 | 278 | 1.11 | 0.09 | 195–210 | 53 | 158 | 1.005 | 0.30 |

the table. As is seen, the C/M5$^+$ ratios vary from 0.02 to 0.30 as the longitude varies from 30° to 210°. It is well known that in the galaxies of the Local Group this ratio is correlated with the metal abundance of those systems. It might be noted that this ratio is equal to 0.7 and 4.4 for the LMC and SMC, respectively (Richer, 1989, in IAU Coll. 106, *Evolution of Peculiar Red Giant Stars*, p. 35).

# On the Molecular Structure of Circumstellar Envelopes Surrounding C Stars

## A. BEATE C. PATZER, JAN MARTIN WINTERS and ERWIN SEDLMAYR

*Institut für Astronomie und Astrophysik, TU Berlin, Germany*

Circumstellar envelopes around cool carbon stars show a rich chemistry as indicated by observations of many different molecular species. Most of the molecules are detected in the shell of the nearby C star IRC +10216. In recent years detailed information on the spatial distribution of molecules has been obtained from high-resolution interferometry and spectroscopy. Within circumstellar envelopes, widely different physical conditions are encountered: high densities and high temperatures prevailing in the inner shell, and cool rarefied conditions predominating in the outer part, which is exposed to the interstellar radiation field.

Based upon the model structure of a stationary dust-driven wind we have investigated the molecular structure of the entire circumstellar shell. The chemical composition of the envelope has been determined by solving a network of chemical rate equations including gas kinetic reactions and photochemical processes. Taking dust shielding and the wavelength dependence of the photo-reactions explicitly into account, the photo rates have been calculated by integration of the mean intensity of the local radiation field over the associated frequency-dependent photo cross section.

Starting from the photosphere with chemical equilibrium abundances as the initial condition, characteristic features of the molecular structure of the circumstellar envelope are reproduced by the model. In general the theoretical molecular distributions tend to be less extended than the observed ones. Possible reasons are an overestimation of the interstellar radiation field and the neglect of molecular shielding effects. Therefore, further investigations will account for these effects.

This work was supported by the BMBF grant 05 3BT 13A 6.

# Maser Mapping of Red Supergiants and the Onset of Bipolar Outflow

## ANITA M. S. RICHARDS[1], JEREMY A. YATES[2], and R. JAMES COHEN[1]

[1] *Jodrell Bank, University of Manchester, Cheshire, U.K.*
[2] *School of Chemistry, University of Bristol, Bristol, U.K.*

Maser emission from SiO, $H_2O$ and OH is found in oxygen–rich red giant circumstellar envelopes at successively larger distances from the star. $H_2O$ emission at 22 GHz has been mapped with MERLIN (the Multi Element Radio Linked Interferometer, http://www.jb.man.ac.uk/merlin/MERLIN.html) with a beamsize of 8 mas. Typically the 22 GHz emitting region is seen to be a thick expanding shell 100 mas across. Small-scale clumpiness is revealed. The shell is resolved into discrete features of velocity extent 1 to 3 $km\,s^{-1}$ and spatial extent 2 to 20 mas. Using this 3–D datacube, constraints are placed on the physical conditions such as kinetic temperature, density and local radiation field. These observations show that the size of the region increases with stellar mass-loss rate and that the mass outflow is undergoing acceleration. The outflow is gravitationally bound to the star at the inner radius of the $H_2O$ masing region, but unbound at its outer radius (Chapman & Cohen 1986, *MNRAS*, 230, 415; Yates & Cohen 1994, *MNRAS*, 270, 529).

In NML Cyg (mass $\sim 50\,M_\odot$), proper motion has been detected in two isolated features outside the rest of the $H_2O$ maser shell which are symmetric in position and velocity with respect to the stellar position (Richards, Yates & Cohen 1996, *MNRAS*, 282, 665). The increase in separation over 9 years corresponds to a transverse velocity of $19 \pm 6$ $km\,s^{-1}$ at a distance of 2 kpc. This is consistent with the observed radial-velocity range of the 22 GHz emission. OH and dust observations show a similar axis of symmetry, and this may be a sign of the onset of bipolar outflow. Axisymmetric structures are often seen in supernova remnants, and also in the planetary nebulae which originate from $\sim 1\,M_\odot$ stars. Results are also given for other supergiant stars currently being observed.

Papers relating to this work will be available by ftp; those interested should write to amsr@jb.man.ac.uk for details.

# The Peculiar Object IRAS 06088+1909

## A. RICHICHI[1], G. CALAMAI[1], F. LISI[1], B. STECKLUM[2], T. HERBST[3] and E. THAMM[4]

[1] Arcetri Observatory, Florence, Italy
[2] Thüringer Landessternwarte, Tautenburg, Germany
[3] Max-Planck Institut für Astronomie, Heidelberg, Germany
[4] Max-Planck Research Unit "Dust in Star Forming Regions" Jena, Germany

Our observation of IRAS 06088+1909 began with two occultation events recorded in the $K$ band (2.2 $\mu$m) during two simultaneous runs at the CAHA, Calar Alto, and TIRGO observatories in October 1994. The difference in latitude of these two sites insures a sufficiently different scan angle to investigate deviations from spherical symmetry. These first occultations were followed by broad-band photometry in the $1-5\,\mu$m range obtained at TIRGO. The interesting characteristics of this star stimulated further observations: (a) imaging and polarimetry in the $V, R, I$ bands; (b) IR speckle interferometry; (c) near-IR low resolution spectroscopy; and (d) near-IR polarimetry. The sequence of observations was concluded by observing the occultation of January 1995, again from both CAHA ($K$ filter) and TIRGO ($L$ filter at 3.8 $\mu$m). Of the four lunar occultations, only the Calar Alto light curve from October 1994 is consistent with a circular stellar disk. In this case, however, the diameter of $0.''007$ would imply a very low photospheric temperature ($T_{\rm eff} < 1500$ K). In all the remaining light curves, the data show that more extended emission is present. This is particularly evident in the $L$-band light curve. Photometric results show that the star is extremely heavily reddened ($V$–$K \approx 14$ mag); this amount of extinction cannot be ascribed to interstellar dust and must be produced locally, presumably in a circumstellar shell. The polarization measurements indicate that there might be a departure from spherical symmetry in the shell. The $K$ band spectrum also indicates that the star is very red, but more importantly, it shows evidence of at least some photospheric lines, indicating that the shell is not completely optically thick at these wavelengths. Speckle interferometry shows the source to be completely unresolved at the diffraction limit of the telescope in the $K$ filter ($0.''13$). The evidence for extended emission around IRAS 06088+1909 is incontrovertible, and our results indicate some resemblance to the case of IRC +10216. IRAS 06088+1909 is probably of a less extreme nature, but it is nevertheless a quite peculiar object that will deserve further investigation.

# Observations, Assignments and Profiles of SiC$_2$ Absorption and Emission Bands in Carbon Stars

PETER J. SARRE[1], MARK E. HURST[1], and TOM LLOYD EVANS[2]

[1] *The University of Nottingham, Nottingham, United Kingdom*
[2] *South African Astronomical Observatory, Cape Town, South Africa*

Although observed in the blue-green spectra of some carbon stars as long ago as 1926, the Merrill-Sanford bands were not identified as arising from SiC$_2$ until 1956. In 1984 it was recognized that the molecule is not linear like C$_3$, but has an unusual T–shape. Following recent laboratory work, extensive SiC$_2$ absorption spectra of a number of stars, including W Pic, RV TrA, and T Mus, have been assigned for the first time. Hot bands involving low-frequency vibrations are generally very strong in typical N–type spectra but were greatly weakened in T Mus for a time in 1994, indicating that the bands were then formed in a much cooler region than those typical of carbon-star photospheres. Of particular interest is the spectrum of IRAS 12311−3509, where SiC$_2$ bands appear in emission, possibly indicating an edge-on disk.

These results are discussed more fully in *ApJ* 471, L107, 1996.

# The Dust around Cool Stars

## IRAKLI SIMONIA and TSITSINO SIMONIA

*Tbilisi Laboratory of the Abastumani Astrophysical Observatory*
*Tbilisi, Republic of Georgia*

The present work deals with some properties of cool, crystalline hydrocarbons in the dust shells around stars of the late types, studied on the basis of laboratory investigations of terrestrial crystalline hydrocarbons. It is common knowledge that solid particles of silicon carbide, graphite, etc., can form in the vicinity of carbon stars. Taking into account the chemical composition of carbon stars and the complex structure of the "star – dust shell" system, one can suggest that the dust around stars of the types mentioned may contain solid carbon particles formed as separate equally-dispersed crystals, as well as complex, unequally dispersed polycrystals. One of the most interesting properties of solid crystalline hydrocarbons is their ability to luminesce when excited by UV radiation. Terrestrial crystalline hydrocarbons of the aromatic series obtained from different grades of petroleum show luminescence only under UV radiation of 3600–3800 Å. Luminescence of such hydrocarbons occurs in the yellow-green part of the visible spectrum, with a maximum wavelength of 5500 Å. When the intrinsic temperature of the hydrocarbons is 270 K and higher, the luminescence of terrestrial hydrocarbons is of the fluorescence type. However, at intrinsic temperatures below 270 K, the luminescence is of the phosphorescence type. When hydrocarbons are heated up to 310 K, they appear to lose their luminescent properties. In the outer layers of detached dust shells surrounding cool stars, crystalline hydrocarbons can be heated to moderate temperatures. If those slightly heated crystalline hydrocarbons happen to receive UV radiation incident from a distant foreign source, those hydrocarbons will luminesce in the corresponding spectrum. This effect can be termed "weak episodic phosphorescence." It may be observed as a blue or violet glow. The results of these laboratory studies of terrestrial crystalline hydrocarbons can be applied in investigations of other physical and chemical properties of crystalline hydrocarbons found in the dust shells of cool stars.

# Observations of the 11 μm Silicon Carbide Feature in Carbon Star Shells

## ANGELA K. SPECK[1], M. J. BARLOW[1], and C. J. SKINNER[2,*]

[1] *University College London, London, U.K.*
[2] *Space Telescope Science Institute, Baltimore MD, U.S.A.*
* *Deceased 1997 October 21*

Silicon carbide (SiC) is known to form in circumstellar shells around carbon stars. SiC can come in two basic types — hexagonal α–SiC or cubic β–SiC. Laboratory studies have shown that both types of SiC exhibit an emission feature in the 11–11.5 μm region. Such a feature can be seen in the spectra of carbon stars. The size and shape of the feature can vary depending on the type, size and shape of the SiC grains. Silicon carbide grains have also been found in meteorites. The aim of the current work is to identify the type(s) of SiC found in circumstellar shells and how they might relate to meteoritic SiC samples. We have used the CGS3 spectrometer at the 3.8-m UKIRT to obtain 7.5–13.5 μm spectra of 31 definite or proposed carbon stars. After flux calibration, each spectrum was fitted using a $\chi^2$–minimization routine. This routine was equipped with the published laboratory optical constants of six different samples of small SiC particles and had the ability to fit the underlying continuum using a range of grain emissivity laws.

It was found that the majority of observed SiC emission features could only be fitted by α–SiC grains. The lack of β–SiC is surprising, in that this is the form most commonly found in meteorites. Included in the sample was the extreme carbon star AFGL 3068, previously known to show the 11 μm SiC feature in absorption. In addition to it, we have discovered three IRAS sources, all of which have been proposed to be carbon stars, that also appear to show the SiC feature in absorption.

# Carbon- and Oxygen-Rich Stars in the IRAS Two-Color Diagram: Results from Hydrodynamical Models of AGB Winds

## M. STEFFEN[1], R. SZCZERBA[2], A. MEN'SHCHIKOV[1,3], and D. SCHÖNBERNER[1]

[1] *Astrophysikalisches Institut Potsdam, Germany*
[2] *Nicolaus Copernicus Astronomical Center, Toruń, Poland*
[3] *Max-Planck-Gesellschaft, Jena, Germany*

Based on detailed stellar-evolution calculations including mass loss on the AGB (Blöcker 1995, *A&A*, 297, 727), we have investigated the structure, dynamics and spectral energy distribution of dusty circumstellar shells around stars in the final stages of their AGB evolution.

The wind is assumed to be driven by radiation pressure on dust grains and subsequent momentum transfer to gas molecules by collisions. Given the fundamental stellar parameters $(M, L, T_{\text{eff}})$ and the mass loss rate $(\dot{M})$, a physical model of this wind is obtained from the self-consistent solution of the radiative transfer problem and the dynamical problem, using a time-dependent two-component (dust/gas) hydrodynamics code.

By this method we can study the dynamical response of the circumstellar dust/gas shell to variations of the stellar parameters and mass loss rate occurring in the course of stellar evolution. We find that the large variations of $L$ and $\dot{M}$ associated with the final thermal pulses near the end of the AGB evolution lead to characteristic, time-dependent signatures in the density structure and can explain the existence of *detached dust shells*.

We present the resulting "loops" in the IRAS two-color diagram for different assumptions concerning the composition (amorphous carbon, graphite, astronomical silicates) of the dust grains and for tracks with different initial stellar mass in comparison with the observed colors of carbon-rich and oxygen-rich stars.

# Energy Distribution in the Spectra of Carbon Stars

## JANIS-IMANTS STRAUME

*Radioastrophysical Observatory, Riga, Latvia*

Carbon stars have a special evolutionary state among cool giants because of the peculiarity of their chemical composition. Spectra of these stars have a very composite structure with a variety of molecular bands.

We have calculated the energy distribution in the spectra of carbon stars on the basis of model atmospheres. The standard opacity sources in the continuum were taken into account as well as absorption in electronic-vibration bands of the molecules CN, $C_2$, CH, CS, SiS, SiH, SiO, CaCl, MgH, AlH, and ZrO.

# On Carbon Star Evolution in the IRAS Two-Color Diagram

## RYSZARD SZCZERBA[1,3], MATTHIAS STEFFEN[2], and KEVIN VOLK[3]

[1] *Nicolaus Copernicus Astronomical Center, Toruń, Poland*
[2] *Astrophysikalisches Institut Potsdam, Germany*
[3] *The University of Calgary, Calgary, Alberta, Canada*

We present a comparison between recent stellar evolutionary calculations by Blöcker (1995, *A&A*, 297, 727) and by Vassiliadis & Wood (1993, *ApJ*, 413, 641) based on the properties of the circumstellar shells predicted by these two sets of models, which mainly differ in the mass-loss rates adopted during evolution on the Asymptotic Giant Branch (AGB). Given the mass-loss rate as a function of time, the evolution of the dusty circumstellar shell and of the resulting synthetic spectrum is computed with a time-dependent radiation hydrodynamics code (see Steffen et al., this volume). Since the formation and further evolution of C stars is a strong function of mass loss on the AGB (Bryan et al. 1990, *ApJ*, 365, 301; Groenewegen & de Jong 1993, *A&A*, 267, 410) it is possible to examine different scenarios for the mass-loss behavior by comparing their predictions to different observational characteristics of AGB stars. We discuss the evolution in the IRAS two-color diagram, from just before the moment of carbon star formation until the end of their evolution on the AGB. We have adopted the widely accepted scenario that carbon stars are formed during thermal pulses but we have tested different assumptions concerning the moment of their formation (core mass, thermal pulse number).

It has been suggested that visual carbon stars represent a period when mass loss is interrupted. According to both mass-loss scenarios mentioned above, the mass-loss rate is strongly modulated by thermal pulses on the upper AGB, leading to a repeating cycle of evolution from visual to infrared carbon stars after the formation of a C-rich star from an O-rich one. This cyclic variability of mass loss, which previous simulations have not taken into account, explains in a natural way the existence of visual carbon stars with O-rich as well as with only C-rich shells. It also explains why the fraction of visual carbon stars is higher than predicted if the star becomes a visual carbon star only once in its evolution.

# Carbon Stars in the Galactic Halo

## ED J. TOTTEN[1] and MICHAEL J. IRWIN[2]

[1] *Queen's University of Belfast, Belfast, Northern Ireland*
[2] *Royal Greenwich Observatory, Cambridge, U.K.*

A byproduct of the APM high-redshift quasar survey was the discovery of $\sim$20 distant (20 − 100 kpc) cool AGB carbon stars (all N–type) at high galactic latitude. We have surveyed the rest of the high latitude SGC sky with $\delta > -18°$ and found 10 more similar carbon stars. Before our work there were only a handful of published faint high-latitude cool carbon stars known (e.g. Margon et al. 1984, *AJ*, 89, 274; Mould et al. 1985, *PASP*, 97, 130) and there has been considerable speculation as to their origin (e.g. Sanduleak 1980, *PASP*, 92, 246; van den Bergh & Lafontaine 1984, *PASP*, 96, 869). Intermediate-age carbon stars (3 − 7 Gyrs) seem unlikely to have formed in the halo in isolation from other star-forming regions, and one possiblity that we are investigating is that they arise from either the disruption of tidally captured dSph galaxies or are a manifestion of the long-sought optical component of the Magellanic Stream. Lack of proper motion rules out the possibility of the majority being dwarf carbon stars (e.g. Warren et al. 1993, *MNRAS*, 261, 185); indeed no N–type carbon stars have been found to be dwarf carbon stars. Optical spectroscopy confirms their carbon star type (they are indistinguishable from cool AGB carbon stars in nearby dwarf galaxies) and hence their probable large distances. We have recently acquired high-resolution ($\sim$1 Å) spectra for all the SGC carbon stars and the majority of the NGC sample and have determined accurate ($\sim$10 km s$^{-1}$) radial velocities. We are extending the survey phase to the remainder of the NGC which, coupled with a program of *JHK* photometry, should enable us to probe the phase-space distribution of the halo carbon stars and hence determine their origin.

# Infrared Observations of Peculiar Carbon Stars

ANA ULLA[1,2], PETER THEJLL[3], TÔNU KIPPER[4],
and UFFE GRÅE JØRGENSEN[5]

[1] *Instituto Astronomico de las Canarias, Tenerife, Spain*
[2] *Laboratorio de Astrofísica Espacial y Física Fundamental
     Madrid, Spain*
[3] *NORDITA, Copenhagen, Denmark*
[4] *Tartu Astrophysical Observartory, Toravere, Estonia*
[5] *Niels Bohr Institute, Copenhagen, Denmark*

We present a uniform and high-quality set of infrared photometric ($JHK$) observations of the 6 peculiar carbon giant stars V Ari, UV Cam, BD +34°911, TU Gem, BD +57°2161 and BD +34°4134. Comparison of the $J$–$H$ and $H$–$K$ colors to other C stars indicates that our sample of stars has smaller color indices and this in turn implies that they are hotter if we use the standard assumption that there is a link between these colors and the temperatures. Furthermore, using standard assumptions we derive estimates of their effective temperatures, gravities, luminosities and distances.

This paper has appeared in A&A 319, 244, 1997.

# The Frequency of Extrinsic and Intrinsic S Stars in the Henize Sample

SOPHIE VAN ECK[1], ALAIN JORISSEN[1],
MICHEL MAYOR[2], STEPHANE UDRY[2],
and MICHEL BURNET[2]

[1] *Institut d'Astronomie et d'Astrophysique*
*Université Libre de Bruxelles, Belgium*
[2] *Observatoire de Genève, Sauverny, Switzerland*

Previous studies have identified two distinct families among S stars: intrinsic S stars exhibiting Tc lines in their spectrum, and extrinsic S stars lacking Tc lines. Extrinsic S stars were found to be binaries, and probably owe their chemical peculiarities to mass transfer in the binary system. On the contrary, intrinsic S stars are thermally-pulsing AGB stars where the third dredge-up brought heavy elements to the surface. The Henize sample of 205 S stars south of declination $-25°$ is especially well suited for inferring the relative frequency of extrinsic/intrinsic S stars, since it is not biased towards low galactic latitudes where intrinsic S stars tend to concentrate. Each star has been measured 3 or 4 times over a period of 3 years with the spectrovelocimeter CORAVEL. The search for binaries is complicated by the fact that Mira-type pulsations are frequent among intrinsic S stars. Fortunately, radial-velocity variations due to atmospheric motions are generally associated with very broad and asymmetric CORAVEL cross-correlation profiles. Furthermore, such a criterion based on the CORAVEL line-width index $Sb$ (which reduces to a luminosity indicator, since it is found to increase with $M_{\rm bol}$) correlates very well with the Tc/no-Tc dichotomy (as derived from high-resolution spectroscopy for the brightest stars of the sample). It thus provides an independent way to distinguish extrinsic ($Sb < 5$ km s$^{-1}$) from intrinsic ($Sb \geq 5$ km s$^{-1}$) S stars. The frequency of extrinsic S stars is found to be $38 \pm 5\%$. The distribution of the standard deviation of the radial velocity for extrinsic stars is flatter than that of intrinsic stars, reflecting the large frequency of binaries among extrinsic stars. The galactic distribution of the Henize sample can be described with a scale height above the galactic plane of 180 pc for intrinsic stars compared to 580 pc for the whole Henize sample (assuming $M_V = -2$ for intrinsic and $M_V = -1$ for extrinsic S stars).

# New Input Data for Synthetic AGB Evolution

## J. WAGENHUBER

*Max-Planck-Institut für Astrophysik, Garching, Germany*

As shown by Groenewegen & de Jong (1993, A&A, 267, 410), detailed knowledge of the secular luminosity variations of stars on the thermal pulsing (TP) AGB is an important ingredient for *synthetic AGB evolution* which enables one to study, apart from other ill-known quantities, the influence of different mass-loss descriptions by producing statistical data that can be compared to observations. This will not be possible by direct stellar evolution calculations for quite some time. Data from 38 AGB sequences (partly with OPAL opacities, nuclear network etc.), comprising three metallicities ($Z = 0.02$, $0.008$ and $10^{-4}$) and initial masses $M_*/M_\odot$ from 0.8 to 7, all starting from the ZAMS and with a very high resolution, have been used to derive fit formulae on a statistical basis of more than 700 TPs. Except for hot-bottom burning, most of the quantities appear to be quite independent of the model physics. The dependence on $Z$ has been taken into account as well as "turn-on" effects which influence the behavior of the first 5–10 TPs of every sequence (this is crucial for low-mass stars which probably do not experience more). The formulae allow reconstructing the slow variations of $L_*$ between TPs *and* modeling the height and shape of the short-term peaks in order to study, for example, their influence if a threshold behavior of the mass loss is assumed, or objects which show rapid period changes presumably due to luminosity variations after a TP (Mattei, this volume).

As an example, the maximum luminosity $L_A$ attained during quiescent hydrogen burning is given by a linear asymptotic relation, a correction which decreases with increasing $\Delta M_c \equiv M_c - M_{c,0}$ ($M_c$ is the actual mass of the hydrogen-exhausted core, $M_{c,0}$ the core mass at the first TP), and a correction for hot-bottom-burning which is the only term that depends on the model, especially on the value of the mixing-length parameter, and the mass of the convective mantle $M_m \equiv M_* - M_c$ (in solar units):

$$L_A = (50\,900 + 5000 \lg \frac{Z}{0.02})(M_c - 0.469) - 10^{2.31 + 1.56 M_{c,0} + (39.5 - 115 M_{c,0})\Delta M_c}$$
$$+ \ 10^{1.38 - 0.33 \lg \frac{Z}{0.02} + (0.411 + 1.20 \Delta M_c) M_m}.$$

# Silicon and Sulphur Chemistry in the Inner Envelopes of Carbon Stars

### KAREN WILLACY and ISABELLE CHERCHNEFF

*Physics Department, UMIST, Manchester, U.K.*

The chemistry of silicon-, sulphur-, and oxygen-bearing molecules is investigated in the inner envelope of a typical carbon-rich AGB star. The effect of pulsation-driven shocks on the gas close to the stellar photosphere is considered. The chemistry is governed by bimolecular and termolecular reactions between neutrals. Thermal equilibrium calculations predict small amounts of SiS and SiO molecules in stellar photospheres for C/O ratios characteristic of carbon stars. On the other hand, radio maps show that these molecules are present with large abundances close to the star. Our model predicts the formation of SiS and SiO in the periodically shocked inner regions and shows that the inner envelope of an AGB star is a region of very active molecule formation.

# Synthetic Colors of Carbon Stars

WALTER WINDSTEIG, ERNST A. DORFI,
SUSANNE HÖFNER, JOSEF HRON,
and FRANZ KERSCHBAUM

*Institut für Astronomie der Universität Wien, Vienna, Austria*

We have calculated a number of synthetic colors and the spectral energy distributions of C-rich circumstellar envelopes of pulsating AGB stars. From the frequency-dependent radiative transfer calculations we obtain the paths within the IRAS two-color diagrams as well as the profiles of various CO lines during the pulsational cycle. These theoretical results are also compared to ground-based and IRAS observations.

This work is supported by the *Fonds zur Förderung der wissenschaftlichen Forschung*, projects S7305–AST and S7308–AST.

# Temperatures of Peculiar G–Type Stars from Narrow-Band Near-Infrared Photometry

ROBERT F. WING[1], ROBERT F. GARRISON[2], and TUBA KOKTAY[2,3]

[1] *Ohio State University, Columbus OH, U.S.A.*
[2] *David Dunlap Observatory, University of Toronto Richmond Hill, Ontario, Canada*
[3] *İstanbul Üniversitesi, İstanbul, Türkiye*

Photometric observations in 6 of the filters of Wing's 8-color narrow-band near-infrared system have been obtained for 16 of the carbon-peculiar stars found by Olsen, Garrison, & Koktay. The observations provide a continuum color index that is free from atomic and molecular blanketing, as well as an explicit measure of CN strength. The continuum color can be calibrated in terms of effective temperature for unreddened stars, as most stars of this sample are believed to be, and these temperatures will be used in a spectroscopic analysis of these stars by Koktay. Approximately half of these stars have abnormally strong CN for G–type main-sequence stars; the remainder show little if any CN.

# A Photometric Search for Dwarf Carbon Stars

## ROBERT F. WING[1] and D. JACK MacCONNELL[2]

[1] *Ohio State University, Columbus OH, U.S.A.*
[2] *Computer Sciences Corp./Space Telescope Science Institute Baltimore MD, U.S.A.*

Most of the known low-luminosity carbon stars have been identified by measuring significant proper motions for stars already known to have C–type spectra. An alternative approach, which we are using in this study, is to examine the spectra of stars already known to show significant motions. This approach requires an efficient method of detecting strong CN or $C_2$ bands in faint proper-motion stars.

We are using the first six filters of Wing's eight-color near-infrared narrow-band photometric system to search for stars with strong CN absorption around 8120 Å among the stars of Luyten's LHS Catalogue. The stars selected for observation are mostly red stars which lie south of $\delta = -20°$ and lack previous classifications. Most of these stars are, of course, M dwarfs, and it is expected that the principal result of this project will be a set of accurate two-dimensional (TiO, CN) classifications for several hundred M dwarfs not previously classified.

After three observing runs with the 1.0-m telescope at CTIO, we have "re-discovered" two previously-known dC stars but have not found additional examples. We have, however, found a number of stars with detectable CN absorption. These stars must either be giants with enormous space velocities or, more likely, dwarfs with enhanced CN. Such stars may be the result of moderate amounts of mass transferred from a former C-rich giant to a main-sequence companion.

# Optical Appearance of Dynamical Models for Circumstellar Dust Shells around Long-Period Variables: AFGL 3068

## JAN MARTIN WINTERS[1], AXEL J. FLEISCHER[1], THIBAUT LE BERTRE[2], and ERWIN SEDLMAYR[1]

[1] *Institut für Astronomie und Astrophysik, TU Berlin, Germany*
[2] *Observatoire de Paris, DEMIRM, Paris, France*

Based on dynamical models of circumstellar dust shells (CDS) around long-period variables (LPVs), including time-dependent hydrodynamics and a detailed treatment of the processes of formation, growth and evaporation of carbon grains, angle- and frequency-dependent radiative transfer calculations have been carried out. The models are completely determined by 6 parameters, comprising the 4 fundamental stellar parameters ($M_*$, $T_*$, $L_*$, $\epsilon_i$) and the pulsation period and velocity amplitude at the base of the atmosphere ($P$, $\Delta u_P$). It turns out that the discrete onion-like structure of the CDS (see Fleischer et al., this volume) decisively influences the shape of the synthetic light curves as well as the calculated spatial intensity profiles and the corresponding spatial spectra. We present a consistent theoretical model for the CDS of the extreme carbon star AFGL 3068. The synthetic light curves, which are multiperiodic on a timescale of approximately 3 pulsation periods, the spectral energy distributions at different phases, and the terminal wind velocity are in excellent agreement with the observed properties of this object. Therefore, the values of the fundamental stellar parameters and the distance to the object, which are not directly accessible by observation, can be derived from this model: $M_* = 0.9\, M_\odot$, $L_*(t_0) = 13\,000\, L_\odot$, $T_*(t_0) = 2100\,\mathrm{K}$, carbon-to-oxygen ratio $\epsilon_C/\epsilon_O = 1.33$ (the other element abundances are assumed to be solar), $P = 696$ d, and $\Delta u = 8.0$ km s$^{-1}$. This model yields a final outflow velocity $v_\infty = 15.7$ km s$^{-1}$, a mass-loss rate $\dot{M} = 1.7 \times 10^{-4} M_\odot$ yr$^{-1}$, a dust-to-gas ratio $\rho^d/\rho^g = 1.6 \times 10^{-3}$, and a distance to AFGL 3068 of $d = 1.3$ kpc. A temporal evolution of the shell is seen in the calculated brightness profiles and in the corresponding visibility functions. Therefore, to obtain a further test of the model, it would be valuable to observe the brightness distribution of AFGL 3068 directly with very high angular resolution ($\approx 0.01$ arcsec).

This poster, including its figures, may be seen at
http://export.physik.tu-berlin.de/Publikationen/

This work was supported by the BMBF (grant 05 3BT13A 6) and by the DFG (grant Se 420/8-1).

# Envelope Pulsations of a $1\,M_\odot$ AGB Star During Thermal Pulses

## A. YA'ARI[1], Y. TUCHMAN[1], and J. WAGENHUBER[2]

[1] Racah Institute, Hebrew University, Jerusalem, Israel
[2] Max-Planck-Institut für Astrophysik, Garching, Germany

Usually, the lower boundary conditions for non-linear hydrodynamic studies of the pulsations of Mira envelopes are taken from a core mass – luminosity relation and kept fixed. However, during thermal pulses (TPs) the luminosity $L_*$ may vary by up to a factor of five on a timescale which is not much longer than the thermal timescale of the whole envelope. Ya'ari & Tuchman (1996, ApJ, 456, 350) have shown that Miras do not pulsate linearly around an equilibrium configuration but develop into an essentially non-linear regime.

Here, $L_*(t)$ as given by a stellar evolution calculation was used instead. In the "rapid dip" following the TP regular pulsations cease and the behavior is reminiscent of a semi-regular variable. In the following peak very violent oscillations may take place, even leading to some mass loss, if $L_*$ is high enough. In one calculated example $L_*$ (averaged over the pulsation cycles) drops from initially 3000 to $700\,L_\odot$ in the dip about 500 years after the TP and rises up to 3800 in the peak another 500 years later. In the quiescent hydrogen burning phase prior to the TP the period $P$ is about 450 days, hence the star would most probably be observed as a Mira. As a response to the rapid luminosity variations $P$ drops to about 100 days in the dip, showing small amplitudes and irregularities in the light curve, and rises to more than 500 days during the peak. This value is maintained for longer than 1000 years. The maxima of $L_*(t)$ in each pulsation cycle even reach up to $7000\,L_\odot$.

# Motions of Carbon Stars

CAHİT YEŞİLYAPRAK[1], ZEKİ ASLAN[1],
ORHAN GÖLBAŞI[1], and TUNCAY ÖZDEMİR[2]

[1] *Akdeniz University, Antalya, Turkey*
[2] *İnönü University, İnönü, Turkey*

Radial velocities and proper motions of the Hipparcos Input Catalogue have been used for a preliminary study of the motions of the variable and 'non-variable' carbon stars. Large uncertainties in distances and the fact that a large fraction of the 'constant' stars are suspected variables make separation into variable and constant carbon stars barely significant. On the other hand, the mean motions of the N– and R–type carbon stars are not the same: for the mean galactic velocity components we obtain $(u, v, w) = (-15 \pm 4, -19 \pm 4, -7 \pm 8)$ km s$^{-1}$ from 155 N–type stars and $(0 \pm 13, -43 \pm 14, 13 \pm 17)$ km s$^{-1}$ from 74 R–type stars. The dispersions about the solutions are 31 and 71 km s$^{-1}$, respectively. If the 'high velocity' stars with residuals exceeding $3\sigma$ are excluded one obtains $(-14 \pm 3, -15 \pm 3, -6 \pm 7)$ km s$^{-1}$ from the N–type stars, and $(-9 \pm 8, -16 \pm 8, -22 \pm 10)$ from the R–type stars. Of the 8 stars excluded from the solution, 6 stars are of type R. The nature of these 'high velocity' stars is discussed. Solutions including differential galactic rotation and the implied mean absolute magnitude are also discussed.

# VRI Observations of S Stars

### SANDRA B. YORKA and TRACY L. HUARD

*Denison University, Granville OH, U.S.A.*

A large sample of "anonymous" stars from Stephenson's *General Catalogue of Galactic S Stars* are characterized by their $V-R$ and $R-I$ colors. $V-R$ colors range from 0.3 to nearly 3, $R-I$ from 0.3 to 2.4; $V-I$ values range from about 1.5 to over 4. A few of the stars have been monitored for variability: Ste 89 and Ste 706 both show evidence of variability in preliminary data. Twenty-eight of the observed stars have IRAS and IRC identifications, and the ratio $F(12\,\mu\mathrm{m})/F(2.2\,\mu\mathrm{m})$ vs. $R-I$ is discussed.

# Liquid and Solid Carbon Particles in Cool White Dwarf Atmospheres

## VICTOR ZUBKO[1,2]

[1] *Institute of Astronomy, N. Copernicus University, Toruń, Poland*
[2] *on leave from the Main Astronomical Observatory, NAS, Kiev, Ukraine*

About ten years ago Zhilyaev & Zubko (1983, *PAZh*, 9, 227; 1984, *Ap&SS*, 105, 99) theoretically predicted that carbon in the condensed form may be present in the atmospheres of cool white dwarfs. In particular, it was shown by the model-atmosphere technique that models with effective temperature $T_{\text{eff}} \leq 6000\,\text{K}$, gravitation $g = 10^8$ cm s$^{-2}$, and chemical abundances H/He $\leq 10^{-5}$, C/He $\leq 0.001$ and C/O $> 1$ may contain both liquid carbon droplets and solid dust grains, forming a stable layer inside the atmosphere. Now there are opportunities to explore the problem from two directions. The first one concerns the observational search for signs of condensed carbon in the spectra of cool white dwarfs which may be obtained with powerful modern ground- and space-based facilities. In any case, the first observations of white dwarfs with the *Hubble Space Telescope* (e.g. Shipman et al. 1995, *AJ*, 109, 1231) are very promising. The second line of attack is the calculation of refined model atmospheres of cool white dwarfs as possible carriers of carbon condensate, on the basis of relevant model-atmosphere codes (e.g. Bergeron, Saumon & Wesemael 1995, *ApJ*, 443, 764; Aslan & Bues, this volume) supplemented by sophisticated calculations of the thermochemical and ionization equilibrium in complicated molecular systems (Schmidt, Bergeron & Fegley 1995, *ApJ*, 443, 274). However, the physical description of condensate particles in the framework of the thermodynamical approach used in early studies seems overly simplified. We propose an improved microphysical model of the carbon condensate layer in a white dwarf atmosphere based on the stationary solutions of an appropriate kinetic equation. The most important physical processes to be taken into account in modelling a condensate layer are the nucleation, growth-evaporation and sedimentation of condensate particles. The coagulation and diffusion of particles through a condensate layer are inefficient and may be neglected. Estimates show that micronic particles prevail in a condensate layer and that the solid carbon particles should be amorphous rather than crystalline. A condensate layer may become optically thick at visual wavelengths as soon as the fraction of condensed carbon exceeds 1%.

It is a pleasure to thank the SOC of IAU Symposium 177, the International Astronomical Union, and N. Copernicus University (Toruń, Poland) for financial support. This paper has appeared in *MNRAS*, 287, 583, 1997.

# Indexes

LOC member Talat Saygaç and Sylvia Önder having tea in the garden of the Istanbul University Observatory (August 1994).

# Author Index

(**Boldface** type indicates papers presented orally by that author)

| | |
|---|---|
| Abia, C. | 89 |
| Ak, H. | 515 |
| Ake, T. B. | **299** |
| Albayrak, B. | 515 |
| Alexander, D. R. | 517 |
| Alksnis, A. | 516 |
| Allard, F. | 517 |
| Andersen, A. C. | 349 |
| Andretta, V. | 522 |
| Aoki, W. | 518 |
| Aringer, B. | 519, 520, 539, 560 |
| Arnould, M. | 459 |
| Aslan, T. | **97** |
| Aslan, Z. | **507**, 515, 592 |
| Asplund, M. | 521 |
| Augason, G. C. | 517 |
| Azzopardi, M. | **51**, 568 |
| Bagnulo, S. | 522 |
| Bakker, E. J. | **217**, 523 |
| Barlow, M. J. | 578 |
| Barthès, D. | 237 |
| Bennett, P. D. | 303 |
| Bidelman, W. P. | 3 |
| Bikmaev, I. | 569, 570 |
| Blöcker, T. | 469, 524 |
| Blommaert, J. A. D. L. | 534 |
| Borysow, A. | 525 |
| Boughaleb, H. | 165 |
| Brewer, J. P. | **59** |
| Brown, A. | 303 |
| Bues, I. | 97 |
| Burnet, M. | 584 |
| Calamai, G. | 575 |
| Cannon, Robert C. | 449 |
| Cannon, Russell | 526 |
| Cherchneff, I. | **331**, 586 |
| Cohen, M. | 517 |
| Cohen, R. J. | 574 |
| Costa, E. | 41 |
| Crabtree, D. R. | 59 |
| Creech-Eakman, M. J. | 527 |
| Croke, B. | 526 |
| Cunha, K. | **103** |
| de Araújo, F. X. | 562 |
| de Jong, T. | 425 |
| Demers, S. | 528 |
| Demircan, O. | 515 |
| Doikov, D. N. | 529 |
| Doikova, E. M. | 529 |
| Dorfi, E. A. | **325**, 538, 587 |
| Doyle, G. | 522 |
| Dzervitis, U. | 530 |
| Eglite, M. | 531 |
| Eglitis, I. | 531 |
| El Eid, M. | 524 |
| Elitzur, M. | **391**, 399, 566 |
| English, L. | 565 |
| Eriksson, K. | 409, 557 |
| Esendimir, A. | 507 |
| Eyer, L. | 171 |
| Feast, M. W. | **207** |
| Ferguson, D. | 565 |
| Fernie, J. D. | **191** |
| Figueras, F. | 564 |
| Fleischer, A. J. | **377**, 590 |
| Foster, G. | 155, 171 |
| Frantsman, Ju. L. | **463** |
| Frogel, J. A. | **41** |
| Frost, C. A. | 449 |
| Garrison, R. F. | 141, 588 |
| Gauger, A. | 532 |
| Gillet, D. | 237 |
| Giridhar, S. | **117** |
| Gölbaşı, O. | 592 |
| Gómez, A. | 564 |
| Gong, Z. G. | **199** |

Gonzalez, G. .................. 523
Green, P. J. ................... **27**
Grenier, S. .................... 564
Grenon, M. .................. **171**
Groenewegen, M.A.T. ... **385**, 533, 534
Güçlü, C. ..................... 545
Gustafsson, B. ..... **409**, **481**, 557
Habison, P. .............. 546, 547
Harper, G. M. ................. 303
Hashimoto, O. ................ 425
Hatzidimitriou, D. ............ 526
Hauschildt, P. H. .............. 517
Helling, C. ............... 535, 536
Herbst, T. ..................... 575
Herwig, F. .................... 524
Hibbins, R. E. ................. 407
Hinkle, K. .................... 537
Höfner, S. ...... 325, 538, 560, 587
Hrivnak, B. J. ................. **293**
Hron, J. ....... 520, 539, 546, 547, 560, 587
Huard, T. L. .................. 593
Hurst, M. E. ............. 407, 576
Iijima, T. ..................... 540
Irwin, M. J. .............. 528, 582
Isern, J. ...................... 89
Ivezić, Ž. ........... 391, **399**, 566
Izumiura, H. .................. **425**
Janča, J. ...................... 541
Jimenez, R. ................... 542
Johnson, H. R. ................ 536
Jørgensen, U. G. ... **349**, 520, 525, 535, 536, 542, 543, 548, 560, 583
Jorissen, A. ... 103, **259**, 269, 459, 549, 584
Joyce, R. R. ................... 544
Kandemir, G. ................. 545
Keady, J. J. .............. 532, 537
Kerr, T. H. .................... 407

Kerschbaum, F. . 539, 546, 547, 587
Kifonidis, K. .................. 469
Kipper, T. ............... 548, 583
Kiselman, D. ................. 409
Knapp, G. .................... 549
Koktay, T. .............. **141**, 588
Kozasa, T. .................... 550
Küçük, İ ...................... 551
Kunkel, W. E. ................. 528
Kuratov, K. .............. 552, 566
Kurtanidze, O. M. .... **13**, 553, 554
Kuz'kov, V. P. ................. 555
Lambert, D. L. ........... 217, 523
Langhoff, S. R. ................ 520
Larsen, F. ..................... 557
Larson, L. .................... 565
Lattanzio, J. C. ............ **7**, **449**
Le Bertre, T. .................. 590
Lèbre, A. ..................... **237**
Lépine, J. R. D. ............... 556
Li, Y. ........................ 199
Lindqvist, M. ................. 557
Linsky, J. L. .................. **303**
Lisi, F. ....................... 575
Little-Marenin, I. R. . **361**, 558, 559
Lloyd Evans, T. ......... **367**, 576
Loidl, R. ................ 547, 560
Lorenz Martins, S. ....... 561, 562
Loup, C. ................ 145, 425
Lu, W. ....................... 293
Lucas, R. ..................... 557
Luri, X. ...................... 564
Luttermoser, D. G. ........... **105**
MacConnell, D. J. ... **37**, 563, 589
Magalhães, A. M. .............. **433**
Mattei, J. A. ........ **155**, 165, 171
Mauron, N. ................... 237
Mayor, M. .............. 269, 584
McClure, R. D. ............... **249**
Mennessier, M.-O. ....... **165**, 564
Men'shchikov, A. ............. 579

Mickelson, M. .................. 565
Miles, J. R. ................... 407
Miroschnichenko, A. S. ........ 566
Molster, F. ................... 567
Morgan, D. .................. 526
Mowlavi, N. .................. **459**
Muratorio, G. ................. 568
Musaev, F. .............. 569, 570
Müyeseroğlu, Z. .............. 515
Netzer, N. .................... 571
Nikolashvili, M. G. ... 13, 554, 572
Noguchi, K. ................... **21**
Nordsieck, K. H. .............. 433
North, P. ..................... **269**
Ohnaka, K. .................... **81**
Olander, N. ................... 409
Olofsson, H. ... 409, **413**, 547, 557
Omont, A. .................... 557
Özdemir, S. ................... 515
Özdemir, T. ................... 592
Parthasarathy, M. ............. **225**
Patzer, A. B. C. ............... 573
Pişmiş, P. .................... 245
Plez, B. .................. **71**, 536
Plonka, M. .................... 541
Prévot, L. .................... 564
Price, S. D. ................... 559
Qian, Z. ...................... 21
Querci, F. R. ................. **499**
Querci, M. .................... 499
Rao, Y. ....................... 21
Richards, A. M. S. ............. 574
Richer, H. B. .................. 59
Richichi, A. ................... 575
Roberts, W. J. ................. 37
Ryde, N. ...................... 481
Sarre, P. J. .............. **407**, 576
Schönberner, D. .... **469**, 524, 579
Schwarz, H. E. ................ 409
Sedlmayr, E. ...**337**, 377, 532, 573, 590

Selam, S. O. .................. 507
Skinner, C. .............. 522, 578
Sloan, G. C. ................... 559
Simonia, I. .................... 577
Simonia, Ts. ................... 577
Smith, V. V. ............ 103, **443**
Sogawa, H. .................... 550
Šolc, M. ...................... 541
Speck, A. K. .................. 578
Stahlberg, J. .................. 469
Stecklum, B. .................. 575
Steffen, M. .......... 469, 579, 581
Stencel, R. E. ................. 527
Straume, J.-I. ................. 580
Sun, J. ....................... 21
Szczerba, R. .............. 579, 581
Thamm, E. .................... 575
Thejll, P. ..................... 583
Torra, J. ...................... 564
Totten, E. J. .................. 582
Tsuji, T. ............. 81, **313**, 518
Tuchman, Y. ................... 591
Udry, S. ...................... 584
Ulla, A. ....................... 583
Valenti, J. .................... 303
van Dishoeck, E. F. ........... 217
Van Eck, S. .............. 259, 584
van Loon, J. Th. .... **145**, 385, 567
Van Winckel, H. ........ **285**, 567
Veteŝník, M. .................. 541
Volk, K. ...................... 581
Waelkens, C. .................. 285
Wagenhuber, J. .......... 585, 591
Wang, G. ..................... 21
Wang, J. ..................... 21
Waters, L.B.F.M. ... 145, 285, 425, 567
Whitelock, P. A. .... 145, **179**, 385
Willacy, K. .................... 586
Williamson II, R. L. ........... 37
Windsteig, W. ................. 587

Wing, R. F. .... **127**, 543, 588, 589
Winters, J. M. .. 337, 377, 573, 590
Wood, P. R. .............. 385, 449
Ya'ari, A. ..................... 591
Yates, J. A. .................... 574
Yeşilyaprak, C. ............... 592
Yorka, S. B. ................... 593
Young, K. .................... 549
Yüce, K. ...................... 515
Začs, L. ................. **277**, 570
Zijlstra, A. A. ............ 145, 385
Zubko, V. ..................... 594

# Object Index

(Non-stellar objects appear towards end of list)

*Stars named by constellation*

$\beta$ And .......................... 133
R And ........................... 160
W And ........................... 160
X And ............................ 160
RR And .......................... 160
ST And .......................... 160
VX And ................... 363, 559
EU And ........................... 85
U Ant .... 396, 410, 416, 419, 420, 425–428, 431
S Aps ........................... 213
X Aqr ........................... 160
V Aql ..................... 409, 411
W Aql ......... 158–160, 163, 265
V1407 Aql ................ 364, 558
V Ari ............................ 583
$\zeta$ Aur ............... 303, 307–311
V Aur ........................... 160
UU Aur ................... 113, 304
$\alpha$ Boo (= Arcturus) ...... 307, 311
R Cam ........................... 160
S Cam ........................... 160
T Cam ........................... 160
U Cam ........................... 419
ST Cam .......................... 364
UV Cam .......................... 583
XX Cam .......................... 137
$\beta$ Cnc .......................... 133
T Cnc ........................... 160
V Cnc ........................... 160
V CVn .................... 436, 438
Y CVn ... 83, 87, 92, 93, 106, 113, 363, 397, 425, 429–431, 566
TT CMa ................... 364, 558
VY CMa .......................... 368

GM CMa ....... 179, 181, 187, 188
R CMi ........................... 160
R Cap .............. 160, 161, 163
IW Car .... 118–121, 123, 124, 228
R Cas ........................... 213
S Cas .................... 160, 162
U Cas ........................... 160
W Cas ........................... 160
X Cas ........................... 160
ST Cas ........................... 82
WY Cas .......................... 265
WZ Cas ......... 3, 106, 460, 484
RU Cen .......................... 119
RV Cen .............. 160, 162, 185
TT Cen ................... 181, 185
UW Cen .......................... 213
V854 Cen .......... 211, 213, 408
S Cep ..... 131, 160, 173, 175, 178, 397, 566
DG Cep ........................... 83
V386 Cep ................ 364, 558
$\xi^1$ Cet ........................... 281
o Cet (= Mira) . 131, 312, 416, 436, 438, 439
W Cet ........................... 160
V CrA ........................... 213
WX CrA .......................... 213
$\epsilon$ CrB ........................... 133
R CrB ..... 77, 137, 191–198, 213, 216, 344
V CrB .................... 160, 548
$\gamma$ Cru ........................... 304
BH Cru .................. 173, 174
$\chi$ Cyg .................... 160, 162
R Cyg ........................... 160
S Cyg ........................... 160
U Cyg .................... 106, 160
V Cyg ........................... 160
RS Cyg .......................... 160
TT Cyg ............. 416, 419, 420
WX Cyg .......................... 160

601

V360 Cyg . 118, 120, 121, 123, 124
V778 Cyg ................ 85, 364
Z Del ........................ 160
$\gamma$ Dra ......................... 304
T Dra ................... 160, 560
RY Dra ........ 178, 363, 396, 397,
515, 566
UX Dra ....................... 515
AG Dra ............. 103, 283, 445
DR Dra .................. 261, 281
R For ..... 160, 179, 181–185, 189,
213, 367, 369, 370, 373
R Gem ....................... 160
T Gem ....................... 160
SS Gem ....... 118, 120, 121, 123
TU Gem ...................... 583
BM Gem ................. 85, 364
$\pi^1$ Gru ................... 265, 416
g Her (= 30 Her) 108, 110, 111, 307
89 Her ............. 209, 287, 418
U Her ................... 213, 419
X Her ........................ 416
UU Her ...................... 198
AC Her ........ 119, 209, 286, 287
TW Hor ................. 107, 133
V Hya ........ 178, 325, 330, 367,
369–373, 375, 416, 549
W Hya .................. 213, 415
IY Hya ....................... 92
T Ind ........................ 106
V366 Lac (= AFGL 2881) ..... 516
R Leo ........................ 416
RW LMi (= CIT 6) .. 416, 516, 557
R Lep .......... 160–163, 185, 367,
369–373
S Lup ........................ 160
R Lyn ........................ 160
$\alpha$ Lyr ........................ 130
R Lyr ........................ 106
S Lyr ............. 160, 265, 364
U Lyr ........................ 160

EP Lyr 117, 118, 120–122, 124, 228
U Mon .... 119, 137, 286, 287, 291
RR Mon ...................... 160
T Mus .............. 367, 368, 576
RZ Nor ...................... 213
V Oph ....................... 160
$\alpha$ Ori ...... 304, 307, 340, 436–438
$o^1$ Ori ....................... 265
R Ori ........................ 160
U Ori ........................ 419
BL Ori ....................... 106
DY Ori .... 118, 120–122, 124, 228
GP Ori ..................... 4, 5
$\chi$ Peg ....................... 137
RZ Peg ...................... 160
IZ Peg .................. 181, 186
Y Per ........................ 160
RZ Per ...................... 160
DY Per ...................... 516
W Pic ........................ 576
TX Psc (=19 Psc) .. 105–111, 113,
114, 213
ST Pup .................. 226, 228
AR Pup ........ 118, 120–122, 228
NP Pup ...................... 364
R Sge ......... 118, 120–122, 228
FG Sge .................. 210, 540
T Sgr .................. 160, 265
RY Sgr ........ 178, 210, 212, 213
ST Sgr ....................... 160
GU Sgr ...................... 213
R Scl  179, 181, 186, 187, 369, 373,
375, 409–411, 420, 425–427, 560
R Sct ......... 119, 137, 242, 375
S Sct .......... 410, 416, 419, 420
$\alpha$ Tau .................... 304, 366
119 Tau (= CE Tau) ..... 133, 137
Z Tau ..................... 160–163
SU Tau ...................... 212
RS Tel ....................... 213
$\alpha$ TrA .............. 303, 305, 306

X TrA .................... 409, 411
RV TrA ....................... 576
τ UMa ........................ 283
S UMa ........................ 160
VY UMa ............ 396, 397, 566
ε Vir ......................... 133
RU Vir ................. 160, 162
SS Vir ............. 106, 160, 162
R Vol ........................ 185

*Bright Star Catalogue*

HR 107 ....................... 270
HR 774 ....................... 283
HR 1105 ...................... 283
HR 4049 ....... 227–229, 232, 287,
   hfill 288, 291
HR 8752 ..................... 137

*Henry Draper Catalogue*

HD   9529 ............... 271, 272
HD 16115 ............... 253, 254
HD 19014 .................... 261
HD 21120 ............... 281, 283
HD 26455 .................... 272
HD 44179 (Red Rectangle) 220–223,
   227, 228, 287, 288, 407, 408, 418
HD 46703 ............... 228, 287
HD 48565 .................... 272
HD 50843 ............... 261, 275
HD 52961 ........... 227–229, 287
HD 56126 .. 217–220, 222, 229, 230,
   237–244
HD 58337 .................... 254
HD 58364 .................... 254
HD 65854 ............... 261, 278
HD 70379 .................... 229
HD 76846 .................... 254
HD 79319 .................... 254
HD 81817 ............... 281, 283

HD 85066 .................... 253
HD 95345 ............... 261, 275
HD 95767 .................... 287
HD 100503 ................... 282
HD 104979 ................... 278
HD 105262 ................... 230
HD 108015 .............. 286, 287
HD 116745 (=ROA 24) .. 229, 232
HD 119185 ......... 261, 275, 278
HD 121447 ................... 282
HD 122547 ................... 254
HD 130255 ................... 278
HD 131356 .............. 286, 287
HD 131670 ................... 278
HD 133656 ................... 230
HD 148897 ................... 555
HD 156074 ......... 252, 254, 554
HD 160538 (=DR Dra) ....... 261
HD 161796 .............. 209, 229
HD 163838 ................... 254
HD 165141 ................... 264
HD 178717 ................... 283
HD 179821 .............. 229, 231
HD 182040 ................... 137
HD 182274 ................... 271
HD 187885 .............. 229, 286
HD 188650 ................... 555
HD 191589 .............. 263, 265
HD 197604 ................... 253
HD 197989 .............. 142, 143
HD 204848 .............. 142, 143
HD 205011 ................... 278
HD 207687 .............. 142, 143
HD 213985 .......... 287–289, 291
HD 214714 ................... 555
HD 218851 ................... 254
HD 223392 ................... 254
HDE 235858 ......... 217, 222, 223
HDE 286436 ................. 254
HDE 332077 ............. 263, 265

*Durchmusterung Stars (pole to pole)*

BD +83°442 .................. 254
BD +57°2161 ................. 583
BD +39°4926 ... 227, 228, 287, 291
BD +38°118 .................. 263
BD +34°911 .................. 583
BD +34°4134 ................. 583
BD +33°1194 ................. 254
BD +30°2637 ................. 254
BD +30°3639 ................. 418
BD +29°95 ................... 253
BD +28°3530 ................. 254
BD +23°601 .................. 254
BD +23°2998 ................. 254
BD +21°64 ................... 254
BD +17°3325 ................. 254
BD +15°726 (= GP Ori) ....... 4
BD +4°2735 .................. 254
BD +2°2446 .................. 254
BD +2°3336 .................. 253
BD +2°4338 .................. 254
BD −21°3873 ................. 445
CPD −61°455 ................. 232
CPD −64°4333 ................ 282

*Hipparcos Input Catalog*

HIP   4284 .................. 176
HIP  26753 .............. 173, 175
HIP  53085 (= V Hya) ........ 178
HIP  63152 (= RY Dra) ....... 178
HIP  94730 (= RY Sgr) ....... 178
HIP  99653 .............. 173, 175
HIP 106583 (= S Cep) ... 173, 175, 178
HIP 109089 ............. 173, 175

*Two-Micron Sky Survey*

IRC −20101 (= GM CMa) ..... 187

IRC +10216 .... 343, 363, 399–403, 408, 414, 416, 417, 419, 424, 492, 533, 537, 573, 575
IRC +10420 ................... 567

*Infrared Astronomy Satellite*

IRAS 05341+0852 ........ 230, 231
IRAS 06088+1909 ............. 575
IRAS 07027−7934 ............. 233
IRAS 07197−1451 (= TT CMa) 364, 558
IRAS 07430+1115 ............. 297
IRAS 08005−2356 .... 217, 222, 223
IRAS 11169−6111 ........ 364, 558
IRAS 12311−3509 ... 367, 368, 375, 576
IRAS 15347−5555 ........ 364, 558
IRAS 16490−4618 ........ 364, 558
IRAS 18095+2704 ... 230, 294–297
IRAS 19385+0155 ............. 297
IRAS 19475+3119 ............. 295
IRAS 19500−1709 ............. 295
IRAS 19545−1122 (= V1407 Aql) 364, 558
IRAS 20004+2955 ............. 295
IRAS 21029+4917 ........ 364, 558
IRAS 22223+4327 ........ 294–297
IRAS 22272+5435 ........ 294–296
IRAS 22512+6100 (= V386 Cep) 364, 558
IRAS 23304+6147 ............. 297

*Air Force Geophysics Laboratory*

AFGL 190 .................... 390
AFGL 618 (= CRL 618) ....... 417
AFGL 2343 ................... 567
AFGL 2688 (= CRL 2688) 208, 417
AFGL 2881 (= V366 Lac) .... 516
AFGL 3068 ............. 578, 590

AFGL 4106 .................... 567

*Stars in globular clusters*

V 1 in $\omega$ Cen ............. 118–124
ROA 24 in $\omega$ Cen (= HD 116745)
..................................... 229, 232
No. 1412 in M4 ................ 232
Barnard 29 in M13 ........... 232

*Miscellaneous stellar*

C*22 ......................... 29
CBS 311 ............... 31, 33, 38
CCS 2–3635 (= GCCCS 3635) .. 39
CIT 6 (= RW LMi) .. 416, 419, 516, 557
CLS 31 ....................... 29
CLS 96 ............. 29, 318, 319
CRL 618 (= AFGL 618) ....... 417
CRL 2688 (= AFGL 2688) . 208, 417
ESO 439-162 .................. 98
G 77–61 . 29, 31, 33, 35, 37, 38, 251
G 99–37 ....................... 97
GCCCS 447 ................... 85
GCCCS 3635 .................. 39
GCGSS 1237 .................. 39
GD 165 B ................ 314, 315
Gl 229 B ................. 314, 315
Kapteyn's star (= HD 33793) .. 40
LHS 69 ....................... 97
LHS 1075 ....... 29, 318, 319, 324
LHS 1126 ................. 97–101
LMC 570 ..................... 388
LMC 1506 ................... 387
LP 328-57 .................... 29
MSB 65 ...................... 83
N55 in SMC ................. 534
NC 83 ........................ 85
NML Cyg ............... 424, 574
PG 0824+289 .......... 31, 33, 38

SAO 34504 ............... 229, 230
SAO 173329 .... 286, 287, 289–291
SAO 244567 (= Hen 1357) 233, 234
SBS 1517+5017 ............... 38
SP 105 ...................... 554
Ste 3795 (= HD 156074) ....... 554
Ste 3801 (viz. GCCCS 3801) .. 554
Ste 89 (viz. GCGSS 89) ....... 593
Ste 706 (viz. GCGSS 706) ..... 593
280 Schjellerup (= WZ Cas) ... 3

*Planetary Nebulae*

Abell 78 ..................... 245
BD +30°3639 ................ 418
Helix nebula (NGC 7293) ..... 420
IRAS 07027-7934 ............. 233
IC 4997 ..................... 233
NGC 2346 ................... 418
NGC 6720 ................... 418
NGC 6781 ................... 418
NGC 7027 ............. 418, 488
SwSt 1 ...................... 233

*Galaxies*

Large Magellanic Cloud (LMC) ...
   41–50, 52, 53, 55, 60, 61, 65, 67,
   135, 136, 145–151, 199–204, 385–
   390, 463, 482, 483, 488, 528, 572
Small Magellanic Cloud (SMC) ...
   42, 50, 52–55, 65, 146, 448,
   463–465, 485, 488, 489, 528,
   534, 572

*Dwarf spheriodal galaxies:*
   Carina ............. 54, 55, 57
   Draco ................. 54, 55
   Fornax ..... 54, 55, 57, 61, 568
   Leo I ................. 54, 55
   Leo II ........ 54, 55, 57, 568
   Sagittarius ............ 55, 57

    Sculptor ............... 54, 55
    Sextans ............... 55, 57
    Ursa Minor ........... 54, 55
IC 1613 .................... 56, 57
M31 ........... 56, 57, 59–68, 489
M33 ....................... 56, 57
M81 .......................... 51
NGC 55 .................... 56, 57
NGC 205 .............. 56, 57, 61
NGC 300 .................. 56, 57
NGC 2403 ............. 51, 56, 57
NGC 6822 ................. 56, 57
WLM ...................... 56, 57

*Asteroid*

Vesta ................... 142, 143

*Other non-stellar*

IC 2220 ...................... 418
M 1-92 ....................... 417
OH 26.5+0.6 ................ 416
OH 127.8–0.0 ............... 416
OH 231.8+4.2 ............... 417
Red Rectangle (see HD 44179)
Roberts 22 ................... 233

# Subject Index

(**Boldface** indicates papers that refer to the topic repeatedly)

AAVSO data ................................. **155, 165, 171, 179**, 501
abundance determinations ....... 33, 72, 75, 94, 103, 109, **117**, 127, **225,
**277**, 570
abundances ............. 4, 32–33, **89**, 121–123, 142, **225**, 238, **277**, **481**,
523, 531, 548, 570
AGB stars (*see also* carbon stars, *etc.*)
   general .......... 21, 32, 51, 75, 90, 103, 259, 303, **313**, 337, **413**, 433,
**443**, 534, 556, 584, 586, 591
   evolution ........ **7**, 13, 21, **41**, 59, 91, **145**, **225**, 328, **449**, **459**, **463**,
**469**, 524, 579, 581, 585
   pulsation ................................................. **199, 463**
   in the LMC ..................................... **41, 145, 199, 385**
   in M31 ............................................. 56–57, **59**, 489
asymmetries ......... 212–213, 325, 329–30, 416, 424, **433**, 557, 567, 575
asymptotic giant branch ... 7, 179, 200, 217, 225, 249, 260, 293, 413, 538,
562, 579, 581, 585
barium stars .......... 5, 30, 32, 103, 135, 137, **249, 259, 269, 277**, 299,
444–445, 447, 501, 555, **564**, 570
binary stars ........ **27**, 38, 103, 117, 163, 228–229, **249, 259, 269, 277,
285, 293**, 299, **303**, 325, 371, 408, 424, 444, 447, 464,
523, 544, 570, 584
bolometric magnitudes ......................... 21, **41**, 91, 463, 519, 584
brown dwarfs ............................................. 35, 72, **313**
$C/M5^+$ ratio ......................................... 57, 62–65, 68, **572**
$C/O$ ratio .......... 3, 5, 27, 33, 71, 76, 81, 84, 94–95, 134, 145, 326–327,
351, 353, **361**, 380–381, 383, 444, 455–456, 460, 462,
488, 490, 517, 530, **531**, 532, 548, 559, 586, 590
carbon isotope ratio ....... 30, 33, 35, 66, 75, **81**, **89**, 222, **349**, 453, 455,
487–488, **518**, 548
carbon stars (*see also* AGB stars, *etc.*)
   C stars .......... **13, 27, 37, 41, 51, 59, 89, 127, 145**, 155–156, 160,
**165, 171, 337, 361**, 388, 390, 396, 416, 419, 444,
**481**, 517, 531, 537, 549, 553, 554, 558, 563, 572, 573,
581, 583, 589
   J stars ................. 65–66, 68, 71, **81**, 93, 257, 355, 425, 430, 458,
488–489, 494, 561

carbon stars (*continued*)

    N stars ............ 4, 24, 30, 71, 76, **81**, **105**, 155–156, **171**, 253, 409, 447, 463, 467, 489, 530, 576, 582, 592

    R stars .......... 4, 13, 18, 24, 28, 30, 71, 76, 107, 155–156, **171**, **249**, 253, 530, 592

    CH stars ........... **27**, 45, 71, 103, **249**, 269, 272, 274, 284, 324, 444, 447, 463, 518, 548, 555

    subgiant CH (sgCH) stars ................. 30, 32, **249**, **269**, 284, 447

    CH–like stars ................................................ 253

    faint high-latitude carbon (FHLC) stars ............ **27**, 544, 554, 582

    (unspecified type) ........ **21**, **179**, **325**, **349**, **367**, 407, 425, 452, 456, 515, 516, 519, 522, 524, 526, 528, 529, 538, 542, 543, 545, 547, 548, 550, 555, 556, 557, 559, 560, 566, 568, 571, 577, 578, 579, 580, 586, 587

    detection ............ **13**, 28, **51**, **59**, 526, 528, 542, 553, 563, 568, 582

    in external galaxies .................... **41**, **51**, **59**, 526, 528, 542, 568

    luminosity ............ 8, 33, 38, **41**, 53, 61–62, 66, 149, **385**, 429, 456, **463**, 542, 556, 583, 592

    space distribution ............................................ **13**, **21**

    spectra .............. **3**, 54, **59**, **81**, 92, **105**, 131, 133, 143, **367**, 528, 548, 568, 582

    variability (*see* light curves, Mira variables, variable stars, *etc.*)

cepheids ............................... 117, 191, 195, 226, 228, 240, 523

chemical evolution of galaxies ..................................... **481**

chemical evolution of the Galaxy .............. 14, 92, 304, 314, 350, 461

chromospheres ..................... 80, **105**, 210, 215, 305–306, 375, 533

circumstellar dust ...... 113, 205, **225**, 285, 287, **361**, 435, 522, 527, 532, 538, 562, 571, 577

circumstellar shells ..... 86, 118, 180, **217**, **367**, **377**, **425**, 533, 537, 556, 575, 578, 579, 581, 590

circumstellar envelopes .... 145, 147, 314, **331**, **337**, **409**, **413**, 501, 545, 550, 561, 573, 586, 587

CNO cycle ............ 66, 93, 120, 229, 233, 356, 445, 450, 452–453, 487

collision-induced absorption (CIA) .................. 100, 316, 319, **525**

color-magnitude diagram .......................... 54, 117–118, 528, 551

convection ................. **7**, 33, 72, 78, 80, 93, 119, 212, 215, **449**, 524

curve of growth ...................................... 82–83, 218–220, 518

diffuse interstellar bands ....................................... **407**, **541**

diffusion .................................................. 8, 284, 524

dredge-up ........... 7, 38, 65–66, 84, 93, 124, 142, **225**, 259, 303, 328, **449**, 482, 494, 523, 524, 542

dust formation ....... 78, 124, **207**, **225**, **313**, **325**, **331**, **337**, **349**, **367**, **377**, 520, 527, 538, 539, 550, 590
dwarf barium stars ........................... 30, 32, 251, **259**, **269**, 447
dwarf carbon stars ......... 20, **27**, **37**, 135–136, 255, 260, **313**, 525, 544, 582, **589**
dwarf galaxies ..................................... 28, **51**, 542, 568, 582
dwarf S stars ................................................................ **37**
dynamic models (*see also* hydrodynamics) .. **71**, 106, 111–112, **377**, 520, 535, 536, 538, 548, 560, **579**, **590**
Eddington limit ........................................ 215, 345, **521**
effective temperature ..... **41**, 75, 127, 138, 234, **313**, 529, 530, 546, 547, 556, 583, 588
emission lines ........ 28, 65, **105**, 197, 221, 227, 233, 239, 244, 305, 330, 371, 375, 383, 408, **409**, 438
excitation temperature ........................................... **217**
Galactic bulge .................................................. 55, 62
Galactic center ...................................... 16, 22, 24, 209, 553
Galactic halo ..................................... 28, **51**, 518, **582**
galactic structure ..................................... 13, 180, 551, 553
globular clusters .............................................. 117, 551
globular cluster stars  34, **117**, 226, 229, 232–233, 324, 458, 525, 542, 554
grain formation  (*see* dust formation)
grism surveys ................................................. 34, 52–53
H$^-$ opacity minimum ....................................... 136–137, 543
H–deficient stars ...................... 71, 76–77, 137, 191, 198, 336, 521
helium-burning reactions ... 7, 66, 117, 147, 179, 186, 229, 233, 249, 252, 255, 262, 354, 356, **443**, **449**, **459**, 464, 487, 490, 492, 524, 556
HIPPARCOS data ........................................... **171**, 592
horizontal branch stars ......................................... 230, 554
hot bottom burning (HBB)  8–9, 72, 80, 145, 225, 231, **449**, **481**, 489, 585
HR diagram ................................... 77, 180, 186, **199**, 303, 343
HST data ............................................ **105**, **303**, 500, 594
hydrocarbon molecules ........................... **97**, 333–334, 532, 577
hydrodynamics ......... 326, 332, 338, 344, **377**, 399, **469**, 524, 579, 581, 590, 591
infrared spectra (*see also* spectral energy distribution, IRAS, ISO) .. **13**, 520, 537, 544, 575
interferometry .................. 72, 213, 308, 440, 503, 557, 573, 574, 575
IRAS low-resolution spectra ................. **361**, 519, 527, 558, 559, 566

IRAS photometry ... 147–149, **165**, 186–187, 227, 237, 369, **385**, 394–395, 398, 401, 427–428, 534, 556, 567, 579, 581, 587, 593
IRAS sources (*see also the* Object Index) .. 146–147, 208, **225**, 293, 361, 363, 367–368, 375, 386, 393, 396, 546, 547, 559, 575, 576, 578
ISO .................... 50, 86, 146, 291, 322, 425, 429–430, 527, 560
IUE data ............... 31, **105**, 234, 251, 265, 304, 306–307, 372, 500
J-silicate stars (*see* silicate carbon stars)
K stars ...................... 71, 76, 81, 103, 134, 138, 253, 275, **303**
lambda Boötis ($\lambda$ Boo) stars ................................. 229, 523
Large Magellanic Cloud (LMC) .... **41**, 52–53, 60–61, 65, 135–136, **145**, 180, **199**, **385**, 457, 463, 482, 488, 526, 528, 572
light curves ....... 117, **155**, 174–175, 178, **179**, 238, **293**, 370, 372, 377, 515, 590
Local Group ............................................. **51**, **59**, 572
M dwarfs ..................... 39, 74, 134–135, 138, 314, 525, 543, 589
M giants .......... 52–53, **59**, 71, 75–76, 81, 84, **105**, 112, 130, 134–135, 137–138, 163, 304, 307, 339, **361**, 398, 438, 517, 543, 546, 553, 558
MK classifications ............................................. 134–135
Magellanic Clouds (see also Large/Small Magellanic Clouds) .... 41, **51**, 90–91, 146, 463, 528
main-sequence precursors ....................... 33, 251, 255, 260, 263
maser emission ................................. 146, 213, **413**, 434, 574
mass loss ............. 86, **145**, 179, 191, **207**, **225**, 313, **325**, **367**, **413**, 426, 431, 433, 440, **469**, 482, 494, 501, 520, 538, 567, 579, 585, 591
mass-loss rate ...... 13, 45, 52, 91–92, 122–123, **145**, 200, 289, 291, **303**, **325**, 343, 383, **391**, 404, **409**, 420–421, 425, 427, 430–431, 470, 474, 477, 522, 533, 537, 538, 546, 547, 550, 556, 558, 562, 571, 574, 579, 581, 590
metallicity ........... 29, 31, 33, 35, 49, 51, 59, 62–64, 71, 103, 134, 261, 282–283, 314, 317, 324, 386, 481, 489, 494, 525, 542, 548, 551, 572, 585
meteorites / meteoritic grains ..................... 8, **349**, 458, 491, 578
Mira variables .... 75–76, 107, 129, 131, 156, 180, 199, 208, 212–213, 244, 312, 321, 332, 344, 362–363, **377**, 389, 416, 419, 433, 438, 501, 527, 537, 538, 539, 546, 559, 562, 584, 591
    carbon Miras .. 131–132, **155**, **165**, 175, 178, **179**, **367**, 424, 516, 537, 547, 548, 549, 560, 590

mixing length ........................................ 8, 72, 80, 199, 524
model atmospheres ...... 33, **71**, **81**, 92, **97**, 108, 121, 128–129, 138, 317,
320, 439, 517, 518, 521, 525, 531, 535, 536, 543, 545,
548, 570, 580, 594
molecular opacities ............ 71, 316, 321, 517, 525, 529, 536, 548, 580
molecules in CS shells ...... **217**, 226, 238, 333, 363, **407**, **413**, 557, 573,
574, 586
molecules in stellar spectra ..... 14, 17, 27, 37, 51–52, 60, **71**, **81**, 89, **97**,
118–119, **127**, 176, 319, 322, **367**, 436–439, 518,
520, 525, 526, 530, 531, 537, 540, 543, 544, 560,
563, 568, 576, 580, 588, 589
Monte Carlo simulations ..................... 30, 261, 283, 439, 561, 562
non-LTE ........................... 78, 80, **105**, 138, 306–307, 438, 517
nuclear reaction rates .............................................. 450
nucleosynthesis ............ **7**, 81, 90, 119, 225, 227, **443**, **449**, **459**, **481**
objective-prism surveys ........ **13**, 28, 37, 42, 51, 54, 526, 553, **563**, 572
observing facilities ................................ **499**, **507**, **552**, **569**
OH/IR stars .......................... 44, 199, 226, 439, 501, 556, 562
opacity distribution function (ODF) .......................... 73, 76–77
opacity sampling (OS) .............................. 73, 76–77, 535, 560
OPAL opacities ........................................... 201, 450, 585
open clusters ............................................. **519**, 551, 553
PAH molecules ...... 13, 227, 230, 233, 237, **331**, 341, 343, 363, 408, 536
photometry
  infrared (*JHK*...) ....... 6, 18, **21**, 30, 33, **41**, 53, 101, 136, **145**, **179**,
211–213, 324, **367**, 534, 543, 555, 556, 566, 575,
582, 583
  medium-bandwidth ............................................... 530
  narrow-band .............. 28, 52, **59**, 74–75, **127**, 517, 542, **588**, **589**
  *UBVRI* .............. **41**, **171**, **191**, 515, 516, 542, 551, 554, 566, 593
  visual .................... **155**, **165**, **171**, 183–184, 370, 372, 374, 501
  Strömgren *uvby* ............................................. 141–142
planetary nebulae ..... 13, 54, 86, 117, 123, 217, 225, 233–234, **245**, 293,
325, 330, 413, 418, 424, 434, 439, **469**, 482, 486,
488–491, 494, 502, 549, 562, 574
polarization ................................. 98, 213, 375, 424, **433**, 575
population types ................................. 47, 231–232, **481**, 518
Population II stars ........................... 30, 117, 232, 249, 493, 518
post–AGB stars ........ **117**, 137, **145**, **207**, **217**, **225**, **237**, **285**, **293**,
**413**, 523, 540, 567

pre– (or proto–) planetary nebulae (PPN) ..... 207, **293**, 364, 434, 439, 549, 567
proper-motion stars ......... 29–30, **37**, 135–136, 544, 554, 582, 589, 592
pulsation ............... 78, 113, 119, 146, 179, **191**, **199**, **207**, 226, 240, 243–244, **293**, 311, 331, **377**, 520, 538, 546, 547, 584, 587, 590, 591
r-process ................................................... 353, 444
R Coronae Borealis (RCB) stars ... 71, 76–77, 137, 177–178, **179**, **191**, **207**, 336, **337**, 367, 408, 431, 516, **521**
RV Tauri stars ..... **117**, 137, 187, 209, 213, 226, 228, 240, 242, 286–287, 291, 294, 375, 523
radial-velocity monitoring ...... 33, 35, 124–125, 196, 209, 237, 242, **249**, **259**, **269**, **277**, **285**, **293**, 584
radio line emission ..... 226, 230, 238, 241, 304, **409**, **413**, 489, 492, 501, 533, 547, 549, 557
Roche lobe overflow ... 31–32, 251, 260, 262, 264, 280, 287, 289, 291, 464
s-process .................... 103, 229, 233, 281, 353, 355, **443**, 460, 524
s-process abundances ....................... 75, 86, 253, 284, 447, **481**
s-process enhancements .. 9, 27, 30, 33, 71, 103, **225**, 237, 252, 261, **277**, 300, **443**, 492, 548
S stars ............. 5, 32, **37**, 59, 65, 71, 76, 90, 130, 135, 137, 145, 155, 158–160, 249–250, **259**, 299–300, **361**, 443, 447, 493, 501, 517, 532, 537, 558, **584**, **593**
SC stars ....................................... 5, 71, **81**, 174, 364, 490
shock waves ........ 72, 119–120, 191, **237**, 331, 344, 375, **377**, 438, 475, 527, 586
silicate carbon stars .................................... 85, 257, 364, 561
Small Magellanic Cloud (SMC) .. 52–53, 65, 75, 146, 448, **463**, 485, 489, **526**, 528, 534, 572
space density ................................................... 20, 31
spallation reactions ................................................. 90
spectral classification ....................................... 14, **141**, 558
spectral energy distribution ... 22–23, 101, 137, 149, 151, **285**, **313**, 377, **385**, 393, 396–397, **399**, 471–472, 522, 525, 533, 534, 556, 579, 580, 587, 590
subgiant stars (see also sgCH stars under carbon stars) ........ 142, 564
Sun .................................................... 105, 142–143
supergiants ........ 122, **127**, 146, 199, 208–209, 217, **225**, 285, 304, 308, 311, 363, 424, 433, 436–437, 491, 501, 563, 567, 574
supernovae ............................. 90, **337**, 352–353, 459, **481**, 574

symbiotic stars .................................... **103**, 299, 445
synthetic spectra ...... 72, 75, 93, 106, 108, 120, 132, 134, 138, 223, 517, 518, 520, 525, 531, 543, 548, 560, 581
technetium (Tc) ....... 9, 32–33, 250, **259**, 287, 299–300, 431, 443, 447, 492, 584
thermal pulses ....... 7–8, 32, 81, 145, 202–203, 227, 250, 259, 426, **463**, **469**, 493, 524, 579, 581, 584, 585, **591**
transition probabilities ............................................. 74
UU Her stars ........................................... 198, 226, 294
variability studies .............. **155**, **165**, **171**, **179**, **191**, **499**, 516, 593
variable stars (see also Miras, etc., and the Object Index)
  irregular ................................. 114, 171, 363, 546, 547, 559
  semi-regular .......... 114, 156, **165**, 171, 177–178, **179**, 199, 362–363, 369, 438, 501, 515, 539, 546, 547, 559, 560
white dwarfs ........ 13, 31–33, 38, **97**, 103, 123, 180, 187, 225, 250–251, 260, 264–265, **277**, 287, 299, 312, 413, 444, 500, **594**
wind ................. 31–32, 44, 211, 234, 251, 262, 264, 279–280, 299, **303**, **313**, **325**, **337**, 384, 413, 434, 475, 482, 521, 536, 538, 549, 571, 574, 579
Wolf–Rayet stars ............................... 231, 303, 336, 459, 461
zero-age main sequence (ZAMS) .......................... **7**, 200, 585